Lecture Notes in Computer Science 11868

More information about this series at http://www.springer.com/series/7412

Aythami Morales · Julian Fierrez ·
José Salvador Sánchez · Bernardete Ribeiro (Eds.)

Pattern Recognition and Image Analysis

9th Iberian Conference, IbPRIA 2019
Madrid, Spain, July 1–4, 2019
Proceedings, Part II

 Springer

Editors
Aythami Morales
Universidad Autónoma de Madrid
Madrid, Spain

Julian Fierrez
Universidad Autónoma de Madrid
Madrid, Spain

José Salvador Sánchez
Universitat Jaume I
Castellón de la Plana, Spain

Bernardete Ribeiro
University of Coimbra
Coimbra, Portugal

ISSN 0302-9743   ISSN 1611-3349   (electronic)
Lecture Notes in Computer Science
ISBN 978-3-030-31320-3   ISBN 978-3-030-31321-0   (eBook)
https://doi.org/10.1007/978-3-030-31321-0

LNCS Sublibrary: SL6 – Image Processing, Computer Vision, Pattern Recognition, and Graphics

This Springer imprint is published by the registered company Springer Nature Switzerland AG
The registered company address is: Gewerbestrasse 11, 6330 Cham, Switzerland

Preface

Now in its ninth edition, IbPRIA has become a key research event in pattern recognition and image analysis on the Iberian Peninsula organized by the national IAPR associations for pattern recognition in Spain (AERFAI) and Portugal (APRP).

Most of the research reported here therefore comes from authors from Spain and Portugal. Out of the 401 authors who presented work to IbPRIA 2019, 29% are from Spain and 20% are from Portugal. More than 50% of the authors are from another 32 countries from around the world, with high representation from countries like Algeria, Brazil, Colombia, India, Italy, or Mexico. Our efforts to strengthen the bonds between the research conducted on the Iberian Peninsula and other countries are patent in the program, which emphasizes interactive poster sessions and includes a special session dedicated to international research cooperation.

On the other hand, we are witnessing a deep transformation in our field, now increasingly dominated by advances occurring in industry. We have also tried to integrate IBPRIA in this vortex by including in the program a number of panel discussions with international research leaders from companies like Google, Microsoft, Telefonica, Vodafone, and Accenture.

IbPRIA 2019 received 137 submissions. The review process for IbPRIA 2019 was diligent and required careful consideration of more than 400 reviews from 100 reviewers who spent significant time and effort in reviewing the papers. In the end 99 papers were accepted, which is an acceptance rate of 72%. To form the final program 30 papers were selected for oral presentations (22% acceptance rate) and 69 as poster presentations.

We hope that this book will result in a valuable resource for the pattern recognition research community. We would like to thank all who made this possible, especially the authors and reviewers.

August 2019

Aythami Morales
Julian Fierrez
José Salvador Sánchez
Bernardete Ribeiro

Organization

Organizing Committee

General Chairs

General Co-chair AERFAI
José Salvador Sánchez   Universitat Jaume I, Castellón, Spain

General Co-chair APRP
Bernardete Ribeiro   University of Coimbra, Portugal

Local Chair

Julian Fierrez   Universidad Autonoma de Madrid, Spain

Program Chairs

Aythami Morales	Universidad Autonoma de Madrid, Spain
Manuel J. Marin	University of Cordoba, Spain
Antonio Pertusa	University of Alicante, Spain
Hugo Proença	University of Beira Interior, Portugal

Local Committee

Ruben Vera-Rodriguez	Universidad Autonoma de Madrid, Spain
Ruben Tolosana	Universidad Autonoma de Madrid, Spain
Javier Hernandez-Ortega	Universidad Autonoma de Madrid, Spain
Alejandro Acien	Universidad Autonoma de Madrid, Spain
Ignacio Serna	Universidad Autonoma de Madrid, Spain
Ivan Bartolome	Universidad Autonoma de Madrid, Spain

Program Committee

Abhijit Das	Griffith University, Australia
Adrian Perez-Suay	University of Valencia, Spain
Ana Mendonça	University of Porto, Portugal
Antonino Furnari	Università degli Studi di Catania, Italy
Antonio Bandera	University of Malaga, Spain
Antonio Javier Gallego Sánchez	University of Alicante, Spain
Antonio Pertusa	University of Alicante, Spain
Antonio-José Sánchez-Salmerón	Universitat Politècnica de València, Spain
António Cunha	UTAD, Spain
António J. R. Neves	University of Aveiro, Portugal

Armando Pinho	University of Aveiro, Portugal
Arsénio Reis	UTAD, Spain
Bilge Gunsel	Istanbul Technical University, Turkey
Billy Mark Peralta Marquez	Pontificia Universidad Catolica de Chile, Chile
Carlo Sansone	University of Naples Federico II, Italy
Catarina Silva	ESTG-IPLEIRIA-PORTUGAL, Portugal
Constantine Kotropoulos	Aristotle University of Thessaloniki, Greece
Cristina Carmona-Duarte	Universidad de Las Palmas de Gran Canaria, Spain
Daniel Acevedo	Universidad de Buenos Aires, Argentina
David Menotti	UFPR - DInf, Panama
Diego Sebastián Comas	UNMDP, Argentina
Enrique Vidal	Universitat Politècnica de València, Spain
Ethem Alpaydin	Bogazici University, Turkey
Fernando Monteiro	Polytechnic Institute of Bragança, Portugal
Filiberto Pla	University Jaume I, Spain
Filip Malmberg	Uppsala University, Sweden
Francesc J. Ferri	University of Valencia, Spain
Francisco Casacuberta	Universitat Politècnica de València, Spain
Francisco Herrera	University of Granada, Spain
German Castellanos	Universidad Nacional de Colombia, Colombia
Giorgio Fumera	University of Cagliari, Italy
Helio Lopes	PUC-RIo, Brazil
Hugo Jair Escalante	INAOE, Mexico
Hugo Proença	University of Beira Interior, Portugal
Ignacio Ponzoni	UNS - CONICET, Argentina
Jacques Facon	Pontifícia Universidade Católica do Paraná, Brazil
Jaime Cardoso	University of Porto, Portugal
Jesus Ariel Carrasco-Ochoa	INAOE, Mexico
Johan Prueba	Centro de Investigacion en Matematicas, Spain
Jordi Vitria	CVC, Spain
Jorge Calvo-Zaragoza	University of Alicante, Spain
Jorge S. Marques	IST/ISR, Portugal
Jose Garcia-Rodriguez	University of Alicante, Spain
Jose Miguel Benedi	Universitat Politècnica de València, Spain
Jose Salvador Sanchez	Universitat Jaume I, Spain
Jose Silvestre Silva	Academia Militar, Spain
João Carlos Neves	IT - Instituto de Telecomunicações, Portugal
João M. F. Rodrigues	Universidade do Algarve, Portugal
Juan Valentín Lorenzo-Ginori	Universidad Central Marta Abreu de Las Villas, Cuba
Kalman Palagyi	University of Szeged, Hungary
Laurent Heutte	Université de Rouen, France
Lawrence O'Gorman	Alcatel-Lucent Bell Labs, USA
Lev Goldfarb	Faculty of CS, UNB, Canada
Luis-Carlos González-Gurrola	Universidad Autonoma de Chihuahua, Mexico

Luís A. Alexandre	UBI and Instituto de Telecomunicações, Portugal
Manuel J. Marín-Jiménez	University of Cordoba, Spain
Manuel Montes-Y-Gómez	Instituto Nacional de Astrofísica, Óptica y Electrónica, Mexico
Marcelo Fiori	Universidad de la República, Uruguay
Marcos A. Levano	Universidad Catolica de Temuco, Chile
Mariella Dimiccoli	Institut de Robòtica i Informàtica Industrial, Spain
Mario Bruno	Universidad de Playa Ancha, Chile
Mark Embrechts	RPI, New York, USA
Martin Kampel	Vienna University of Technology, Austria
Matilde Santos	Universidad Complutense de Madrid, Spain
Michele Nappi	Università di Salerno, Italy
Miguel Angel Guevara Lopez	University of Minho, Portugal
Moises Diaz	Universidad del Atlantico Medio, Spain
Nicolaie Popescu-Bodorin	University of S-E Europe Lumina, Romania
Nicolas Perez De La Blanca	University of Granada, Spain
Niusvel Acosta-Mendoza	Advanced Technologies Application Center (CENATAV), Cuba
Paolo Rosso	Universitat Politècnica de València, Spain
Paulo Correia	Instituto de Telecomunicacoes - Instituto Superior Tecnico, Portugal
Pedro Cardoso	Universidade do Algarve, Portugal
Pedro Latorre Carmona	Universidad Jaume I, Castellon de la Plana, Spain
Pedro Real Jurado	Institute Mathematics of Seville University (IMUS), Spain
Rafael Medina-Carnicer	Cordoba University, Spain
Ramón A. Mollineda Cárdenas	University Jaume I, Spain
Rebeca Marfil	University of Malaga, Spain
Ricardo Torres	Institute of Computing, University of Campinas, Brazil
Roberto Alejo	Tecnológico Nacional de México, Campus Toluca, Mexico
Sebastian Moreno	Universidad Adolfo Ibañez, Chile
Sergio A. Velastin	Universidad Carlos III de Madrid, Spain
Turki Turki	King Abdulaziz University, Saudi Arabia
V. Javier Traver	Universitat Jaume I, Spain
Ventzeslav Valev	Bulgarian Academy of Sciences, Bulgaria
Vitaly Kober	CICESE, Mexico
Vitomir Struc	University of Ljubljana, Slovenia
Xiaoyi Jiang	University of Münster, Germany

Contents – Part II

Biometrics

What Is the Role of Annotations in the Detection
of Dermoscopic Structures? 3
   Bárbara Ferreira, Catarina Barata, and Jorge S. Marques

Keystroke Mobile Authentication: Performance of Long-Term Approaches
and Fusion with Behavioral Profiling. 12
   Alejandro Acien, Aythami Morales, Ruben Vera-Rodriguez,
   and Julian Fierrez

Incremental Learning Techniques Within a Self-updating Approach
for Face Verification in Video-Surveillance 25
   Eric Lopez-Lopez, Carlos V. Regueiro, Xosé M. Pardo,
   Annalisa Franco, and Alessandra Lumini

Don't You Forget About Me: A Study on Long-Term Performance
in ECG Biometrics 38
   Gabriel Lopes, João Ribeiro Pinto, and Jaime S. Cardoso

Face Identification Using Local Ternary Tree Pattern Based Spatial
Structural Components. 50
   Rinku Datta Rakshit, Dakshina Ranjan Kisku, Massimo Tistarelli,
   and Phalguni Gupta

Catastrophic Interference in Disguised Face Recognition 64
   Parichehr B. Ardakani, Diego Velazquez, Josep M. Gonfaus,
   Pau Rodríguez, F. Xavier Roca, and Jordi Gonzàlez

Iris Center Localization Using Geodesic Distance and CNN 76
   Radovan Fusek and Eduard Sojka

Low-Light Face Image Enhancement Based on Dynamic Face
Part Selection 86
   Adel Oulefki, Mustapha Aouache, and Messaoud Bengherabi

Retinal Blood Vessel Segmentation: A Semi-supervised Approach 98
   Tanmai K. Ghosh, Sajib Saha, G. M. Atiqur Rahaman, Md. Abu Sayed,
   and Yogesan Kanagasingam

Quality-Based Pulse Estimation from NIR Face Video with Application
to Driver Monitoring   108
    Javier Hernandez-Ortega, Shigenori Nagae, Julian Fierrez,
    and Aythami Morales

Handwriting and Document Analysis

Multi-task Layout Analysis of Handwritten Musical Scores   123
    Lorenzo Quirós, Alejandro H. Toselli, and Enrique Vidal

Domain Adaptation for Handwritten Symbol Recognition: A Case of Study
in Old Music Manuscripts   135
    Tudor N. Mateiu, Antonio-Javier Gallego, and Jorge Calvo-Zaragoza

Approaching End-to-End Optical Music Recognition
for Homophonic Scores   147
    María Alfaro-Contreras, Jorge Calvo-Zaragoza, and José M. Iñesta

Glyph and Position Classification of Music Symbols
in Early Music Manuscripts   159
    Alicia Nuñez-Alcover, Pedro J. Ponce de León,
    and Jorge Calvo-Zaragoza

Recognition of Arabic Handwritten Literal Amounts Using Deep
Convolutional Neural Networks   169
    Moumen El-Melegy, Asmaa Abdelbaset, Alaa Abdel-Hakim,
    and Gamal El-Sayed

Offline Signature Verification Using Textural Descriptors   177
    Ismail Hadjadj, Abdeljalil Gattal, Chawki Djeddi, Mouloud Ayad, Imran
    Siddiqi, and Faycel Abass

Pencil Drawing of Microscopic Images Through Edge
Preserving Filtering   189
    Harbinder Singh, Carlos Sánchez, Gabriel Cristóbal, and Gloria Bueno

Line Segmentation Free Probabilistic Keyword Spotting and Indexing   201
    Killian Barrere, Alejandro H. Toselli, and Enrique Vidal

Other Applications

Incremental Learning for Football Match Outcomes Prediction   217
    José Domingues, Bernardo Lopes, Petya Mihaylova,
    and Petia Georgieva

Frame by Frame Pain Estimation Using Locally Spatial
Attention Learning   229
  Jun Yu, Toru Kurihara, and Shu Zhan

Mosquito Larvae Image Classification Based on DenseNet and Guided
Grad-CAM.   239
  Zaira García, Keiji Yanai, Mariko Nakano, Antonio Arista,
  Laura Cleofas Sanchez, and Hector Perez

Towards Automatic Rat's Gait Analysis Under Suboptimal
Illumination Conditions   247
  Ana F. Adonias, Joana Ferreira-Gomes, Raquel Alonso, Fani Neto,
  and Jaime S. Cardoso

Impact of Enhancement for Coronary Artery Segmentation Based on Deep
Learning Neural Network.   260
  Ahmed Ghazi Blaiech, Asma Mansour, Asma Kerkeni,
  Mohamed Hédi Bedoui, and Asma Ben Abdallah

Real-Time Traffic Monitoring with Occlusion Handling.   273
  Mauro Fernández-Sanjurjo, Manuel Mucientes, and Víctor M. Brea

Image Based Estimation of Fruit Phytopathogenic Lesions Area   285
  André R. S. Marcal, Elisabete M. D. S. Santos, and Fernando Tavares

A Weakly-Supervised Approach for Discovering Common Objects
in Airport Video Surveillance Footage.   296
  Francisco Manuel Castro, Rubén Delgado-Escaño, Nicolás Guil,
  and Manuel Jesús Marín-Jiménez

Standard Plenoptic Camera Calibration for a Range of Zoom
and Focus Levels   309
  Nuno Barroso Monteiro and José António Gaspar

Going Back to Basics on Volumetric Segmentation of the Lungs in CT:
A Fully Image Processing Based Technique   322
  Ana Catarina Oliveira, Inês Domingues, Hugo Duarte, João Santos,
  and Pedro H. Abreu

Radiogenomics: Lung Cancer-Related Genes Mutation Status Prediction   335
  Catarina Dias, Gil Pinheiro, António Cunha, and Hélder P. Oliveira

Learning to Perform Visual Tasks from Human Demonstrations   346
  Afonso Nunes, Rui Figueiredo, and Plinio Moreno

Serious Game Controlled by a Human-Computer Interface for Upper Limb
Motor Rehabilitation: A Feasibility Study.   359
Sergio David Pulido, Álvaro José Bocanegra, Sandra Liliana Cancino,
and Juan Manuel López

Weapon Detection for Particular Scenarios Using Deep Learning   371
Noelia Vallez, Alberto Velasco-Mata, Juan Jose Corroto,
and Oscar Deniz

Hierarchical Deep Learning Approach for Plant Disease Detection   383
Joana Costa, Catarina Silva, and Bernardete Ribeiro

An Artificial Vision Based Method for Vehicle Detection and Classification
in Urban Traffic   394
Camilo Camacho, César Pedraza, and Carolina Higuera

Breaking Text-Based CAPTCHA with Sparse Convolutional
Neural Networks.   404
Diogo Daniel Ferreira, Luís Leira, Petya Mihaylova,
and Petia Georgieva

Image Processing Method for Epidermal Cells Detection and Measurement
in *Arabidopsis Thaliana* Leaves   416
Manuel G. Forero, Sammy A. Perdomo, Mauricio A. Quimbaya,
and Guillermo F. Perez

User Modeling on Mobile Device Based on Facial Clustering and Object
Detection in Photos and Videos   429
Ivan Grechikhin and Andrey V. Savchenko

Gun and Knife Detection Based on Faster R-CNN for Video Surveillance . . .   441
M. Milagro Fernandez-Carrobles, Oscar Deniz, and Fernando Maroto

A Method for the Evaluation and Classification of the Orange Peel Effect
on Painted Injection Moulded Part Surfaces   453
Atae Jafari-Tabrizi, Hannah Luise Lichtenegger, and Dieter P. Gruber

A New Automatic Cancer Colony Forming Units Counting Method   465
Nicolás Roldán, Lizeth Rodriguez, Andrea Hernandez, Karen Cepeda,
Alejandro Ondo-Méndez, Sandra Liliana Cancino Suárez,
Manuel G. Forero, and Juan M. Lopéz

Deep Vesselness Measure from Scale-Space Analysis of Hessian
Matrix Eigenvalues   473
Ricardo J. Araújo, Jaime S. Cardoso, and Hélder P. Oliveira

Segmentation in Corridor Environments:
Combining Floor and Ceiling Detection   485
Sergio Lafuente-Arroyo, Saturnino Maldonado-Bascón,
Hilario Gómez-Moreno, and Cristina Alén-Cordero

Development of a Fire Detection Based on the Analysis of Video Data
by Means of Convolutional Neural Networks   497
Jan Lehr, Christian Gerson, Mohamad Ajami, and Jörg Krüger

Towards Automatic and Robust Particle Tracking
in Microrheology Studies   508
Marina Castro, Ricardo J. Araújo, Laura Campo-Deaño,
and Hélder P. Oliveira

Study of the Impact of Pre-processing Applied to Images Acquired
by the Cygno Experiment   520
G. S. P. Lopes, E. Baracchini, F. Bellini, L. Benussi, S. Bianco,
G. Cavoto, I. A. Costa, E. Di Marco, G. Maccarrone, M. Marafini,
G. Mazzitelli, A. Messina, R. A. Nobrega, D. Piccolo, D. Pinci, F. Renga,
F. Rosatelli, D. M. Souza, and S. Tomassini

Author Index   531

Contents – Part I

Best Ranked Papers

Towards a Joint Approach to Produce Decisions and Explanations
Using CNNs. 3
   Isabel Rio-Torto, Kelwin Fernandes, and Luís F. Teixeira

Interactive-Predictive Neural Multimodal Systems. 16
   Álvaro Peris and Francisco Casacuberta

Uncertainty Estimation for Black-Box Classification Models: A Use Case
for Sentiment Analysis. 29
   José Mena, Axel Brando, Oriol Pujol, and Jordi Vitrià

Impact of Ultrasound Image Reconstruction Method on Breast Lesion
Classification with Deep Learning. 41
   Michal Byra, Tomasz Sznajder, Danijel Korzinek,
   Hanna Piotrzkowska-Wroblewska, Katarzyna Dobruch-Sobczak,
   Andrzej Nowicki, and Krzysztof Marasek

Segmentation of Cell Nuclei in Fluorescence Microscopy Images
Using Deep Learning. 53
   Hemaxi Narotamo, J. Miguel Sanches, and Margarida Silveira

Food Recognition by Integrating Local and Flat Classifiers 65
   Eduardo Aguilar and Petia Radeva

Machine Learning

Combining Online Clustering and Rank Pooling Dynamics
for Action Proposals 77
   Nadjia Khatir, Roberto J. López-Sastre, Marcos Baptista-Ríos,
   Safia Nait-Bahloul, and Francisco Javier Acevedo-Rodríguez

On the Direction Guidance in Structure Tensor Total Variation
Based Denoising. 89
   Ezgi Demircan-Tureyen and Mustafa E. Kamasak

Impact of Fused Visible-Infrared Video Streams on Visual Tracking 101
   Stéphane Vujasinović, Stefan Becker, Norbert Scherer-Negenborn,
   and Michael Arens

Model Based Recursive Partitioning for Customized Price
Optimization Analytics   113
   Jorge M. Arevalillo

3D Reconstruction of Archaeological Pottery from Its Point Cloud   125
   Wilson Sakpere

Geometric Interpretation of CNNs' Last Layer   137
   Alejandro de la Calle, Javier Tovar, and Emilio J. Almazán

Re-Weighted ℓ_1 Algorithms within the Lagrange Duality Framework:
Bringing Interpretability to Weights.   148
   Matías Valdés and Marcelo Fiori

A Note on Gradient-Based Intensity Normalization   161
   Manuel G. Forero, Carlos Arias-Rubio,
   José de Anchieta C. Horta-Júnior, and Dolores E. López

Blind Robust 3-D Mesh Watermarking Based on Mesh Saliency
and QIM Quantization for Copyright Protection   170
   Mohamed Hamidi, Aladine Chetouani, Mohamed El Haziti,
   Mohammed El Hassouni, and Hocine Cherifi

Using Copies to Remove Sensitive Data: A Case Study on Fair Superhero
Alignment Prediction.   182
   Irene Unceta, Jordi Nin, and Oriol Pujol

Weighted Multisource Tradaboost   194
   João Antunes, Alexandre Bernardino, Asim Smailagic,
   and Daniel Siewiorek

A Proposal of Neural Networks with Intermediate Outputs   206
   Billy Peralta, Juan Reyes, Luis Caro, and Christian Pieringer

Addressing the Big Data Multi-class Imbalance Problem
with Oversampling and Deep Learning Neural Networks   216
   V. M. González-Barcenas, E. Rendón, R. Alejo, E. E. Granda-Gutiérrez,
   and R. M. Valdovinos

Reinforcement Learning and Neuroevolution in Flappy Bird Game   225
   André Brandão, Pedro Pires, and Petia Georgieva

Pattern Recognition

Description and Recognition of Activity Patterns Using Sparse
Vector Fields   239
   Ana Portêlo, Andrea Cavallaro, Catarina Barata, and Jorge S. Marques

Instance Selection for the Nearest Neighbor Classifier:
Connecting the Performance to the Underlying Data Structure 249
*Vicente García, Josep Salvador Sánchez, Alberto Ochoa-Ortiz,
and Abraham López-Najera*

Modified DBSCAN Algorithm for Microscopic Image Analysis of Wood . . . 257
Aurora L. R. Martins, André R. S. Marcal, and José Pissarra

Automatic Detection of Tuberculosis Bacilli from Microscopic Sputum
Smear Images Using Faster R-CNN, Transfer Learning and Augmentation. . . 270
Moumen El-Melegy, Doaa Mohamed, and Tarek ElMelegy

Detection of Stone Circles in Periglacial Regions of Antarctica
in UAV Datasets. 279
Pedro Pina, Francisco Pereira, Jorge S. Marques, and Sandra Heleno

Lesion Detection in Breast Ultrasound Images Using a Machine Learning
Approach and Genetic Optimization 289
*Fabian Torres, Boris Escalante-Ramirez, Jimena Olveres,
and Ping-Lang Yen*

Evaluating the Impact of Color Information in Deep Neural Networks. 302
Vanessa Buhrmester, David Münch, Dimitri Bulatov, and Michael Arens

Diatom Classification Including Morphological Adaptations Using CNNs . . . 317
Carlos Sánchez, Noelia Vállez, Gloria Bueno, and Gabriel Cristóbal

Deep Learning of Visual and Textual Data for Region Detection Applied
to Item Coding. 329
*Roberto Arroyo, Javier Tovar, Francisco J. Delgado, Emilio J. Almazán,
Diego G. Serrador, and Antonio Hurtado*

Deep Learning Versus Classic Methods for Multi-taxon
Diatom Segmentation 342
*Jesús Ruiz-Santaquitaria, Anibal Pedraza, Carlos Sánchez,
José A. Libreros, Jesús Salido, Oscar Deniz, Saúl Blanco,
Gabriel Cristóbal, and Gloria Bueno*

Estimation of Sulfonamides Concentration in Water Based
on Digital Colourimetry. 355
*Pedro H. Carvalho, Sílvia Bessa, Ana Rosa M. Silva, Patrícia S. Peixoto,
Marcela A. Segundo, and Hélder P. Oliveira*

Characterization of Cardiac and Respiratory System of Healthy Subjects
in Supine and Sitting Position. 367
Angel D. Ruiz, Juan S. Mejía, Juan M. López, and Beatriz F. Giraldo

Automatic Fault Detection in a Cascaded Transformer Multilevel Inverter
Using Pattern Recognition Techniques...................................... 378
    Diego Salazar-D'antonio, Nohora Meneses-Casas, Manuel G. Forero,
    and Oswaldo López-Santos

Collision Anticipation via Deep Reinforcement Learning
for Visual Navigation 386
    Eduardo Gutiérrez-Maestro, Roberto J. López-Sastre,
    and Saturnino Maldonado-Bascón

Spectral Band Subset Selection for Discrimination of Healthy Skin
and Cutaneous Leishmanial Ulcers 398
    Ricardo Franco-Ceballos, Maria C. Torres-Madronero,
    July Galeano-Zea, Javier Murillo, Artur Zarzycki, Johnson Garzon,
    and Sara M. Robledo

Data Augmentation of Minority Class with Transfer Learning for
Classification of Imbalanced Breast Cancer Dataset Using Inception-V3..... 409
    Manisha Saini and Seba Susan

Image Processing and Representation

Single View Facial Hair 3D Reconstruction 423
    Gemma Rotger, Francesc Moreno-Noguer, Felipe Lumbreras,
    and Antonio Agudo

From Features to Attribute Graphs for Point Set Registration 437
    Carlos Orrite, Elias Herrero, and Mauricio Valencia

BELID: Boosted Efficient Local Image Descriptor 449
    Iago Suárez, Ghesn Sfeir, José M. Buenaposada, and Luis Baumela

A Novel Graph-Based Approach for Seriation of Mouse Brain
Cross-Section from Images... 461
    S. Sarbazvatan, R. Ventura, F. F. Esteves, S. Q. Lima, and J. M. Sanches

Class Reconstruction Driven Adversarial Domain Adaptation
for Hyperspectral Image Classification.................................... 472
    Shivam Pande, Biplab Banerjee, and Aleksandra Pižurica

Multi-label Logo Classification Using Convolutional Neural Networks 485
    Antonio-Javier Gallego, Antonio Pertusa, and Marisa Bernabeu

Non-destructively Prediction of Quality Parameters of Dry-Cured Iberian
Ham by Applying Computer Vision and Low-Field MRI 498
    Juan Pedro Torres, Mar Ávila, Andrés Caro, Trinidad Pérez-Palacios,
    and Daniel Caballero

Personalised Aesthetics with Residual Adapters   508
  Carlos Rodríguez-Pardo and Hakan Bilen

An Improvement for Capsule Networks Using Depthwise
Separable Convolution.   521
  Nguyen Huu Phong and Bernardete Ribeiro

Wave Front Tracking in High Speed Videos Using a Dynamic
Template Matching   531
  Samee Maharjan

An Efficient Binary Descriptor to Describe Retinal Bifurcation Point
for Image Registration   543
  Sarder Tazul Islam, Sajib Saha, G. M. Atiqur Rahaman, Deep Dutta,
  and Yogesan Kanagasingam

Aggregation of Deep Features for Image Retrieval Based
on Object Detection.   553
  Juan Ignacio Forcén, Miguel Pagola, Edurne Barrenechea,
  and Humberto Bustince

Impact of Pre-Processing on Recognition of Cursive Video Text.   565
  Ali Mirza, Imran Siddiqi, Syed Ghulam Mustufa, and Mazahir Hussain

Image Feature Detection Based on Phase Congruency by Monogenic Filters
with New Noise Estimation   577
  Carlos Jacanamejoy Jamioy, Nohora Meneses-Casas,
  and Manuel G. Forero

Texture Classification Using Capsule Networks   589
  Bharat Mamidibathula, Satakarni Amirneni, Sai Shravani Sistla,
  and Niharika Patnam

Automatic Vision Based Calibration System for Planar Cable-Driven
Parallel Robots   600
  Andrés García-Vanegas, Brhayan Liberato-Tafur,
  Manuel Guillermo Forero, Antonio Gonzalez-Rodríguez,
  and Fernando Castillo-García

3D Non-rigid Registration of Deformable Object Using GPU   610
  Junesuk Lee, Eung-su Kim, and Soon-Yong Park

Focus Estimation in Academic Environments Using Computer Vision   620
  Daniel Canedo, Alina Trifan, and António J. R. Neves

Author Index   629

Biometrics

What Is the Role of Annotations in the Detection of Dermoscopic Structures?

Bárbara Ferreira, Catarina Barata$^{(\boxtimes)}$ ⓘ, and Jorge S. Marques ⓘ

Institute for Systems and Robotics, Instituto Superior Técnico, Lisboa, Portugal
`ana.c.fidalgo.barata@tecnico.ulisboa.pt`

Abstract. There has been an increasing demand for computer-aided diagnosis systems to become self-explainable. However, in fields such as dermoscopy image analysis this comes at the cost of asking physicians to annotate datasets in a detailed way, such that they simultaneously identify and manually segment regions of medical interest (dermoscopic criteria) in the images. The segmentations are then used to train an automatic detection system to reproduce the procedure. Unfortunately, providing manual segmentations is a cumbersome and time consuming task that may not be generalized to large amounts of data. Thus, this work aims to understand how much information is really needed for a system to learn to detect dermoscopic criteria. In particular, we will show that given sufficient data, it is possible to train a model to detect dermoscopic criteria solely using *global annotations* at the image level, and achieve similar performances to that of a fully supervised approach, where the model has access to *local annotations* at the pixel level (segmentations).

Keywords: Skin cancer · Dermoscopic structures · Supervised model · Weakly supervised model · corr-LDA

1 Introduction

Annual reports, such as [13], show that the incidence rates of skin cancer, in particular of its most aggressive form (melanoma), have been steadily increasing for the past decades. Although dermoscopy has been shown to be a powerful imaging technique, the diagnosis of dermoscopy images remains subjective a challenge for untrained dermatologists [2]. This has favored the development of computer-aided diagnosis (CAD) systems that can perform a preliminary screening of the dermoscopy images and work as second opinion tool, from which inexperienced doctors may learn [11,12].

One of the most challenging aspects of CAD systems is the selection of discriminative features to characterize the dermoscopy images [4]. Earlier works based their image descriptors on the medical ABCD rule, which amounted to compute abstract asymmetry, border, color, and texture descriptors, while most recent methods avoid the process of feature design and rely on deep learning

© Springer Nature Switzerland AG 2019
A. Morales et al. (Eds.): IbPRIA 2019, LNCS 11868, pp. 3–11, 2019.
https://doi.org/10.1007/978-3-030-31321-0_1

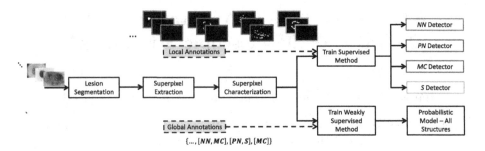

Fig. 1. Block diagram of the training of the proposed models. The colors represent the different dermoscopic structures: NN - negative network (green); PN - pigment network (red); MC - milia-like cysts (blue); and S - streaks (orange). (Color figure online)

architectures to learn the most discriminative features. In both cases, the features lack medical meaning, making it hard to: (i) understand the diagnosis and (ii) be accepted by the medical community. This issue has fostered a promising line of research, where the goal is to develop clinically oriented systems that extract features that have a medical meaning, namely dermoscopic structures and colors [8]. However, this is a challenging field, due to the subtlety of several of the dermoscopic criteria and the need for extensively annotated datasets. These datasets must not only comprise the diagnosis of a skin lesion, but also detailed information of the observed dermoscopic criteria, such as segmentations at the pixel level, which we will refer to as *local annotations*. Until very recently, a dataset of this kind was not publicly available, which fomented the proposal of weakly supervised methods [3,10] that could deal with the only available information (*global annotations, i.e.*, labels at the image level), and still be able to localize the criteria in the skin lesions. Figure 1, shows the difference between *local* and *global annotations*.

Very recently, the International Skin Imaging Collaboration (ISIC), released a dataset that contained more than 2000 images, each with *local annotations* for a small set of relevant dermoscopic structures (pigment network, streaks, milia-like cysts, and negative network) [9]. This dataset has the potential to push forward the development of clinically inspired systems. Nonetheless, several dermoscopic criteria are still missing and the number of examples for most of the structures is small.

For dermatologists to provide *local annotations* is a cumbersome and time consuming task. Thus, it is important to understand if this information is really needed or if it is possible to develop a clinically inspired method, solely using *global labels*. This paper addresses the aforementioned issue and performs a comparison between a fully supervised system, which learns to detect dermoscopic criteria from *local annotations*, and a weakly-supervised system that only uses *global annotations* to perform the same task.

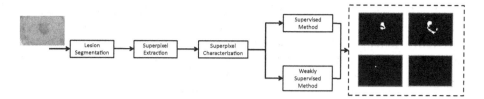

Fig. 2. Block diagram for testing the proposed models and desired output. The colors-lined masks represent the localization of the dermoscopic structures: negative network (green); pigment network (red); milia-like cysts (blue); streaks (orange). (Color figure online)

2   Proposed Framework

The goal of this work is to compare the performance of a supervised learning system against a weakly supervised one on the task of localizing four different dermoscopic structures (pigment network, negative network, streaks, and milia-like cysts) in skin lesions. The block diagram for training and testing the methods is shown in Figs. 1 and 2. The first diagram clearly states the main different between the supervised and weakly supervised strategies: the first relies on labels at the pixel level, called *local annotations*, thus has access to very detailed information, while the latter relies on less informative text labels at the image level (*global annotations*). The second diagram shows that we expect to obtain the same output with both methods, *i.e.*, a set of binary masks with the location of the structures.

Below, we succinctly describe the steps that are common to the two methods. The supervised method is detailed in Sect. 3, while the weakly supervised approach is described in Sect. 4.

2.1   Lesion Segmentation and Superpixel Extraction

Skin lesions are segmented in order to separate them from the surrounding healthy skin. In this work we use manual segmentation masks performed by experts [9].

The following step is to divide the lesion into small and homogeneous regions, called superpixels, such that each region can be analyzed independently and classified regarding the presence/absence of each dermoscopic structure. The algorithm used to compute the superpixels is SLIC0 [1,9].

2.2   Superpixel Characterization

The superpixel algorithm leads to the segmentation of the skin lesion into N regions that must be characterized by a feature vector $r_n \in \mathbb{R}^f$, such that an image d may be represented by the following set $\mathbf{r}^d = \{r_1^d, ..., r_N^d\} \in \mathbb{R}^{f \times N^d}$, where N^d is the number of superpixels of that image. We rely on color and texture information to characterize the superpixels, in particular:

Color features: This property is characterized using the mean color vector in the HSV space (μ_{HSV}).

Texture features: Three types of descriptors are used to characterize the texture of a region. The mean contrast (μ_c), the mean contrast $\times$ anisotropy (μ_{ca}), and statistics computed using the directional filters from [5]. Both contrast and anisotropy are computed from the second moment matrix estimated at each pixel $M(x, y)$

$$a(x,y) = 1 - \frac{\lambda_2}{\lambda_1}, \quad c(x,y) = 2\sqrt{\lambda_1 + \lambda_2}, \tag{1}$$

where λ_1, λ_2 are the eigenvalues of $M(x, y)$ [7]. The directional filters are applied at different orientations $\theta_i \in [0, \pi]$, $i = 0, \ldots, 9$, with the impulse response for any direction θ_i given by

$$h_{\theta_i}(x,y) = G_1(x,y) - G_2(x,y), \tag{2}$$

where G_k is a Gaussian filter:

$$G_k(x,y) = C_k \exp\left\{-\frac{x'^2}{2\sigma_{x_k}^2} - \frac{y'^2}{2\sigma_{y_k}^2}\right\}, k = 1, 2. \tag{3}$$

C_k is a normalization constant and the values of (x', y') are related with (x, y) by a rotation of amplitude θ_i.

$$\begin{aligned} x' &= x \cos\theta_i + y \sin\theta_i, \\ y' &= y \cos\theta_i - x \sin\theta_i. \end{aligned} \tag{4}$$

We compute the output of the directional filters (2) for all the directions and keep the maximum and minimum at each pixel (x, y). The regions are described by the mean and standard deviation of these values (μ_M, σ_M, μ_m, and σ_m).

3 Supervised Model

In the supervised context, the model has access to all detailed information during the training phase. In the case of this work, this means that to be able to train a supervised model, we need to have access to detailed ground-truth annotations at the pixel or superpixel level (*local annotations*, as shown in Fig. 3). The annotations used in this work are at the superpixel level.

Given the *local annotations*, the problem of localizing dermoscopic structures becomes a simple classification problem, where our goal is to classify each superpixel into one of four possible classes: negative network, pigment network, milia-like cysts, and streaks. Although this could be treated as a multi-class problem, some of the superpixels have more than one label (see Fig. 3). Thus, we will train a separate classifier for each of the classes, as shown in Fig. 1. During the test phase, each of the classifiers is separately used to label the superpixels.

The classification algorithm used in this work is SVM (support vector machines) with an radial basis function (RBF) kernel.

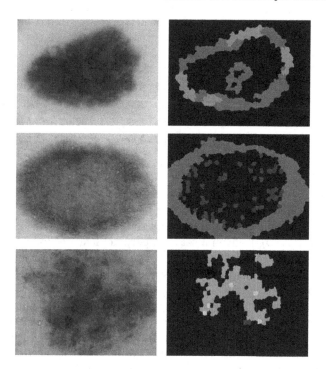

Fig. 3. *Local annotations*: negative network (green); pigment network (red); milia-like cysts (blue); and streaks (orange). The remaining colors identify superpixels that have multiple labels. (Color figure online)

4   Weakly Supervised Model

The supervised approached described in Sect. 3 corresponds to the ideal scenario, where we have access to detailed annotations of the presence and location of the dermoscopic structures that we want to detected. However, until the recent release of challenge related datasets, such as ISIC 2017 [9], this kind of information was unavailable. Most of times, one has only access to the dermoscopy images and a set of *global annotations* at the image level, which dermoscopic structures are present. To address this limitation and still be able to localize the dermoscopic structures, Barata et al. [3] proposed a framework based on the correspondence latent Dirichlet allocation (corr-LDA) model [6].

Corr-LDA belongs to the family of generative algorithms for image captioning [6]. The main idea of this method is that we can represent an image as a distribution over a set of K latent variables z, called topics. Each of the topics allow us to simultaneously express: (i) a distribution over superpixel features $p(r_n|z_k, \Omega_k)$, where Ω_k is the set of parameters of the distribution associate to z_k; and (ii) a multinomial distribution over the possible *global annotations* $p(w|z_k, \beta_k)$ with parameter β_k. Thus, through the topics, we are able to find

Table 1. ISIC 2017 statistics: PN - pigment network; MC - milia-like cysts; NN - negative network; and S - streaks.

Set	Local annotations					Global annotations				
	# SPixel	% PN	% MC	% NN	% S	# Images	% PN	% MC	% NN	% S
Train	460272	16.92	1.01	0.71	0.46	2000	56.40	28.40	6.25	5.80
Val.	31946	10.41	1.02	1.03	0.04	150	42.67	28.67	7.33	1.33
Test	193730	10.38	0.66	1.12	0.07	600	55.50	24.50	7.50	1.50

the relationship between superpixel features r_n and the *global annotations*, and consequently label each superpixel.

Training the corr-LDA model amounts to estimating the set of parameters $\{\Omega_1, \ldots, \Omega_K, \beta_1, \ldots, \beta_K\}$, given a set of superpixel features for different images $\mathcal{R} = \{\mathbf{r}^1, \ldots, \mathbf{r}^D\}$ and their corresponding *global annotations* (negative network, pigment network, milia-like cysts, and streaks). We use the strategy described in [3,6], to train our model. On the test phase, we apply the estimated model and compute the probability $p(w|r_n)$ (see [3]), to determine the probability of each global label. Since some of the superpixels may be associated with more than one label (recall Fig. 3), we experimentally determined a threshold on the probability of each annotation, such that a dermoscopic structure was assigned to a superpixel if its probability was greater than the threshold.

5   Experimental Results

5.1   Dataset and Performance Metrics

The supervised and weakly supervised approaches were evaluated using the ISIC 2017 challenge dataset [9], which comprises 2750 images divided into training, validation, and test sets. The training set was used to estimate parameters of the supervised (SVM) and weakly supervised (corr-LDA) models, while the validation set was used to selected the best model and respective hyperparameters (the kernel width in RBF and the number of topics K in corr-LDA). The test set was used to evaluate and compare the models.

Each of the images was segmented by experts, who also annotated each of the superpixels into one or more of the following dermoscopic structures: negative network, pigment network, milia-like cysts, and streaks. Table 1 shows the number of superpixels per set, as well as the percentage that is associated to each structure. Since the weakly supervised model uses image-level annotations, we have defined that an image receives the *global annotation* of a specific structure if it has at least one superpixel with that label. For computational reasons it was necessary to define an additional *global annotation* called "without structure" to deal with images that do not exhibit any structure. The proportion of *global labels* per type of structure is also shown in Table 1.

Table 2. Experimental results for the supervised and weakly supervised models. In **bold** we highlight the most interesting scores.

Dermoscopic structure	Supervised			Weakly supervised		
	SE	SP	BACC	SE	SP	BACC
Pigment network	**84.6%**	**69.2%**	**76.9%**	**73.3%**	**76.0%**	**74.7%**
Milia-like cysts	62.6%	60.3%	61.5%	0.6%	98.6%	49.6%
Negative network	67.6%	70.8%	69.2%	3.4%	99.3%	51.3%
Streaks	71.4%	73.0%	72.2%	25.7%	94.5%	60.1%

These statistics show that there is a significant imbalance in the dataset, with pigment network being the most common annotation both at the superpixel and image level. To overcome this limitation, we used two strategies: (i) in the supervised approach, we have assigned different weights to the classes during the training of the classifiers, based on their distribution; (ii) in the weakly supervised approach we have artificially augmented the data, such that there were at last 500 images for each type of *global annotation*.

The metrics used to evaluate the models are the sensitivity (SE), specificity (SP), and balanced accuracy ($BACC$)

$$SE = \frac{TP}{TP+FN}, \ SP = \frac{TN}{TN+FP}, \ BACC = \frac{SE+SP}{2}, \tag{5}$$

where TN, TP, FN, and FP are the total number of superpixels that are respectively true negatives, true positives, false negatives, and false positives.

5.2 Results

The scores for the supervised and weakly supervised models are shown in Table 2. In the case of the supervised model, the classification of the superpixels w.r.t to each dermoscopic structure is achieved with different degrees of success. In particular, it seems that milia-like cysts and negative network are harder to identify than the other two structures.

The weakly supervised model exhibits low SE for milia-like cysts, negative network, and streaks. However, the most noteworthy result is that of pigment network. The performance for this structure is similar to that of the supervised approach at the cost of approximately 0.43% for the annotations: recall that the supervised method relies on more than 460K *local annotations* at the superpixel level, while the weakly supervised uses 2K *global annotations* at the image level. This suggests that given a sufficient number of images with a specific dermoscopic structure, one does not require very detailed annotations as a weakly supervised model seems to be able to achieve a similar performance. Moreover, the adopted region features may also play a role, as pigment network is a highly directional structure with sharp transitions between the lines and the background. This properties may be well described by the directional filters. The performance for streaks, a structure that shares properties with pigment network, seems to support this claim, as with only 5.8% of the images with this annotation, corr-LDA still achieves a $SE = 25.7\%$ and a $SP = 94.5\%$ and the supervised model is the second best. Since providing detailed annotations is a cumbersome task, this result opens new possibilities for training models with less information.

Negative network and especially milia-like cysts seem to be the structures that achieve the worse performances in both supervised and weakly-supervised approaches. This suggests that the features used to characterize the superpixels may not efficiently represent these two structures. Thus, a future direction of improvement will be to replace the features described in Sect. 2.2 with more powerful descriptors, namely those based on deep neural networks.

6   Conclusions

This paper compares two models for the detection and localization of four dermoscopic structures (pigment network, milia-like cysts, negative network, and streaks) in superpixels of dermoscopy images. The first model was fully supervised, thus was trained using more than 4.60×10^5 *local annotations* at the superpixel level, while the second model relied on a weakly-supervised algorithm (corr-LDA), trained using 2×10^3 *global annotations* at the image level. The experimental results surprisingly showed that in the case of pigment network, the weakly supervised model is able to achieve a similar performance to that of the supervised one, with a significantly smaller amount of ground truth information. However, the supervised model achieved better performances for the remaining structures, suggesting that better features and more data is required. Nonetheless, the most relevant output of this work is that given a sufficient number of images, we may not need detailed information to be able to detect dermoscopic structures.

Acknowledgments. This work was supported by the FCT project and plurianual funding: [PTDC/EEI PRO/0426/2014], [UID/EEA/50009/2019].

References

1. Achanta, R., Shaji, A., Smith, K., Lucchi, A., Fua, P., Susstrunk, S.: SLIC superpixels compared to state-of-the-art superpixel methods. IEEE Trans. Pattern Anal. Mach. Intell. **34**(11), 2274–2282 (2012)
2. Argenziano, G., Soyer, H.P., De Giorgi, V., et al.: Interactive Atlas of Dermoscopy. EDRA Medical Publishing & New Media (2000)
3. Barata, C., Celebi, M.E., Marques, J.S.: Development of a clinically oriented system for melanoma diagnosis. Pattern Recogn. **69**, 270–285 (2017)
4. Barata, C., Celebi, M.E., Marques, J.S.: A survey of feature extraction in dermoscopy image analysis of skin cancer. IEEE J. Biomed. Health Inform. **23**(3), 1096–1109 (2018)
5. Barata, C., Marques, J.S., Rozeira, J.: A system for the detection of pigment network in dermoscopy images using directional filters. IEEE Trans. Biomed. Eng. **59**(10), 2744–2754 (2012)

6. Blei, D., Jordan, M.: Modeling annotated data. In: 26th Annual International ACM SIGIR Conference on Research and Development in Informataion Retrieval, pp. 127–134. ACM (2003)
7. Carson, C., Belongie, S., Greenspan, H., Malik, J.: Blobworld: image segmentation using expectation-maximization and its application to image querying. IEEE Trans. Pattern Anal. Mach. Intell. **24**(8), 1026–1038 (2002)
8. Celebi, M.E., Codella, N., Halpern, A.: Dermoscopy image analysis: overview and future directions. IEEE J. Biomed. Health Inform. **23**(2), 474–478 (2019)
9. Codella, N.C.F., Gutman, D., Celebi, M.E., et al.: Skin lesion analysis toward melanoma detection: a challenge at the 2017 international symposium on biomedical imaging (ISBI), hosted by the international skin imaging collaboration (ISIC). In: IEEE 15th International Symposium on Biomedical Imaging (ISBI 2018), pp. 168–172 (2018)
10. Madooei, A., Drew, M.S., Hajimirsadeghi, H.: Learning to detect blue-white structures in dermoscopy images with weak supervision. IEEE J. Biomed. Health Inform. **23**(2), 779–786 (2018)
11. Oliveira, R., Papa, J., Pereira, A., Tavares, J.: Computational methods for pigmented skin lesion classification in images: review and future trends. Neural Comput. Appl. **29**(3), 1–24 (2016). https://doi.org/10.1007/s00521-016-2482-6
12. Pathan, S., Prabhu, K.G.S., Siddalingaswamy, P.C.: Techniques and algorithms for computer aided diagnosis of pigmented skin lesions - a review. Biomed. Signal Process. Control **39**, 237–262 (2018)
13. Siegel, R.L., Miller, K.D., Jemal, A.: Cancer statistics, 2019. CA: Cancer J. Clin. **69**, 7–34 (2019)

Keystroke Mobile Authentication: Performance of Long-Term Approaches and Fusion with Behavioral Profiling

Alejandro Acien$^{(\boxtimes)}$, Aythami Morales, Ruben Vera-Rodriguez, and Julian Fierrez

BiDA Lab, School of Engineering, Universidad Autonoma de Madrid,
C/ Francisco Tomas y Valiente 11, 28049 Madrid, Spain
{alejandro.acien,aythami.morales,ruben.vera,
julian.fierrez}@uam.es

Abstract. In this paper we evaluate the performance of mobile keystroke authentication according to: (1) data availability to model the user; and (2) combination with behavioral-based profiling techniques. We have developed an ensemble of three behavioral based-profile authentication techniques (WiFi, GPS Location, and App usage) and a Keystroke state-of-the-art recognition approach. Algorithms based on template update are employed for profiling systems meanwhile bidirectional recurrent neuronal networks with a Siamese training setup is used for the keystroke system. Our experiments are conducted on the semi-uncontrolled UMDAA-02 database. This database comprises smartphone sensor signals acquired during natural human-mobile interaction. Our results show that it is necessary 6 days of usage data stored to achieve the best performance in average. The template update allows to improve the equal error rate of keystroke by a relative 20%–30% performance.

Keywords: Mobile authentication · Biometric recognition · Behavioral pattern · Behavioral-based profiling · Keystroke dynamics

1 Introduction

In the last decade smartphones have become a vital gadget for an important percentage of the world population. Recent reports reveal that mobile lines exceeded the world population in 2018 [1]. Moreover, more than 90% of citizens decline to go out without their smartphones due to the need for contact with their friends or work responsibilities among others reasons [2]. During our daily routines, smartphones become a sort of data hubs storing a wide variety of sensitive information: from personal information (e.g. photos, videos, chats messages) stored by ourselves, behavioral traits (e.g. touch gestures, GPS location, WiFi connections, keystroke patterns) stored by the smartphones during the user interaction, up to critical information (e.g. bank transactions, account's passwords, contacts list). Due to this capacity of storing sensitive information, according to [3] more than a half of population would be willing to pay 500$ and the 30% would pay up to 1000$, regardless the price of the device, in order to recover the smartphone information when stolen.

© Springer Nature Switzerland AG 2019
A. Morales et al. (Eds.): IbPRIA 2019, LNCS 11868, pp. 12–24, 2019.
https://doi.org/10.1007/978-3-030-31321-0_2

However, some surveys have shown that about 34% or more smartphone users did not use any form of authentication mechanism on their mobile devices [4]. Among the reasons for this, inconvenience of use was cited to be the main reason. They find out that mobile device users considered unlock screens unnecessary in 24% of the situations and they spend up to 9% of time they use their smartphone unlocking the screens despite of many modern smartphones have fingerprint and face recognition algorithms.

In order to deal with this problem of convenience of use, the research community is developing transparent biometric authentication mobile systems [5]. These approaches analyze behavioral information (e.g. touch gestures, keystroke patterns) stored by the smartphone during normal user-device interaction and check the user's identity in the background. This way, the smartphone will be able to assist in user authentication avoiding to disturb the owner with traditional authentication mechanisms (e.g. passwords, swipe patterns).

Despite some of these biometric authentication mobile systems work really well achieving a good performance under certain conditions (e.g. limited number of users, supervised scenarios), these systems are not usually tested in a real life scenario in which a new user installs the authentication system in the device and starts using it. In that moment, the amount of behavioral data available for the biometric authentication will be scarce and the performance may be low. The device will need a traditional authentication mechanism until it has enough behavioral biometric information to check the identity of the user by itself with good performance.

The aim of this paper is to analyze how the performance of these mobile biometric authentication systems evolve according to the amount of behavioral information available from the owner. Our experiments include up to four different information channels (Keystroke, GPS location, WiFi signals, and App Usage) and the fusion of all of them to train a reliable authentication system by employing each time more user's information to authenticate. Finally, we will analyze how much information these biometric systems need to work with a good performance. For this, our experiments are conducted on the UMDAA-02 mobile database [6], a challenging mobile dataset acquired under unsupervised conditions.

The rest of this paper is organized as follows: Sect. 2 makes an overview of the state-of-the-art works related and links with this work. Section 3 describes the architecture followed to implement the different systems proposed. Section 4 explains the experimental protocol, describing the database and the experiments performed. Section 4.3 presents and analyzes the results achieved and Sect. 5 summarizes the conclusions and future work.

2 Background and Related Works

Authentication systems based on keystroke dynamics have been widely studied in computer keyboards, achieving very good results in fixed text [7] (i.e. the keystroke sequence of the input authentication system is prefixed) and free text [8] (the input can be any kind of keystroke sequence). The feature set usually employed in keystroke recognition is generated using the elapsed time of press and release events between consecutive keys [9]: hold time, inter-key latency, press and release latency. In the authentication stage; Manhattan distances, DTW and digraphs achieve the best results in most of the cases for fixed text scenarios [7, 10, 11], whereas binary classifiers (SVM, KNN), Hierarchical Trees and Recurrent Neuronal Networks work better in free text [12–14].

Regarding keystroke authentication in smartphones, similar architectures have been applied with little adaptations. In [15], they take advantage of the hand postures while holding the device during typing as discriminative information, and combining this with time features they reduce the error rates up to 36.8% in a fixed text scenario with binary classifiers (SVM, NB, and KNN). In [16], Monaco *et al.* proposed Partially Observable Hidden Markov Models (POHMM) as an extension of the traditional Hidden Markov Models (HMMs), but with the difference that each hidden state is conditioned on an independent Markov chain. The algorithm is motivated by the idea that typing events depend both on past events and also on a separate process. More recently, [17] proposed a Siamese Long Short-Term Memory network architecture in which the keystroke authentication is performed by calculating the Euclidean distance between two embedding vectors (the outputs of the Siamese model).

The WiFi networks detected by our smartphone provide useful information about when and where we go, and hence, they can detect possible variations in our daily routines. This discriminative information is considered as behavioral biometric and it could help in the mobile authentication process. In this assumption, [18] explores a WiFi authentication system based on templates. They store in a template the time and the name of the WiFi networks detected during the training process, then they test by comparing the template with the new WiFi networks detected and compute a kind of confidence score.

Regarding Geo-location based authentication approaches, Mahub *et al.* [19] developed a modified HMM to characterize the mobile trace histories, they suggest that the human mobility can be described as a Markovian Motion, and they predict the new user location exploiting the sparseness of the data and past locations. In [13], they classify mobile user location with SVM by using the latitude and longitude as features and calibrating the scores with logistic regression. They also implement App Usage based authentication by ranking the top 20 mobile applications most visited by the user that appear in the training set. The classification process is performed by comparing these top ranks of more used applications with the new test data and calculating a similarity score. In the other hand, [20] suggests that the unknown applications and unforeseen events have more impact in App Usage authentication than the top N-apps, and they should be incorporated in the models by adopting smoothing techniques with HMMs. They are capable of detecting an intrusion in less than 3 min of application usage with only 30 min of historical data to train.

Finally, how to integrate all these different modalities in a multimodal mobile authentication architecture is not trivial [21]. Due to many differences between the architectures proposed for each biometric trait, the fusion is usually done at decision level. For example in [13], they fused at decision level web browsing, application usage, GPS location, and keystroking data using information from slice time windows. They suggest that the performance increases according to the size of the time window. In [22], they merge also at decision level touch dynamics, power consumption, and physical movements modalities with a dataset captured under supervised conditions. In

[17], they merge up to 8 modalities (keystroke dynamics, GPS location, accelerometer, gyroscope, magnetometer, linear accelerometer, gravity, and rotation sensors) at score level with a Siamese Long Short-Term Memory network architecture and 3 s window. The fusion approach enhances the performance more than 20% compared to each modality separately. In [23], they designed a mobile authentication app that collects data from WiFi, Bluetooth, accelerometer, and gyroscope sources during natural user interaction and fused them at score level achieving up to 90% of accuracy in the best scenario.

Previous works fusing different modalities [13, 17, 22, 23] have focused their approaches on obtaining time windows from the different modalities and then, they carry out the fusion with the architectures previously trained for each user. However, this does not represent a realistic scenario because biometric information is not always available at the beginning and therefore, the lack of these biometric information could decrease the performance.

The major contributions of this paper are: (i) a performance analysis of user mobile authentication based on keystroke biometrics traits and 3 behavioral-based profiling techniques (GPS location, WiFi, and App usage) separately and the fusion of all them at score level for a multimodal approach, and (ii) a study of the performance evolution of these authentication systems across the time according to the amount of user biometric information available in each moment.

3   Systems Description

In this paper we will analyze 4 mobile sources of information: Keystroking, GPS Location, App Usage, and WiFi. According to the literature, keystroke patterns are related to the neuromotor skills of the people based on, for instance, time differences between consecutive keys, which are directly related to muscles activation/deactivation timing [24]. On the other hand, GPS location, WiFi, and App Usage belong to behavioral based-profiling systems that describe daily habits and manners from the user according to the services they use or the places they visit [5]. In the next subsection we describe the approach followed for each of the 4 systems taking into account the above definitions.

3.1   Keystroke System

In keystroking, the discriminative user information is allocated in the temporal relationships of press and release events between two or more consecutive keys. For this reason, we decided to implement a Recurrent Neural Network (RNN) algorithm for keystroking authentication. To the best of our knowledge, RNN has demonstrated to be one of the best algorithms to deal with temporal data and works well with free-text keystroke patterns [14, 17]. The feature set chosen is as follows: (i) Hold Latency (HL): the elapsed time between press and release key events; (ii) Inter-key Latency (IL): the elapsed time between releasing a key and pressing the next key; and (iii) Press Latency

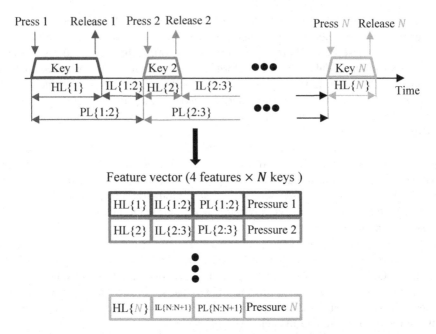

Fig. 1. Example of feature extraction for a keystroke sequence of N keys. The number in brackets shows the key corresponding to each feature.

(PL): the elapsed time between two consecutive press events. Additionally, we add the pressure as another feature to provide more behavioral user information (see Fig. 1 for details).

Our RNN model has a fixed length input N. To handle keystroke sequences of varying length, we concatenate them until we have the length necessary to feed the RNN, as proposed in [17]. The longer sequence we choose, the better performance the RNN model usually achieves. However, the system has to wait until the user has pressed enough number of keys to authenticate the user. So there is a trade-off between the performance and the authentication time delay.

The architecture of the RNN model that achieved our best results is depicted in Fig. 2. That RNN consists of two LSTM layers of 32 units with batch normalization and dropout rate of 0.5 between layers to avoid overfitting. We suggest that the next keys typed are as relevant as past keys, therefore, in order to consider forward and backward time relationships between consecutive keys, we decided to set up the LSTM layers in a bidirectional mode (duplicating the number of neurons in each layer, one for each forward and backward direction). The output of the RNN model is an embedding vector of 64 units' size (32×2), this embedding vector is a feature representation of the input keystroke sequence that we will use to distinguish a keystroke sequence from genuine and impostor users. By training the RNN model in a Siamese setup, the RNN model will learn discriminative information of the keystroke sequence and transform this information into an embedding space where keystroke sequences of the same user (genuine samples) are close, and far in the opposite case.

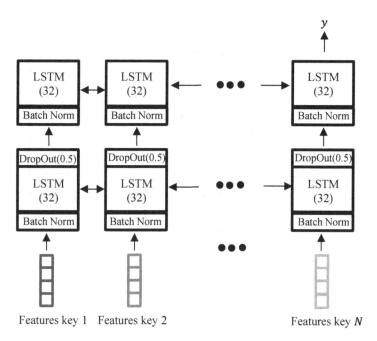

Fig. 2. Architecture of the Bidirectional RNN model proposed. The output of the model y is a embedding vector of 64 (32×2) features (the bidirectional mode duplicates the number of neurons in each layer).

In this setup, the RNN model has two inputs (the two keystroke sequences to compare) and outputs two embedding vectors (see Fig. 3 for details). By calculating the Euclidean distance between this pair of embedding vectors we will obtain a score between 0 and α, where 0 means that both keystroke sequences belong to the same user and α means that they come from different users.

For this, the contrastive loss is defined to regulate large or small distances depending on the label y_{ij} associated with the pair of samples [17]. Let's define X_i and X_j as both inputs of the Siamese model, the Euclidean distance between the pairs $d(X_i, X_j)$ is defined as:

$$d(X_i, X_j) = \|f(X_i) - f(X_j)\| \tag{1}$$

where $f(X_i)$ and $f(X_j)$ are the outputs (embedding vectors) of the RNN Model. Finally, with the contrastive loss, the RNN model will learn to make this distance small for genuine pairs and large for impostor pairs according to the label y_{ij}:

$$Loss = (1 - y_{ij}) \frac{d^2(X_i, X_j)}{2} + y_{ij} \frac{\max^2(0, \alpha - d(X_i, X_j))}{2} \tag{2}$$

where the label y_{ij} is set to 0 for genuine pairs and 1 for impostor pairs and $\alpha > 0$ is called the margin (the maximum margin between genuine and impostor distances).

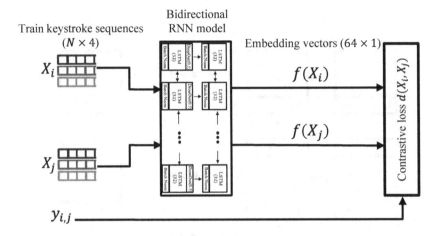

Fig. 3. Siamese keystroke setup for training. N is the number of keys in each sequence.

3.2 Behavioral-Based Profiling Systems

For WiFi, GPS location, and App Usage behavioral-based profiling systems we employ a template-based matching algorithm that has proved to work well according to [18, 20]. This algorithm consists in user's templates that record the time stamps and the frequency of the events occurred during the daily routines of the user. These events are the WiFi networks detected, the latitude and longitude of a location or the name of the app for WiFi, GPS location, and App Usage systems respectively.

Table 1 shows an example of a template for the WiFi system. First of all, we divided the 24 h of the day in M time slots of fixed duration. For example, for $M = 48$ we will have 48 slots of 30 min length (24/48 = 0.5 h = 30 min). Once the size of the time slots is set, the template records for each WiFi network the time slot it belongs and the name of the network. The frequency column shows the number of sessions that WiFi network was detected in the same time slot. In other words, the template-based algorithm describes the daily routines of the user during a period of time according to the events detected by their smartphone when he/she unlocks the smartphone.

Table 1. Example of a WiFi user template generated according the data captured during a week.

Event (WiFi network)	Time slot	Frequency
Network 1	4	7
Network 2	10	3
Network 3	10	1
Network 1	15	7
Network 4	24	5

Table 2. General UMDAA-02 dataset information.

Description	Statistics
Gender	36M/12F
Age	22–31 years
Avg. Days/User	10 days
Avg. Sessions/User	248 sessions
Avg. Sensors/Session	5.2 Sensors
Avg. Sessions/Day	26 sessions

Finally, we test the system by comparing the new sessions with the user template. We match the new events detected with the events of the template for each time slot and calculate a confidence score as:

$$score = \sum_{i=1}^{S} f_i^2 \qquad (3)$$

where f_i is the frequency of the event stored in the user template that matches with the test event i in the same time slot and S is the total number of events detected in that test session. For example, if the test session includes the WiFi networks of '*Network 2*' and '*Network 3*' during the tenth time slot, the score confidence will be $1^2 + 3^2 = 10$ (according to the template showed in Table 1). Based on this, a higher score in the test session implies higher confidence for authentication.

4 Experiments

4.1 Database

The experiments were conducted with the UMDAA-02 database [6] that comprises more than 140 GB smartphone sensor signals collected during natural user-device interaction. Table 2 summarize the characteristics of the database. The users were mainly students from the university of Maryland, they used a smartphone provided by the researchers as their primary device during their daily life (unsupervised scenario) over a period of two months. A huge range of smartphone sensors were captured: touchscreen (i.e. touch gestures and keystroking), gyroscope, magnetic field, GPS location, and WiFi networks, among others. Information related to mobile user's behavior like lock and unlock time events, start and end time stamps of calls, and app usage are also stored.

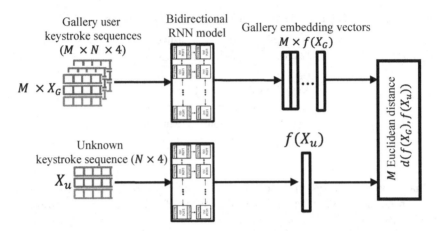

Fig. 4. Siamese keystroke setup for testing. The number M of gallery samples in test varies according to the number of keystroke sessions employed for testing. The number of keys in each sequence N is set to 20.

The structure of the database is divided in sessions (i.e. the elapsed time between the user unlocks the screen until the next lock). For each session, the device stores in a folder all the sensor signals employed in that session. For example, if the user unlocks the smartphone to check the email inbox maybe there are no GPS locations or keystroke signals but WiFi and swipes gestures could be provided. The amount of data and the kind of signals acquired vary according to the user's behavior. This reason motivated us to analyze the temporal performance evolution of our systems at session level instead of fixed time slots like days or weeks. Some users could provide a large amount of information in only one day whereas it could be scarce in other users.

4.2 Experimental Protocol

For the behavioral based-profiling systems (WiFi, GPS Location, and App Usage), we train the templates with the first M sessions acquired for each user and using the remaining sessions as genuine test sessions. Sessions from the others users are considered as impostor data.

The RNN model for keystroke recognition is trained in a Siamese setup, which had showed to perform very well with short time sequences like signatures or smartphone time signals [17, 25]. For this, we train de RNN model with pairs of keystroking sequences from train users (80% of the users). Regarding the training details, the best results were achieved with a learning rate of 0.005, Adam optimizer was used with $\beta_1 = 0.9$, $\beta_2 = 0.999$ and $\varepsilon = 10^{-8}$ respectively, batch size of 512 pairs and the margin set to $\alpha = 1.5$.

For testing, Fig. 4 shows the details of the setup in which we compare the first M keystroke sequences of each test user (commonly named gallery samples) with new keystroke sequences that belong to the same user (genuine samples) or other test users

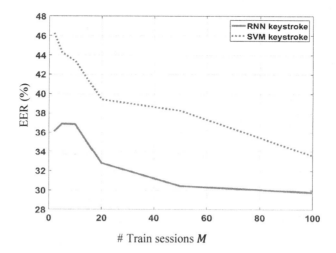

Fig. 5. Evaluation of keystroke performance with bidirectional RNN and SVM models across the number of sessions employed to train.

(impostor samples). Our idea is to build a RNN model able to generalize, distinguishing keystroke sequences from any kind of users.

As we commented before, the keystroke data in UMDAA-02 is stored by sessions and the length of the keystroke sequences vary depending on the session and user, but the RNN model has a fixed length input that we set to $N = 20$, the average of the keystroke sequence length of the database. To avoid zero padding or truncating, we concatenate consecutively keystroke sequences from the user sessions to build the input keystroke sequence of the RNN model so this sequence will belong to only one session in average but could be more or less. In this assumption, we have a total of 8615 keystroke sequence in total for all users in the database.

Finally, to study the temporal evolution of the performance across the time (sessions in this paper) in the keystroke system, we will increase the number of gallery sequences assuming that each gallery sequence is a new genuine user session and then, we test the unknown sample comparing it to all gallery sequences and averaging the M resulting distances (see Fig. 4).

4.3   Results and Discussion

We first compare the bidirectional RNN model for keystroke proposed in this paper with a SVM model, following the traditional workflow of global feature extraction and classification. For this, we extract again the same time features as in the RNN model (HL, IL, PL, and Pressure) and then we compute the global features for each time feature: mean, median, standard deviation, 1 percentile, 99 percentile, and 99-1 percentile. According to this protocol, for each keystroke session we have a feature vector of size 24 (6×4). Then, we train a SVM for each user using his/her first M keystroke sessions as genuine samples and M samples from other users as impostor ones.

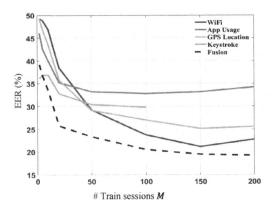

Fig. 6. Evaluation performance for all modalities and the fusion of all across the number of sessions employed to train.

Figure 5 shows the EER curves for keystroke systems using both SVM and RNN algorithms versus the number of sessions employed to train the models. EER refers to Equal Error Rate, the value where False Acceptance Rate (percentage of impostors classified as genuine) and False Rejection Rate (percentage of genuine users classified as impostors) are equal. The results suggest that RNN networks work better than SVM in all cases, even when there are few user samples available for training the RNN algorithm. However, the EER for SVM tends to drop faster than RNN as more user data is available.

Finally, we evaluate the performance of each biometric system individually and the fusion of all of them. Figure 6 shows the performance versus number of sessions employed for training. The results show that the best individual system in terms of EER is the WiFi system, achieving less than 23% of EER with 150 sessions to train the templates. However, the keystroke system drops faster and performs better with few user data. We think that this occurs because the keystroke system (bidirectional RNN) was previously trained with data from other users and learnt discriminative keystroke patterns, being able to authenticate new users with few samples.

The dotted curve shows the fusion of all systems at score level. To get the best of the fusion scheme, we weighted the systems by giving higher weights to the best ones (WiFi and GPS location). The best results are around 19% of EER with more than 150 sessions to train. According to Table 2 (last row), the authentication systems need more than 6 days in average to authenticate users to achieve the best performance possible.

5   Conclusions and Future Work

In this paper we evaluate the performance of mobile keystroke authentication according to: (1) data availability to model the user; and (2) combination with behavioral-based profiling techniques. We have developed an ensemble of three behavioral-based profile authentication techniques (WiFi, GPS Location and App usage) and a keystroke state-of-the-art recognition approach. The results showed that even though behavioral based-profile systems tend to work better with large amounts of training data, the performance gets worse when the amount of data to model the user is scarce. We therefore suggest that behavioral profile systems work well at long terms, when the smartphone has stored enough data to train the templates.

On the other hand, a keystroke system based on bidirectional RNN seems to work better with few samples. We suggest that this happens due to the pre-training phase of the RNN model with development users in a Siamese setup.

Although the keystroke system works better than the others with few samples, the performance is not competitive for large amounts of training data. For future work, we propose to improve the keystroke performance by employing transfer learning techniques and adapting the RNN model to each user when the amount of user data is enough.

Acknowledgments. This work was financed by projects: BIBECA (RTI2018-101248-B-I00 from MICINN/FEDER) and BioGuard (Ayudas Fundacion BBVA).

References

1. Mobile World Congress (2018). https://elpais.com/tecnologia/2018/02/27/actualidad/1519725291_071783.html. Accessed 01 Apr 2019
2. Impacts of Cell Phone Addiction. https://ifpgod.wordpress.com/about/impacts-of-cell-phone-addiction/. Accessed 01 Apr 2019
3. Would you be willing to be in danger to get your cell phone back? http://primerasnoticias.com/2014/05/correr-peligro-recuperar-movil/. Accessed 01 Apr 2019
4. Harbach, M., Von Zezschwitz, E., Fichtner, A., De Luca, A., Smith, A.: It's a hard lock life: a field study of smartphone (un) locking behavior and risk perception. In: 10th Symposium on Usable Privacy and Security (SOUPS), pp. 213–230 (2014)
5. Patel, V.M., Chellappa, R., Chandra, D., Barbello, B.: Continuous user authentication on mobile devices: recent progress and remaining challenges. Proc. IEEE Signal Process. Mag. **33**, 49–61 (2016)
6. Mahbub, U., Sarkar, S., Patel, V.M., Chellappa, R.: Active user authentication for smartphones: a challenge data set and benchmark results. In: Proceedings of the IEEE 8th International Conference on Biometrics Theory, Applications and Systems, New York, USA (2016)
7. Morales, A., Fierrez, J., et al.: Keystroke biometrics ongoing competition. IEEE Access **4**, 7736–7746 (2016)
8. Tappert, C.C., Cha, S.H., Villani, M., Zack, R.S.: A keystroke biometric system for long-text input. In: Optimizing Information Security and Advancing Privacy Assurance: New Technologies, pp. 32–57. IGI Global (2012)
9. Neal, M., Balagani, K., Phoha, V., Rosenberg, A., Serwadda, A., Karim, M.E.: Context-aware active authentication using touch gestures, typing patterns and body movement (No. AFRL-RI-RS-TR-2016-076). Louisiana Tech University, Ruston United States (2016)
10. Monaco, J.V.: Robust keystroke biometric anomaly detection. arXiv preprint arXiv:1606.09075 (2016)
11. Montalvão, J., Freire, E.O., Bezerra Jr., M.A., Garcia, R.: Contributions to empirical analysis of keystroke dynamics in passwords. Pattern Recogn. Lett. **52**, 80–86 (2015)

12. Ceker, C., Upadhyaya, S.: User authentication with keystroke dynamics in long-text data. In: Proceedings of the IEEE 8th International Conference on Biometrics Theory, Applications and Systems (BTAS), pp. 1–6 (2016)
13. Fridman, L., Weber, S., Greenstadt, R., Kam, M.: Active authentication on mobile devices via stylometry, application usage, web browsing, and GPS location. IEEE Syst. J. **11**(2), 513–521 (2017)
14. Xiaofeng, L., Shengfei, Z., Shengwei, Y.: Continuous authentication by free-text keystroke based on CNN plus RNN. Procedia Comput. Sci. **147**, 314–318 (2019)
15. Buschek, D., De Luca, A., Alt, F.: Improving accuracy, applicability and usability of keystroke biometrics on mobile touchscreen devices. In: Proceedings of the 33rd Annual ACM Conference on Human Factors in Computing Systems, Seoul, Republic of Korea (2015)
16. Monaco, J.V., Tappert, C.C.: The partially observable hidden Markov model and its application to keystroke dynamics. Pattern Recogn. **76**, 449–462 (2018)
17. Deb, D., Ross, A., Jain, A.K., Prakah-Asante, K., Prasad, K.V.: Actions speak louder than (pass) words: passive authentication of smartphone users via deep temporal features. In: Proceedings of the 12th IAPR International Conference on Biometrics, Crete, Greece (2019)
18. Li, G., Bours, P.: Studying WiFi and accelerometer data based authentication method on mobile phones. In: Proceedings of the 2nd International Conference on Biometric Engineering and Applications, Amsterdam, Netherlands (2018)
19. Mahbub, U., Chellappa, R.: PATH: person authentication using trace histories. In: Proceedings of the Ubiquitous Computing, Electronics and Mobile Communication Conference. IEEE, New York (2016)
20. Mahbub, U., Komulainen, J., Ferreira, D., Chellappa, R.: Continuous authentication of smartphones based on application usage. IEEE Transactions on Biometrics, Behavior, and Identity Science **1**(3), 165–180 (2018)
21. Fierrez, J., Morales, A., Vera-Rodriguez, R., Camacho, D.: Multiple classifiers in biometrics. Part 2: trends and challenges. Inf. Fusion **44**, 103–112 (2018)
22. Liu, X., Shen, C., Chen, Y.: Multi-source interactive behavior analysis for continuous user authentication on smartphones. In: Proceedings of Chinese Conference on Biometric Recognition, Urumchi, China (2018)
23. Li, G., Bours, P.: A mobile app authentication approach by fusing the scores from multi-modal data. In: Proceedings of 21st International Conference on Information Fusion, Cambridge, UK (2018)
24. Giancardo, L., et al.: Computer keyboard interaction as an indicator of early Parkinson's disease. In: Sci. Rep. vol. 8 (2016)
25. Tolosana, R., Vera-Rodriguez, R., Fierrez, J., Ortega-Garcia, J.: Exploring recurrent neural networks for on-line handwritten signature biometrics. IEEE Access **6**, 1–11 (2018)
26. Taigman, Y., Yang, M., Ranzato, M., Wolf, L.: Closing the gap to human-level performance in face verification. In: Proceedings of the IEEE Computer Vision and Pattern Recognition (CVPR) (2014)

Incremental Learning Techniques Within a Self-updating Approach for Face Verification in Video-Surveillance

Eric Lopez-Lopez[1](✉) ⓘ, Carlos V. Regueiro[1] ⓘ, Xosé M. Pardo[2] ⓘ,
Annalisa Franco[3] ⓘ, and Alessandra Lumini[3] ⓘ

[1] Computer Architecture Group, CITIC, Universidade da Coruña, A Coruña, Spain
{eric.lopez,carlos.vazquez.regueiro}@udc.es
[2] Centro de Investigación en Tecnoloxías Intelixentes (CiTIUS),
Universidade de Santiago de Compostela, Santiago de Compostela, Spain
xose.pardo@usc.es
[3] DISI - Department of Computer Science and Engineering,
Università di Bologna, Bologna, Italy
{annalisa.franco,alessandra.lumini}@unibo.it

Abstract. Data labelling is still a crucial task which precedes the training of a face verification system. In contexts where training data are obtained online during operational stages, and/or the genuine identity changes over time, supervised approaches are less suitable.

This work proposes a face verification system capable of autonomously generating a robust model of a target identity (genuine) from a very limited amount of labelled data (one or a few video frames). A self-updating approach is used to wrap two well known incremental learning techniques, namely Incremental SVM and Online Sequential ELM.

The performance of both strategies are compared by measuring their ability to unsupervisedly improve the model of the genuine identity over time, as the system is queried by both genuine and impostor identities. Results confirm the feasibility and potential of the self-updating approach in a video-surveillance context.

Keywords: Face verification · Video-surveillance ·
Incremental learning · Self-updating

1 Introduction

The aim of face verification in video-surveillance (FViVS) is to determine whether the faces captured in a sequence of video frames belong to a target (genuine identity). In addition to the general difficulties found in face verification using still photos, such as pose variations, illumination conditions, or occlusions, video face verification also incorporates its own issues (e.g. motion blur, low-resolution).

Depending on the treatment received by the video data three different scenarios emerge [13]. First, in Video-to-Video face verification (V2V) a system is

© Springer Nature Switzerland AG 2019
A. Morales et al. (Eds.): IbPRIA 2019, LNCS 11868, pp. 25–37, 2019.
https://doi.org/10.1007/978-3-030-31321-0_3

queried using a sequence of video frames in order to find the same target identity in another part of the video [23]. Second, in Still-to-Video (S2V) a system is queried with a still photo in order to find the same identity in a video [1,5]. And finally, in Video-to-Still (V2S) verification tasks the goal is to find a target identity in a set of still images using a video query [4,8].

Offline learning is commonly used when a large number of labelled face images are available for training in advance, and offers a good coverage of the (stationary) target domain. However, due to the high computational complexity required for retraining, it is not adequate for purpose where the regularly update of the classifier is needed. Conversely, online learning presents an efficient alternative by updating the classifier knowledge upon the arrival of new data. When the availability of labels is scarce, or relevant visual changes are expected in the target's appearance after the modelling, online approaches are advantageous. It also learns to remove patterns that become extraneous and redundant over time. Thus, online learning has two components namely incremental and decremental learning, though here, we will only address the first one.

Hybrid approaches begin with an offline supervised learning, and then enhance the model over time in a semi-supervised or unsupervised way [4,22]. Online approaches build and update their models in a semi-supervised or unsupervised way [20]. In biometry, both approaches are commonly referred as template updating. Here, we use self-updating to refer to methods where the decision whether to update or not is also driven by themselves.

Deserving a special mention, deep learning techniques have boosted face verification in terms of performance. Notwithstanding, their requirements of huge amount of training data hinder the applicability to real scenarios. In FViVS (either V2V or S2V/V2S) this limitation is due to the difficulty in curating and annotating such large video datasets as needed. In order to circumvent these difficulties, solutions like fine-tuning [21], or the utilisation of pre-trained networks as feature extractors have been proposed [19]. Nevertheless a general solution to transfer face recognition is still far from being reached [18]. Similarly, despite its growing interest in the scientific community, the adaptation of deep learning techniques to semi or unsupervised scenarios remains something pendent [24].

This work proposes a FViVS system capable of autonomously generates a robust model of a target identity, when starting with a minimum template. This template is unsupervisedly improved over time, as new samples of the target and different identities (impostors) are presented. To achieve this behaviour, incremental learning methods were selected. The scenario where a controlling agent selects one video frame that contains the target face, and the system is able to create the complete model by itself, epitomises an illustrative case of use. The main contributions are:

1. The application of a self-updating strategy to FViVS.
2. A comparison between two classification approaches designed for incremental learning (Incremental SVM [15] and Online Sequential ELM [16]) within a self-updating framework.
3. A study of the relevance of the initial template in the self-updating strategy.

The rest of the paper is organised as follows: Sect. 2 presents common strategies used to develop an unsupervised face verification system. Section 2.1 formally defines the hypothesis and the elements of a self-updating system. Sections 4 and 5 describe the experimental setting and the performed experiments. Finally, Sect. 6 exposes the conclusions of this work.

2 Unsupervised Face Verification in Videos

Traditionally, the use of template updating methods have been focused in two similar but different tasks: (i) the adaptation of a previously trained model to mitigate the impact of changes in either environments or facial appearance [4,7], and, (ii) the gradual improvement of a template when the amount of labelled data is low [25]. This work try to provide insights on the second challenging task.

The absence of labels entails the necessity for somehow inferring this information and solving the dilemma of updating or not (Sect. 2.1). In the literature this is usually referred as the *stability-plasticity* or the *exploitation-exploration* dilemmas [9,11]. In the specific case of videos, the possibility of exploiting a time series of images will help in the task of this inference (Sect. 2.2).

2.1 Self-updating

Firstly proposed in the scope of natural language processing [25], this approach relies on the output of the model to infer the labels to perform the template updating. The updating will be performed whenever the target identity is verified [6,8]. Consequently, once the initial model is created (using a quite limited amount samples), the labelling (i.e. supervision) requirements is null.

In contrast, the main concern is how to determine the adequate **threshold** of the confidence value assigned to each label. Each new sample, labelled as belonging to the genuine identity and which confidence value is above the threshold, is used to update the template. While a high confidence threshold avoids the template corruption by outliers, it also prevents the system from accepting new valuable samples that differ from the ones contained in the template. Conversely, lower confidence thresholds can ease the acceptance and the subsequent addition of diverse information at the risk of corrupting the model with false positives.

2.2 Temporal Coherence

Often remarked as one of the keys to the actual development of an unsupervised learning method [8,20,24], the idea behind temporal coherence is something quite intuitive for humans. For example, if one of colleagues puts on a wig and sunglasses in front of you, it is natural to assume that the identity of this person is still the same despite his drastic look change. In videos, this idea is exploited assuming that successive frames tend to contain very similar information [2].

In FViVS, the exploitation of temporal coherence is performed with the help of a face tracker. This way, we can assume that an output video sequence of a face tracker belongs to the same identity despite changes in pose or illumination that could potentially damage the performance of a frame by frame recogniser.

3   Self-updating for FViVS

The idea of self-updating methods is to rely on the current model (M^t) at time t, to make the decision about updating itself when a video query has been identified (at time t) as belonging to the same identity. This way, unlabelled samples are gathered over time in order to improve the model without supervision.

Taking this into account the considered scenario assumes that initially ($t = 0$) a controlling agent selects one or a few video frames of the target identity (genuine) from a sequence given by a face tracker to create the *template*. It is also assumed the availability of a bunch of negative samples (impostors) from the domain of operation (in the literature this set is often called Universal Model, UM [4]) necessary to compare against the genuine information we are retrieving. The set of both the genuine template and the UM compose the set D^0.

Over time ($t = 1, 2, 3...$), the system is queried with new video sequences from unknown identities (both genuine and impostor) from the Cohort Model, CM [4]. If the model M^t accepts the query sequence, the sequence will be added to D^t in order to generate D^{t+1} and create the model M^{t+1}. In the opposite case, D^{t+1} remains the same so as M^{t+1}. The hypothesis of self-updating systems consist on assuming that this procedure will allow to improve performance.

3.1   Decision Rules for Self-updating

Since using a self-updating strategy gives to the models the power of deciding the label of a video sequence, we need to define three different rules:

- The *Frame Decision Rule (FDR)*. This rule assigns a score to every frame of the query video sequence. It corresponds to the outcome provided by the selected model (Sect. 3.2) and, consequently, dependant on it.
- The *Sequence Decision Rule (SDR)*. This rule is the actual implementation of the exploit of the temporal coherence described in Sect. 2.2. It is assumed that even if some frames of the sequence are not recognised we could still use the fact that the whole sequence belongs to a same identity in order to reject or accept it.

  In practical terms, this rule assigns an unique score to every query video sequence based on the individual scores given by the FDR to each frame of their frames. It is computed using the **median** of the scores assigned by the FDR to each frame of the sequence (which is the equivalent of a majority voting). Identities will be verified by fixing a **threshold**. The cautiousness or greediness in this fixation is directly related with the *stability-plasticity* dilemma.
- The *Update Rule (UR)*. This rule marks how and when the model will be updated. In our case, whenever the identity is verified. Since it is planned to use only incremental methods, the update will consist in perform a partial fit using the actual query sequence.

3.2   Selected Incremental Learning Methods

Any classification method can be used within a self-update strategy. A self-updating method is used as a 'wrapper' one that in practice converts a supervised classification method into an unsupervised one.

In this work we compare two different incremental learning techniques within this strategy. The advantage of the incremental methods is that they provide a natural way of performing the template updating:

- **Incremental Support Vector Machines (I-SVM).** [15] Solves the widely known problem of the Linear Support Vector Machines [3] by using the Stochastic Gradient Descent approach to incrementally find the hyper-plane parameters of the solution:

$$\mathbf{w_0} \cdot \mathbf{x} + b_0 = 0$$

where $\mathbf{w_0}$ and b_0 are the parameters of the hyper-plane and $\mathbf{x}$ represents a vector in the feature space.
- **Online Sequential Extreme Learning Machine (OS-ELM).** [16] Derived by the well known ELM neural network classifier [12], this approach is specifically adapted to be able to compute and update the weight values sequentially as more data is becoming available ('chunk-by-chunk' or one-by-one).
  In our case, a sigmoid function is used as activation function and the number of hidden nodes is empirically fixed at $\tilde{N} = 80$.

4   Experimental Setting

In this section, the experimental setup is explained. First, the database and the face detection algorithms are described. Then, the protocol for testing and the metrics used are presented.

4.1   FACE COX Database

The FACE COX database [13] gathers video frames of a total of 1000 users. There are 3 video sequences captured by 3 different cameras (cam1, cam2 and cam3) and a high quality still photo of each user. The faces of the subjects, who walked along a S-path, were captured on fixed cameras with varying pose, illumination, scale, and amount of blur. Each camera recorded a part of the path, without temporal overlapping between them.

As it has been explained in Sect. 3, in a self-update strategy the update is performed after each video query. In this dataset, the number of sequences of a same user is quite limited (3 sequences per user). Therefore, a priori, this dataset would allow a maximum of 3 updates (without taking into account the testing needs). In order to mitigate this limitation, each video sequence was divided in a number of sub-sequences, while being ware of their temporal order.

Table 1. FACE COX database user and camera division learning.

	Genuine				Impostor			
	still	cam1	cam2	cam3	still	cam1	cam2	cam3
Train	0	0	0	0	300	300	300	300
Gallery	0	700	700	0	0	0	0	0
Probe	0	0	0	700	0	0	0	700

4.2   Face Detection and Feature Extractor

A face detection over each frame of the sequence is performed in order to isolate and correctly align the region of the face from the rest of the background using the tool provided in the Dlib library [14]. After that, we use the power of the pre-trained ResNet Convolutional Neural Network [10] implementation provided by the Dlib library [14] for feature extraction. This implementation achieves an accuracy of 99.38% in the LFW dataset and has shown to have very good properties in terms of robustness to non-identity related variations [17].

4.3   Training and Testing Protocol

Inspired by the protocol proposed by FACE COX database, we have created different subsets (Table 1):

- The **train subset** contains face images used as negative samples to train each method. In our experiments this subset is conformed by the images of 300 users taken from each 3 cameras.
- The **gallery subset** that contains the images used to create the initial template as well as the ones used to simulate the video queries (both genuine and impostor). To build this set in our experiments we will use the images from the other 700 users taken from cam1 and cam2. Each video sequence taken from each camera were divided in **5** different sub-sequences given a total of 10 possible queries.
- The **probe subset** contains the images used to perform the testing of the system. The testing will be performed after each query of the learning phase to follow the evolution of the model. To build this subset we will use images taken from cam3 from the same 700 users used to build the *gallery subset*. In this case we have divided each user sequence into **10** sub-sequences.

The identities that are present in the *train subset* will not be present in the other two subsets. This way, the identities of the training subset will conform the Universal Model (UM) and the identities in the *gallery* and the *probe subset* will make up the Cohort Model (CM). In the experiments, each identity will have a specific CM that will contain its data and the data of the 10 'most similar' (using a SVM metric [17]) impostors.

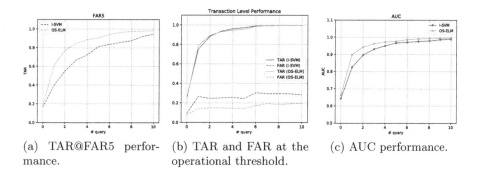

(a) TAR@FAR5 perfor-   (b) TAR and FAR at the   (c) AUC performance.
mance.   operational threshold.

Fig. 1. Supervised performance of I-SVM and OS-ELM. Performance is measured after each query (in this case only genuine ones) presented to the system.

It is important to note that the *train subset* will be used as the validation set that will help us to fix the decision threshold of the SDR (see Sect. 3.1). The value of this *operational threshold* will be set to 10% FAR of the model created using the initial template.

4.4   Metrics

The metrics used to evaluate our system were the Area Under the Curve (AUC) of the Receiver Operating Characteristic (ROC) generated by the True Acceptance Rate (TAR) and False Acceptance Rate (FAR) when we vary the decision threshold. Besides, measurements of TAR at FAR of 5% (TAR@FAR5) and Transaction Level performance are also provided [4].

Performance is measured using using the *probe subset*, with a distribution of 10 sub-sequences per genuine identity and 1 sub-sequence per impostor identity. Then, the results obtained for each identity are averaged between the 700 identities from the *probe subset*.

5   Experiments and Results

The high degrees of freedom of the self-updating strategy forced us to be cautious during the testing. The first step we have taken is to establish a baseline or 'upper-limit' in the achievable performance. That is the case where the system is updated in a supervised manner (Sect. 5.1).

Afterwards, we have moved to measure the performance evolution by actually using the self-update strategy in unsupervised conditions (Sect. 5.2). Both genuine and impostor sequences are presented. The system needs to distinguish between them and update or not consequently. Finally, we highlight the importance of a good initial template for achieving good final performance (Sect. 5.3).

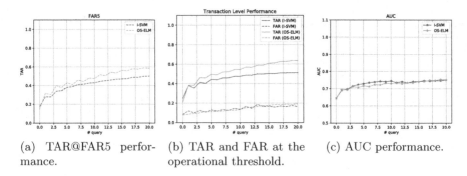

(a) TAR@FAR5 perfor-   (b) TAR and FAR at the   (c) AUC performance.
mance.   operational threshold.

Fig. 2. Unsupervised performance of I-SVM and OS-ELM using a 1 frame initial template. Performance is measured after each query presented to the system.

5.1 Supervised Learning (Baseline)

The supervised case represents the upper-limit since labels query labels are provided to the system. Consequently, it does not have to decide between accepting or not. The initial template is composed by just one frame, the first one of the sequence which is used to create M^0. From this point, the system is queried with 10 different queries from the genuine identity.

Performance is measured on the *probe subset* using the members of the CM of each identity, at the initial step ($t = 0$) and after each query ($t = 1, 2, ..., 10$). This means that the CM is conformed by the genuine identity and the 10 most similar impostors (see Sect. 4.3). This has been done in order to have comparable results of this experiment with the following made under unsupervised conditions.

As it can be seen in Fig. 1a and c, both methods are able to achieve quite high performance (+0.90 TAR@FAR5 and +0.95 AUC) showing an overall good supervised modelling. However, it is important to note that the OS-ELM method shows the best behaviour in two important aspects. First, Fig. 1a shows a quicker improvement in performance, proving that this method is able to build a more robust model with the same data. This effect is specially visible during the first steps, when the genuine information is more limited.

Second, when the performance is measured for a given operational threshold (Fig. 1b), OS-ELM shows a more steady FAR over time than I-SVM. This can be specially relevant in the unsupervised learning due to the fact that an increasing FAR means that the probability of accepting impostors during the training will increase as well, and thus the risk of corrupting the model.

5.2 Unsupervised Learning (1 Frame Template)

Here we start testing the unsupervised capabilities of the two methods. The philosophy of the experiment is similar to the former one, requiring now the use of the SDR to distinguish between genuine and impostor identities. Thus, after the generation of M^0 using the initial template (1 frame), the system is queried

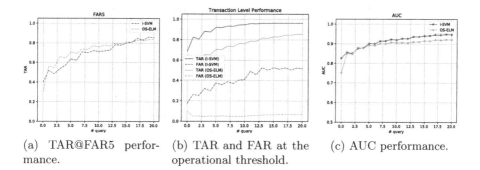

(a) TAR@FAR5 performance.   (b) TAR and FAR at the operational threshold.   (c) AUC performance.

Fig. 3. Unsupervised performance of I-SVM and OS-ELM using a 5 frames initial template. Performance is measured after each query presented to the system.

by 10 genuine sequences and 10 impostor sequences, both of them belonging to the *gallery subset* (identities of the CM). For each genuine query (odd query, $t = 1, 3, ..., 19$) we will have an impostor query (even query, $t = 2, 4, ..., 20$) afterwards. Each impostor query belongs to a different identity.

It can be noted in Fig. 2a and c an important drop in performance of both methods with respect to the supervised cases. An explanation can be found on the fact that initial performance (≈ 0.2 TAR) is too poor to make a reliable decision, as Fig. 2b seems to proof. While FAR at the operational threshold remains mostly the same, TAR drops compared with the supervised case. Nevertheless, the self-updating strategy stills manages to achieve an important improvement specially in TAR@FAR5, +0.32 in I-SVM and +0.39 in OS-ELM (see Table 2). Overall, OS-ELM shows a slightly better performance respect I-SVM in every performance measurement.

One explanation for the moderate performance showed by both models in this experiment, could be found on the high requirements that were demanded. Specially because the template which was built with just one video frame. It could be affirmed that such a poor initial performance does not allow a system to start accepting/rejecting the right information. In the next section, we will repeat the same experiment with the difference that the template is built with 5 frames instead of just one.

5.3 Unsupervised Learning (5 Frame Template)

This experiment is the same that the previous one but changing the initial template from 1 frame to the 5 first frames As it can be seen in Fig. 3a and c, performance is significantly improved respect to the one-frame template. In both cases, TAR@FAR5 reaches values of 0.84.

Nevertheless, despite having a pretty similar performance, Fig. 3b shows an important difference between both methods. While OS-ELM maintains a steady FAR during all the experiment, I-SVM increases it over time. This could possibly means that I-SVM has a more unstable performance for a given threshold during

Table 2. Summary of TAR@FAR5 performances values obtained.

Model	Template	Initial	Final Superv.		Final Unsuperv.	
			Value	Improv.	Value	Improv.
I-SVM	1 frame	0.170	0.926	+0.756	0.498	+0.328
	5 frames	0.415	0.946	+0.531	0.846	+0.431
OS-ELM	1 frame	0.187	0.981	+0.794	0.583	+0.396
	5 frames	0.297	0.983	+0.686	0.848	+0.551

the online training. Nevertheless, since every impostor query has a different identity, this malfunctioning is not reflected too much in TAR@FAR5. Unlike the accepted genuine sequences, the impact of the accepted impostors is not constructive. This may have greater impact in the case where a same impostor is repeatedly querying the system.

5.4   Discussion

In Table 2, a summary of the experiments conducted in this work is presented. We have added the case of an initial template of 5 frames in supervised conditions in order to see the complete picture. Overall it can be said that the self-updating strategy is able to improve performance in every experiment. It is important to remark the extremely low labelled conditions (1 or 5 low quality video frames) in which our experiments were conducted have not been able to avoid this improvement. On the other hand, our experiments show that the self-updating strategy is quite sensible to the initial performance of the model (which in our case is expressed in the necessity of more genuine data to create the initial template). This fact makes our systems move from about 0.50 to 0.85 TAR@FAR5.

One final appointment to mention is how the systems are affected by our decisions when defining the self-updating strategy. In this case, for the sake of simplicity, a fixed threshold in the SDR was established in order to decide whether to update or not. This threshold was fixed selecting the point of M^0 ROC curve that corresponds to a 10% FAR. Nevertheless, we cannot assure that this ROC point will be stable with time. Therefore, the fixed threshold benefits the classification methods that preserve (or even decrease) the ROC point of functioning (as it can be seen in Fig. 3b where, unlike OS-ELM, I-SVM's FAR constantly increases).

6   Conclusions

In this work, the unsupervised FViVS problem using a self-update strategy has been explored. The case of study starts with a surveillance agent selecting one frame from a video sequence, and then the autonomous video-surveillance system try to detect the same identity within the same or a different video sequence, while incrementally build a robust model of the target identity.

Experiments showed that a self-updating strategy seems to be viable to build the identity model without the necessity of labels, or at least capable of improving initial performance. In addition, the importance of a correct decision rule is highlighted during the online training as well as its correlation with the classification method at hand. This fact makes OS-ELM performance stands out respect to I-SVM.

For future work, the aim is to perform a deeper study including more classification techniques and an extended experimental assessment. It would also be interesting to explore the behaviour of this approach in life-long learning conditions in order to study its robustness to unwanted drifts.

Acknowledgements. This work has received financial support from the Spanish government (project TIN2017-90135-R MINECO (FEDER)), from The Consellería de Cultura, Educación e Ordenación Universitaria (accreditations 2016–2019, EDG431G/01 and ED431G/08), and reference competitive groups (2017–2020 ED431C 2017/69, and ED431C 2017/04), and from the European Regional Development Fund (ERDF). Eric López had received financial support from the Xunta de Galicia and the European Union (European Social Fund - ESF).

References

1. Bashbaghi, S., Granger, E., Sabourin, R., Bilodeau, G.A.: Dynamic ensembles of exemplar-SVMs for still-to-video face recognition. Pattern Recogn. **69**, 61–81 (2017). https://doi.org/10.1016/j.patcog.2017.04.014
2. Becker, S.: Implicit learning in 3D object recognition: the importance of temporal context. Neural Comput. **11**(2), 347–374 (1999). https://doi.org/10.1162/089976699300016683
3. Cortes, C., Vapnik, V.: Support-vector networks. Mach. Learn. **20**(3), 273–297 (1995). https://doi.org/10.1007/BF00994018
4. De la Torre, M., Granger, E., Radtke, P.V., Sabourin, R., Gorodnichy, D.O.: Partially-supervised learning from facial trajectories for face recognition in video surveillance. Inf. Fusion **24**, 31–53 (2015). https://doi.org/10.1016/j.inffus.2014.05.006
5. Dewan, M.A.A., Granger, E., Marcialis, G.L., Sabourin, R., Roli, F.: Adaptive appearance model tracking for still-to-video face recognition. Pattern Recogn. **49**, 129–151 (2016). https://doi.org/10.1016/j.patcog.2015.08.002
6. Didaci, L., Marcialis, G.L., Roli, F.: Analysis of unsupervised template update in biometric recognition systems. Pattern Recogn. Lett. **37**, 151–160 (2014). https://doi.org/10.1016/j.patrec.2013.05.021
7. Ditzler, G., Roveri, M., Alippi, C., Polikar, R.: Learning in nonstationary environments: a survey. IEEE Comput. Intell. Mag. **10**(4), 12–25 (2015). https://doi.org/10.1109/MCI.2015.2471196
8. Franco, A., Maio, D., Maltoni, D.: Incremental template updating for face recognition in home environments. Pattern Recogn. **43**(8), 2891–2903 (2010). https://doi.org/10.1016/j.patcog.2010.02.017

9. Grossberg, S.: Nonlinear neural networks: Principles, mechanisms, and architectures. Neural Netw. **1**(1), 17–61 (1988). https://doi.org/10.1016/0893-6080(88)90021-4

10. He, K., Zhang, X., Ren, S., Sun, J.: Deep residual learning for image recognition. In: Computer Vision and Pattern Recognition (CVPR), pp. 770–778 (2016). https://doi.org/10.1109/CVPR.2016.90

11. Hoens, T.R., Polikar, R., Chawla, N.V.: Learning from streaming data with concept drift and imbalance: an overview. Prog. Artif. Intell. **1**, 89–101 (2012). https://doi.org/10.1007/s13748-011-0008-0

12. Huang, G.B., Zhu, Q.Y., Siew, C.K.: Extreme learning machine: theory and applications. Neurocomputing **70**(1), 489–501 (2006). https://doi.org/10.1016/j.neucom.2005.12.126

13. Huang, Z., et al.: A benchmark and comparative study of video-based face recognition on cox face database. IEEE Trans. Image Process. **24**(12), 5967–5981 (2015). https://doi.org/10.1109/TIP.2015.2493448

14. King, D.E.: Dlib-ml: a machine learning toolkit. J. Mach. Learn. Res. **10**, 1755–1758 (2009). https://doi.org/10.1145/1577069.1755843

15. Kivinen, J., Smola, A.J., Williamson, R.C.: Online learning with kernels. IEEE Trans. Signal Process. **52**(8), 2165–2176 (2004). https://doi.org/10.1109/TSP.2004.830991

16. Liang, N., Huang, G., Saratchandran, P., Sundararajan, N.: A fast and accurate online sequential learning algorithm for feedforward networks. IEEE Trans. Neural Netw. **17**(6), 1411–1423 (2006). https://doi.org/10.1109/TNN.2006.880583

17. López-López, E., Pardo, X.M., Regueiro, C.V., Iglesias, R., Casado, F.E.: Dataset bias exposed in face verification. IET Biom. **8**(4), 249–258 (2019). https://doi.org/10.1049/iet-bmt.2018.5224

18. Masi, I., Wu, Y., Hassner, T., Natarajan, P.: Deep face recognition: A survey. In: Conference on Graphics, Patterns and Images (SIBGRAPI), pp. 471–478 (2018). https://doi.org/10.1109/SIBGRAPI.2018.00067

19. Pernici, F., Bartoli, F., Bruni, M., Del Bimbo, A.: Memory based online learning of deep representations from video streams. In: Computer Vision and Pattern Recognition (CVPR), pp. 2324–2334 (2018). https://doi.org/10.1109/CVPR.2018.00247

20. Pernici, F., Bimbo, A.D.: Unsupervised incremental learning of deep descriptors from video streams. In: International Conference on Multimedia Expo Workshops (ICMEW), pp. 477–482 (2017). https://doi.org/10.1109/ICMEW.2017.8026276

21. Sohn, K., Liu, S., Zhong, G., Yu, X., Yang, M., Chandraker, M.: Unsupervised domain adaptation for face recognition in unlabeled videos. In: International Conference on Computer Vision (ICCV), pp. 5917–5925 (2017). https://doi.org/10.1109/ICCV.2017.630

22. Villamizar, M., Sanfeliu, A., Moreno-Noguer, F.: Online learning and detection of faces with low human supervision. Vis. Comput. **35**(3), 349–370 (2019). https://doi.org/10.1007/s00371-018-01617-y

23. Wang, R., Shan, S., Chen, X., Gao, W.: Manifold-manifold distance with application to face recognition based on image set. In: Computer Vision and Pattern Recognition (CVPR), pp. 1–8 (2008). https://doi.org/10.1109/CVPR.2008.4587719

24. Wang, X., Gupta, A.: Unsupervised learning of visual representations using videos. In: International Conference on Computer Vision (ICCV), pp. 2794–2802 (2015). https://doi.org/10.1109/ICCV.2015.320

25. Yarowsky, D.: Unsupervised word sense disambiguation rivaling supervised methods. In: Annual Meeting on Association for Computational Linguistics (ACL), pp. 189–196. Association for Computational Linguistics, Stroudsburg, PA, USA (1995). https://doi.org/10.3115/981658.981684

Don't You Forget About Me: A Study on Long-Term Performance in ECG Biometrics

Gabriel Lopes[1], João Ribeiro Pinto[1,2]([✉]) , and Jaime S. Cardoso[1,2]

[1] Faculdade de Engenharia, Universidade do Porto, Porto, Portugal
jtpinto@fe.up.pt
[2] Centre for Telecommunications and Multimedia, INESC TEC, Porto, Portugal

Abstract. The performance of biometric systems is known to decay over time, eventually rendering them ineffective. Focused on ECG-based biometrics, this work aims to study the permanence of these signals for biometric identification in state-of-the-art methods, and measure the effect of template update on their long-term performance. Ensuring realistic testing settings, four literature methods based on autocorrelation, autoencoders, and discrete wavelet and cosine transforms, were evaluated with and without template update, using Holter signals from THEW's E-HOL 24 h database. The results reveal ECG signals are unreliable for long-term biometric applications, and template update techniques offer considerable improvements over the state-of-the-art results. Nevertheless, further efforts are required to ensure long-term effectiveness in real applications.

Keywords: Biometrics · Electrocardiogram · Identification · Template update

1 Introduction

The ability to identify or recognize another human being is of utmost importance. For access control and other purposes, security systems are based on unique credentials. Several methods are used such as PIN codes, passwords, ID cards, or keys [1]. These credentials are susceptible to be copied or counterfeit [2], and it is very easy for the person to forget, share, or lose them. The alternative is for systems to use something that is characteristic and belongs only to one person: a biometric trait [3]. Biometric traits currently include fingerprint, voice, face, and electrocardiogram (ECG), and only require the individual to be present when access is requested [4]. Relative to the traditional techniques, biometric systems are considered more secure, as traits are more difficult to counterfeited or steal than extrinsic credentials [5].

Current off-the-person ECG acquisition techniques, aiming towards increased simplicity, usability, and comfort, make ECG-based biometric systems effortless

A. Morales et al. (Eds.): IbPRIA 2019, LNCS 11868, pp. 38–49, 2019.
https://doi.org/10.1007/978-3-030-31321-0_4

for the user [6,7]. However, when considering long-term usage, the performance decays over time [8]. This applies to human-machine interfaces that require frequent identity control, especially those that are information-sensitive.

Variability on the input biometric data, biometric trait's aging, and variations caused by the subject's interaction with the sensor contribute to large intrasubject variability [5]. This makes stored individual templates to quickly lose representativity, resulting in poor recognition performance and placing serious challenges on long-term recognition. In biometrics in general, long-term identification requires frequent update of the templates to maintain acceptable performance over time [9], and avoid security flaws in information-sensitive systems.

Previously, Labati *et al.* [10] have studied the performance decay over time on their proposed algorithm for ECG-based authentication. In this paper, we extend their work, by studying the effect of ECG permanence and variability in biometric identification and evaluating the impact of template update on long-term performance in four state-of-the-art algorithms for identification.

2   State-of-the-art Review

In the literature, it is difficult to find a strong and widely accepted rule for template update. Most methods are based on heuristics and empirically determined thresholds, which are highly dependent on the data and application settings. For example, Komeili *et al.* [8], for authentication, have set the acceptance threshold equal to the point of zero false acceptance rate, thus ensuring updates with only genuine samples.

Nevertheless, it is possible to identify some common mechanisms that may vary depending on different factors: these include the choice of the update criterion (based on thresholds or graphs), the update periodicity (online or offline), the selection mechanism, and the template update working mode system (supervised or semi-supervised). The taxonomy of template update (see Fig. 1) divides the existing techniques into two categories: supervised and semi-supervised.

Supervised methods are offline methods in which label attribution is given by a supervisor. These contain the *Clustering* subcategory, which includes the *MDIST* that aims to search for the templates that minimize the intra distance among all the samples in the database (*i.e.*, the most similar) and *DEND* that aims to search for the templates that exhibit large intra-class variations resort to the dendrogram (*i.e.*, the most different) [12]. The second subcategory comprises Editing-based methods, which are independent of the number of templates and give focus on the whole collected training set T. A subset $E \in T$ is generated, maintaining the classification performance offered by T. The best subsets were obtained by reviewing the structure of the data (needs to be done to each subject) [9,13]. All the algorithms (based on k-Nearest Neighbors) must be representative of T and can be roughly described as *incremental* when the E starts empty and grows, or *decremental* when E starts equal to T and in each iteration some instances are deleted, until some criterion is reached [9].

Semi-supervised methods merge labeled (in biometrics correspond to the initial training samples) and unlabeled (correspond to the samples available during

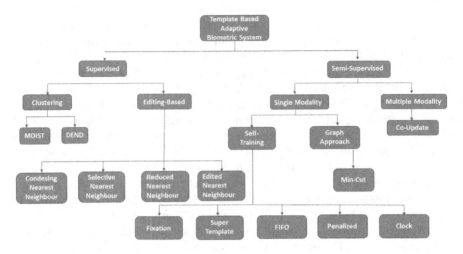

Fig. 1. Dendrogram representing the taxonomy of template update techniques (based on [11]).

system operation) data to improve the system's performance. This category comprises the *Single Modality* (for unimodal biometric systems) and *Multiple Modality* (for systems using more than one biometric trait) subcategories. The *Single Modality* subcategory includes the *Self-Training* approaches such as *FIFO* (first-in-first-out), *Fixation*, *Super Template* (X composed by N templates x) where new genuine date is always fused to a common single template [14] updated online during the execution of continuous verification, *Penalized template update* method based on the mean of the past ECG's and the actual ECG [15] and *clock method* where the current template is tested against all the others stored in the database [16].

Generally, a new, unknown trait measurement is used for template update if its score (returned by the biometric recognition system) is above a set threshold. Hence, the future performance of the system relies heavily on the chosen threshold value [11].

The update threshold is commonly estimated using enrollment templates or training data. When training data are scarce or when using short enrollments, this leads to some problems: as important intrasubject variability is missed because use only the patterns similar to the templates stored; online methods are dependent on the order of the sequence of input data; vulnerability to large intra-class variations; since the algorithm normally looks for the minimal cost (high score), it can be stuck in local maximum and always use highly confident data for updating.

Semi-supervised methods also include *Graph* approaches. These commonly define a graph where the nodes are either labeled (the identity is known) or unlabeled (unknown identity) data, and the edges (which can have different weights) are the similarity between those samples [11,17]. To be considered a

Table 1. Graph-based template update methods and their respective loss and regularizer functions (based on [17]).

Method	Source	Loss	Regularizer
Min cut	[18]	$\sum_{i \in L}(y_i - y_{i\|L})^2$	$\frac{1}{2}\sum_{ij} w_{ij}(y_i - y_j)^2$
Gaussian random fields and harmonic function	[19]	$\sum_{i \in L}(f_i - y_i)^2$	$f^T \Delta f$
Local and global consistency	[20]	$\sum_{i=1}^{n}(f_i - y_i)^2$	$D^{-\frac{1}{2}}\Delta D^{\frac{1}{2}}$
Tikhonov regularization	[21]	$\frac{1}{K}\sum_i(f_i - y_i)^2$	$\gamma f^T S f$
Manifold regularization	[22]	$\frac{1}{l}\sum_{i=1}^{l} V(x_i, Y_i, f)$	$\gamma_A \|f\|_k^2 + \gamma_I \|f\|_I^2$
Graph Kernel from the spectrum of Laplacian	[23]	$\min \frac{1}{2}w^T W$	$\exp(-\frac{\sigma}{2}\lambda)$
Spectral graph transducer	[19]	$\min c(f - \gamma)^T C(f - \gamma)$	$f^T L f$
Local learning regularization	[24]	$\min \frac{1}{k}\sum_{i=1}^{k}(y_i - f_k(x_i))^2$	$\frac{\gamma}{k}\|f_k\|^2$

graph-based semi-supervised method, it must estimate a function f, approximate the known Y on the labeled nodes, and include two terms to turn the graph smooth: a loss function and a regularizer. These two terms are what define each approach (as can be seen in Table 1) [17], among which the most common in biometrics is min-cut graphs [11].

Considering the topic of template update is still to be adequately addressed on ECG biometrics, this work studies the effect of ECG permanence and variability in long-term identification performance. Furthermore, it aimed to evaluate the effect of template update techniques, on the performance of several state-of-the-art methods.

3   Methods

3.1   Implemented Identification Methods

To fully and objectively evaluate the effects of ECG variability on the performance of biometric algorithms, a study was conducted on four literature methods, described below.

Plataniotis *et al.* [25] proposed an ECG biometric recognition method using a non-fiducial approach. Signals are preprocessed using a bandpass filter (0.5–40 Hz), followed by feature extraction with autocorrelation (AC) and dimensionality reduction using discrete cosine transform (DCT). The fifteen most relevant features were selected, and Euclidean distance was used for classification.

Tawfik *et al.* [26] used a bandpass filter (1–40 Hz) in the preprocessing phase. QRS complexes (most stable part of ECG) were cut from the signal using a 0.35 sec window. The average ensemble QRS was computed and features were extracted using DCT technique (thirty most relevant features were selected). A multilayer perceptron (MLP) is used for classification.

Belgacem *et al.* [27] also preprocessed signals with a bandpass filter (1–40 Hz). The QRS complexes were located and cut from the signal, and the average QRS was computed. The feature extraction resorted to Discrete Wavelet Transform (DWT). From all DWT decomposition levels, only the most relevant were selected, and a Random Forest is used for classification. The authors used this technique for authentication.

Eduardo *et al.* [28] used a Finite Impulse Response (5–20 Hz) filter for preprocessing. Heartbeats were cut with a fixed length of $[-200, 400]$ ms around each R peak. Outliers were detected and removed using DMEAN ($\alpha = 0.5$ and $\beta = 1.5$, with Euclidean distance). For decision, the k-nearest neighbors (kNN) classifier was used with $k = 3$ and cosine distance.

Standard sample wise normalization was performed (following Eq. 1) for all methods except that of Eduardo *et al.* [28], which required $[-1, 1]$ *min-max* normalization (Eq. 2), where x represents the input signal and $\tilde{x}$ the normalized signal.

$$\tilde{x}[g] = \frac{x[g] - \overline{x[g]}}{\sigma(x[g])} \tag{1}$$

$$\tilde{x}[g] = 2 \left(\frac{x[g] - \min(x[g])}{\max(x[g]) - \min(x[g])} \right) - 1 \tag{2}$$

3.2   Template Update Methods

FIFO (first-in-first-out) is the most common strategy and, computationally, is very light. Here, the database is updated using new samples whose score is above or below a threshold (whether the score represents similarity or dissimilarity, respectively), or between two threshold values (discarding previously stored sample) [8,29]. The score of a new sample can either be output by a classifier, or be a measure of distance or similarity between that sample and the stored templates [30].

In this work, the training data were used to search for threshold values. Among all training samples, 75% were used to train a model, which was used to obtain scores for the remaining data samples. Comparing the scores with several thresholds, the error at each threshold was analyzed (Fig. 2) to find one that simultaneously maximizes true positives and minimizes false positives.

Fixation consists on fixing certain templates, allowing only the remaining stored samples to be updated [31]. In this work, 25, 50, or 75% of the enrollment templates of the individual are fixed, while the rest of the samples are free to be updated. This ensures some initial, labeled information of the subjects remains on the system over time.

An adaptation of this technique was explored. Here, $n + j \times n$ samples were fixed, where $n \in [1, 2, 3]$ is the number of fixed initial templates, and j increases

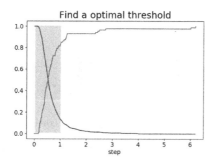

Fig. 2. Illustration of the search for the ideal threshold. The values were chosen near the intersection, inside the yellow zone. (Color figure online)

over time. In this work, $j \in [0, 6]$ increased by one at each testing moment ($j \in [0, 6]$), which allowed the system to fix more and more samples over time, thus storing information on the subject's variability over time. In a real system with potentially endless use, the parameters n and j should be carefully chosen to avoid the eventual fixation of the entire template gallery.

4   Experimental Settings

4.1   Dataset

For evaluation, the ECG signals used were from the E-HOL-03-0202-003 database[1] (most commonly designated as E-HOL 24 h). This database consists of a study of 202 healthy subjects (only 201 were provided), recorded using three leads at 200 Hz sampling frequency, after an initial resting supine period of 20 min. From the available data of 201 subjects, thirteen were discarded due to saturation or unacceptable noise (subjects 1043, 9003, 9005, 9020, 9021, 9022, 9025, 9046, 9061, 9064, 9071, 9082 and 9105), similar to what was done by Labati *et al.* [14]. From each of the remaining 188 subjects, only the lead most closely resembling Lead I ECG was selected, to approximate off-the-person settings.

4.2   Experiments

In order to fit the used data, some changes were introduced to the original methods. On the method from Eduardo *et al.*, the cutoff frequencies of the bandpass filter were changed to 1 and 40 Hz, to retrieve important information on higher frequencies; outlier removal was reparametrized with $\alpha = 1.2$ and $\beta = 1.5$. The autoencoder had the topology $[120, 60, 40, 20, 40, 60, 120]$ and was trained using the Adam optimizer with learning rate 0.01. Classification used $k = 1$.

---

[1] THEW. Available on: http://thew-project.org/Database/E-HOL-03-0202-003.html.

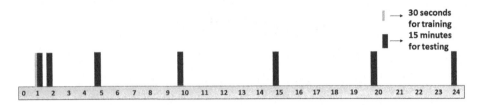

Fig. 3. Schema illustrating the use of each E-HOL record for training and testing (in orange - training segment; in blue - each test segment). (Color figure online)

For the method of Belgacem *et al.*, DWT feature extraction was performed in four decomposition levels, due to lower data sampling rate, and $cd4$, $cd3$, $cd2$, and $cd1$ coefficients were used as features.

Data were divided into train and test sets. The training phase used the last 30 s (mimicking short enrollments on real-life applications) of the first 60 min (avoiding unrealistic calm after the initial resting period) of each subject. Five-second overlap was used to obtain 26 samples from each 30 s of training data. To study performance over time, testing was performed over seven time points (see Fig. 3): one immediately after enrollment, another after one hour, and regularly until the end of the records. From each point, from 15 min of data, thirty 30 s samples are extracted, and batches are built with one sample from each of the 188 subjects.

5   Results

5.1   Without Template Update

After implementing, for identification, the method proposed by Labati *et al.* [10] (replicating their evaluation conditions), it was possible to conclude that the ECG signal is not fully permanent over 24 h. However, similarly to what was stated by Labati *et al.*, the results are relatively good over the first two hours (see Fig. 4a), although permanence was not verified.

The performance results at each test hour, obtained through the weighted average of the corresponding batches, for the state-of-the-art methods can be found in Fig. 4b. It was found that the performance is mostly acceptable in the first test point, but performance decays significantly over time and variability changes considerably over the day.

Moreover, a minimum around the 15th h occurs independently of the chosen method. Considering that most of the records start between 8–12 am, after 15 h the subjects must be sleeping. In this perspective, it appears that the ECG is most different from normal when the subject is asleep.

5.2   With Template Update

Considering the previous results, template update was applied to the methods, in an effort to avoid performance decay over time. Figure 5 presents the results using the FIFO technique, with diverse thresholds.

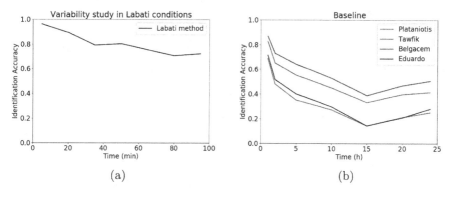

Fig. 4. Identification performance over time corresponding to (a) the Labati *et al.* method, and (b) the implemented state-of-the-art methods.

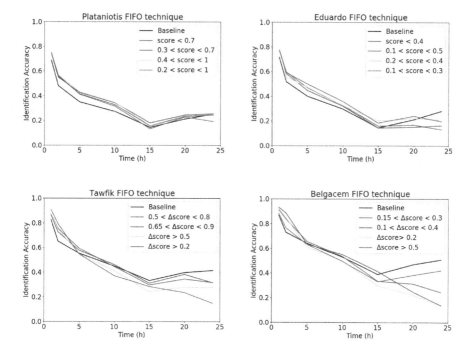

Fig. 5. Comparative FIFO methods with different thresholds using different identification methodologies.

For the methods of Plataniotis *et al.* and Eduardo *et al.*, the best results were obtained using two thresholds, respectively, {0.3, 0.7} (4.7% accuracy improvement) and {0.1, 0.3} (+5.7% accuracy), improving all performance results until the 15th h. However, for the method of Belgacem *et al.*, the performance worsens with template update after the first two hours (best results obtained when the

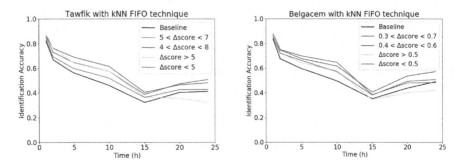

Fig. 6. Results using FIFO update with different thresholds.

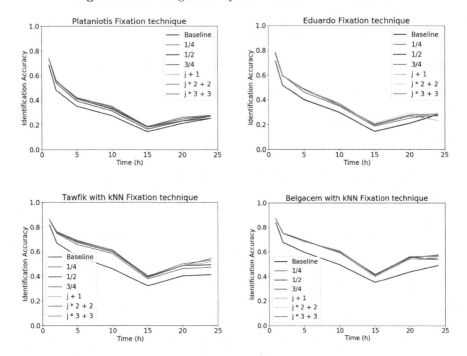

Fig. 7. Results using fixation update (the corresponding value represents the number of samples that was fixated per subject).

difference between the highest and second highest scores $\Delta score \in [0.15, 0.3]$). The same was verified for the method of Tawfik *et al.* which, in the first two hours, offered best results with $\Delta score > 0.2$. In general, using two thresholds instead of one offered the best results.

Considering this, it appeared that the Random Forest and MLP classifiers are not suitable for these kinds of template/model update. This was confirmed after a repetition of the evaluation of these methods, with kNN replacing the classifiers (see Fig. 6). With kNN, the template update was able to reduce the performance decay over time, improving accuracy, on average, by 7.9% and 9.2%, respectively, for the methods of Belgacem *et al.* and Tawfik *et al.*

As for the Fixation technique, the obtained results were more promising (see Fig. 7). This template update technique brought performance improvements for all methods. The fixation technique that offered the best results was $j \times 3 + 3$, improving the baseline identification accuracy, on average, by 10.0%.

6   Conclusion

This work studied how the ECG variability effects the performance of state-of-the-art biometric algorithms, and how template update could mitigate performance decay over time. The results have shown long-term identification performance in ECG biometrics is generally weak, despite the promising results often presented in the literature.

Template update techniques proved successful in enhancing the long-term performance of state-of-the-art methods, especially when using template fixation techniques. However, further efforts are needed for the study and development of more advanced techniques, with special focus on supervised techniques, so that ECG-based biometric systems can offer reliable performances over long periods.

Acknowledgements. This work was financed by the ERDF – European Regional Development Fund through the Operational Programme for Competitiveness and Internationalization - COMPETE 2020 Programme and by National Funds through the Portuguese funding agency, FCT–Fundação para a Ciência e a Tecnologia within project "POCI-01-0145-FEDER-030707", and within the PhD grant "SFRH/BD/137720/2018". Data used for this research was provided by the Telemetric and Holter ECG Warehouse of the University of Rochester (THEW), NY.

References

1. Prabhakar, S., Pankanti, S., Jain, A.K.: Biometric recognition: security and privacy concerns. IEEE Secur. Priv. Mag. **1**(2), 33–42 (2003)
2. Hadid, A., Evans, N., Marcel, S., Julian, F.: Biometrics systems under spoofing attack: an evaluation methodology and lessons learned. IEEE Signal Process. Mag. **32**(5), 20–30 (2015)
3. Agrafioti, F., Bui, F.M., Hatzinakos, D.: Secure telemedicine: biometrics for remote and continuous patient verification. J. Comput. Netw. Commun. **2012**, 11 p. (2012). Article ID 924791
4. Akhtar, Z., Micheloni, C., Foresti, G.: Biometric liveness detection: challenges and research opportunities. IEEE Secur. Priv. **13**(5), 63–72 (2015)
5. Jain, A.K., Ross, A., Prabhakar, S.: An introduction to biometric recognition. IEEE Trans. Circ. Syst. Video Technol. **14**(1), 4–20 (2004)
6. Pinto, J.R., Cardoso, J.S., Lourenco, A., Carreiras, C.: Towards a continuous biometric system based on ECG signals acquired on the steering wheel. Sensors **17**(10), 2228 (2017)
7. Pinto, J.R., Cardoso, J.S., Lourenco, A.: Evolution, current challenges, and future possibilities in ECG Biometrics. IEEE Access **6**, 34746–34776 (2018)

8. Komeili, M., Armanfard, N., Hatzinakos, D.: Liveness detection and automatic template updating using fusion of ECG and fingerprint. IEEE Trans. Inf. Forensics Secur. **13**(7), 1810–1822 (2018)

9. Freni, B.: Template editing and replacement: novel methods for biometric template selection and update. Ph.D. thesis, University of Cagliari, Italy (2010)

10. Labati, R., Sassi, R., Scotti, F.: ECG biometric recognition: permanence analysis of QRS signals for 24 hours continuous authentication. In: 2013 IEEE International Workshop on Information Forensics and Security (WIFS), Guangzhou, China, pp. 31–36. IEEE, November 2013

11. Rattani, A.: Adaptive biometric system based on template update procedures. Ph.D. thesis, University of Cagliari (2010)

12. Lumini, A., Nanni, L.: A clustering method for automatic biometric template selection. Pattern Recogn. **39**(3), 495–497 (2006)

13. Hutchison, D., Anagnostopoulos, G.: Structural, syntactic, and statistical pattern recognition. In: Proceedings of Joint IAPR International Workshop, SSPR & SPR 2008, Orlando, USA. Springer, Heidelberg, 4–6 December 2008. https://doi.org/10.1007/978-3-540-89689-0

14. Labati, R.D., Piuri, V., Sassi, R., Scotti, F., Sforza, G.: Adaptive ECG biometric recognition: a study on re-enrollment methods for QRS signals. In: 2014 IEEE Symposium on Computational Intelligence in Biometrics and Identity Management (CIBIM), Orlando, FL, USA, pp. 30–37. IEEE, December 2014

15. Chun, S.Y.: Small scale single pulse ECG-based authentication using GLRT that considers T wave shift and adaptive template update with prior information. In: 2016 23rd International Conference on Pattern Recognition (ICPR), Cancun, Mexico, pp. 3043–3048. IEEE, December 2016

16. Scheidat, T., Makrushin, A., Vielhauer, C.: Automatic Template Update Strategies for Biometrics. Technical report, Otto-von-Guericke University of Magdeburg, Germany (2007)

17. Zhu, X.: Semi-Supervised Learning Literature Survey. University of Wisconsin-Madison, USA (2006)

18. Blum, A., Chawla, S.: Learning from labeled and unlabeled data using graph mincuts. In: Proceedings of International Conference on Machine Learning (ICML-2001) (2001)

19. Zhu, X., Ghahramani, Z., Lafferty, J.: Semi-supervised learning using gaussian fields and harmonic functions. In:Twentieth International Conference on Machine Learning (ICML-2003), Washington DC, p. 8 (2003)

20. Zha, Z., Mei, T., Wang, J., Wang, Z., Hua, X.: Graph-based semi-supervised learning with multiple labels. J. Vis. Commun. Image Represent. **20**(2), 97–103 (2009)

21. Belkin, M., Matveeva, I., Niyogi, P.: Regularization and semi-supervised learning on large graphs. In: Shawe-Taylor, J., Singer, Y. (eds.) COLT 2004. LNCS (LNAI), vol. 3120, pp. 624–638. Springer, Heidelberg (2004). https://doi.org/10.1007/978-3-540-27819-1_43

22. Sindhwani, V., Niyogi, P., Belkin, M.,Keerthi, S.: Linear manifold regularization for large scale semi-supervised learning. In: Proceedings of the 22nd ICML Workshop on Learning with Partially Classified Training Data, Bonn, Germany, p 4, August 2005

23. Chapelle, O., Schölkopf, B., Zien, A. (eds.): Semi-supervised Learning. Adaptive Computation and Machine Learning. MIT Press, Cambridge (2006)

24. Kokkinos, Y., Margaritis, K.: Local learning regularization networks for localized regression. Neural Comput. Appl. **28**(6), 1309–1328 (2017)

25. Plataniotis, K., Hatzinakos, D., Lee, J.: ECG biometric recognition without fiducial detection. In: 2006 Biometrics Symposium: Special Session on Research at the Biometric Consortium Conference, Baltimore, MD, USA, pp. 1–6. IEEE, September 2006

26. Tawfik, M.M., Selim, H., Kamal, T.: Human identification using time normalized QT signal and the QRS complex of the ECG. In: 2010 7th International Symposium on Communication Systems, Networks Digital Signal Processing (CSNDSP 2010), pp. 755–759, July 2010

27. Belgacem, N., Nait-Ali, A., Fournier, R., Bereksi-Reguig, F.: ECG based human authentication using wavelets and random forests. Int. J. Crypt. Inf. Secur. **2**(2), 1–11 (2012)

28. Eduardo, A., Aidos, H., Fred, A.: ECG-based biometrics using a deep autoencoder for feature learning - an empirical study on transferability. In: Proceedings of the 6th International Conference on Pattern Recognition Applications and Methods, Porto, Portugal, pp. 463–470 (2017)

29. Coutinho, D.P., Fred, A.L.N., Figueiredo, M.A.T. : ECG-Based continuous authentication system using adaptive string matching. In: Proceedings of the International Conference on Bio-inspired Systems and Signal Processing, Rome, Italy, pp. 354–359 (2011)

30. Lourenco, A., Silva, H., Fred, A.: Unveiling the biometric potential of finger-based ECG signals. Comput. Intell. Neurosci. **2011**, 8 p. (2011). Article ID 720971

31. Guerra-Casanova, J., Sánchez-Ávila, C., Sierra, A.S., del Pozo, G.B.: Score optimization and template updating in a biometric technique for authentication in mobiles based on gestures. J. Syst. Softw. **84**(11), 2013–2021 (2011)

Face Identification Using Local Ternary Tree Pattern Based Spatial Structural Components

Rinku Datta Rakshit[1], Dakshina Ranjan Kisku[2(✉)],
Massimo Tistarelli[3], and Phalguni Gupta[4]

[1] Department of Information Technology, Asansol Engineering College,
Vivekananda Sarani, Kanyapur, Asansol 713305, West Bengal, India
rakshit_rinku@rediffmail.com
[2] Department of Computer Science and Engineering,
National Institute of Technology Durgapur, Mahatma Gandhi Road,
A-Zone, Durgapur 713209, West Bengal, India
drkisku@cse.nitdgp.ac.in
[3] Computer Vision Lab, DAP, University of Sassari, 07041 Alghero, SS, Italy
tista@uniss.it
[4] Department of Computer Science and Engineering, IIT Kanpur,
Kanpur 208016, UP, India
pg@cse.iitk.ac.in

Abstract. This paper reports a face identification system which makes use of a novel local descriptor called Local Ternary Tree Pattern (LTTP). Exploiting and extracting distinctive local descriptor from a face image plays a crucial role in face identification task in the presence of a variety of face images including constrained, unconstrained and plastic surgery images. LTTP has been used to extract robust and useful spatial features which use to describe the various structural components on a face. To extract the features, a ternary tree is formed for each pixel with its eight neighbors in each block. LTTP pattern can be generated in four forms such as LTTP–Left Depth (LTTP-LD), LTTP–Left Breadth (LTTP-LB), LTTP–Right Depth (LTTP-RD) and LTTP–Right Breadth (LTTP-RB). The encoding schemes of these patterns are very simple and efficient in terms of computational as well as time complexity. The proposed face identification system is tested on six face databases, namely, the UMIST, the JAFFE, the extended Yale face B, the Plastic Surgery, the LFW and the UFI. The experimental evaluation demonstrates the most promising results considering a variety of faces captured under different environments. The proposed LTTP based system is also compared with some local descriptors under identical conditions.

Keywords: Face identification · Local descriptor · Ternary tree ·
Cosine similarity · Sum of absolute differences · Classifier

1 Introduction

The outcomes of face recognition in the last decades exhibit enormous improvements in person identification while constrained face images are used with little variations. However, it becomes difficult when a face identification system is presented with more

A. Morales et al. (Eds.): IbPRIA 2019, LNCS 11868, pp. 50–63, 2019.
https://doi.org/10.1007/978-3-030-31321-0_5

twisted and unconstrained face images for identification. Further, plastic surgery images also make the identification more challenging. Now, face identification has moved on to unconstrained scenarios with more non-deterministic factors. Face recognition [7] is a continuing research process in computer vision and it has been attained significant attention due to extensive use in surveillance, law enforcement and information security. The wide applications of face recognition have been motivated the researchers due to its dynamicity in reliability and robustness. Recognizing a face in a controlled environment is not a difficult task, however, it raises an adverse situation in unconstrained scenarios. The success of face recognition relies on the choice of robust descriptors and features which can deal with uncertainties of a face image. Applying local descriptors to face recognition is a powerful approach and it has successfully used many face recognition systems. Among three modes of face recognition such as verification, identification and matching, identification seems to be a difficult mode where an unidentified probe face is compared with all face templates of the asserted identity in the database. In face identification, to find the identity of an unknown person, the input face image is compared with all face templates of the registered persons present in the database. The face identification is more demanding in crime inquisition, law enforcement, identification of a suspicious person in public places like school, bank, railway station, airport and border of a country. The main objective of this paper is to propose a robust face identification system which can deal with real life situations efficiently and at the same time it ensures the robustness of the system. To achieve this, a novel local descriptor LTTP is applied for extracting discriminatory facial features.

1.1 Related Works

To handle the challenging situations in face recognition, a large number of local descriptors [1–3, 5, 10, 12, 16, 18, 19] have been employed in many works. The existing local descriptors are mainly exploited for face verification, however, not experimented for face identification. Local descriptors have the competency to extract discriminatory and stable information subjected to extract the features information effectively from each structural component of a face image.

Some local descriptors such as Local Binary Pattern (LBP) [3, 12], Multi scale-Local Binary Pattern (MS-LBP) [10], Multi block-Local Binary Pattern (MB-LBP) [18], Local Texture Pattern (LTP) [16], Local Derivative Pattern (LDP) [19], Local Vector Pattern (LVP) [5], Local Graph Structure (LGS) [2] and Symmetric Local Graph Structure (SLGS) [1] are used for face recognition. Among them, LBP [3, 12] is known to be a widely used local descriptor to encode the local relationship of the pixels. The operator works with the eight neighbors of a pixel using the value of the center pixel as a threshold. A binary pattern is generated for every pixel of the face image and produces a transformed image (TI). Histogram generated from the transformed image (TI) is used as a feature vector. The LBP operator is able to extract micro-patterns from face images and these micro-patterns are invariant to monotonic grey scale transformations. The scale variation is an important factor which affects the face recognition system due to variations of texture in a different scale. To handle this situation, MS-LBP [10] is used to improve performance. The MS-LBP is able to extract micro-structures from the face images at different scales. Its accuracy is better than LBP operator due to enhance size of the neighborhood. The MB-LBP [18] is another variant which shows enhanced performance over LBP by computing the average value of a sub-region of the face image. LTP [16] uses a constant to threshold the pixels into three values. After thresholding, a ternary pattern is generated for every pixel and formed a transformed image. The LTP operator is found to be more discriminative and less sensory to noise in identical regions. On the contrary, LDP [19] uses the local derivative to encode directional pattern. To extract discriminatory features from a high-order derivative space, LVP can be used. To increase the robustness of the structure of micropatterns, LVP encodes different pair-wise directions of vectors and generate an invariant facial description. Some existing local descriptors which use graphs structures can also be found useful for providing compatible performance with LBP and its variants. Such as Local Graph Structures (LGS) [2] and Symmetric Local Graph Structure (SLGS) [1] are used local directional graph for a pixel to generate a binary pattern. These local descriptors have successfully been applied in face recognition due to their distinctive feature representation.

1.2   Major Contributions

The local descriptors are very powerful tools which can able to extract diverse and contrasting features from a face. The proposed work uses an efficient local descriptor called Local Ternary Tree Pattern (LTTP). It has the ability to identify the strength through the structural patterns which are derived from the face image. The descriptor is computationally not very expensive. Further, the identification performance under complicated environments is found to be outstanding. With structural uniqueness of LTTP, face identification can achieve its dominance phase.

The rest of the paper is organized as follows: Sect. 2 describes the proposed Local Ternary Tree Pattern (LTTP). Section 3 presents the proposed face identification framework which uses Local Ternary Tree Pattern in its feature extraction phase. The experimental results are provided in the next section. Finally, a conclusion is drawn in the last section.

2   Local Ternary Tree Pattern (LTTP)

The Local Ternary Tree Pattern (LTTP) is used effectively to extract the features which represent the invariant local textures of a face image in the way of extracting structural components on the face. Unlike LBP, the LTTP generates a ternary tree for each pixel of an image with its 8 neighbors and generates four patterns rather than one.

To extract the discriminatory and structural components from a face image, an image $I(P)$ of size $h \times v$ is divided into a number of smaller sub-regions with dimension $h' \times v'$, where $h' \ll h$ and $v' \ll v$. Then LTTP operator is applied on each pixel of the gray scale face image and generates transformed value. To determine the transformed value for a pixel, a ternary tree is formed by considering the pixel as root of the tree covering eight neighbors in a 3×3 region. Then binary labeling of edges is performed. During labeling, the root node is compared with its child nodes and computes the differences between the root node and child nodes. If the difference is found positive or equal to 0, then assign 1 to the edge between the root node (source pixel) and child node (neighborhood pixel), otherwise assign 0 to that edge. Finally, binary labels (0 or 1) of edges of the ternary tree are concatenated together using four rules to form 8-bit binary pattern which is then converted to a decimal number and assigned to the target pixel. The four concatenation rules are: Left oriented–Depth First Traversal (LTTP-LD), Left oriented–Breadth First Traversal (LTTP-LB), Right oriented-Depth First Traversal (LTTP-RD), and Right oriented-Breadth First Traversal (LTTP-RB). A ternary tree with a target pixel '9' is shown in Fig. 1. To encode the LTTP-LD pattern, at first starts from root node, goes to the depth in left side then goes to depth in right side and one by one concatenate all eight labels of edges. According to Fig. 1, the sequence of edges to generate LTTP-LD is $9 \rightarrow 9$, $9 \rightarrow 8$, $9 \rightarrow 6$, $6 \rightarrow 5$, $6 \rightarrow 11$, $6 \rightarrow 7$, $9 \rightarrow 8$, $8 \rightarrow 10$ (A $\rightarrow$ B, B $\rightarrow$ E, A $\rightarrow$ C, C $\rightarrow$ F, C $\rightarrow$ G, C $\rightarrow$ H, A $\rightarrow$ D, D $\rightarrow$ I). To encode the LTTP-LB pattern, at first starts from root node, goes to breadth from left side to right side, covers all levels of the tree and one by one concatenate all eight labels of edges. According to Fig. 1, the sequence of edges to generate LTTP-LB is $9 \rightarrow 9$, $9 \rightarrow 6$, $9 \rightarrow 8$, $9 \rightarrow 8$, $6 \rightarrow 5$, $6 \rightarrow 11$, $6 \rightarrow 7$, $8 \rightarrow 10$ (A $\rightarrow$ B, A $\rightarrow$ C, A $\rightarrow$ D, B $\rightarrow$ E, C $\rightarrow$ F, C $\rightarrow$ G, C $\rightarrow$ H, D $\rightarrow$ I). To encode the LTTP-RD pattern, at first starts from root node, goes to the depth in right side then goes to depth in left side and one by one concatenate all eight labels of edges. According to Fig. 1, the sequence of edges to generate LTTP-RD is $9 \rightarrow 8$, $8 \rightarrow 10$, $9 \rightarrow 6$, $6 \rightarrow 7$, $6 \rightarrow 11$, $6 \rightarrow 5$, $9 \rightarrow 9$, $9 \rightarrow 8$ (A $\rightarrow$ D, D $\rightarrow$ I, A $\rightarrow$ C, C $\rightarrow$ H, C $\rightarrow$ G, C $\rightarrow$ F, A $\rightarrow$ B, B $\rightarrow$ E). Lastly, to encode the LTTP-RB pattern, at first starts from root node, goes to breadth from right side to left side, covers all levels of the tree and one by one concatenate all eight labels of edges. According to Fig. 1, the sequence of edges to generate LTTP-RB is $9 \rightarrow 8$, $9 \rightarrow 6$, $9 \rightarrow 9$, $8 \rightarrow 10$, $6 \rightarrow 7$, $6 \rightarrow 11$, $6 \rightarrow 5$, $9 \rightarrow 8$ (A $\rightarrow$ D, A $\rightarrow$ C, A $\rightarrow$ B, D $\rightarrow$ I, C $\rightarrow$ H, C $\rightarrow$ G, C $\rightarrow$ F, B $\rightarrow$ E). The encoding scheme of LTTP considers the direct relationship of a target pixel with its neighbors as well as the relationship between the pixels that form the local ternary tree of the target pixel. This encoding scheme enables the LTTP to generate unique face representation and subsequently the improved identification performance.

To produce the decimal number by applying LTTP operator for a pixel $I(P_t)$, a binomial weight 2^q is multiplied to each label of edges of the ternary tree and then all multiplied labeled values are added.

The LTTP code for a target pixel $I(P_t)$ is given by

$$LTTP(I(P_t)) = \sum_{q=0}^{L-1} f(P_r - P_c) * 2^q \quad where \quad f(x) = \begin{cases} 1, & if \quad x \geq 0 \\ 0, & if \quad x < 0 \end{cases} \quad (1)$$

where P_r denotes the gray value of root node (source pixel) and P_c denotes the gray value of child node (neighboring pixel). In Eq. (1), L is the total number of neighboring pixels of the target pixel. The pictorial representation of basic LTTP operator is shown in Fig. 1.

The four different types of encoding schemes in LTTP would contribute a large number of distinguishable features to the feature set considering different environmental conditions including unconstrained and post-surgery.

3   Proposed Face Identification Framework

The aim of this paper is to present a robust face identification system which uses the proposed local descriptor – LTTP in its feature extraction phase to improve the overall performance of the system under degraded, unconstrained and post-surgery scenarios. The system consists of two main phases – feature extraction and classification.

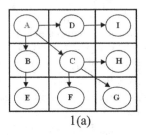

1(a)

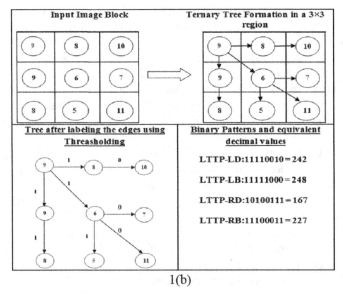

1(b)

Fig. 1. Pictorial representation of Local Ternary Tree Pattern (LTTP).

The face identification [13, 14] is a paradigm of one to many matching that compares a probe image with all templates of face images present in a face database to infer the identity of the probe face. Suppose, a probe face image I_p is considered for testing and a full set of gallery images G is considered for training. The feature vector F (I_p) is generated for a probe image I_p. Then the objective function of the proposed face identification system can be defined as:

$$f(I_p) = \min_i(D(F(I_p), F(G_i))) \quad where \quad i = 1, 2, 3, \dots, N \qquad (2)$$

Where $D(.)$ is the distance function used to measure the similarity between two feature vectors $F(I_p)$ and $F(G_i)$. $F(G_i)$ is the feature vector of ith gallery image.

The robust and discriminatory facial feature extraction is the key to the success of a face identification system. This phase transforms a raw face image into a transformed image and generates the feature vector. In this phase, the proposed Local Ternary Tree Pattern (LTTP) is applied on all pixels of a face image to generate the transformed image (TI) and finally, the feature vector is generated from the transformed image by flattening it. The proposed LTTP is capable to handle different challenges of a face image like changes in photometric condition, facial expression, head pose, facial accessories (makeup glasses etc.), imaging modality and unconstrained environment. An input face image and its transformed patterns generated using LTTP variants are shown in Fig. 2.

The classification plays an important role in the proposed face identification system. Classification phase can be divided into two sub-phases such as matching and identity generation.

During matching, the template generated from a probe face image is compared with all templates produced from gallery face images and computes their similarity scores. The cosine similarity measure (CS) [11] and the sum of absolute differences (SAD) [4] are used to measure the similarity between the feature of the probe image and the feature of the gallery images.

During identity generation, the similarity scores produced by matching are used to generate a rank order list for each probe image. The gallery feature which has the maximum similarity to the probe feature is selected and the identity of the corresponding gallery image is considered as the identity of the probe image.

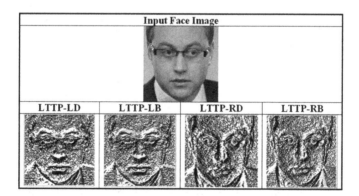

Fig. 2. An input image and transformed images generated by applying LTTP operator.

4   Evaluation

The proposed face identification system is evaluated on six challenging face databases, namely, the UMIST [8], the extended Yale face B [17], the JAFFE [16], the Plastic Surgery [15], the LFW [6] and the UFI [9] databases. Identification accuracy at Rank-1 strategy is determined using two distance metrics viz. cosine similarity measure (CS) [11] and the sum of absolute differences (SAD) [4]. During experiments, when the system takes a probe face as input, the Rank-1 strategy retrieves a face image corresponding to highest proximity, from database. In this experiment, the LTTP is used for facial feature extraction from the face image. However, prior to feature extraction, no image enhancement operation is applied. During training and testing, each face image is partitioned into a number of image blocks of size 3×3. Then, LTTP is applied to each block and obtain local textural and structural features which are concatenated together with other local features to form a global feature set. To generate probe and gallery sets, a random partition is used in the databases prior to training and testing. However, a different strategy other than random partition is adopted for the LFW and the UFI databases where a subset of images of the entire database is taken into consideration for evaluation and the subset is divided biasedly, to obtain the probe and gallery sets. Finally, the performance of the novel operator LTTP for face identification is compared with other local descriptors such as LBP [3, 12], LGS [2] and LTP [16] in identical scenarios with the same databases. A brief overview of six databases is listed in Table 1 and sample images of these databases are shown in Fig. 3. The experimental results are shown in Table 2.

Table 1. Brief description of six face databases

Name of database	No. of subjects	Total no. of images	Image format	Resolution	Description
JAFFE	10	213	.tiff	121×146	Facial expression variations
UMIST	20	564	.pgm	92×112	Pose variations (profile to frontal)
Extended Yale face B	38	2432	.pgm	40×46	Illumination variations
LFW	5749	13233	.pgm	64×64	Unconstrained face images
UFI	605	4921	.pgm	128×128	Unconstrained face images
Plastic surgery	53 (obtained) 900 (original)	106 (obtained) 1800 (original)	.jpg	238×273	Post-surgery variations

The identification accuracy (*IA*) is computed using Eqs. (3) and (4). For a probe image I_p, the Identification Accuracy '*IA*' of the proposed system is defined as

$$IA = \left(\frac{1}{|M|} \sum_{p=1}^{|M|} \sum_{i=1}^{|N|} \Delta(\Phi(I_p), \Phi(G_i), R(I_p, G_i))\right) \times 100 \qquad (3)$$

where $|M|$ is the size of probe set, $|N|$ is the size of the gallery, and $\Delta(.)$ computes k^{th} best match for the probe image I_p as

$$\Delta(\Phi(I_p), \Phi(G_i), R(I_p, G_i)) = \begin{cases} 1, & if \quad \Phi(I_p) = \Phi(G_i) \quad and \quad R(I_p, G_i) = k \\ 0, & else \end{cases} \quad (4)$$

$\Phi(.)$ is a function which returns the class of an image and $R(I_p, G_i)$ returns the ranked position of a gallery image G_i with respect to the test image I_p from the probe set.

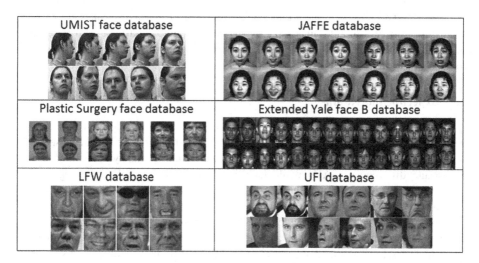

Fig. 3. Sample images from the UMIST [8], the JAFFE [16], the extended Yale face B [17], the Plastic Surgery [15], the LFW [6] and the UFI [9] Face Databases.

4.1 Experimental Design and Protocol

Experiments are conducted with face images obtained from different databases and each of these databases is partitioned into two sets – probe and training. From the UMIST database [8] one image per subject (total 20 face images) having pose variation is considered in the probe set. Some of the subjects having eyeglasses are also considered in the probe set. Rest of the images are used for training. From the JAFFE database [16], one image per subject (total 10 face images) having varied facial expressions are included in the probe set and the remaining 203 face images are considered for the training set. In the extended Yale face B database [17], 8 images per subject (total 304 face images) with different illumination conditions are included in the probe set and remaining 2128 images are included in the training set. In the Plastic Surgery face database [15], 15 to 22 face images labeled as post-surgery are included in the probe set and 53 face images labeled as pre-surgery are considered for the training set. It is due to the difficulty in separation of pre-surgery and post-surgery images and unavailability of plastic surgery face images, only upto 53 images are considered for the experiment. The face images present in the LFW database [6] are captured in unconstrained environments. The accuracy is highly affected due to illumination variation, facial expression changes, occlusion (external or self), pose variations and facial accessories like scarf, mask, hat or sunglasses. The database contains 13233 face images of 5749 individuals. We conducted the experiment only on 1,680 subjects with the criteria of having two or more than two face images. One image per subject is considered in the probe set and the rest of the images are included in the training set. The UFI database [9] is another unconstrained database which contains face images of 605 subjects. The presence of various backgrounds in face images makes the task of identification more difficult. During the experiment on the database, several probe sets depending upon different variations present in face images are created and 4316 images are used in the training set.

4.2   Experimental Results

The UMIST [8] face database covers a range of poses from profile to frontal views. From experimental outcomes, it has been observed that the proposed LTTP based face identification gives 100% accuracy at Rank-1 using cosine similarity (CS) and the sum of absolute differences (SAD) while handling pose variations of face images. The experimental results reveal that the LTTP is robust to frontal as well as different pose variations of the face images.

The Extended Yale Face B database [17] containing face images having a number of illumination effects and it makes the database challenging. The identification accuracy determined on this database achieves 100% at Rank-1 using cosine similarity (CS) and the sum of absolute differences (SAD) while LTTP operator is used as a feature descriptor in face identification. The operator is capable to handle the effects of illumination variations of face images.

The identification accuracy determined on the JAFFE [16] database achieves 100% at Rank-1 using cosine similarity (CS) and the sum of absolute differences (SAD) while the database covers 7 facial expressions. The outcomes reveal that LTTP operator could capture variations in facial appearance due to expression changes very efficiently.

The post-surgery variations on a face create a challenging situation in identification. The Plastic Surgery face database [15] contains pre-surgery and post-surgery face images of 900 subjects. However, the evaluation uses images of 53 subjects as the rest of the images from 847 subjects are not found due to some difficulty. The proposed face identification system has achieved 100% accuracy at Rank-1 using cosine similarity (CS) and the sum of absolute differences (SAD) as classifiers.

The proposed face identification achieves 100% and 99.66% accuracies on the LFW [6] and the UFI [9] unconstrained face databases respectively while Rank-1 strategy is employed. The outcomes reveal that the proposed LTTP shows the robustness when face images are captured under unconstrained environments. The experimental results are shown in Table 2.

Table 2. Identification accuracies (%) determined at Rank-1 on six face databases.

Name of databases	LTTP-LD		LTTP-LB		LTTP-RD		LTTP-RB	
	CS	SAD	CS	SAD	CS	SAD	CS	SAD
UMIST	100	100	100	100	100	100	100	100
JAFFE	100	100	100	100	100	100	100	100
Extended YALE face B	100	100	100	100	100	100	100	100
Plastic surgery	100	100	100	100	100	100	100	100
LFW	100	100	100	100	100	100	100	100
UFI	99.66	99.66	99.66	99.66	99.66	99.66	99.66	99.66

Most of the works on the LFW database have been reported in face verification. Since face identification is a demanding task in law enforcement and in an unconstrained environment, a low-quality probe face may not be able to provide a sufficient description of a face image, therefore this motivates to check the effect of different environmental conditions such as pose, facial expression, lighting conditions, occlusions, background, accessories and photographic quality explicitly during experiments. The effect of different environmental conditions are analyzed separately in the LFW database. Several experiments are conducted on the LFW database to check the robustness of the proposed descriptor under different environmental conditions separately as the unconstrained environment is the collection of various situations that affect the appearance of a face image to a large extent. It is enlisted best identification accuracy at Rank-1 in Table 2 while a subset of images is selected biasedly from a number of subsets. The experimental outcomes reveal that the proposed LTTP operator shows the robustness in the probe set of the LFW database where mainly pose variations, facial expression variations and illumination variations are considered.

In law enforcement, to determine the identity of a subject based on one or more probe images, a top 200 ranked list may be retrieved from the gallery and which suffices for analyzers [20]. This motivates to conduct further experiments on the LFW database to test the combined effect of different environmental conditions. The experiments are conducted on the dataset containing twenty and more than twenty images per subject and generate identification accuracy up to Rank-50 shown in Table 3.

Table 3. Identification accuracies (%) on the LFW dataset having the criteria of twenty images per subject at Rank-1, Rank-10, Rank-20, Rank-30, Rank-40 and Rank-50.

Rank	LTTP-LD		LTTP-LB		LTTP-RD		LTTP-RB	
	CS	SAD	CS	SAD	CS	SAD	CS	SAD
1	50.90	56.36	45.45	49.09	32.72	32.72	25.45	36.36
10	89.09	89.09	81.81	87.27	72.72	78.18	76.36	78.18
20	94.54	96.36	92.72	94.54	80	90.90	83.63	92.72
30	98.18	98.18	96.36	98.18	89.09	98.18	89.09	96.36
40	98.18	100	96.36	100	92.72	98.18	92.72	96.36
50	100	100	96.36	100	96.36	98.18	94.54	98.18

A number of experiments are conducted on the UFI face database and enlist the best identification accuracy at Rank-1 achieved among different probe sets of the database, in Table 2. During experiments, the effects of different environmental conditions like pose, facial expression, illumination, occlusion, image quality and background are experimented explicitly and analyze the effect of different conditions. If an image with pose variation is present in the probe set, but not an image with pose variations in the training set, then it is very difficult to identify the probe image at Rank-1. Presence of auxiliary background in the probe image or images captured in extreme illumination or self-occlusion or low-quality image affects the performance of the identification system adversely. The UFI is an unconstrained database which contains images captured with no restrictions over environmental conditions. To analyze the effect of different environmental conditions, different probe sets out of 605 subjects are created. The probe set which contains images with facial expression variations; illumination variations are identified correctly at Rank-1. The probe set which contains images with pose variations and self-occlusions are identified correctly if such kinds of variations are present in the training set. In these cases, 100% accuracy is achieved. If low-quality images and images captured in extreme illumination are present in the probe set, it affects the performance of the identification system adversely. In these cases, special preprocessing approaches are required before feature extraction.

To check the combined effect of different environmental conditions (unconstrained environment), further experiments are conducted on the UFI database in an unbiased way, where one image per subject (total 605 subjects) is considered together in a probe set and 4316 images in the training set. The experimental results are shown in Table 4.

Table 4. Identification accuracies (%) on the UFI dataset having images of all subjects in probe set at Rank-1, Rank-10, Rank-20, Rank-30, Rank-40, Rank-50.

Rank	LTTP-LD		LTTP-LB		LTTP-RD		LTTP-RB	
	CS	SAD	CS	SAD	CS	SAD	CS	SAD
1	44.13	50.24	42.64	49.09	39.83	42.97	38.51	43.30
10	62.31	67.6	60.49	67.27	54.04	58.84	53.88	61.15
20	67.76	74.04	66.11	72.39	59.50	66.77	59.17	67.10
30	69.75	76.85	69.25	75.86	64.46	69.91	63.47	71.40
40	72.72	79.50	71.40	78.67	67.10	72.56	67.27	73.71
50	73.71	80.99	73.55	80.16	69.25	74.21	69.75	75.20

4.3 Comparative Study

This section presents a comparison of LTTP based system with some existing local descriptors such local binary pattern (LBP) [3, 12], local graph structure (LGS) [2] and local texture pattern (LTP) [16] which are simulated and tested on the same databases. The identification accuracy achieved by LTTP is compared with the accuracies obtained by LBP, LGS, and LTP at the corresponding ranks. Table 5 shows the comparison summary on six face databases.

When evaluation is performed on the JAFFE database, it can be seen from Table 5 that the proposed LTTP yields 100% accuracy at Rank-1 using both CS and SAD metrics, LBP produces 90% accuracy using the CS classifier and 100% accuracy using the SAD classifier, also, LGS achieves 100% accuracy using both the classifiers and LTP achieves 70% accuracy using both the classifiers. The experimental outcomes reveal that the proposed LTTP performs well in facial expression variations. When evaluation is performed on the UMIST face database, both LTTP and LBP operators achieve 100% accuracy using both the classifiers, the LGS achieves 100% accuracy using the SAD classifier and 95% accuracy using the CS classifier and LTP achieves 90% accuracy using the CS and 80% accuracy using the SAD classifier. Thus, the proposed descriptor performs well in case of pose variations. When the same identification system is experimented on the extended Yale face B database, the LBP achieves 62.5% accuracy using the CS and 74.67% accuracy using the SAD, the LGS achieves 70% accuracy using the CS and 76.97% accuracy using the SAD, the LTP achieves 38.15% accuracy using the CS and 44.07% accuracy using the SAD, while LTTP achieves 100% accuracy using both classifiers. The outcomes on the extended Yale face B database reveal that the proposed LTTP is more robust than existing local descriptors in case of illumination variations. When the same identification framework is evaluated on the Plastic Surgery face database, the proposed LTTP performs well. The LBP achieves 46.66% accuracy using the CS and 53.33% accuracy using the SAD, the LGS achieves 73.33% accuracy using both the CS and SAD classifiers, and the LTP achieves 13.33% accuracy using both the CS and SAD classifiers, while LTTP achieves 100% accuracy using both classifiers. When the same identification framework tested on the LFW face database the LBP achieves 17.77% accuracy using the CS and 15.55% accuracy using the SAD, the LGS achieves 28.88% accuracy using the CS and 31.11% accuracy using the SAD, the LTP achieves 4.44% accuracy using the CS and 6.66% accuracy using SAD, while LTTP achieves 100% accuracy using both CS and the SAD. When the UFI database is used in the experiment, the LBP achieves 47% accuracy using CS and 48% accuracy using SAD, the LGS achieves 50.5% accuracy using CS and 55.5% accuracy using SAD, the LTP achieves 10% using CS and 11.5% using SAD, while LTTP achieves 99.66% accuracy using both the CS and SAD classifiers. From Table 5 we can say that the proposed LTTP is much more robust than existing local descriptors in case of the pose variations, facial expression variations, illumination variations, post-surgery variations and unconstrained environments.

Table 5. Comparison of the proposed LTTP with state-of-the-art local descriptors at Rank-1 strategy.

Name of face databases	Name of local descriptors							
	LBP		LGS		LTP		LTTP	
	CS	SAD	CS	SAD	CS	SAD	CS	SAD
JAFFE	90	100	100	100	70	70	100	100
UMIST	100	100	95	100	90	80	100	100
Extended YALE face B	62.5	74.67	70	76.97	38.15	44.07	100	100
Plastic surgery	46.66	53.33	73.33	73.33	13.33	13.33	100	100
LFW	17.77	15.55	28.88	31.11	4.44	6.66	100	100
UFI	47	48	50.5	55.5	10	11.5	99.66	99.66

5 Conclusion and Future Works

This paper has presented a promising work on face identification which has used a novel local descriptor (LTTP) for facial feature extraction. The proposed descriptor LTTP, as well as the face identification system, have validated through a variety of face images captured under extensive experiments including plastic surgery faces. The LTTP is a ternary tree based local descriptor where a pixel is represented by a ternary tree with its eight neighbors. Then using a thresholding rule a binary pattern is generated for the target pixel. The experiments illustrated that the use of LTTP in feature extraction phase of the face identification system has enabled it with greater discriminative power. The proposed LTTP is more robust than other local descriptors because its ternary tree structure is able to capture more discriminatory information from a face. The LTTP descriptor might have a huge impact if it is used for sustainable facial recognition systems. However, a few more tests on vibrant databases such as low resolution as well as heterogeneous face images will establish its usefulness as an unconquerable feature representation tool.

References

1. Abdullah, M.F.A., Sayeed, M.S., Muthu, K.S., Bashier, H.K., Azman, A., Ibrahim, S.Z.: Face recognition with symmetric local graph structure (SLGS). Expert Syst. Appl. **41**(14), 6131–6137 (2014)
2. Abusham, E.E.A., Bashir, H.K.: Face recognition using local graph structure (LGS). In: Jacko, J.A. (ed.) HCI 2011. LNCS, vol. 6762, pp. 169–175. Springer, Heidelberg (2011). https://doi.org/10.1007/978-3-642-21605-3_19
3. Ahonen, T., Hadid, A., Pietikäinen, M.: Face recognition with local binary patterns. In: Pajdla, T., Matas, J. (eds.) ECCV 2004. LNCS, vol. 3021, pp. 469–481. Springer, Heidelberg (2004). https://doi.org/10.1007/978-3-540-24670-1_36
4. Alsaade, F.: Fast and accurate template matching algorithm based on image pyramid and sum of absolute difference similarity measure. Res. J. Inf. Technol. **4**(4), 204–211 (2012)
5. Fan, K.C., Hung, T.Y.: A novel local pattern descriptor—local vector pattern in high-order derivative space for face recognition. IEEE Trans. Image Process. **23**(7), 2877–2891 (2014)
6. Huang, G.B., Mattar, M., Berg, T., Learned-Miller, E.: Labeled faces in the wild: a database for studying face recognition in unconstrained environments. In: Workshop on Faces in 'Real-Life' Images: Detection, Alignment, and Recognition (2008)
7. Jain, A.K., Li, S.Z.: Handbook of Face Recognition. Springer, New York (2011). https://doi.org/10.1007/978-0-85729-932-1
8. Kisku, D.R., Mehrotra, H., Gupta, P., Sing, J.K.: Robust multi-camera view face recognition. Int. J. Comput. Appl. **33**(3), 211–219 (2011)
9. Lenc, L., Král, P.: Unconstrained facial images: database for face recognition under real-world conditions. In: Lagunas, O.P., Alcántara, O.H., Figueroa, G.A. (eds.) MICAI 2015. LNCS (LNAI), vol. 9414, pp. 349–361. Springer, Cham (2015). https://doi.org/10.1007/978-3-319-27101-9_26
10. Chan, C.-H., Kittler, J., Messer, K.: Multi-scale local binary pattern histograms for face recognition. In: Lee, S.-W., Li, S.Z. (eds.) ICB 2007. LNCS, vol. 4642, pp. 809–818. Springer, Heidelberg (2007). https://doi.org/10.1007/978-3-540-74549-5_85

11. Nguyen, H.V., Bai, L.: Cosine similarity metric learning for face verification. In: Kimmel, R., Klette, R., Sugimoto, A. (eds.) ACCV 2010. LNCS, vol. 6493, pp. 709–720. Springer, Heidelberg (2011). https://doi.org/10.1007/978-3-642-19309-5_55

12. Ojala, T., Pietikainen, M., Maenpaa, T.: Multiresolution gray-scale and rotation invariant texture classification with local binary patterns. IEEE Trans. Pattern Anal. Mach. Intell. **24** (7), 971–987 (2002)

13. Rakshit, R.D., Nath, S.C., Kisku, D.R.: An improved local pattern descriptor for biometrics face encoding: a LC–LBP approach toward face identification. J. Chin. Inst. Eng. **40**(1), 82–92 (2017)

14. Rakshit, R.D., Nath, S.C., Kisku, D.R.: Face identification using some novel local descriptors under the influence of facial complexities. Expert Syst. Appl. **92**, 82–94 (2018)

15. Singh, R., Vatsa, M., Bhatt, H.S., Bharadwaj, S., Noore, A., Nooreyezdan, S.S.: Plastic surgery: a new dimension to face recognition. IEEE Trans. Inf. Forensics Secur. **5**(3), 441–448 (2010)

16. Suruliandi, A., Meena, K., Rose, R.R.: Local binary pattern and its derivatives for face recognition. IET Comput. Vis. **6**(5), 480–488 (2012)

17. UCSDRepository (2001). http://vision.ucsd.edu/~leekc/ExtYaleDatabase/ExtYaleB.html

18. Zhang, L., Chu, R., Xiang, S., Liao, S., Li, S.Z.: Face detection based on multi-block LBP representation. In: Lee, S.-W., Li, S.Z. (eds.) ICB 2007. LNCS, vol. 4642, pp. 11–18. Springer, Heidelberg (2007). https://doi.org/10.1007/978-3-540-74549-5_2

19. Zhang, B., Gao, Y., Zhao, S., Liu, J.: Local derivative pattern versus local binary pattern: face recognition with high-order local pattern descriptor. IEEE Trans. Image Process. **19**(2), 533–544 (2010)

20. Best-Rowden, L., Han, H., Otto, C., Klare, B.F., Jain, A.K.: Unconstrained face recognition: Identifying a person of interest from a media collection. IEEE Trans. Inf. Forensics Secur. **9** (12), 2144–2157 (2014)

Catastrophic Interference in Disguised Face Recognition

Parichehr B. Ardakani[1]([✉]), Diego Velazquez[1], Josep M. Gonfaus[2],
Pau Rodríguez[3], F. Xavier Roca[1], and Jordi Gonzàlez[1]

[1] Computer Vision Center, Univ. Autònoma de Barcelona, 08193 Bellaterra, Spain
pbehjati@cvc.uab.es
[2] Visual Tagging Services, Parc de Recerca UAB, 08193 Bellaterra, Spain
[3] Element AI, Montreal 6650, Canada

Abstract. It is commonly known the natural tendency of artificial neural networks to completely and abruptly forget previously known information when learning new information. We explore this behaviour in the context of Face Verification on the recently proposed Disguised Faces in the Wild dataset (DFW). We empirically evaluate several commonly used DCNN architectures on Face Recognition and distill some insights about the effect of sequential learning on distinct identities from different datasets, showing that the catastrophic forgetness phenomenon is present even in feature embeddings fine-tuned on different tasks from the original domain.

Keywords: Neural network forgetness · Face recognition ·
Disguised Faces

1 Introduction

Deep Convolutional Neural Networks (DCNNs) have achieved remarkable success in various cognitive applications such as image recognition, facial detection, signal processing, on supervised, unsupervised and reinforcement learning tasks through feature representations at successively higher, more abstract layers. Computational complexity and the time needed to train large networks is one of the major challenges for convolutional networks. It is common to pretrain a DCNN on a large dataset and then use the trained network as an initialization or as a fixed feature extractor for a particular application [24]. A major downside of such DCNNs is the inability to retain previous knowledge while learning new information. This problem is called Catastrophic forgetting.

Catastrophic forgetting is a term, often used in connectionist literature, to describe a common problem with many traditional artificial neural network models. It refers to forgetting what has been learned upon learning new or different

---

P. B. Ardakani and D. Velazquez—Both authors contributed equally to this work.

© Springer Nature Switzerland AG 2019
A. Morales et al. (Eds.): IbPRIA 2019, LNCS 11868, pp. 64–75, 2019.
https://doi.org/10.1007/978-3-030-31321-0_6

information. For instance, when a network is first trained to convergence on one task, and then trained on a second task, it forgets how to perform the former.

There are some approaches to improve the performance of models when learning new information that benefit from previously learned information, for example fine-tuning [6], where the old task parameters are adjusted to adapt to a new task. Other approaches well known are feature extraction [5] where the parameters of the old network are unchanged and the parameters of the outputs of one or more layers are used to extract features for the new task. There is also a paradigm called joint train [4] in which parameters of old and new tasks are trained together to minimize the loss in all tasks.

Overcoming the problem of catastrophic forgetting is an important step. Some methods have already been developed to overcome this problem [8,17,27]. But even with these and other methods, the problem of catastrophic forgetting is still a key problem within the Artificial Intelligence (AI) community and it is time to move towards algorithms that can learn multiple tasks over time [25].

The novel approaches have been developed specifically for the tasks of visual face recognition and verification in order to boost performance on public datasets such as Labeled Faces in the wild (LFW) [11]. However, the performance on completely unconstrained datasets like Youtube Face (YTF) [30], and UMDFaces [1] remains subpar at low false alarm rates. These datasets contain significant variations in illumination, pose, expression, aging and tend to have low resolution and clutter filled images. This indicates that the problem of face recognition is far from *solved*. The recently announced Disguised Faces in the Wild (DFW) dataset aims to study another covariate of the face verification pipeline - *disguises*.

Disguise and impersonation are part of a sub-field of face recognition where the subjects are non-cooperative and are actively trying to deceive the system. A disguise involves both intentional or unintentional changes on a face through which one can either obfuscate his/her identity. This means that the subject is trying to adopt a new identity in order to hide his/her own. A subject might impersonate someone else's identity. Obfuscation increases the inter-class variations whereas impersonation reduces the inter-class dissimilarity, thereby affecting face recognition/verification task and making it non-trivial. This is a very challenging face verification problem and has not been studied in a comprehensive way, primarily due to the unavailability of such a dataset. The aim of a face verification system in such cases is to identify a given subject under varying disguises while rejecting impostors trying to look like the subject of interest in an uncontrolled setting. From the point of view of an automated computer vision method, it is important to extract rich face features in-order to distinguish among the identities and verify them correctly.

In this paper, we explore catastrophic behaviour in the context of Face Verification on the DFW dataset. We empirically evaluate several commonly used DCNN architectures on Face Recognition and distill some insights about the effect of sequential learning on distinct identities from different datasets which are explained in the following sections.

2   Related Work

In this section we briefly review some recent related work and proposed methods on face recognition/verification and catastrophic forgetting.

Disguised faces recognition focuses on recognizing the identity of disguised faces and impersonators. There is limited research focus on this topic. MiRA-Face [31] uses two CNNs networks one for aligned input and the other for unaligned input to perform generic face recognition. Then, Principal Component Analysis (PCA) is used to find the transformation matrix for face recognition adaptation. Another work is Deep Disguise Recognizer (DDRNET) [14] uses an Inception Network along with Center loss [29] followed by classification using a similarity metric. DisguisedNet [28] proposed a Siamese-based approach using the pretrained VGG-Face [20] and after that, cosine similarity is applied for performing classification of the learned features. AEFRL [26], performs face detection and alignment on the input images using Multi-task Cascaded Convolutional Networks(MT-CNN) [32] followed by horizontal flipping. An ensemble of five networks is used to obtain features for original and flipped images. The concatenation of these features are used to perform classification using cosine similarity. UMDNets [2] is another work which uses All-in-One [21] to align the images using facial landmarks. They performs feature extraction using two networks, followed by independent score computation. Then, classification is performed by averaging the scores obtained via the two feature sets. Table 1 provides a list of the proposed approaches on DFW dataset for face verification.

The problem of catastrophic forgetting is a big issue in machine learning and artificial intelligence if our goal is to build a system that learns through time, and is able to deal with more than a single problem. According to [18], without this capability we will not be able to build truly intelligent systems, we can only create models that solve isolated problems in a specific domain. There are some recent works that tried to overcome this problem, e.g., domain adaptation that uses the knowledge learned to solve one task and transfers it to help learning another, but those two tasks have to be related. This approach was used in [12] to avoid the problem of catastrophic forgetting, in order to do so they use two properties. The first property was to keep the decision boundary unchanged and the second one, was that the feature extractor from the source data by the target network should be present in a position close to the features extracted from the source data by the source network. As was shown in the experiments, by keeping the decision boundaries unchanged new classes cannot be learned, making this approach unable to deal with related tasks that present a different number of classes. Early attempts to alleviate catastrophic forgetting often consists of a memory system that store previous data and replays the sampled old examples with the new data [22], and similar approaches are still used today [16]. [23] learns a generative model to capture the data distribution of previous tasks, and both generated samples and real samples from the current task are used to train the new model so that the forgetting can be alleviated for continual learning.

In our work, we will show that the intrinsic forgetness property of neural networks not only present when performing classification on new problems but also when extracting features even for tasks whose domain is the same.

Table 1. Different approaches to face verification

Model	Brief description
AEFRL	MTCNN + 4 Networks for feature extraction + Cosine distance
DDRNET	Weighted linear combination of ensemble of 3 CNNs
DisguiseNet	VGG siamese architecture + Weighted Loss + Cosine Distance
MiRA-Face	MTCNN + RSA + Ensemble of CNNs Text follows
UMDNets	All-in-One + average across scores obtained by 2 networks

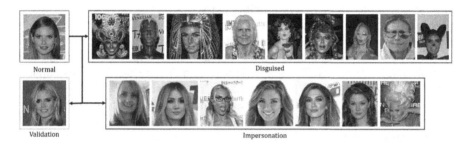

Fig. 1. Some example images of Disguised Faces in the wild (DFW) dataset. The dataset contains four kinds of images: normal, validation, disguise and impersonator (figure taken from [15])

3   Methodology

In this section we describe the DFW dataset [15] along with the evaluation protocols it presents. We comment on the different architectures we use for our experiments and explain the identity overlap between the datasets used for training and the DFW test set.

3.1   Dataset

In our experiments we evaluate the performance of different models in the face verification task using the DFW dataset. This dataset has been created to address the issue of disguised and imposter faces in the wild. The dataset consists of 11,157 images belonging to 1000 different subjects. Images pertaining to 400 subjects form the training set, while the remaining 600 subjects constitute the test set. Each subject has at least five face images, and can have four types of images: (i) *normal*, (ii) *validation*, (iii) *disguised* and (iv) *impersonator*.

Normal images correspond to non-disguised frontal face images. Validation images are used to generate a non-disguised pair within a subject. Disguised images correspond to a face image of the same subject having intentional or unintentional disguise. Impersonator images correspond to a face image of individuals who intentionally or unintentionally look similar to a different subject. An example of each is shown in Fig. 1.

Three verification protocols have been provided with the DFW dataset to understand and evaluate the effect of disguises on face recognition.

- **Protocol-1 (Impersonation)** evaluates the capacity of the system to differentiate genuine users from impersonators. Genuine pairs for this protocol are created by combining a genuine image and a validation image from the same subject. Impostors pairs are created by combining impersonator images with normal, validation and disguised images from the same subject. This protocol is made up of 25,046 possible pairs.
- **Protocol-2 (Obfuscation)** evaluates the robustness of the system when it comes to detecting when a subject is unintentionally or intentionally trying to hide his identity. The genuine set for this protocol comprises pairs formed by (normal, validation), (validation, disguise) and (disguise$_1$, disguise$_2$) images from the same subject. Where disguised$_n$ corresponds to the nth disguised image of a subject. Impostor pairs are generated by creating cross-subject pairs, combining normal, validation and disguised images of one subject with their counterpart from another subject. This protocol consists of 9,041,283 possible pairs.
- **Protocol-3 (Overall Performance)** is a the combination of the previous two and evaluates the overall performance of the system. A valid genuine or impostor pair for this protocol can be any genuine or impostor pair from protocols 1 and 2. This protocol comprises 9,066,329 possible pairs.

3.2   Neural Network Architectures

In order to carry out the experiments we used three neural network architectures: (i) VGG-Face [20] (ii) ResNet-50 [9] (iii) Se-ResNet-50 [10]. The training and testing details will be explained in the following section.

VGG-Face. In our first experiment, we use a pretrained implementation of the VGG-Face CNN which is one of the top performing deep learning models for face recognition, this will act as our baseline for the rest of the experiments. The network was trained on the VGG-FACE dataset [20].

Table 2. Datasets used for the training of each model. The last column refers to the number of different identities present in the training set of each dataset that can also be found in the DFW test set.

Dataset	VGG	Resnet50	Resnet50-ft	Senet	Overlapping identities
VGG-Face	✓				203 (33%)
VGG-Face2		✓	✓	✓	122 (20%)
MS-Celeb-1M			✓	✓	348 (58%)
DFW (non-overlapping)					143 (24%)

ResNet-50. In the next experiment, we use two residual networks for our face verification system, concretely two Resnet-50. One network is trained on MS-Celeb-1M [7] and then fine-tuned VGG-Face2 [3], while the other one is just trained on VGG-Face2. The architecture comprises 50 convolutional layers followed by a fully connected layer of dimension 2048.

Se-ResNet-50. Lastly we use a pretrained Se-Resnet-50 in our last experiment. This network is trained on MS-Celeb-1M. The only difference between the architecture of this model and ResNet-50 is that 'Squeeze-and-Excitation' (SE) block is added to the convolutional layers of the ResNet-50 followed by an embedding of 256 dimension. SE block can be used with any standard architecture. The SE block tries to use global information to selectively emphasize informative features and suppress less useful once.

3.3   Dataset Overlap

The datasets that were used to pretrain the models we are evaluating present overlapping identities with the DFW test set. Despite containing the same identities, the face images do not need to be the same. Studying how each architecture performs when evaluated on these identities will provide us with insight into the ability of statistical models to retain and generalize previously acquired knowledge when fitting a new distribution. Table 2 shows which dataset was used to train each of the models we evaluate. Note that there are also identities from DFW that overlap in more than one dataset: VGG-Face $\cap$ MS-Celeb-1M $= 145$, and VGG-Face2 $\cap$ MS-Celeb-1M $= 71$.

4   Experiments and Results

In this section we present the different experiments and results obtained on every DFW protocol over every overlapping set. We also present some hard examples and an embedding visualization.

Table 3. Verification accuracy (%) of the different approaches and our results (last 4 rows). Models are evaluated on protocol-1 (P1), protocol-2 (P2) and protocol-3 (P3). *Senet + Resnet50-ft* represents an embedding of these two models

Algorithm	GAR-P1		GAR-P2		GAR-P3	
	1% FAR	0.1% FAR	1% FAR	0.1% FAR	1% FAR	0.1% FAR
Baseline (VGG-Face)	55.29	28.91	34.32	17.58	36.25	19.35
AEFRL	96.08	57.64	87.82	77.06	87.90	75.54
DDRNET	84.20	51.26	71.04	49.28	71.43	49.08
MIRA-Face	95.46	51.09	90.65	80.56	90.62	79.26
UMDNets	94.28	53.27	86.62	74.69	86.75	72.90
DisguiseNet	1.34[a]	1.34[b]	66.32	28.99	60.89	23.25
Resnet50	81.18	49.92	75.63	55.16	75.92	54.26
Resnet50-ft	83.70	53.45	77.91	58.37	78.00	56.98
Senet	86.72	50.92	78.93	60.39	79.07	58.92
Senet+Resnet50-ft	86.89	55.63	80.71	63.02	80.89	61.12

[a]GAR@0.95%FAR
[b]The smallest FAR value is 0.95% for DisguiseNet

4.1   Performance on DFW

First, we evaluate and compare the different models on the standard dataset. We use the Genuine Acceptance Rate (GAR) at False Acceptance Rate of 1% and 0.1% (FAR), as defined in the original paper [15]. Table 3 shows the results obtained by several algorithms in each of the DFW evaluation protocols. The top performing methods do so well because they use models pretrained with over 5M images and fine-tune them on the DFW dataset for the face verification task.

Figure 2 shows the results of our experiments on each DFW protocol. It is clear that the models obtain competitive results despite none of them being specifically trained for this task, or fine-tuned in the DFW training set. This is, of course, due to the aforementioned identity overlap and the high capacity of the models used.

4.2   Dataset Overlapping Study

As presented on Sect. 3.3 the datasets that were used to pretrain the models have overlapping identities with the DFW test set.

Model performance can vary significantly when evaluated on different subsets of the data, mainly due to the difficulty of the image pairs from each subset. Despite this, the overall performance is directly correlated with the model capacity and the quantity of images seen during training. Table 4 presents the performance of every evaluated model across different overlapping sets of identities. Scores on overlapping sets of identities *seen* by the architecture during training are presented in bold. It is easy to understand that the models will perform better on these subsets of the data.

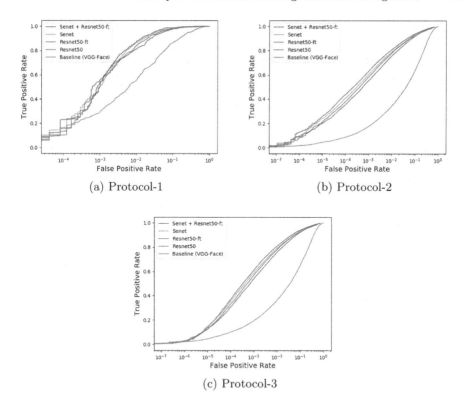

(a) Protocol-1   (b) Protocol-2

(c) Protocol-3

Fig. 2. ROC Curves for every evaluated model on each DFW protocol

Catastrophic forgetting in neural networks occurs because of the stability-plasticity dilemma [13]. The model requires sufficient plasticity to acquire new tasks, but large weight changes will cause forgetting by distributing previously learned representations. A concrete example of catastrophic forgetting is when a network is training on new tasks or categories, a neural network tends to forget the information learned in the previous trained tasks from different domains. This usually means a new task will likely override the weights that have been learned in the past, and thus degrade the model performance for the past tasks. In this work we show that as the domain of two task remains unchanged, the weight changes are small, therefore the improvement ratio of the fine-tuned ResNet over the original model (*Resnet50-ft* vs *Resnet50*) remains constant (~3%) across different overlapping sets. This effect indicates that the fine-tuned network is not able to retain specific knowledge from the first distribution it was trained on (the Ms-Celeb-1M dataset). If this were not the case, the fine-tuned network would perform much better than the original model on this overlapping set. Therefore, the overall improvement seems to arise solely from the increase in *seen* images.

Due to the intrinsic forgetness property of statistical models learning multiple task from mutually exclusive domains, without forgetting all but one of them, is unfeasible. However, this experiment shows that even when the domain of the learned tasks are the same, the catastrophic forgetness problem persists. Therefore, the forgetness problem seems to not only affect the fully connected layers acting as classifiers, but also the deepest layers in charge of feature extraction.

Table 4. Performance (GAR@1%FAR) of every evaluated model across different overlapping sets of subjects. The scores in bold indicate the performance of the model on identities *seen* during training

Overlapping set	VGG	Resnet50	Resnet50-ft	Senet
VGG-Face	**39.63**	73.79	74.71	76.25
VGG-Face2	36.59	**77.70**	**80.65**	**84.70**
Ms-Celeb-1M	35.34	73.97	**75.64**	**76.66**
VGG-Face ∩ Ms-Celeb-1M	**41.28**	76.85	**80.03**	**79.94**
VGG-Face2 ∩ Ms-Celeb-1M	28.59	**77.64**	**79.70**	**81.16**
None Overlapping	35.14	75.84	77.60	78.79
All sets	36.25	75.92	78.00	79.07

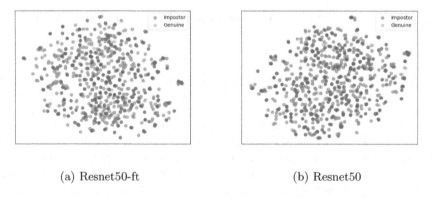

(a) Resnet50-ft   (b) Resnet50

Fig. 3. Embedding representation of Genuine and impostor subjects that overlap with the MS-Celeb-1M dataset

4.3   Face Embedding Representation

The forgetness property can also be analyzed by projecting the face image embeddings of impostors and genuine subjects into a 2d space using t-SNE [19]. All the face images were created by padding the provided face coordinates and resizing the resulting bounding box maintaining aspect ratio with shorter side of 256 and then center cropped to 224×224.

Note on Fig. 3, how the embeddings of both Resnet50 architectures struggle similarly to separate impostors from genuine subjects further evidencing our hypothesis stating that the fine-tuned architecture has forgotten the faces it was originally trained on. The apparently random distribution of both embeddings also demonstrate the high level of complexity that the face verification task represents.

Fig. 4. The pairs consistently misclassified by every model. These are genuine pairs that were labeled as an impostor pair (false negatives). All the pairs have been extracted from Protocol-3, since it is the one that comprises every possible pair.

4.4   Visualizing Hard Examples

To shed some light into the difficulty of the face verification task in the DFW dataset, we show some examples of pairs commonly misclassified by every architecture on Protocol-3. Figure 4 shows some hard genuine pairs. Note how often the misclassified genuine pairs represent drastic changes in face structure, pose and texture. This makes it notoriously hard, even for humans, to correctly classify these pairs.

5   Conclusions

In this study we show that the intrinsic forgetness property of neural networks is not only present when performing classification but also when extracting features for similar tasks sharing the same domain. After the fine-tuning process, even powerful architectures like Resnet50 will fail to remember the distribution they first learned.

In our experiments, we observe that the model that has been pretrained on MS-Celeb-1M and then fine-tuned on VGG-dataset-2, has a relatively constant improvement across different overlapping subsets of identities. This behaviour indicates that the model has forgotten some specifics about the previously fitted distribution to accommodate a new one. The consistent gain in accuracy across different overlapping subsets is solely due to the larger amount of *seen* images.

Acknowledgements. Authors acknowledge the support of the Spanish project TIN2015-65464-R (MINECO FEDER), the 2016FI_B 01163 grant of Generalitat de Catalunya, and the COST Action IC1307 iV&L Net. We also gratefully acknowledge the support of NVIDIA Corporation with the donation of a Tesla K40 GPU and a GTX TITAN GPU, used for this research.

References

1. Bansal, A., Nanduri, A., Castillo, C.D., Ranjan, R., Chellappa, R.: UMDFaces: an annotated face dataset for training deep networks. In: 2017 IEEE International Joint Conference on Biometrics (IJCB), pp. 464–473. IEEE (2017)
2. Bansal, A., Ranjan, R., Castillo, C.D., Chellappa, R.: Deep features for recognizing disguised faces in the wild. In: Proceedings of the IEEE Conference on Computer Vision and Pattern Recognition Workshops, pp. 10–16 (2018)
3. Cao, Q., Shen, L., Xie, W., Parkhi, O.M., Zisserman, A.: VGGFace2: a dataset for recognising faces across pose and age. In: 2018 13th IEEE International Conference on Automatic Face & Gesture Recognition (FG 2018), pp. 67–74. IEEE (2018)
4. Caruana, R.: Multitask learning. Mach. Learn. **28**(1), 41–75 (1997)
5. Donahue, J., et al.: DeCAF: a deep convolutional activation feature for generic visual recognition. In: International Conference on Machine Learning, pp. 647–655 (2014)
6. Girshick, R., Donahue, J., Darrell, T., Malik, J.: Rich feature hierarchies for accurate object detection and semantic segmentation. arXiv preprint arXiv:1311.2524 (2014)
7. Guo, Y., Zhang, L., Hu, Y., He, X., Gao, J.: MS-Celeb-1M: a dataset and benchmark for large-scale face recognition. In: Leibe, B., Matas, J., Sebe, N., Welling, M. (eds.) ECCV 2016. LNCS, vol. 9907, pp. 87–102. Springer, Cham (2016). https://doi.org/10.1007/978-3-319-46487-9_6
8. Gutstein, S., Stump, E.: Reduction of catastrophic forgetting with transfer learning and ternary output codes. In: 2015 International Joint Conference on Neural Networks (IJCNN), pp. 1–8. IEEE (2015)
9. He, K., Zhang, X., Ren, S., Sun, J.: Deep residual learning for image recognition. In: Proceedings of the IEEE Conference on Computer Vision and Pattern Recognition, pp. 770–778 (2016)
10. Hu, J., Shen, L., Sun, G.: Squeeze-and-excitation networks. In: Proceedings of the IEEE Conference on Computer Vision and Pattern Recognition, pp. 7132–7141 (2018)
11. Huang, G.B., Mattar, M., Berg, T., Learned-Miller, E.: Labeled faces in the wild: a database forstudying face recognition in unconstrained environments. In: Workshop on Faces in 'Real-Life' Images: Detection, Alignment, and Recognition (2008)
12. Jung, H., Ju, J., Jung, M., Kim, J.: Less-forgetting learning in deep neural networks. arXiv preprint arXiv:1607.00122 (2016)
13. Kemker, R., McClure, M., Abitino, A., Hayes, T.L., Kanan, C.: Measuring catastrophic forgetting in neural networks. In: Thirty-Second AAAI Conference on Artificial Intelligence (2018)
14. Kohli, N., Yadav, D., Noore, A.: Face verification with disguise variations via deep disguise recognizer. In: Proceedings of the IEEE Conference on Computer Vision and Pattern Recognition Workshops, pp. 17–24 (2018)

15. Kushwaha, V., Singh, M., Singh, R., Vatsa, M., Ratha, N., Chellappa, R.: Disguised faces in the wild. In: Proceedings of the IEEE Conference on Computer Vision and Pattern Recognition Workshops, pp. 1–9 (2018)
16. Lesort, T., Caselles-Dupré, H., Garcia-Ortiz, M., Stoian, A., Filliat, D.: Generative models from the perspective of continual learning. arXiv preprint arXiv:1812.09111 (2018)
17. Li, Z., Hoiem, D.: Learning without forgetting. IEEE Trans. Pattern Anal. Mach. Intell. **40**(12), 2935–2947 (2018)
18. Liu, B.: Lifelong machine learning: a paradigm for continuous learning. Front. Comput. Sci. **11**(3), 359–361 (2017)
19. van der Maaten, L., Hinton, G.: Visualizing data using t-SNE. J. Mach. Learn. Res. **9**(Nov), 2579–2605 (2008)
20. Parkhi, O.M., Vedaldi, A., Zisserman, A.: Deep face recognition. In: British Machine Vision Conference (2015)
21. Ranjan, R., Sankaranarayanan, S., Castillo, C.D., Chellappa, R.: An all-in-one convolutional neural network for face analysis. In: 2017 12th IEEE International Conference on Automatic Face & Gesture Recognition (FG 2017), pp. 17–24. IEEE (2017)
22. Robins, A.: Catastrophic forgetting, rehearsal and pseudorehearsal. Connection Sci. **7**(2), 123–146 (1995)
23. Rusu, A.A., et al.: Progressive neural networks. arXiv preprint arXiv:1606.04671 (2016)
24. Sharif Razavian, A., Azizpour, H., Sullivan, J., Carlsson, S.: CNN features off-the-shelf: an astounding baseline for recognition. In: Proceedings of the IEEE Conference on Computer Vision and Pattern Recognition Workshops, pp. 806–813 (2014)
25. Silver, D.L., Yang, Q., Li, L.: Lifelong machine learning systems: beyond learning algorithms. In: 2013 AAAI Spring Symposium Series (2013)
26. Smirnov, E., Melnikov, A., Oleinik, A., Ivanova, E., Kalinovskiy, I., Luckyanets, E.: Hard example mining with auxiliary embeddings. In: Proceedings of the IEEE Conference on Computer Vision and Pattern Recognition Workshops, pp. 37–46 (2018)
27. Van Merriënboer, B., et al.: Blocks and fuel: frameworks for deep learning. arXiv preprint arXiv:1506.00619 (2015)
28. Vishwanath Peri, S., Dhall, A.: DisguiseNet: a contrastive approach for disguised face verification in the wild. In: Proceedings of the IEEE Conference on Computer Vision and Pattern Recognition Workshops, pp. 25–31 (2018)
29. Wen, Y., Zhang, K., Li, Z., Qiao, Y.: A discriminative feature learning approach for deep face recognition. In: Leibe, B., Matas, J., Sebe, N., Welling, M. (eds.) ECCV 2016. LNCS, vol. 9911, pp. 499–515. Springer, Cham (2016). https://doi.org/10.1007/978-3-319-46478-7_31
30. Wolf, L., Hassner, T., Maoz, I.: Face recognition in unconstrained videos with matched background similarity. IEEE (2011)
31. Zhang, K., Chang, Y.L., Hsu, W.: Deep disguised faces recognition. In: Proceedings of the IEEE Conference on Computer Vision and Pattern Recognition Workshops, pp. 32–36 (2018)
32. Zhang, K., Zhang, Z., Li, Z., Qiao, Y.: Joint face detection and alignment using multitask cascaded convolutional networks. IEEE Signal Process. Lett. **23**(10), 1499–1503 (2016)

Iris Center Localization Using Geodesic Distance and CNN

Radovan Fusek$^{(\boxtimes)}$ and Eduard Sojka

FEECS, Department of Computer Science, Technical University of Ostrava,
17. listopadu 2172/15, 708 00 Ostrava-Poruba, Czech Republic
{radovan.fusek,eduard.sojka}@vsb.cz

Abstract. In this paper, we propose a new eye iris center localization method for remote tracking scenarios. The method combines the geodesic distance with CNN-based classification. Firstly, the geodesic distance is used for fast preliminary localization of the regions possibly containing the iris. Then a convolutional neural network is used to carry out the final decision and to refine the final position of the iris center. In the first step, the areas that do not appear to contain the eyeball are quickly filtered out, which makes the whole algorithm fast even on less powerful computers. The proposed method is evaluated and compared with the state-of-the-art methods on two publicly available datasets focused to the remote tracking scenarios (namely BioID [9], GI4E [15]).

Keywords: CNN · Iris detection · Geodesic distance · Deep learning

1 Introduction

In the area of recognition of eye movements, the remote and head-mounted eye-tracker systems have been widely deployed in recent years. The head-mounted eye-tracker systems are represented by the devices that are very often attached to the user's head. These systems can be used to obtain accurate information on the eye movements, such as gaze direction, or iris and pupil positions. However, these systems are more intrusive for the users than the remote eye-tracker systems. The remote trackers can be created by a single camera or by multiple cameras located away from the user. For example, these kinds of trackers are used inside the vehicle cockpits to recognize fatigue of the driver or blinking frequency. The remote systems can also be used for iris and pupil localization, however, due to the fact that the images provided by the remote systems have usually a low resolution, recognition of the eye parts represents a challenging task.

In this paper, we propose a method for localization of iris center for the remote tracking scenarios. The method is based on the geodesic distance combined with a convolutional neural network (CNN). In [6], the authors show that the geodesic distance can be used for pupil localization. We experimented with that method and we observed detection shortcomings, which became the motivation for this paper. However, we found that the method can be useful, especially,

© Springer Nature Switzerland AG 2019
A. Morales et al. (Eds.): IbPRIA 2019, LNCS 11868, pp. 76–85, 2019.
https://doi.org/10.1007/978-3-030-31321-0_7

for fast detecting the coarse position of iris. Our new method runs in two steps. In the first step, we use the ideas presented in [6] for preliminarily estimating the candidate areas. The final determination of iris position is done by making use of CNN in the second step. The second step extends and improves the original method, which is the main contribution of this paper. The presented experiments show that the proposed method outperforms the original method [6] and the state-of-the-art methods in this area.

The rest of the paper is organized as follows. The previously presented papers from the area of eye analysis are mentioned in Sect. 2. In Sect. 3, the main steps of the proposed method are described. In Sect. 4, the results of experiments are presented.

2 Related Work

In the area of iris and pupil detection, many different approaches have been presented. In [13], a method designed for head-mounted eye-tracking systems for pupil localization was proposed. The main steps include: removing the corneal reflection, pupil edge detection using a feature-based technique, and the ellipse fitting step using RANSAC. Swirski et al. [14] presented the method that is based on a Haar-like feature detector to roughly estimate the pupil location in the first step. In the next step, the potential pupil region is segmented using k-means clustering to find the largest black region. In the final step, the edge pixels of region are used for ellipse fitting using RANSAC. Exclusive Curve Selector or $ExCuSe$ was proposed in [2]. This method is based on the histogram analysis combined with the Canny edge detector and ellipse estimation using the direct least squares method. In [8], another pupil detection method known as SET is proposed. The method is based on thresholding, segmentation, border extraction using the convex hull method, and selection of the segment with the best fit. In [5], another approach known as $ElSe$ is presented. The method uses edge filtering, ellipse evaluation, and pupil validation. Another method for determining the iris centre in low-resolution images is proposed in [7]. In the first step, the coarse location of iris centre is determined using a novel hybrid convolution operator. In the second step, the iris location is further refined using boundary tracing and ellipse fitting. In [10], the pupil localization method based on the training process and the Hough regression forest was proposed. The method based on a convolutional neural network is proposed in [3,4]. An evaluation of the state-of-the-art pupil detection algorithms is presented in [1].

3 Proposed Method

In many iris or pupil detection methods, the coarse position of iris or pupil is localized in the first step. For example, a circle-shaped (due to the shape of pupil) convolution filter is used in [7]. In [14], the approximate pupil region is localized using a Haar-like center-surround feature.

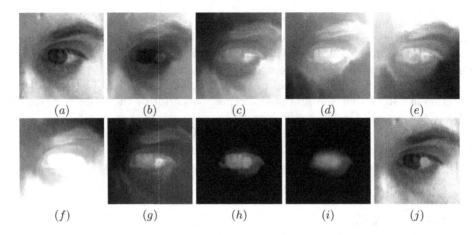

Fig. 1. The steps of eyeball and iris center localization using Geodesic distances. The input image (a). The visualization of the distance function from the centroid (b) and from particular corners (c, d, e, f). The mean of all corner distances (g). The difference (h) between (g) and (b) (only the non-zero distances are shown). The result of convolution step (i). The final position of iris center (j). The values of distance function are depicted by the level of brightness.

In this paper, we adopt the coarse localization of iris (eyeball) presented in [6]. For convenience of the reader, we briefly mention this approach. The approach is based on the geodesic distance that is used in the following way. Suppose that the image of eye region (Fig. 1(a)) is obtained beforehand (e.g. using facial landmarks or eye detector). In the first step, the geodesic distance is computed from the centroid (the point located in the center of the eye region) to all other points inside the eye image (Fig. 1(b)). The geodesic distance between two points computes the shortest curve that connects both points along the image manifold. Since the values of distance function are high in the area of eyebrow, this step is useful for its removing. It can be clearly seen that the areas with low distances represent the potential location of pupil and iris.

In the next step, the geodesic distance is also computed from each image corner to all other points inside the image (Fig. 1(c–f)). Then, the mean of all corner distances is calculated (Fig. 1(g)). Thereafter, for automatic extraction of eyeball area, the difference between Fig. 1(g) and (b) is carried out. In the image that shows this difference (Fig. 1(h)), it can be seen that the eyebrow area is removed and the potential area of iris is localized. In [6], the authors used the convolution with the Gaussian kernel in the last step (Fig. 1(i)). Then, the final iris position is determined as the location with the maximum value. In Fig. 1(j), the iris center position obtained using this approach is shown. In this particular case, it can be seen that the method fails to find the correct pupil and iris center (position) due to the fact that the iris is gently off-centered. Figure 1(a) is taken from the GI4E dataset [16] that contains many similar off-center iris and pupil images. We observed that these kinds of images cause difficulties for the method

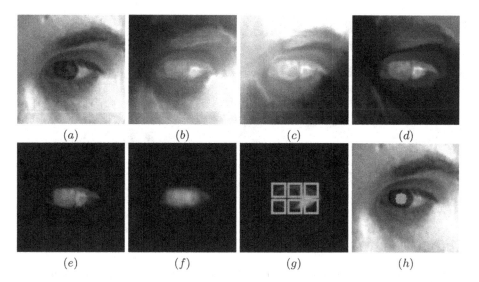

Fig. 2. The steps of iris center localization using the proposed approach. The input image (a). The visualization of the distance function from the two corners (top left (b) and bottom right (c)). The mean of two corner distances (d). An example of extracted preliminary iris region (e) using the difference step between (d) and Fig. 1(b). The result of convolution step (f). An example of cropped images (windows) that are used as an input for the CNN-based detector (g). The final position of iris center obtained using the proposed approach (h). The values of distance function are depicted by the level of brightness.

that was presented in [6] due to the fact that the final detection is based on finding one point only with a maximum distance, which does not seem to be reliable enough.

In contrast to the approach from [6], the main steps of our new approach are as follows. In the first step, the candidates for iris center are quickly determined. In the second step, the most probable centre is determined among the candidates by making use of a traditional convolutional neural network. Rapidly filtering out the points that do not have a chance to become the iris center speeds up the whole algorithm, which is often required. In addition to this, the first step also contributes to the successfulness of recognition since the neural network is asked to decide only certain specific pixel configurations in image. In the subsequent paragraphs, this general idea is presented in more details.

In the first step, we follow the approach presented in [6] that has been briefly repeated at the beginning of this section. Since, in the case of the method presented here, the goal of the first step is only to determine the candidates (not to determine the final position of the iris center directly), we may simplify the algorithm presented in [6], which is desirable since the first step should be fast. We do the following: Instead of measuring the distances from the four corners, which was done in the original method, we compute the distances only from two

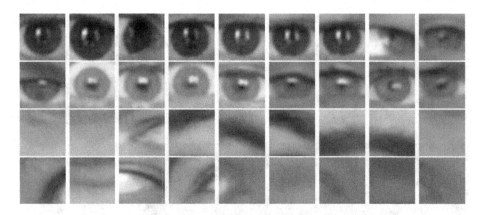

Fig. 3. An example of iris and non-iris images.

cornes with the hope that the subsequent use of CNN will compensate for this simplification. We use the top left and bottom right corner, see Fig. 2(b), (c). For the same reason, a smaller kernel size may be used in convolution smoothing the difference between the distances from the center and the mean of the distances from the corners (see Fig. 2 again), i.e. less aggressive smoothing is used. We note that the expectations we mention here will also be confirmed experimentally in Sect. 4.

Before carrying out the second step, suppose that the CNN-based classifier is trained with a sufficient amount of training iris and non-iris images (Fig. 3). In the second step, the distance differences produced in the first step are subjected to thresholding. It means that the position is verified by CNN only if the distance value is big enough at that point; a window (centered at the point that is being verified) of the gray-scale image is used by CNN (Fig. 2(g)). Finally, the location with the best response of CNN-based detector represents the final iris position (Fig. 2(h)).

The main advantages of this approach can be summarized as follows. Since, the original method uses only the maximum distance value for determining the final position (i.e. feature vector with one value), the combination with CNN-based detector has a positive effect on detection accuracy due to the fact that the model of iris is now described using a more sophisticated feature vector. With the use of coarse iris localization, the CNN classification is carried out only in the neighborhood of points with high distance values to fine-tune the position of iris. This step positively influences the speed of the whole method. Moreover, a smaller number of negative training images can be used if the iris position is approximately detected in advance (CNN will decide only certain specific situations).

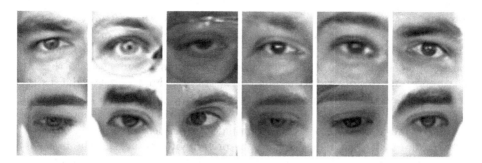

Fig. 4. Examples of eye images used in experiments. The BioID images are in the first row. The GI4E images are in the second row.

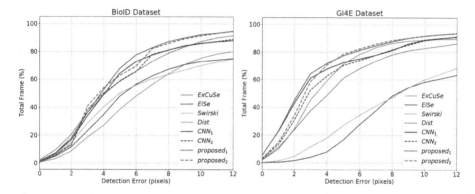

Fig. 5. The cumulative distribution of detection error. The error that is calculated as the Euclidean distance (in pixels) is in the x-axis. The y-axis shows the percentage of frames with the detection error smaller or equal to a specific error. The names of datasets are placed above the pictures.

4 Experiments

As we described in the previous section, after detection of the approximate iris area based on the geodesic distance, the potential points that are selected using the appropriate threshold are further evaluated with the use of CNN. Based on our experiments, we observed that 85% of all points in the eye image can be discarded based on their low distance values. It means that we examine only 15% of all points in the image (the locations with the highest distance values) using CNN. Since we would like to keep a fast computational time of the approach, we use a general architecture of LeNet [12] network for CNN. The network consists of two convolutional layers with the depth of 6 and 16, respectively, and a 5×5 filter size with a 1×1 stride. Each of the layers is followed by a rectified linear activation function. Thereafter, a max pooling layer with a window size of 2×2 and with a 2×2 stride is added; the last two layers are fully connected. We used stochastic gradient descent with the learning rate of 0.01 annealed to 0.0001

To compute the recognition score (confidence), we use the soft-max layer, and 32×32 grayscale images are used as an input. The implementation of CNN is based on Dlib [11]. The training set consists of 4600 iris images and 4600 non-iris images that were manually extracted from our eye image data (Fig. 3). It is important to note that the number of training images is low due to the fact that the geodesic distance is used to find the preliminary iris location, and the CNN-based detector is used to refine the final iris position. Therefore, the negative training data were obtained around the iris location only.

We examine two configurations of the presented approach. In the first configuration, we use the CNN detector that evaluates the neighborhood of every point after the distance thresholding (15% of all points). The method with this configuration is denoted as $proposed_1$ in the following experiments. We also created a faster version of our method in which only every fourth point is examined after distance thresholding. This method is referred to as $proposed_2$. The size of extracted area around each point is 32×32 pixels in both variants.

To compare the proposed algorithm to the state-of-the-art methods, we have chosen the following methods. Namely *ElSe*, *ExCuSe*, *Swirski*, the original distance method (denoted as *Dist*), and two CNN-based iris detectors: CNN_1 and CNN_2. In the first CNN-based detector (CNN_1), we used a sliding window technique applied to the entire input eye image with one pixel stride, and the stride of four pixels is used in the second detector (CNN_2). The size of sliding window is 32×32 pixels in both variants (i.e. 32×32 grayscale images are used as an input). The architecture and training process of networks are the same as in the proposed method. It is worth mentioning that *ElSe*, *ExCuSe*, and *Swirski* were primarily developed to work with images acquired by head-mounted cameras, however, the experiments in [1] show that the methods can be used in the images captured with the use of remote sensors as well. We also experimented with the parameters of particular methods. For *ElSe*, we directly used the setting for remotely acquired images published by the authors of the algorithm.

To evaluate the methods, we used two public datasets; BioID [9] and GI4E [15]. The BioID dataset contains 1521 images with the resolution of 384×286 pixels. The GI4E database contains 1339 images with the resolution of 800×600. From both datasets, the eye regions are selected based on the provided ground truth data of eye corner positions. It is important to mention that the eye images from datasets are purposely extracted with the eyebrow to test the methods in complicated conditions. The size of each extracted eye image (from both datasets) is 100×100 pixels in the following experiments. Example images of the GI4E and BioID datasets that are used for experiments are shown in Fig. 4.

In Table 1, the detection results and average times of methods are shown. We note that the average time for processing one eye region was measured on an Intel core i3 processor (3.7 GHz) with NVIDIA GeForce GTX1050. The errors are calculated as the Euclidean distance between the ground truth of iris center and the center provided by the particular detection method. In Fig. 5, we also provide the resulting plots of detection results. In the plots, the cumulative

Table 1. The detection results of methods.

	BioID Mean error (pixels)	GI4E Mean error (pixels)	Time per region (ms)
$proposed_1$	4.97	4.09	18
$proposed_2$	5.36	4.35	9
$Dist$	5.51	5.58	10
CNN_1	6.41	4.65	240
CNN_2	6.34	4.92	15
$ElSe$	10.50	12.72	16
$ExCuSe$	11.00	7.10	8
$Swirski$	10.43	11.10	10

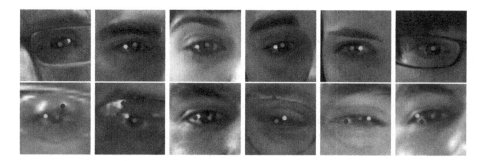

Fig. 6. Examples of images in which the proposed method performs better compared to other tested methods. The results of methods are distinguished by color: $proposed_2$ - red, CNN_2 - blue, $Dist$ - cyan. The first row: GI4E dataset, the second row: BioID dataset. (Color figure online)

distribution of detection error is shown (i.e. the figures show the percentage of frames with the detection error smaller or equal to a specific value).

Based on the results, we can conclude that the proposed method achieved very stable results and outperforms all methods in the images of both datasets. For BioID datasets, the average detection error of proposed method ($proposed_1$) is 4.97 pixels. It means that the presented method also outperforms the original method ($Dist$) in the area of detection accuracy (4.97 vs. 5.51). The faster variant of our method ($proposed_2$) also achieved promising results (5.36). It is worth mentioning that the CNN-based detectors achieved good detection score (6.41 and 6.34), however, the detection time is unnecessarily long in the first variant of CNN (CNN_1). The situation is better in the second faster variant of CNN detector (CNN_2), unfortunately, the detection error is bigger than in the faster variant of proposed approach (6.34 vs 5.36). Based on the results in Fig. 5, it can be observed that the proposed method is able detect approximately 90% of all frames with detection error smaller than 8 pixels. Even in the case of GI4E datasets, the proposed detectors achieved smaller errors than all tested methods (4.09 and 4.35). This situation can also be seen in Fig. 5.

In summary, our results show that the proposed method outperforms the main competitors: the original method presented in [6] and the iris detectors based on CNN. The proposed method that combines CNN with the distance-based preprocessing also achieved the promising time needed for processing one eye region (9 ms in $proposed_2$). Figure 6 shows several cases in which our method works better compared to other tested methods (namely, the main competitors: CNN_2 and $Dist$). Based on the results in Fig. 6, it may be said that the common errors are caused by the presence of glasses and reflections. However, the proposed method is better in such cases than the other tested methods.

5   Conclusion

In this paper, we proposed a new approach for iris center localization. The approach combines the geodesic distance with a convolutional neural network. Firstly, the geodesic distance is used to determine the areas possibly containing the iris. CNN is then used for the final decision. The proposed approach was evaluated and compared with the state-of-the-art methods on two publicly available datasets. Based on the experimental results, we can conclude that the proposed method achieved better recognition performance and a reasonable computational time when compared to the existing methods. We leave the deeper experiments with another architectures of CNN for future work.

Acknowledgments. This work was partially supported by Grant of SGS No. SP2019/71, VŠB - Technical University of Ostrava, Czech Republic.

References

1. Fuhl, W., Geisler, D., Santini, T., Rosenstiel, W., Kasneci, E.: Evaluation of state-of-the-art pupil detection algorithms on remote eye images. In: Proceedings of the 2016 ACM International Joint Conference on Pervasive and Ubiquitous Computing: Adjunct, UbiComp 2016, pp. 1716–1725. ACM, New York (2016). https://doi.org/10.1145/2968219.2968340. http://doi.acm.org/10.1145/2968219.2968340
2. Fuhl, W., Kübler, T., Sippel, K., Rosenstiel, W., Kasneci, E.: ExCuSe: robust pupil detection in real-world scenarios. In: Azzopardi, G., Petkov, N. (eds.) CAIP 2015. LNCS, vol. 9256, pp. 39–51. Springer, Cham (2015). https://doi.org/10.1007/978-3-319-23192-1_4
3. Fuhl, W., Santini, T., Kasneci, G., Kasneci, E.: PupilNet: convolutional neural networks for robust pupil detection. CoRR abs/1601.04902 (2016). http://arxiv.org/abs/1601.04902
4. Fuhl, W., Santini, T., Kasneci, G., Rosenstiel, W., Kasneci, E.: PupilNet v2.0: convolutional neural networks for CPU based real time robust pupil detection. CoRR abs/1711.00112 (2017). http://arxiv.org/abs/1711.00112
5. Fuhl, W., Santini, T.C., Kübler, T.C., Kasneci, E.: Else: ellipse selection for robust pupil detection in real-world environments. CoRR abs/1511.06575 (2015). http://arxiv.org/abs/1511.06575

6. Fusek, R.: Pupil localization using geodesic distance. In: Bebis, G., et al. (eds.) ISVC 2018. LNCS, vol. 11241, pp. 433–444. Springer, Cham (2018). https://doi.org/10.1007/978-3-030-03801-4_38

7. George, A., Routray, A.: Fast and accurate algorithm for eye localisation for gaze tracking in low-resolution images. IET Comput. Vis. **10**(7), 660–669 (2016). https://doi.org/10.1049/iet-cvi.2015.0316

8. Javadi, A.H., Hakimi, Z., Barati, M., Walsh, V., Tcheang, L.: Set: a pupil detection method using sinusoidal approximation. Front. Neuroeng. **8**, 4 (2015). https://doi.org/10.3389/fneng.2015.00004. https://www.frontiersin.org/article/10.3389/fneng.2015.00004

9. Jesorsky, O., Kirchberg, K.J., Frischholz, R.W.: Robust face detection using the hausdorff distance. In: Bigun, J., Smeraldi, F. (eds.) AVBPA 2001. LNCS, vol. 2091, pp. 90–95. Springer, Heidelberg (2001). https://doi.org/10.1007/3-540-45344-X_14

10. Kacete, A., Royan, J., Seguier, R., Collobert, M., Soladie, C.: Real-time eye pupil localization using Hough regression forest. In: 2016 IEEE Winter Conference on Applications of Computer Vision (WACV), pp. 1–8, March 2016. https://doi.org/10.1109/WACV.2016.7477666

11. King, D.E.: Dlib-ml: a machine learning toolkit. J. Mach. Learn. Res. **10**, 1755–1758 (2009)

12. Lecun, Y., Bottou, L., Bengio, Y., Haffner, P.: Gradient-based learning applied to document recognition. Proc. IEEE **86**(11), 2278–2324 (1998). https://doi.org/10.1109/5.726791

13. Li, D., Winfield, D., Parkhurst, D.J.: Starburst: a hybrid algorithm for video-based eye tracking combining feature-based and model-based approaches. In: 2005 IEEE Computer Society Conference on Computer Vision and Pattern Recognition (CVPR 2005) - Workshops, pp. 79–79, June 2005. https://doi.org/10.1109/CVPR.2005.531

14. Świrski, L., Bulling, A., Dodgson, N.: Robust real-time pupil tracking in highly off-axis images. In: Proceedings of the Symposium on Eye Tracking Research and Applications, ETRA 2012, pp. 173–176. ACM, New York (2012). https://doi.org/10.1145/2168556.2168585. http://doi.acm.org/10.1145/2168556.2168585

15. Villanueva, A., Ponz, V., Sesma-Sanchez, L., Ariz, M., Porta, S., Cabeza, R.: Hybrid method based on topography for robust detection of iris center and eye corners. ACM Trans. Multimedia Comput. Commun. Appl. **9**(4), 25:1–25:20 (2013). https://doi.org/10.1145/2501643.2501647. http://doi.acm.org/10.1145/2501643.2501647

16. Zhang, X., Sugano, Y., Fritz, M., Bulling, A.: Appearance-based gaze estimation in the wild. In: 2015 IEEE Conference on Computer Vision and Pattern Recognition (CVPR), pp. 4511–4520, June 2015. https://doi.org/10.1109/CVPR.2015.7299081

Low-Light Face Image Enhancement
Based on Dynamic Face Part Selection

Adel Oulefki$^{(\boxtimes)}$ ⓘ, Mustapha Aouache, and Messaoud Bengherabi

Centre de Développement des Technologies Avancés - CDTA,
PO. Box 17, Baba Hassen, 16303 Algiers, Algeria
aoulefki@cdta.dz
http://www.cdta.dz

Abstract. A common challenge faced by face recognition community is struggling to circumvent face images that are acquired under low-light situation. The present work aims to couple the power of the popular CLAHE algorithm for face preprocessing with a Fuzzy inference system in such a way to correct the annoyance of non-uniform illumination of face images in a targeted and a precise manner. Due to the particularity of the low-light illumination problem. Firstly, the input face image is divided into two equal sub-regions. Subsequently, the degree of brightness in each sub-region and in the whole face is used for dynamic decision of whether to normalize. In the case where only one region of the face undertakes the CLAHE-Fuzzy approach is applied. Thus, the left and right face regions are grouped back followed by further processing like a blur removal and contrast enhancement (smoothing). Visual results showed that more facial features appeared in comparison with other approaches for enhancement. Besides, we quantitatively validate the accuracy of the developed Partial Fuzzy Enhancement Approach (PFEA) with four different metrics. The effectiveness of PFEA technique has been demonstrated by presenting extensive experimental results using Extended Yale-B, CMU-PIE, Mobio, and CAS-PEAL databases.

Keywords: Face image enhancement ·
Partial Fuzzy Enhancement Approach (PFEA) · Blending images

1 Introduction

1.1 Motivation and Research Objectives

During the past decades, automatic face recognition has received widespread attention from research communities. However, very large intra-subject variations such as facial expressions, occlusion [9], aging, and outdoor illumination make the task of recognition more challenging, especially in real-world applications [1, 29]. Usually, face acquisition is realized by an outdoor camera, where the subjects may not be cooperative to match with a face database. Furthermore, images are so difficult and unclear since they are acquired under unpredictable

© Springer Nature Switzerland AG 2019
A. Morales et al. (Eds.): IbPRIA 2019, LNCS 11868, pp. 86–97, 2019.
https://doi.org/10.1007/978-3-030-31321-0_8

varying lighting conditions. For this reasons, subject recognition involves a pre-processing stage which is commonly known as face image enhancement. This step aims to improve the face image quality in such way to bring out the occluded face characteristic. Particularly, finding a good trade-off between brightness and contrast is important for enhancing face images [19]. Despite the large variety of the existing face enhancement approaches in literature, yet, face images in conjunction with illumination variations still suffer from unclarity [12]. In order to compensate the illumination, Fuzzy [21], CLAHE [27] and multi-resolution pyramid [20] approaches have been harnessed.

1.2   Literature Review

During the last decade, the link between image prepocessing and pattern recognition area has been at the center of much attention from scholarly literature [15]. There are a large number of published methods (e.g., Iratni et al. [14]; Oulefki et al. [19,21]) where they involved image analysis stage for the seek of face image enhancement. Moreover, enhancement strategies might include spatial [26,30] and frequency domain [6,13] to improve images. For example, Du et al. [6] presented discrete wavelets transform (DWT) in which they applied HE to the low frequency and accentuate the high frequency coefficients. Jobson et al. [16] improved the multiscale retinex (MSR) method which is the (HE) Histogram Equalization extension of the previous single-scale center/surround retinex. This last cancels lots of the low frequency information by subdividing the given image to a smoothed version of itself. Due to its fast implementation [36]. HE is one of the basic method for improving and adjusting the contrast of images. Despite its accuracy, it remains an uneven particularly when it processed under varying lighting conditions.

Pizer et al. [24] discussed and summarized Contrast-Limited Adaptive Histogram Equalization (CLAHE). The standard CLAHE computes distinct histograms blocks corresponding to a divided section of the given image then readjust the image's brightness of each part. However, CLAHE has a tendency to over-amplify/enhance images which led to loss of information in some local region.

Generally, CLAHE is lacking the nice property of balancing overall patterns and image sharpness as stated in [32]. To overcome the limitations of CLAHE, many variants have been proposed in literature [4]. Toward the end of improving CLAHE performance for face enhancements under low light constraints, we propose in this paper a hybrid CLAHE-FUZZY enhancement pipeline. This method will produce optimal contrast without losing any local information of the face image which is most important for recognition and detection human faces. The proposed method consists of two stages of processing to increase the potentiality homogeneity regions of an image and to preserve the local details in the images. The details of the proposed method are presented in the next section.

1.3   Contributions

In this paper, an innovative spatial-frequency domain enhancement named Fuzzy
Partial Enhancement Approach (FPEA) has been proposed seeking the correction of face images illumination in a balanced way. The FPEA compensation is
based on the intensity of too dark, or too bright sides of face images. Firstly, the
input face image is split into two areas of face images right and left sides. Then,
the corresponding intensity of dark, normal, or bright pixels in the gray-scale
channel of each part are calculated. Afterward, FPEA correction is applied to
compensate the illumination of the affected part only (low or high intensity).
Meanwhile, laplacian pyramid decomposition is used to reconstruct the facial
image, in such way to smooth the line that separates the two face parts, thus
blending the local area around the separation [20]. This latter approach provides the ability to deal with separate frequency while locally preserving the
face information. The compensation in the FPEA method is adaptively applied
based on the face affected part. Moreover, a region-based FPEA method that
entails applying the FPEA algorithm in two sub-regions of the image reduces the
computing time. In addition, the experimental results based on the four most
widely used face databases proved the efficiency of the FPEA.

The remainder of this paper is organized as follows: Sect. 1 reviews the state-
of-the-art face image enhancement. In Sect. 2, the proposed method is discussed
in detail, followed by Sect. 3 which discusses in depth the quality assessment and
database used in this paper, and Sect. 4, Sect. 5 presents the experimental results
of quantitative and visual measurements respectively along with a discussion.
Finally, Sect. 6 presents the conclusion by interpreting overall results.

2   The Proposed Enhancement Framework

Imperfections such as over/less enhancements are one of the drawbacks of using
the conventional image enhancement methods. Moreover, artifacts effect could be
generated also when applied to given images. Thus, the proposed approach aims
to cut down the aforesaid issues is proposed. On one side, it compensates both
brightness and the contrast of the input image, without ignoring hidden details
that will appear after using our approach. Fuzzy-Reasoning Model (FRM) [21]
along with Contrast Limited Adaptive Histogram Equalization (CLAHE) [24] to
compromise in a more precise manner the problem of illumination variation on
face images. Firstly, we apply Viola–Jones Algorithm for detecting faces. The
cropped input face image (I) is then divided into two equal parts labeled (A)
and (B). This two face parts denote the left and the right part, respectively (See
Fig. 1). At that point, we calculate the intensity from each part including the
input cropped face image (I).

After that, we perform enhancement first by using CLAHE. Following
CLAHE stage, we carry out FRM operator as an exponentiation to produce
a factor in such a way that adjust illumination at the part of face that has been
affected by the lack of luminance.

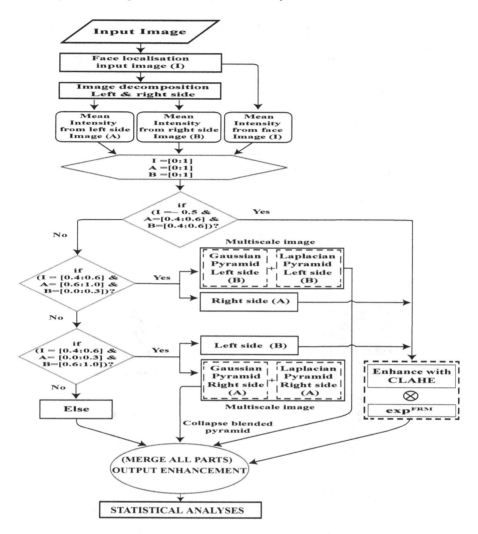

Fig. 1. The pipeline of the proposed PFEA for face illumination enhancements

According to FRM model, FRM generates an adaptive enhancement which corrects and improves non-uniform illumination and low contrasts of an image by comprising luminance component as given in Eq. 1:

$$FRM = I^{exp(\overline{\alpha}_2 - \overline{\alpha}_2)} \tag{1}$$

where I is the input image, and $\overline{\alpha}_0$ represent the mean intensity of I. $\overline{\alpha}_f$ is the resulting value from the Fuzzy system. The FRM rules used and its description be found in [21].

Prior to this, in order to detect face pixels that are affected by the lack of illumination. The intensity varies as a function of the relative direction to the illumination at the acquisition stage. The regions that could be enhanced depend on the poor intensity that are presented. First, we normalized histogram to calculate the changes of intensity of input image, left side and right as represented in Fig. 1 by the triplet $[I, A, B]$.

After studious testing we have found that intensity belonging to face region with a poor illumination are located when $A = [0.6, 1]$ and $B = [0.0, 0.3]$ for the right side and vice versa for the left one.

Finally, in order to have smooth blending, multi-resolution pyramid approach [3] applied on the none affected area in a given band that should be recombined to obtain the final face image enhanced mosaic (see Fig. 2). In other words, for a greater realism of the enhanced image the multi-resolution pyramid approach is carried out only for the part that is not affected by lack of illumination for making a realistic face looking. This is proven by the results of visual assessments. The aim is to seamlessly stitch together both right and left images into a face image mosaic by smoothing the boundary in a scale-dependent way to avoid boundary artifacts.

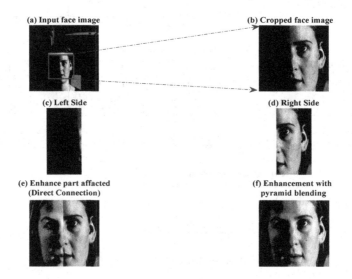

Fig. 2. The spline applied to a sample face image from CMU-PIE database. When the left half of (c) is joined to the right half of face (d) without a spline, the boundary is clearly visible (e). However, by applying our approach. No boundary is visible when the multi-resolution spline is used (f)

3   Quality Assessment and Databases Used

We have chosen four performance metrics for this paper. No-reference Image Quality Assessments (IQAs) tested in this paper are summarized in Table 1.

Table 1. Image Quality Assessments (IQAs) used in the present work.

Acronym	Description
EME [7]	$EME = \chi(\frac{1}{k_1 k_2} \sum\limits_{i=1}^{k_1} \sum\limits_{j=1}^{k_2} 20 ln \left[\frac{I_{max;k,l}^w}{I_{min;k,l}^w + c} \right])$ (2) (I) denotes the input image of (N * M) divided into ($k_1 \times k_2$) blocks; l(i; j) of size ($l_1 \times l_2$), I_{min} and I_{max} are the maximum and minimum values of the pixels in each block
SDME [22]	$SDME_{k_1,k_2} = -\frac{1}{k_1 k_2} \sum\limits_{i=1}^{k_1} \sum\limits_{j=1}^{k_2} 20 ln \left[\frac{I_{max;k,l} - 2 I_{center;k,l} + I_{min;k,l}}{I_{max;k,l} + 2 I_{center;k,l} + I_{min;k,l}} \right]$ (3) where, $I(i,j)$ is the gray value of pixel (i,j), and $I(m,n)$ is the gray value of adjacent pixel (i,j) in the block (window) of 3×3
CPP [23]	$CPP = \dfrac{\sum_{i=0}^{M} \sum_{j=0}^{N} \left(\sum_{(m,n) \in R_3^{(i,j)}} \lvert I(i,j) - I(m,n) \rvert \right)}{MN}$ (4) Indicate the estimation by averaging the intensity difference between a pixel and its adjacent pixel
NIQE	Practical implementation available in [18]

For the IQA without mathematical description (NIQE), the exact details of its implementation and more details can be found in the giving reference [18]. The data-sets used for this project consisted of CMU-PIE, E-Yale-B, CAS-PEAL and Mobio. These face databases were selected to represent a typical implementation of our approach along with specification of images included faces affected by illumination invariant.

4 Experiments

In this section, the performance of our technique (PFEA) compared with the state-of-the-art image improvement approaches such us HE [2,25], BPDFHE [34], NMHE [26], LIME [5,11], FRM [21], and AdaptGC [14] using CMU-PIE [31], mobio [17], CAS-PEAL [8] and E-Yale-B [10] databases. We use four IQAs to compare the performance of all these approaches including EME, SDME, CPP, and NIQE to indicate good performance in terms of correlation with human visual assessment [28,37]. To display each quantitative metric in one single figure. We selected a bar plot to present the comparisons by calculating the means of EME, SDME, CPP and NIQE of the proposed and the stat-of-the-art of all face images from EYale-B to CMU-PIE, through CAS-PEAL and Mobio data set. While the color symbols blue, red, orange, and white, having been selected to represent E-Yale-B, CMU-PIE, Mobio and CAS-PEAL data-sets respectively. Moreover we include the information of Confident Interval (CI) (95%) [33] to refer to the level C of a confidence which provides the probability that the range CI obtained by the approaches employed includes the true mean value of the IQAs used over the entire database.

4.1   IQAs Results and Discussion

As can be seen in the Fig. 3, the means of EME of the improved face images using the proposed PFEA is greater than the five enhancement approaches picked for comparison, this means that the best enhancement effect is achieved by the developed approach. Meanwhile, NMHE provides the least indicator enhancement followed by PDFHE and HE, whereas AdaptGC maintaining competitive.

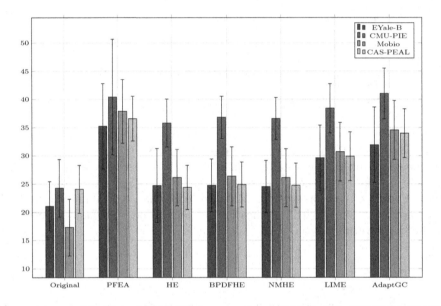

Fig. 3. Means of EME, and CI (95%) computed from PFEA, HE, BPDFHE, NMHE, LIME and AdaptGC by taking original images' EME as reference of respectively EYale-B, CMU-PIE, mobio and CAS-PEAL face Data bases

In order to assess the effects of the enhancement quality using SDME metric, the experimental results on Fig. 4 shows that the mean of HE, BPDFHE, NMHE are lower than those of the other methods. The provided results when performing PFEA methods along with the other enhancement methods assessed in this paper, proved that the PFEA produce a best result.

In terms of CPP metric where it is plotted in Fig. 5, we can say that across the different quality measurement the results are just about the same of all datasets, with a slight superiority of the PFEA.

The summary statistics of NIQE metrics which by definition a smaller score indicates better perceptual quality. In Fig. 6, it is obvious that across the different enhancements used the PFEA provide the best results as the mean NIQE are just about the same of all datasets.

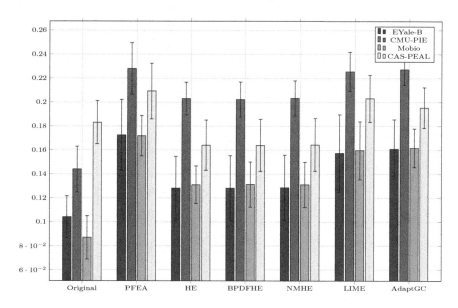

Fig. 4. Means of SDME, and CI (95%) computed from PFEA, HE, BPDFHE, NMHE, LIME and AdaptGC by taking original images' SDME as reference of respectively EYale-B, CMU-PIE, mobio and CAS-PEAL face Data bases

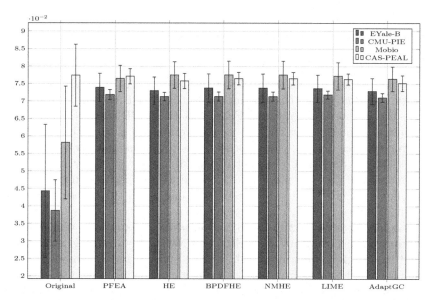

Fig. 5. Means of CPP, and CI (95%) computed from PFEA, HE, BPDFHE, NMHE, LIME and AdaptGC by taking original images' CPP as reference of respectively EYale-B, CMU-PIE, mobio and CAS-PEAL face Data bases

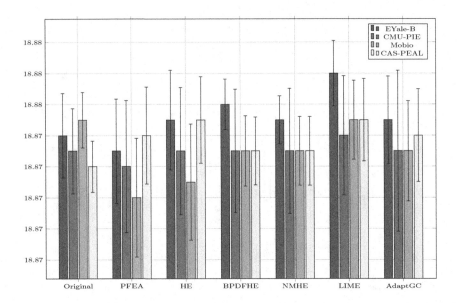

Fig. 6. Means of NIQE, and CI (95%) computed from PFEA, HE, BPDFHE, NMHE, LIME and AdaptGC by taking original images' NIQE as reference of respectively EYale-B, CMU-PIE, mobio and CAS-PEAL face Data bases

5 Qualitative Results and Discussion

Samples of low-light face images before and after enhancement results are shown in Fig. 7. We picked up disparate low-light face images from already stated databases of E-Yale-B, CMU-PIE, mobio, and CAS-PEAL. On this spot, we compare the proposed algorithm PFEA with HE, BPDFHE, NMHE, LIME, and AdaptGC. For comparison approaches, we applied firstly Viola and Jones [35] to crop the face region. As we carried out both spacial and frequency tone to enhance both the dark and over light regions of face image, the visual comparisons demonstrate that the proposed PFEA can effectively face enhance images which differ from the darkest to the lightest ones even the lowest or highest-dark/light whatever the region affected. For example, the results of AdaptGC includes some saturated regions and the noises also come out. Whereas, face image improvement using HE, BPDFHE and NMHE does not add some sharpness and fix the poor blurriness. While the results of the proposed PFEA as shown in the third column from the left to right can enhance the images without such saturated region also removes some unnecessary disturbances by smoothing the face images. Another advantage of applying PFEA is the ability to extract significant facial features (eye contours, eyebrows, lip contours, and nose tip) also producing balanced resulting visually.

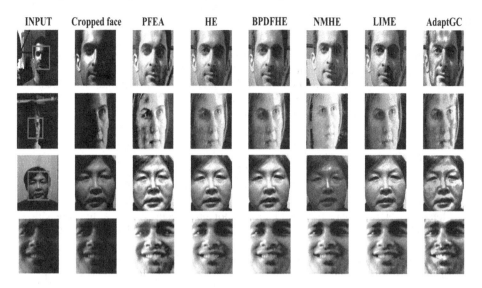

Fig. 7. Visual comparisons using PFEA, HE, BPDFHE, NMHE, LIME and AdaptGC test images: (from top to bottom) low-light face images from E-YaleB, CMU-PIE, CAS-PEAL, and mobio

6 Conclusion

In this work, we proposed an efficient method for face enhancement. Both, the key to the low-light/hard-dark image enhancement and the affected part of the face map are estimated. The proposed PFEA uses fuzzy logic to handle the inexactness of face images in a better way compared to stat of the art techniques, resulting in higher quality performance. It can also improve the affected face image parts, regarding both the brightness and darkest sides. It also automatically takes into account the enhancement of the image contrast. The proposed algorithm was tested on different face images from EYale-B, CMU-PIE, mobio, and CAS-PEAL databases. The performance of proposed PFEA algorithm was evaluated and compared in terms of EME, SDME, CPP and NIQE metrics. Furthermore, PFEA was compared to state-of-the-art image enhancements methods using both qualitative and quantitative performance metrics. Experiments demonstrate that the proposed method significantly eliminate the washed-out appearance and adverse artifacts induced by several existing methods. This method is simple and suitable for consumer electronic products. We are investigating the proposed method for improving face recognition performances, where the preliminary results are promising.

References

1. Abbadi, B., Oulefki, A., Mostefai, M.: Development and implementation of a new dynamic face detection operator. Int. J. Comput. Appl. **49**(20) (2012)
2. Agaian, S., Roopaei, M., Shadaram, M., Bagalkot, S.S.: Bright and dark distance-based image decomposition and enhancement. In: 2014 IEEE International Conference on Imaging Systems and Techniques (IST) Proceedings, pp. 73–78. IEEE (2014)
3. Burt, P.J., Adelson, E.H.: A multiresolution spline with application to image mosaics. ACM Trans. Graph. **2**(4), 217–236 (1983)
4. Chang, Y., Jung, C., Ke, P., Song, H., Hwang, J.: Automatic contrast-limited adaptive histogram equalization with dual gamma correction. IEEE Access **6**, 11782–11792 (2018)
5. Dong, X., et al.: Fast efficient algorithm for enhancement of low lighting video. In: 2011 IEEE International Conference on Multimedia and Expo, pp. 1–6. IEEE (2011)
6. Du, S., Ward, R.: Wavelet-based illumination normalization for face recognition. In: IEEE International Conference on Image Processing, ICIP 2005, vol. 2, p. II-954. IEEE (2005)
7. Gao, C., Panetta, K., Agaian, S.: No reference color image quality measures. In: 2013 IEEE International Conference on Cybernetics (CYBCO), pp. 243–248. IEEE (2013)
8. Gao, W., et al.: The CAS-PEAL large-scale chinese face database and baseline evaluations. IEEE Trans. Syst. Man Cybern.-Part A: Syst. Hum. **38**(1), 149–161 (2008)
9. García-Montero, M., Redondo-Cabrera, C., López-Sastre, R., Tuytelaars, T.: Fast head pose estimation for human-computer interaction. In: Paredes, R., Cardoso, J.S., Pardo, X.M. (eds.) IbPRIA 2015. LNCS, vol. 9117, pp. 101–110. Springer, Cham (2015). https://doi.org/10.1007/978-3-319-19390-8_12
10. Georghiades, A.S., Belhumeur, P.N., Kriegman, D.J.: From few to many: illumination cone models for face recognition under variable lighting and pose. IEEE Trans. Pattern Anal. Mach. Intell. (6), 643–660 (2001)
11. Guo, X., Li, Y., Ling, H.: Lime: low-light image enhancement via illumination map estimation. IEEE Trans. Image Process. **26**(2), 982–993 (2017)
12. Hassaballah, M., Aly, S.: Face recognition: challenges, achievements and future directions. IET Comput. Vis. **9**(4), 614–626 (2015)
13. Hu, H.: Illumination invariant face recognition based on dual-tree complex wavelet transform. IET Comput. Vis. **9**(2), 163–173 (2014)
14. Iratni, A., Aouache, M., Adel, O.: Adaptive gamma correction-based expert system for nonuniform illumination face enhancement. J. Electron. Imaging **27**(2), 023028 (2018)
15. Jain, A.K.: Fundamentals of Digital Image Processing. Prentice Hall, Englewood Cliffs (1989)
16. Jobson, D.J., Rahman, Z., Woodell, G.A.: A multiscale retinex for bridging the gap between color images and the human observation of scenes. IEEE Trans. Image process. **6**(7), 965–976 (1997)
17. McCool, C., et al.: Bi-modal person recognition on a mobile phone: using mobile phone data. In: 2012 IEEE International Conference on Multimedia and Expo Workshops, pp. 635–640. IEEE (2012)
18. Mittal, A., Soundararajan, R., Bovik, A.C.: Making a "completely blind" image quality analyzer. IEEE Signal Process. Lett. **20**(3), 209–212 (2013)

19. Mustapha, A., Oulefki, A., Bengherabi, M., Boutellaa, E., Algaet, M.A.: Towards nonuniform illumination face enhancement via adaptive contrast stretching. Multimed. Tools Appl. **76**(21), 21961–21999 (2017)
20. Ogden, J.M., Adelson, E.H., Bergen, J.R., Burt, P.J.: Pyramid-based computer graphics. RCA Eng. **30**(5), 4–15 (1985)
21. Oulefki, A., Mustapha, A., Boutellaa, E., Bengherabi, M., Tifarine, A.A.: Fuzzy reasoning model to improve face illumination invariance. SIViP **12**(3), 421–428 (2018)
22. Panetta, K., Zhou, Y., Agaian, S., Jia, H.: Nonlinear unsharp masking for mammogram enhancement. IEEE Trans. Inf. Technol. Biomed. **15**(6), 918–928 (2011)
23. Peli, E.: Contrast in complex images. JOSA A **7**(10), 2032–2040 (1990)
24. Pizer, S.M., Johnston, R.E., Ericksen, J.P., Yankaskas, B.C., Muller, K.E.: Contrast-limited adaptive histogram equalization: speed and effectiveness. In: Proceedings of the First Conference on Visualization in Biomedical Computing, pp. 337–345, May 1990. https://doi.org/10.1109/VBC.1990.109340
25. Pizer, S.M., et al.: Adaptive histogram equalization and its variations. Comput. Vis. Graph. Image Process. **39**(3), 355–368 (1987)
26. Poddar, S., Tewary, S., Sharma, D., Karar, V., Ghosh, A., Pal, S.K.: Nonparametric modified histogram equalisation for contrast enhancement. IET Image Proc. **7**(7), 641–652 (2013)
27. Reza, A.M.: Realization of the contrast limited adaptive histogram equalization (CLAHE) for real-time image enhancement. J. VLSI Signal Process. Syst. Signal Image Video Technol. **38**(1), 35–44 (2004)
28. Samani, A., Panetta, K., Agaian, S.: Quality assessment of color images affected by transmission error, quantization noise, and noneccentricity pattern noise. In: 2015 IEEE International Symposium on Technologies for Homeland Security (HST), pp. 1–6. IEEE (2015)
29. Savchenko, A.V.: Deep convolutional neural networks and maximum-likelihood principle in approximate nearest neighbor search. In: Alexandre, L.A., Salvador Sánchez, J., Rodrigues, J.M.F. (eds.) IbPRIA 2017. LNCS, vol. 10255, pp. 42–49. Springer, Cham (2017). https://doi.org/10.1007/978-3-319-58838-4_5
30. Sheet, D., Garud, H., Suveer, A., Mahadevappa, M., Chatterjee, J.: Brightness preserving dynamic fuzzy histogram equalization. IEEE Trans. Consum. Electron. **56**(4) (2010)
31. Sim, T., Baker, S., Bsat, M.: The CMU pose, illumination, and expression (PIE) database. In: Proceedings of Fifth IEEE International Conference on Automatic Face Gesture Recognition, pp. 53–58. IEEE (2002)
32. Sun, X., Xu, Q., Zhu, L.: An effective Gaussian fitting approach for image contrast enhancement. IEEE Access (2019)
33. Tan, S.H., Tan, S.B.: The correct interpretation of confidence intervals. Proc. Singapore Healthcare **19**(3), 276–278 (2010)
34. Tzimiropoulos, G., Zafeiriou, S., Pantic, M.: Subspace learning from image gradient orientations. IEEE Trans. Pattern Anal. Mach. Intell. **34**(12), 2454–2466 (2012)
35. Viola, P., Jones, M.J., Snow, D.: Detecting pedestrians using patterns of motion and appearance. Int. J. Comput. Vis. **63**(2), 153–161 (2005)
36. Wang, Y., Chen, Q., Zhang, B.: Image enhancement based on equal area dualistic sub-image histogram equalization method. IEEE Trans. Consum. Electron. **45**(1), 68–75 (1999). https://doi.org/10.1109/30.754419
37. Wharton, E., Panetta, K., Agaian, S.: Human visual system based similarity metrics. In: 2008 IEEE International Conference on Systems, Man and Cybernetics, pp. 685–690. IEEE (2008)

Retinal Blood Vessel Segmentation: A Semi-supervised Approach

Tanmai K. Ghosh[1], Sajib Saha[2(✉)], G. M. Atiqur Rahaman[1], Md. Abu Sayed[1], and Yogesan Kanagasingam[2]

[1] Computational Color and Spectral Image Analysis Lab. Computer Science and Engineering Discipline, Khulna University, Khulna, Bangladesh
`tanmai1532@cseku.ac.bd, gmatiqur@gmail.com, sayed4931@gmail.com`
[2] Australian e-Health Research Centre,
Commonwealth Scientific and Industrial Research Organization (CSIRO),
Perth, WA, Australia
`{Sajib.Saha,Yogi.Kanagasingam}@csiro.au`

Abstract. Segmentation of retinal blood vessels is an important step in several retinal image analysis tasks. State-of-the-art papers are still incapable to segment retinal vessels correctly, especially, in presence of pathology. In this paper an innovative descriptor named **R**obust **F**eature **D**escriptor (RFD) is proposed to describe vessel pixels more uniquely in the presence of pathology. For accurate segmentation of blood vessels, the method combines both supervised and unsupervised approaches. Extensive experiments have been conducted on three publicly available datasets namely DRIVE, STARE and CHASE_DB1; and the method has been compared with other state-of-the-art methods. The proposed method achieves an overall segmentation accuracy of 0.961, 0.960 and 0.955 respectively on DRIVE, STARE and CHASE_DB1 datasets, which are better than the state-of-the-art methods in comparison. The sensitivity, specificity and area under curve (AUC) of the method are respectively 0.737, 0.981, 0.859 on DRIVE dataset; 0.805, 0.972, 0.889 on STARE dataset; and 0.763, 0.969, 0.866 on CHASE_DB1 dataset.

Keywords: Retinal image · Vessel segmentation ·
Multi-scale line detector · Robust Feature Descriptor · Random forest

1 Introduction

The segmentation of retinal blood vessels plays an important role in various retinal images analysis that includes automatic pathology detection and registration of retinal images [1]. For automatic detection of many eye-related diseases such as diabetic retinopathy, hypertension, and vein occlusion [2], the blood vessels segmentation is widely used as a preprocessing step. Use of retinal blood vessels rather than other alternative features is more authentic for image registration [3]. The segmentation of retinal blood vessels and depiction of morphological structures of retinal blood vessels such as length, width, tortuosity

© Springer Nature Switzerland AG 2019
A. Morales et al. (Eds.): IbPRIA 2019, LNCS 11868, pp. 98–107, 2019.
https://doi.org/10.1007/978-3-030-31321-0_9

and/or branching pattern and angles are widely used for diagnosis, screening, treatment, and analysis of various cardiovascular and ophthalmologic diseases like polygenic disease, hypertension, arterial sclerosis and choroidal neovascularization [4]. It is noted that, the retinal vascular tree like structure is found to be unique for every individual. Hence, the segmented vessel structure can be used for biometric authentication [4]. Manual segmentation of retinal blood vessels is very challenging and tedious task even for the specialists. Moreover, the segmentation result varies from observer to observer. That is why we need automated methods. Over the last two decades, a tremendous number of algorithms and processes were introduced. Despite the fact, still there are challenges to address. Some of the important challenges for blood vessel segmentation are listed below [5].

1. Segmenting retinal blood vessels in the presence of central vessel reflex.
2. Segmenting blood vessels presenting in crossover and bifurcation regions.
3. Segmenting the merging of close vessels.
4. Segmenting the small and thin vessels.
5. Segmenting the blood vessels in the pathological region (Dark lesion and bright legion).

Nyugen et al. [5] recently introduced an efficient approach to solve many of the challenges mentioned above. Despite being efficient, it still lacks in accurately segmenting blood vessels in the presence of pathology. In this work we augment Nguyen et al.'s method [5] by incorporating robust description and supervised learning steps with it. Results show that blood vessels can be detected more accurately even with the presence of pathology by the proposed method.

2   Literature Review

A wide number of approaches have been introduced relating to the automated segmentation of retinal blood vessels in the last two decades, and here we briefly discuss the most recent and relevant ones. These methods can be broadly divided into two categories - supervised and unsupervised.

Roychowdhury et al. [6] proposed a supervised method that presents a novel three stage blood vessel segmentation algorithm. At the first stage, two binary images are extracted from the green channel and morphologically reconstructed enhanced image. Then, common regions of the binary images are extracted as the major vessels. In the second stage, Gaussian Mixture Model (GMM) is used to classify the remaining pixels. In the final stage, combination of the major portions of blood vessels with the classified vessels is performed. The proposed algorithm is evaluated on three publicly available datasets DRIVE, STARE, and CHASE_DB1 respectively. Lupascu et al. [7] also proposed a supervised method based on AdaBoost. A 41-D feature vector is constructed for each pixel in the field of view (FoV) of the image. Finally, AdaBoost classifier is used to classify the pixels as vessels or non-vessels based on the extracted features. The method is evaluated on DRIVE dataset only.

Zhao et al. [8] proposed an unsupervised method to segment the retinal vessels based on level set and region growing. Firstly, preprocessing is performed using the contrast-limited adaptive histogram equalization and a 2D Gabor wavelet to enhance the vessels. To smooth the image and preserve vessel boundaries, an anisotropic diffusion filter is used. Finally, extraction of retinal vessels is done by the region growing method and a region-based active contour model with the implementation of level set. The final segmentation is achieved by combining the results. Method evaluation is performed on the publicly available DRIVE and STARE databases. Ricci et al. [9] also proposed an unsupervised segmentation method based on basic line operators (Ricci-line). Though the method was a major breakthrough, it has some drawbacks such poor segmentation result in the presence of central vessel reflex, the possibility of merging close vessels, at bifurcation and crossover regions. Nguyen et al. [5] proposed a method based on the line detector of varying length for minimizing the limitations of Ricci's method. The method has significant contribution and therefore, it can segment the vessels – (1) in presence of central vessel reflex, (2) at bifurcation and crossover regions, and (3) in presence of merging of close vessels. However, the method fails to segment blood vessels accurately in the presence of pathological lesions. An example of such misclassification is shown in Fig. 1.

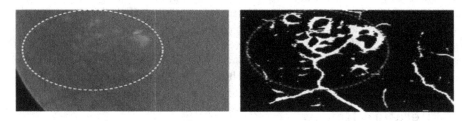

Fig. 1. Example misclassification of pathology pixels as vessel by Nguyen et al.'s [5] method. Left – portion of a color fundus image with pathology, right – segmentation using Nguyen et al.'s method. (Color figure online)

3   Proposed Method

The proposed method has been inspired by the multi-scale line detector approach by Nguyen et al. [5]. In an aim to augment Nguyen et al.'s method and to perform blood vessel segmentation more accurately in the presence of pathology, the proposed method performs blood vessel segmentation in two steps. In the first step a preliminary segmentation is performed relying on multi-scale line detector approach [5]. In the second step fine segmentation is performed relying on robust feature description, and supervised learning. A diagram of the proposed system is shown in Fig. 2.

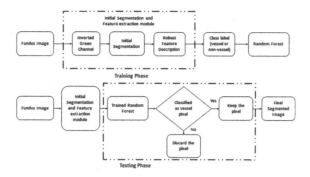

Fig. 2. Diagram of the proposed system. Operations shown within the dotted box are performed pixel-wise.

3.1 Preliminary Segmentation

The multi-scale line detector approach [5] is applied. A window of size $w \times w$ is taken centered at each pixel and 12 lines of varying length (1 to window size, w) and oriented at twelve different directions of $15°$ angular difference is considered. The raw response value is calculated for the varying length of line for each pixel, and which are then standardized. Line responses are computed likewise at varying scales, and they are finally linearly combined. The multi-scale line detector is computed on the inverted green channel image as recommend in [5]. An example segmentation produced at this step is shown in Fig. 3.

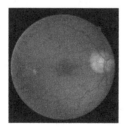 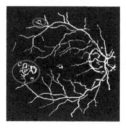

Fig. 3. Example segmentation relying on multi-scale line detector approach [5]. Left – original color fundus image from DRIVE dataset [10], right – vessel segmentation by Nguyen et al.'s method. Misclassified pathology pixels are circled in blue. (Color figure online)

3.2 Fine Segmentation

In the fine segmentation stage, a supervised approach is incorporated to remove misclassified pixels. A novel descriptor named **R**obust **F**eature **D**escriptor (RFD) is proposed. A random forest classifier is finally trained to classify each pixel as true vessel or not depending on RFD.

3.2.1   Robust Feature Descriptor (RFD)

In order to extract useful information surrounding the pixel, RFD relies on Hear wavelet responses (Fig. 4) likewise in [11]. However, in different to [11], here wavelet responses are computed at one scale, which is determined by the Euclidean distance between the optic disc and macula centers. At the same time, instead of using local gradient information for each keypoint or pixel of interest, a global orientation is used. The global orientation is computed based on optic disc and macula centers.

Orientation Assignment. Prior to computing Haar wavelet responses, we identify a reproducible orientation of the image, which is then used to rotate the image. For that purpose, we first compute the centers of the optic disc and macula relying on the method proposed by Rust et al. in [12]. Let, (X_M, Y_M) and (X_{OD}, Y_{OD}) are the coordinates of the macula and optic disk center respective, then the reproducible orientation of the image θ is computed as, $\theta = tan^{-1} \frac{Y_M - Y_{OD}}{X_M - X_{OD}}$.

Image Resizing. Prior to computing wavelet responses, we also resize the image. We compute the Euclidean distance $E_i = \sqrt{(X_M - X_{OD})^2 + (Y_M - Y_{OD})^2}$ between the optic disc and macula centers. Then the image resizing factor, s is determined as the ratio of E_i and E_{avg}, where E_{avg} is the average Euclidean distance between optic disc and macula and centres computed on 1000 selected images from EyePACS (http://www.eyepacs.com/).

Descriptor Components. A square region of size 36×36 around the pixel of interest is considered. This region is further split up into smaller 4×4 square sub-regions. For each sub-region, we compute Haar wavelet responses in the X and Y directions. The wavelet responses and their absolute values are summed up over each subregion and a 4-D vector is formed $v = (\sum d_x, \sum d_y, \sum |d_x|, \sum |d_y|)$, where d_x, d_y are respectively the wavelet responses in the X and Y directions. The responses are computed at 3×3 regularly spaced intervals using a 4×4 window.

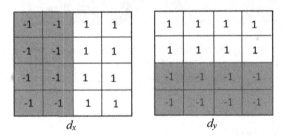

Fig. 4. Haar wavelets

The responses are then weighted with a Gaussian of $\sigma = 12$ centered at the pixel of interest. Vectors computed over all the sub-regions are then concatenated to form the descriptor of length 64 to represent the pixel. The descriptor is finally normalized to have unit length (Fig. 5).

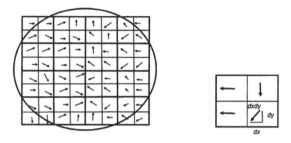

Fig. 5. Feature description process of RFD.

3.2.2 Random Forest Classifier

A random forest classifier [13] is trained to classify a pixel as vessel or not. The training algorithm for random forests applies the general technique of bagging to tree learners. Given a training set $X = x_1, x_2, \ldots, x_n$ with responses $Y = y_1, y_2, \ldots, y_n$ bagging repeatedly (K times) selects random sample with replacement of the training set and fits trees to these samples: For $k = 1, \ldots, K$:

1. Sample, with replacement, n training examples from X, Y; call these X_k, Y_k.
2. Train a classification or regression tree f_k on X_k, Y_k.

After training, predictions for unseen samples x' are made by taking the majority vote in the case of classification trees (Fig. 6).

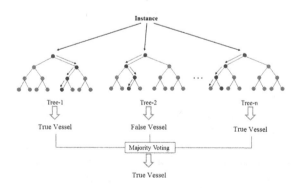

Fig. 6. Random forest classifier to classify a pixel as true vessel or not.

RFDs are computed for all the pixels determined as vessels by Nguyen et al.'s method [5]. Ground truth labels of these pixels determined by experienced grader were made available while training the classifier. Once trained it classified a given pixel as vessel or not vessel depending on its RFD.

4   Experiments and Results

Experiments were conducted on three publicly available datasets: DRIVE [10], STARE [14], and CHASE_DB1 [15]. A summary of these datasets is provided in Table 1. 90% of these images are used for training and rest 10% are used for testing.

Table 1. Summary of the datasets used for the experiments

Datasets	No of images	Image resolution	Pathology information
DRIVE	40	565×584	7 images with diabetic retinopathy (DR), rest 33 without DR
STARE	20	700×605	10 images contain pathology and rest 10 images are normal
CHASE_DB1	28	999×960	Not available

Sensitivity, specificity, accuracy and area under ROC curve (AUC) as computed in [16] and defined below are used to measure the performance of the proposed and the state-of-the-art methods quantitatively.

Sensitivity, $SN = \frac{TP}{TP+FN}$, Specificity, $SP = \frac{TN}{TN+FP}$, Accuracy, Acc $= \frac{TP+TN}{TP+TN+FP+FN}$, Area Under Curve, AUC $= \frac{Sensitivity+Specificity}{2}$.

Here, true positive (TP) refers to a pixel classified as vessel in both in the ground truth and the segmented image, false positive (FP) refers to a pixel classified as a vessel in segmented image but it is recognized as a non-vessel in the ground truth, true negative (TN) refers to a pixel classified as non-vessel in both in the ground truth and the segmented image, false negative (FN) refers to a pixel classified as a non-vessel in segmented image but is recognized as a vessel in the ground truth [16]. Some sample outputs produced by the proposed method and Nguyen et al.'s method is shown in Fig. 7.

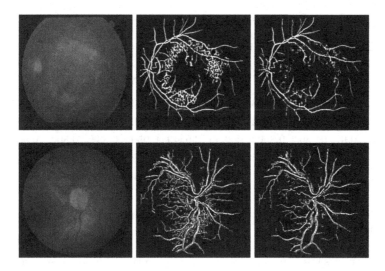

Fig. 7. Sample outputs of the proposed and Nguyen et al.'s method. Original image (first column), multiscale segmented image (second column) and segmented image by proposed method (third column).

Table 2 summarizes the sensitivity, specificity, accuracy and AUC of the proposed method with state-of-the-art methods on DRIVE and STARE dataset.

Table 2. Comparison of performance on DRIVE and STARE datasets

Methods	Dataset							
	DRIVE				STARE			
	Acc	AUC	SE	SP	Acc	AUC	SE	SP
Supervised methods								
Lupascu et al. [7]	0.959	–	0.720	–	–	–	–	–
Marin et al. [17]	0.945	0.843	0.706	0.980	0.952	0.838	0.694	0.982
Roychowdhury et al. [6]	0.952	0.844	0.725	0.962	0.951	0.873	0.772	0.973
Unsupervised methods								
Zhao et al. [8]	0.948	–	0.735	0.979	0.951	–	0.719	0.977
Budai et al. [18]	0.957	0.816	0.644	0.987	0.938	0.781	0.580	0.982
Nguyen et al. [5]	0.941	–	–	–	0.932	–	–	–
Proposed	**0.961**	**0.859**	**0.737**	**0.981**	**0.960**	**0.889**	**0.805**	**0.972**

In Table 3, a comparison of performance between Nyugen et al.'s method and our proposed method in a new dataset CHASE_DB1 is shown.

Table 3. Comparison of performance on CHASE_DB1

Performance (CHASE_DB1)								
Nyugen et al.					Proposed			
Image title	Acc	AUC	SE	SP	Acc	AUC	SE	SP
Image_05R	0.9433	0.8613	0.7625	0.9602	0.953	0.855	0.737	0.973
Image_06R	0.9316	0.8362	0.7241	0.9484	0.950	0.829	0.687	0.971
Image_09R	0.9364	0.8904	0.8359	0.9449	0.959	0.884	0.802	0.966
Image_11R	0.9437	0.8955	0.8417	0.9494	0.959	0.897	0.827	0.966
Average	**0.934**	**0.870**	**0.791**	**0.950**	**0.955**	**0.866**	**0.763**	**0.969**

5 Conclusion

In this paper, a semi-supervised technique for retinal blood vessels segmentation is proposed. The method augments the multi-scale line detector approach of Nguyen et al. [5] by incorporating robust description and supervised learning steps with it. An innovative descriptor named robust feature descriptor is proposed to describe retinal pixels of interest. The descriptor extracts rich texture information around the pixel of interest so that the pixel is true vessel or not can be determined. Experimental results show that the proposed method produces higher accuracy than the state-of-the-art methods, with comparable or higher sensitivity, specificity, and AUC. For DRIVE dataset an accuracy of 0.961 is observed, for STARE and CHASE_DB1 datasets accuracies are respectively 0.960 and 0.955. For Nguyen et al.'s method these values are respectively 0.941, 0.932, and 0.934. Future work will focus on determining more reformed pixel patterns to compute the descriptor and outlining more effective segmentation model. Ensemble learning could also be a way for enhancing the performance of the classifiers.

References

1. Saha, S.K., Xiao, D., Frost, S., Kanagasingam, Y.: A two-step approach for longitudinal registration of retinal images. J. Med. Syst. **40**(12), 277 (2016)
2. Dharmawan, D.A., Ng, B.P.: A new two-dimensional matched filter based on the modified Chebyshev type I function for retinal vessels detection. In: 39th Annual International Conference of the IEEE Engineering in Medicine and Biology Society (EMBC), pp. 369–372 (2017)
3. Saha, S.K., Xiao, D., Bhuiyan, A., Wong, T.Y., Kanagasingam, Y.: Color fundus image registration techniques and applications for automated analysis of diabetic retinopathy progression: a review. Biomed. Signal Process. Control **47**, 288–302 (2019)
4. Fraz, M.M., et al.: Blood vessel segmentation methodologies in retinal images–a survey. Comput. Methods Programs Biomed. **108**(1), 407–433 (2012)

5. Nguyen, U.T., Bhuiyan, A., Park, L.A., Ramamohanarao, K.: An effective retinal blood vessel segmentation method using multi-scale line detection. Pattern Recogn. **46**(3), 703–715 (2013)
6. Roychowdhury, S., Koozekanani, D.D., Parhi, K.K.: Blood vessel segmentation of fundus images by major vessel extraction and subimage classification. IEEE J. Biomed. Health Inform. **19**(3), 1118–1128 (2015)
7. Lupascu, C.A., Tegolo, D., Trucco, E.: FABC: retinal vessel segmentation using AdaBoost. IEEE Trans. Inf Technol. Biomed. **14**(5), 1267–1274 (2010)
8. Zhao, Y.Q., Wang, X.H., Wang, X.F., Shih, F.Y.: Retinal vessels segmentation based on level set and region growing. Pattern Recogn. **47**(7), 2437–2446 (2014)
9. Ricci, E., Perfetti, R.: Retinal blood vessel segmentation using line operators and support vector classification. IEEE Trans. Med. Imaging **26**(10), 1357–1365 (2007)
10. DRIVE Homepage. https://www.isi.uu.nl/Research/Databases/DRIVE//. Accessed 08 July 2018
11. Bay, H., Ess, A., Tuytelaars, T., Van Gool, L.: Speeded-up robust features (SURF). Comput. Vis. Image Underst. **110**(3), 346–359 (2008)
12. Rust, C., Häger, S., Traulsen, N., Modersitzki, J.: A robust algorithm for optic disc segmentation and fovea detection in retinal fundus images. Curr. Dir. Biomed. Eng. **3**(2), 533–537 (2017)
13. Breiman, L.: Random forests. Mach. Learn. **45**(1), 5–32 (2001)
14. STARE Homepage. http://cecas.clemson.edu/~ahoover/stare/. Accessed 29 Nov 2018
15. CHASE_DB1 Homepage. https://blogs.kingston.ac.uk/retinal/chasedb1/. Accessed 15 Nov 2018
16. Fan, Z., Lu, J., Wei, C., Huang, H., Cai, X., Chen, X.: A hierarchical image matting model for blood vessel segmentation in fundus images. IEEE Trans. Image Process. **28**, 2367–2377 (2018)
17. Marin, D., Aquino, A., Gegundez-Arias, M.E., Bravo, J.M.: A new supervised method for blood vessel segmentation in retinal images by using gray-level and moment invariants-based features. IEEE Trans. Med. Imaging **30**(1), 146 (2011)
18. Budai, A., Bock, R., Maier, A., Hornegger, J., Michelson, G.: Robust vessel segmentation in fundus images. Int. J. Biomed. Imaging **2013**, 11 (2013)

Quality-Based Pulse Estimation from NIR Face Video with Application to Driver Monitoring

Javier Hernandez-Ortega[1]($\boxtimes$)(iD), Shigenori Nagae[2], Julian Fierrez[1](iD),
and Aythami Morales[1](iD)

[1] Universidad Autonoma de Madrid, Madrid, Spain
{javier.hernandezo,julian.fierrez,aythami.morales}@uam.es
[2] OMRON Corporation, Kyoto, Japan
shigenori.nagae@omron.com

Abstract. In this paper we develop a robust heart rate (HR) estimation method using face video for challenging scenarios with high variability sources such as head movement, illumination changes, vibration, blur, etc. Our method employs a quality measure Q to extract a remote Plethysmography (rPPG) signal as clean as possible from a specific face video segment. Our main motivation is developing robust technology for driver monitoring. Therefore, for our experiments we use a self-collected dataset consisting of Near Infrared (NIR) videos acquired with a camera mounted in the dashboard of a real moving car. We compare the performance of a classic rPPG algorithm, and the performance of the same method, but using Q for selecting which video segments present a lower amount of variability. Our results show that using the video segments with the highest quality in a realistic driving setup improves the HR estimation with a relative accuracy improvement larger than 20%.

Keywords: Remote Plethysmography · Driver monitoring ·
Heart rate · Quality assessment · Face biometrics · NIR video

1 Introduction

Traffic accidents have become one of the main non-natural causes of death in today's society. The World Health Organization (WHO) published a report in 2018 [18] declaring that 1.35 millions of people die annually all over the world due to traffic accidents, even becoming the main cause of death among young population (those under 30 years old).

Some types of traffic accidents can not be predicted by any manner because they occur due to external factors such as bad weather, roads in poor condition, mechanical issues, etc. However, there is still a high amount of accidents caused by human factors that can be avoided [14]. For example, fatigue is one of the most common causes of accidents, and it is also one of the most preventable.

© Springer Nature Switzerland AG 2019
A. Morales et al. (Eds.): IbPRIA 2019, LNCS 11868, pp. 108–119, 2019.
https://doi.org/10.1007/978-3-030-31321-0_10

Drivers experiencing fatigue have a decrease in their visual perception, reflexes, and psychomotor skills, and they may even fall asleep while driving.

In order to reduce the number of accidents, driver monitoring has attracted a lot of research attention in the recent years [3,7,12]. A driver monitoring system must be able to detect the presence of signals related to fatigue, allowing to take preventive actions to avoid a possible accident. Some of these actions are recommending the driver to stop in a rest area until he is fully recovered, and displaying acoustic and luminous warnings inside the car to keep the driver awake until he can stop.

Driver monitoring systems may follow different ways for achieving their target. Some of them use information about the way the driver is conducting the car, i.e. movements of the steering wheel, status of the pedals, etc. [11]. Physiological signals such as the heart rate (HR), the blood pressure, the brain activity, etc., can also be used to detect fatigue in the driver [10].

A monitoring system capable of estimating physiological components such as the heart rate, or the blood pressure, may present additional benefits. These systems could be able not only to detect signs of fatigue, but also changes in the driver's general health condition. This kind of monitoring systems allow to acquire and process health information daily and non-intrusively. The captured data can be used to help doctors to make better diagnostics, or even for recommending the driver to visit a practitioner if a potential health issue is detected.

The accurate extraction of physiological signals in a real driving scenario is still a challenge. There exist different approaches depending of the acquisition method, i.e. contact-based and image-based, each one with its own strengths and weaknesses. In this paper we focus in improving the performance of an image-based method by introducing a quality assessment algorithm [2]. The target of this algorithm is selecting the video sequences more favorable to a specific heart rate estimation method, in a kind of quality-based processing [6].

The rest of this paper is organized as follows: Sect. 2 introduces driver monitoring techniques, with focus in remote photoplethysmography and its challenges. Section 3 describes the proposed system. Section 4 summarizes the dataset used. Section 5 describes the evaluation protocol and the results obtained. Finally, the concluding remarks and the future work are drawn in Sect. 6.

2   Driver Monitoring Techniques

Early research in driver monitoring was mostly based on acquiring accurate physiological signals from the drivers using contact sensors (e.g. ECG, EEG, or EMG), but this approach may result uncomfortable and impractical in a realistic driving environment. Some parameters that can be obtained this way are the heart rate, respiration, brain activity, muscle activation, corporal temperature, etc. Some works related to this approach are [10] and [16].

Contactless approaches are more convenient for its use in real driver monitoring without bothering the driver with cables and other uncomfortable devices. Regarding this approximation, computer vision techniques result really practical since they use images acquired non-invasively from a camera mounted inside

Table 1. Selection of works related to pulse extraction and/or driver monitoring using contact sensors or images.

Method	Type of data	Parameters extracted	Performance	Target
Brandt et al. [4]	RGB and NIR Video	Head Motion and Eye Blinking	N/A	Driver Fatigue
Shin et al. [16]	ECG	Heart Rate	N/A	Driver Fatigue
Jo et al. [9]	NIR Video	Head Pose and Eye Blinking	Accuracy = 98.55%	Driver Drowsiness and Distraction
Poh et al. [15]	RGB Video	Heart and Breath Rate, HR Variab.	RMSE = 5.63%	Physiological Measurement
Jung et al. [10]	ECG	Heart Rate	N/A	Driver Drowsiness
Tasli et al. [17]	RGB Video	Heart Rate, HR Variab.	MAE = 4.2%	Physiological Measurement
McDuff et al. [13]	RGB-CO Video	Heart and Breath Rate, HR Variab	Correlation = 1.0	Physiological Measurement
Chen et al. [5]	RGB and NIR Video	Heart Rate	RMSE = 1.65%	Physiological Measurement
Present Work	**NIR Video**	**Heart Rate**	**MAE = 8.76%**	**Driver Monitoring**

the vehicle. These images can be processed to analyze physiological parameters using remote photoplethysmography (rPPG). With this technique it is possible to estimate the heart rate, the oxygen saturation, and other pulse related information using only video sequences [15].

2.1 Remote Photoplethysmography

Photoplethysmography (PPG) [1] is a low-cost technique for measuring the cardiovascular Blood Volume Pulse (BVP) through changes in the amount of light reflected or absorbed by human vessels. PPG is often used at hospitals to measure physiological parameters like the heart rate, the blood pressure, or the oxygen saturation. PPG signals are usually be measured with contact sensors often placed at the fingertips, the chest, or the feet. This type of contact measurement may be suitable for a clinic environment, but it can be uncomfortable and inconvenient for daily driver monitoring.

In recent works like [5,13,15,17] remote photoplethysmography techniques have been used for measuring physiological signals from face video sequences captured at distance. These works used signal processing techniques for analyzing the images, and looking for slight color and illumination changes related with the BVP. However, using these methods in a real moving vehicle is not straightforward due to all the variability sources present in this type of video sequences. A selection of works related to driver monitoring and photoplethysmography is shown in Table 1.

2.2 Challenges and Proposed Approach

A moving vehicle is not a perfect environment for obtaining high accuracy when using rPPG algorithms. Images acquired in this scenario may present external illumination changes, low illumination levels, noise, movement of the driver, occlusions, and vibrations of the camera due to the movement of the vehicle. All these factors can make the performance of the rPPG algorithms to drop significantly [2].

In this work we propose a system for pulse estimation for driver monitoring that tries to overcome some of these challenges. We use a NIR camera with active infrared illumination mounted in the dashboard of a real moving car. The NIR spectrum band is highly invariant to ambient light, providing robustness against this external source of variability at a low cost. This also allowed us to extend the application of heart rate estimation to very low illumination environments, e.g. night conditions.

Regarding to the presence of other variability factors such as movement or occlusions, a quality-based approach to rPPG could be adequate [6]. With a short-time analysis, small video segments without enough quality for extracting a robust rPPG signal could be discarded without affecting the global performance of pulse estimation. To accomplish this target, we have proposed a quality metric for short segments of rPPG signals.

Summarizing, in this work: (i) we performed pulse estimation using NIR active illumination to be robust to external illumination variability; (ii) we proposed a quality metric for classifying short rPPG segments and deciding which ones can be used and which ones should be discarded in order to obtain a robust heart rate estimation; and (iii) we compared the performance of a classic rPPG algorithm and our quality-based approach.

3 Proposed System

In this section, we describe the improvements we have done to a baseline rPPG-based heart rate estimation system to increase its performance in a real driving scenario. Classic rPPG systems drastically degrade when facing the variability sources mentioned in the previous sections. This performance problem is caused by the low quality of the extracted rPPG signals which may be affected (in their totality or only in some fragments) by variability sources that the rPPG method does not know how to deal with.

Having this into mind, we thought that computing a quality measure for knowing the amount of variability in each temporal segment of a rPPG signal could be useful for deciding which segments are more suitable for extracting a robust heart rate estimation.

In the next subsection we describe the vanilla rPPG system we used to obtain the baseline results. This method corresponds to the system shown in Fig. 1. In the second subsection we describe the addition of a quality metric to the baseline system. That approach is shown in Fig. 2. In the third subsection we describe how we have obtained the groundtruth of the heart rate for our experiments.

3.1 Baseline rPPG System

The basic method is based in the one used in [8], and consists of the next three main steps:

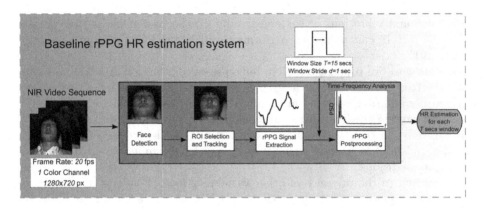

Fig. 1. Architecture of the baseline rPPG system for HR estimation. Given a facial NIR video, the face is detected and the rPPG signal is extracted from the ROI. The raw rPPG signal is windowed and postprocessed in order to obtain an individual HR estimation for each video segment.

- **Face detection and ROI tracking:** The first step consists in detecting the face of the driver on the first frame of the NIR video. We used the Matlab implementation of the Viola-Jones algorithm. This algorithm is known to perform reasonably well and in real time when dealing with frontal faces, as in our case. After the recognition stage we selected the left cheek as the Region Of Interest (ROI), since it is a zone lowly affected by objects like hats, glasses, beards, or mustaches. The next step consisted of detecting corners inside the ROI for tracking them over time using the Kanade-Lucas-Tomasi algorithm, also implemented in Matlab. If at some point of the video the ROI is lost, the face will be redetected, and after that also the ROI and the corners.
- **rPPG signal extraction:** For each frame from the video, we calculated its raw rPPG value as the averaged intensity of the pixels inside the ROI. The final output for each video sequence is a rPPG temporal signal composed by the concatenation of these averaged intensities.
- **rPPG postprocessing:** We wanted to estimate a HR value each d seconds. In order to achieve that target, we extracted windows of T seconds from the rPPG signal, with a stride of d seconds between them. The length of the window (T) is configurable in order to perform a time dependent analysis. For each window we postprocessed the raw rPPG signal and we obtained an estimation of the HR. This postprocessing method consists of three filters:
  - Detrending filter: this temporal filter is employed for reducing the stationary part of the rPPG signal, i.e. eliminating the contribution from environmental light and reducing the slow changes in the rPPG level that are not part of the expected pulse signal.
  - Moving-average filter: this filter is designed to eliminate the random noise on the rPPG signal. That noise may be caused by imperfections on the

sensor and inaccuracies in the capturing process. This filter consists in a moving average of the rPPG values (size 3).

- Band-pass filter: we considered that a regular human heart rate uses to be into the 40–180 beats per minute (bpm) range, which corresponds to signals with frequencies between 0.7 Hz and 3 Hz approximately. All the rPPG frequency components outside that range are unlikely to correspond to the real pulse signal so they are discarded.

After this processing stage we transformed the signal from the time domain to the frequency domain using the Fast Fourier Transform (FFT). Then, we estimated its Power Spectral Density (PSD) distribution. Finally, we searched for the maximum value in that PSD. The frequency correspondent to that maximum is the estimated HR of that specific video segment.

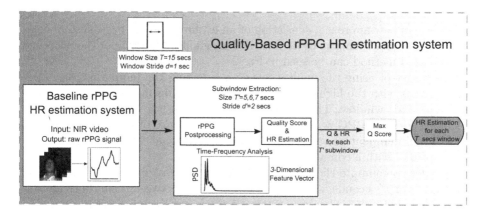

Fig. 2. Architecture of the quality-based rPPG system for HR estimation. We extracted some features from subwindows of the postprocessed rPPG signal. These features were used to compute a quality metric for estimating the presence of noise, head motion, or external illumination variability in the rPPG signal. For each T seconds window we selected the T' seconds subwindow with the highest quality.

3.2   Proposed Quality-Based Approach

The baseline method is able to obtain robust HR estimations in controlled scenarios without too much variability or noise in the recordings. However, the raw rPPG signals acquired in a realistic driver monitoring scenario use to have high variations due to external illumination changes, and frequent movements of the driver's head. There are also other sources of noise, e.g. noise inherent to the acquisition sensor.

All the mentioned factors make the performance of the baseline rPPG algorithm to dramatically fall. In order to make it as robust as possible, we decided

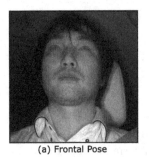 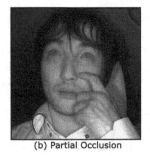 

(a) Frontal Pose   (b) Partial Occlusion   (c) Lateral Pose

Fig. 3. Images extracted from the OMRON database: (a) shows a image with a low level of variability. A high quality rPPG signal could be extracted from a video composed by this type of images; (b) and (c) show examples of images with high variability, such as occlusions and head rotation respectively.

to develop an new approach, consisting of an evolution of the basic system combined with a quality metric of the raw rPPG signals. A scheme of the proposed quality-based method can be seen in Fig. 2.

The target of using the quality metric is selecting the temporal subwindow of T' seconds with the highest quality from all the subwindows available inside each T seconds window. The criteria for determining the best quality consists in looking for the rPPG segment with the less presence of noise, head motion, and external illumination variability, i.e. the rPPG signal closest to one that has been captured with a contact sensor.

In order to compute the quality level, we divided each window into several subwindows of T' seconds, with a stride of d' seconds between them (both parameters are configurable). Then we performed the processing of the rPPG signal in the same way done in the baseline system. From each processed rPPG subwindow we extracted the three features shown in Table 2 (left), and we combined them to obtain a single numerical quality measure (Q) representative of how close is the rPPG signal of each subwindow to one acquired in perfect conditions.

Finally, from each T seconds window, we selected the segment of T' seconds with the highest Q, and we estimate the user's HR with that rPPG segment. This way we discarded the rPPG fragments that may be more affected by variability. This value of the HR is used as the final HR estimation for the whole T seconds window.

4   Dataset

4.1   OMRON Database

We tested our method with a self-collected dataset called OMRON Database. The data in the dataset is composed by Near Infrared (NIR) active videos of the driver's faces, recorded with a camera mounted in a car dashboard. The images were captured at a sampling rate of 20 fps, and a resolution of 1280×720

Table 2. Left: Features extracted to compute the quality of the rPPG postprocessed signals. Right: Final configuration of the parameters of the quality-based method. *From each window of T seconds, we extracted subwindows of $T' = 5, 6, and 7 s$ of duration, and we selected the one with the highest Q value.*

Feature	Description	Parameter	Value
Signal Noise Ratio (SNR)	Power of the maximum value in the PSD and its two first harmonics, divided by the rest of the power.	Window Size T	7 seconds
Bandwidth (BW)	Bandwidth containing the 99% of the power, centered in the maximum value of the PSD.	Window Stride d	1 second
Ratio Peaks (RP)	Power of the highest peak in the PSD divided by the power of second highest peak.	Subwindow Size T'	5, 6, and 7 seconds*
		Subwindow Stride d'	2 seconds
		Feature Vector	SNR, BW, and RP

pixels. The PPG signals used for the groundtruth were captured using a BVP fingerclip sensor with a sampling rate of 500 Hz, and then downsampled to 20 Hz to synchronize them with the images from the camera.

The dataset is comprised of 7 male users, with different ages, skin tones and some of them wearing glasses. Each participant was in front of the camera during a single session with a different duration for each one. The sessions went from 20 min to 60 min long. The full database contains 400, 000 images with an average of 57, 000 images for each subject. The recordings try to represent a real driving scenario inside a moving car. They present different types of variability such as head movement, occlusions, car vibration, or external illumination. These variations mean different levels of quality in the estimated rPPG signals. Examples of images from this database can be seen in Fig. 3.

5   Evaluation

In this section we compare the performance of the heart rate estimations obtained using the quality-based rPPG method with the performance obtained using the baseline rPPG method.

5.1   Setting Quality Parameters and Features

The quality-based method has several parameters to be configured: the window size T, the subwindow size T', the window stride d, and the subwindow stride d'. It is also necessary to decide which features to extract from the rPPG signals, as they must contain information about the quality level of each subwindow T'.

For this work we extracted 3 different features that can give us information about how close/far is a rPPG signal from the one captured in perfect conditions. The features and their descriptions can be seen in Table 2 (left). The final quality metric is computed as the arithmetic mean of these 3 features after normalizing them to the [0, 1] range using a *tanh*-based normalization [6].

Table 3. Results of HR estimation for the rPPG baseline method and the quality-based approach. The results comprehend the individual Mean Absolute Errors (MAE) with respect to the groundtruth HR. The mean value and the standard deviation of the MAE for the whole evaluation data has been also computed. The relative improvement of the MAE is shown between parentheses.

MAE [bpm] Video Number	1	2	3	4	5	6	7	8	9	10	11
Baseline Method	11.9	15.9	7.5	12.6	11.2	12.5	9.6	8.0	9.2	9.6	10.1
Proposed Method	9.1	9.9	7.5	8.1	10.9	8.2	7.9	8.7	5.7	5.8	10.5

MAE [bpm] Video Number	12	13	14	15	16	17	18	19	Mean	Std
Baseline Method	8.8	14.6	10.3	13.1	7.9	11.1	9.8	15.2	11.0	2.4
Proposed Method	9.3	7.6	7.0	7.9	8.4	9.1	11.2	12.7	8.7 (21%)	1.7 (29%)

Based on our own previous rPPG experiments, we decided to test values of T going from 5 s to 15 s, with 1 s of increment for the loop. From our previous work [8] we know that $T = 5$ s was the lowest value that gave good HR estimation with favorable conditions, and using windows longer than 15 s did not show to improve the results.

For setting the subwindow duration T', we decided to test values going from 5 s (limited by the minimum possible T size), to the correspondent T value in each case. We also incremented the T' values using a step of 1 s. The stride d is set to 1 s in order to give an estimation of the HR for each second of the input video. The stride d', i.e. the temporal step between each subwindow, took values going from a minimum of 1 s to a maximum of 5 s (when possible), with 1 s of increment. After this initial configuration experiments the best results were obtained for the parameters shown in Table 2 (right).

To compute the performance of heart rate estimations we decided to use the Mean Absolute Error (MAE) between the groundtruth heart rate in beats per minute (bpm), and the one estimated with the rPPG algorithm.

5.2   Results

For the final evaluation of both methods (baseline and proposed quality-based), we processed 19 NIR videos of 1 min duration each one from the OMRON Database. We used the configuration of parameters shown in Table 2 (right) for both methods (only T and d in the case of the baseline system). We first computed the Mean Absolute Error (MAE) for each NIR video separately. We did this to have an idea of which videos are working better and which ones are working worse. We also computed the mean and standard deviation of the MAE for the whole evaluation dataset.

As can be seen in Table 3, using the quality-based rPPG approach we obtained a MAE value averaged across videos of 8.7 beats per minute (bpm), and a standard deviation of 1.7 bpm for the whole evaluation dataset. Compared to this result, the baseline system (without the quality approach) obtained a MAE

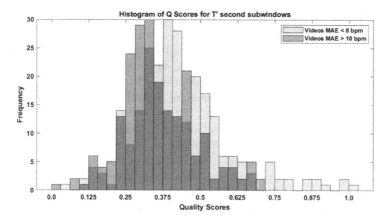

Fig. 4. Quality Scores obtained from T' seconds windows. We have selected those videos with a mean MAE under 8 bpm as representative of high quality videos, and those with a mean MAE over 10 bpm as low quality videos. The histograms show two different distributions, with the high quality videos presenting a higher mean value of the quality score Q.

of 11.0 bpm and a MAE standard deviation of 2.4 bpm for the 19 min. This difference in the performance represents a relative improvement of a 21% in the mean value, and of the 29% in the standard deviation of the Mean Absolute Error.

Table 3 also shows the MAE values for each NIR video of the evaluation dataset, both using the baseline and the quality-based methods. It can be seen that for some specific videos the baseline result is obtaining a more accurate estimation of the HR, but in general, the MAE values obtained using the quality-based approach are lower. The specific cases in which the quality-based method is working worse coincide with those videos with long sequences with high variability, what makes difficult to find clean segments.

In Fig. 4 we are showing the quality scores Q we obtained for a selection of the evaluation videos. We decided to show the distribution of Q scores from those videos with a MAE value (obtained with the quality-based method) lower than 8 bpm, and those with a MAE value higher than 10 bpm. The histograms show two different distributions, with the best performing videos (i.e., MAE ¡ 8 bpm) presenting a higher mean value of the quality score Q.

The results of this section evidenced that, at least with the data from the OMRON Dataset, the quality metric Q has shown to be an effective way to discard segments of video that may impact negatively to the general performance in rPPG, and therefore obtaining an improvement of the global accuracy of HR estimation.

6   Conclusion and Future Work

In this paper we developed a method for improving heart rate (HR) estimation using remote photoplethysmography (rPPG) in challenging scenarios with multiple sources of high-variability and degradation. Our method employs a quality measure to extract a rPPG signal as clean as possible from a specific face video segment, trying to obtain a more robust HR estimation.

Our main motivation is developing robust technology for contactless driver monitoring using computer vision. Therefore, in our experiments we employed Near Infrared (NIR) videos acquired with a camera mounted in a car dashboard. This type of videos present a high number of variability sources such as head movement, external illumination changes, vibration, blur, etc. The target of the quality metric Q we have proposed consists in estimating the amount of presence of those factors. Even though our experimental framework is around driver monitoring, our methods may find application in other high-variability face-based human-computer interaction scenarios such as mobile video-chat.

We have compared the performance of two different methods for HR estimation using rPPG. The first one consisted in a classic rPPG algorithm. The second method consisted in the same algorithm, but using the quality measure Q for selecting which video segments present a lower amount of variability. We used those segments for extracting rPPG signals and their associated HR estimations. The quality metric Q showed to be a reliable estimation of the amount of variability. We achieved better performance in HR estimation using the video segments with the highest possible quality, compared to using all the video frames indistinctly.

Our solution is based on defining the quality Q as a combination of handcrafted features. As future work, other definitions of quality could be also investigated. A different set of features that may correlate more accurately to the presence of noise factors in the rPPG signal can be studied. Training a Deep Neural Network (DNN) for extracting Q from the video sequences is also an interesting possibility. This type of networks may be able of estimating the quality level by learning which factors are more relevant for obtaining robust rPPG signals directly from training data. However, the lack of labeled datasets makes it difficult to train DNNs from scratch, so it would be also beneficial to acquire a larger database. This new database may contain a higher number of users, and it may also present more challenging conditions for testing our quality-based rPPG algorithm, e.g. variant ambient illumination, motion, blur, occlusions, etc.

Acknowledgements. This work was supported in part by projects BIBECA (RTI2018-101248-B-I00 from MICINN/FEDER), and BioGuard (Ayudas Fundacion BBVA). The work was conducted in part during a research stay of J. H.-O. at the Vision Sensing Laboratory, Sensing Technology Research Center, Technology and Intellectual Property H.Q., OMRON Corporation, Kyoto, Japan. He is also supported by a Ph.D. Scholarship from UAM.

References

1. Allen, J.: Photoplethysmography and its application in clinical physiological measurement. Physiol. Meas. **28**(3), R1 (2007)
2. Alonso-Fernandez, F., Fierrez, J., Ortega-Garcia, J.: Quality measures in biometric systems. IEEE Secur. Priv. **10**(6), 52–62 (2012)
3. Awasekar, P., Ravi, M., Doke, S., Shaikh, Z.: Driver fatigue detection and alert system using non-intrusive eye and yawn detection. Int. J. Comput. Appl. **180**, 1–5 (2019). (0975–8887)
4. Brandt, T., Stemmer, R., Rakotonirainy, A.: Affordable visual driver monitoring system for fatigue and monotony. In: IEEE International Conference on Systems, Man and Cybernetics, pp. 6451–6456 (2004)
5. Chen, J., et al.: RealSense = real heart rate: illumination invariant heart rate estimation from videos. In: Image Processing Theory Tools and Applications (IPTA) (2016)
6. Fierrez, J., Morales, A., Vera-Rodriguez, R., Camacho, D.: Multiple classifiers in biometrics. Part 2: trends and challenges. Inf. Fusion **44**, 103–112 (2018)
7. Flores, M.J., Armingol, J.M., de la Escalera, A.: Real-time warning system for driver drowsiness detection using visual information. J. Intell. Robot. Syst. **59**(2), 103–125 (2010)
8. Hernandez-Ortega, J., Fierrez, J., Morales, A., Tome, P.: Time analysis of pulse-based face anti-spoofing in visible and NIR. In: IEEE CVPR Computer Society Workshop on Biometrics (2018)
9. Jo, J., et al.: Vision-based method for detecting driver drowsiness and distraction in driver monitoring system. Opt. Eng. **50**(12), 127202 (2011)
10. Jung, S.J., Shin, H.S., Chung, W.Y.: Driver fatigue and drowsiness monitoring system with embedded electrocardiogram sensor on steering wheel. IET Intell. Transport Syst. **8**(1), 43–50 (2014)
11. Kang, H.B.: Various approaches for driver and driving behavior monitoring: a review. In: IEEE International Conference on Computer Vision Workshops (2013)
12. Lal, S.K., et al.: Development of an algorithm for an EEG-based driver fatigue countermeasure. J. Saf. Res. **34**(3), 321–328 (2003)
13. McDuff, D., Gontarek, S., Picard, R.W.: Improvements in remote cardiopulmonary measurement using a five band digital camera. IEEE Trans. Biomed. Eng. **61**(10), 2593–2601 (2014)
14. Pakgohar, A., Tabrizi, R.S., Khalili, M., Esmaeili, A.: The role of human factor in incidence and severity of road crashes based on the CART and LR regression: a data mining approach. Proc. Comput. Sci. **3**, 764–769 (2011)
15. Poh, M.Z., McDuff, D.J., Picard, R.W.: Advancements in noncontact, multiparameter physiological measurements using a webcam. IEEE Trans. Biomed. Eng. **58**(1), 7–11 (2011)
16. Shin, H.S., Jung, S.J., Kim, J.J., Chung, W.Y.: Real time car driver's condition monitoring system. In: IEEE Sensors, pp. 951–954 (2010)
17. Tasli, H.E., Gudi, A., den Uyl, M.: Remote PPG based vital sign measurement using adaptive facial regions. In: IEEE International Conference on Image Processing (ICIP), pp. 1410–1414 (2014)
18. WHO: Global status report on road safety (2018). https://www.who.int/violence_injury_prevention/road_safety_status/2018/en/

Handwriting and Document Analysis

Multi-task Layout Analysis
of Handwritten Musical Scores

Lorenzo Quirós[✉], Alejandro H. Toselli, and Enrique Vidal

Pattern Recognition and Human Language Technologies Research Center,
Universitat Politècnica de València, Camino de Vera, s/n, 46022 Valencia, Spain
{loquidia,ahector,evidal}@prhlt.upv.es

Abstract. Document Layout Analysis (DLA) is a process that must be performed before attempting to recognize the content of handwritten musical scores by a modern automatic or semiautomatic system. DLA should provide the segmentation of the document image into semantically useful region types such as staff, lyrics, etc. In this paper we extend our previous work for DLA of handwritten text documents to also address complex handwritten music scores. This system is able to perform region segmentation, region classification and baseline detection in an integrated manner.

Several experiments were performed in two different datasets in order to validate this approach and assess it in different scenarios. Results show high accuracy in such complex manuscripts and very competent computational time, which is a good indicator of the scalability of the method for very large collections.

Keywords: Document Layout Analysis ·
Text region detection and classification · Semantic segmentation ·
Music document processing · Music score images

1 Introduction

Thousands of handwritten documents are available in libraries and other institutions around the world, but most of them are not searchable or even browsable by modern digital means due to the lack of available digital transcripts.

Music constitutes one of the main vehicles for cultural transmission. Hence handwritten musical scores have been preserved over the centuries and, thus, it is important that they can be studied, analyzed, and performed.

This kind of manuscripts has a very complex layout and, because of the huge amount of documents available, it is intractable to provide accurate transcripts in a totally manual manner. Consequently, automatic or semi-automatic transcription systems have been developed to accelerate this process.

Those systems are often divided into two sub-processes: Handwritten Text Recognition (HTR) [12,14]—for the lyrics and other textual regions of the document—, and Handwritten Music Recognition (HMR) [2,5]—for the musical content of the document (e.g. the staff).

© Springer Nature Switzerland AG 2019
A. Morales et al. (Eds.): IbPRIA 2019, LNCS 11868, pp. 123–134, 2019.
https://doi.org/10.1007/978-3-030-31321-0_11

For that reason, before attempting to recognize the content written in a musical document, we must perform Document Layout Analysis (DLA) on it, i.e. divide it into relevant regions which can be processed by HTR, HMR or any other system available for a specific region. Those regions must be physically segmented from the document (region segmentation) and labeled accordingly (region classification). Also, the baselines (the imaginary lines upon which the lines of text rest) should be detected before performing HTR.

In this work, we present a system based on Artificial Neural Networks, which is able to segment the document into relevant regions, to provide the label associated to each region and to detect the baselines on text regions. It is an integrated approach where regions and baselines are segmented, labeled and detected in a single process.

The rest of the paper is organized as follows: first in Sect. 2 we review the current state of the art regarding music layout analysis. Section 3 provides an overview of the layout analysis technologies used. In Sect. 4 we present in detail the corpora used in the experiments, the evaluation measures and the system setup. Then, the results are discussed in Sect. 5. Finally, we draw some conclusions and outline possible extensions for future work in Sect. 6.

2   Related Work

Although the importance of Document Layout Analysis process before any recognition step is clear, only a few studies have tackled the task for handwritten music scores. Most of them have focused on pixel-wise classification of the different symbols or elements present in the staff (staff line, symbol, text) [3,4,7,13], where the staff itself is supposed to be segmented previously or later by another method.

Other methods focus on separating music and lyrics sections by searching local minima points on binarized images [1] or using projection profiles and Hidden Markov Models [6], but they are restricted to documents where all regions follow a vertical order (no horizontal split is allowed).

Approaches based on detailed pixel-level classification or symbol-level classification are not scalable due to the cumbersome process of ground-truth generation. Also, approaches restricted by the vertical order of the document fail in many complex scenarios. For this reason we propose an integrated approach which is able to separate the different regions in a document using a region-level ground-truth instead of a pixel-level without any vertical nor horizontal restriction. It follows the ideas previously introduced for DLA of handwritten text images [11].

3   Framework Description

This works extents on the ideas successfully applied to Handwritten Document Layout Analysis [11,12]. Here we show the applicability of those methods to the complex task of Layout Analysis in handwritten music scores.

Following similar formulation as in [11], DLA of music documents is defined as a two task problem:

- *Task-1*: Region segmentation and labeling
- *Task-2*: Baseline detection.

Task-1 consists in classifying the input image into a set of regions and assign each one to the correct class (e.g. heading, paragraph, staff, etc). On the other hand, *Task-2* consists in obtaining the baseline of each text line present in the regions where some text is expected (e.g. heading, paragraph).

The proposed method consists of a set of two main stages used to solve the multi-task problem formulated previously in an integrated manner.[1] In the first stage (called Pixel-level classification) an Artificial Neural Network (ANN) is used to classify the pixels of the input image ($x \in \mathbb{R}^{w \times h \times \gamma}$, with height h, width w and γ channels) into a defined number of regions of interest (text, illustration, staff, etc) and baselines. In the second stage (called Region and baseline consolidation), a contour extraction algorithm is used to consolidate the pixel-level classification into a set of simplified regions delimited by closed polygons. Then a similar process is carried inside each region where a line of text is expected to extract the baselines. In contrast to [11], here we restrict the search of baselines to only those regions where text is expected to be (e.g. a region of type "lyrics" is expected to contain text while "staff" does not).

Stage 1, pixel-level classification: given an input image x, we can define a multi-task variable[2] $y = [y^1, y^2]$, where $y^t = (y_{ij}^t), 1 \leq i \leq w, 1 \leq j \leq h, t \in \{1, 2\}$ and $y_{ij}^t \in \{1, \ldots, K^t\}$ with $K^t \in \mathbb{N}^+$ being the finite number of classes associated with the t-th task. The solution of this problem for some test instance x is given as the following optimization problem:

$$\hat{y} = \arg\max_{y} p(y \mid x) \tag{1}$$

where the conditional distribution $p(y \mid x)$ is usually unknown and has to be estimated from training data $D = \{(x, y)\}_{n=1}^{N} = \{(X, Y)\}$.

In our case *Task-2* ($t = 2$) is a binary classification problem, then $K^2 = 2$ (background, baseline), and *Task-1* ($t = 1$) is a multi-class problem where K^1 is equal to the number of different types of regions in the specific corpus, plus one for the background.

In this work, the conditional distribution $p(y \mid x)$ is estimated under naive Bayes assumption for each pixel in the image by *M-net*, the Conditional Adversarial Network presented in [11]. Under this assumption, the optimization problem formulated in Eq. (1) can be computed element by element:

- *Task-1*:

$$y_{i,j}^{*1} = \arg\max_{y \in \{1, \ldots, K^2\}} \mathcal{M}_{i,j,y}(x), \quad 1 \leq i \leq w, 1 \leq j \leq h \tag{2}$$

---

[1] Notice that both stages work together on both tasks.
[2] For convenience, each task will be represented mathematically as a superscript over the variables (e.g. v^t).

– *Task-2*:

$$y_{i,j}^{*2} = \underset{y\in\{0,1\}}{\arg\max} \mathcal{M}_{i,j,y}(\boldsymbol{x}), \quad 1 \leq i \leq w, 1 \leq j \leq h \qquad (3)$$

where $\mathcal{M}(\cdot)$ is the output of the latest layer of *M-net*.

Stage 2, region and baseline consolidation: let a test instance $\boldsymbol{x}$ and its pixel level classification $\boldsymbol{y}^*$ obtained in the previous stage be given. First, the contour extraction algorithm presented by Suzuki et al. [15] is used for each region type over $\boldsymbol{y}^1$ to determine the vertices of its contour (we call it a region-contour). Then, for each region-contour found, which belongs to a region where a line of text is expected, we apply the same extraction algorithm over $\boldsymbol{y}^2$ to find the contours where baselines are expected to be (baseline-contour), but restricted to the area defined by the region polygon.

Finally, the baseline detection algorithm presented in [11] is used to detect the baseline of each baseline-contour found.

4 Experimental Setup

4.1 Corpus

CAPITÁN: is an archive of manuscripts of Spanish and Latin American music from the 16-th to 18-th centuries. These manuscripts were written using the so-called *white mensuralnotation*, which in many aspects differs from the modern Western musical notation. Furthermore, this archive was written following the slightly different Hispanic notation of that time, increasing its historical and musicological interest. The archive is managed by the Department of Musicology of the Spanish National Research Council of Barcelona, which kindly allowed the use of the archive for research purposes.

On this work, we carried out our experiments on a subset of 96 pages of the archive, using 50 pages for training and 46 for test as defined in [6].

The dataset has been annotated manually into the following layout elements:

– **header:** title of the piece that might appear at the beginning of a piece (top of the first page).
– **staff:** Represents the regions that contains a set of horizontal lines and spaces where each one represent a different musical pitch. This region type does not contain text lines. Hence, no baselines.
– **lyrics:** words that are sung appear below their corresponding staff.

Main statistics of the dataset are presented in Table 1, and an example of an annotated page can be seeing in Fig. 1.

VORAU-253: is a music manuscript referred to as Cod. 253 of the Vorau Abbey library, which was provided by the Austrian Academy of Sciences. It is written in German Gothic notation and dated around year 1450. This manuscript is interesting because of the complexity of its layout, where staff, text and decorations are intertwined to compose the structure of the document (see Fig. 2(a)).

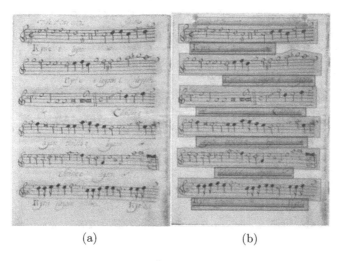

(a)   (b)

Fig. 1. Example of a page of the CAPITÁN dataset. (a) Original image. (b) Annotated layout, green = header, pink = staff, blue = lyrics. Better seeing in color. (Color figure online)

Table 1. Main characteristics of the CAPITÁN dataset.

Region	#Regions			#Lines		
Name	Train	Test	Total	Train	Test	Total
Header	5	4	9	5	4	9
Staff	300	276	576	—	—	—
Lyrics	289	253	542	290	255	545

On this work, we carried out our experiments on a subset of 228 pages of the archive, using 128 randomly selected pages for training and 100 for test.

The dataset has been annotated manually into the following layout elements:

– **staff:** Represents the regions that contains a set of horizontal lines and spaces where each one represent a different musical pitch. This region type does not contain text lines. Hence, no baselines.
– **lyrics:** words that are sung appear below their corresponding staff, and other text in the document.
– **drop-capital:** a decorated letter that might appear at the beginning of a word or text line.

Main statistics of the dataset are presented in Table 2, and an example of an annotated page can be seeing in Fig. 2.

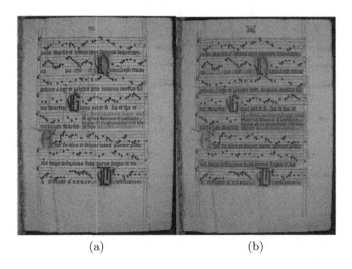

(a)   (b)

Fig. 2. Example of a page of the VORAU-253 dataset. (a) Original image. (b) Annotated layout, green = drop-capital, pink = staff, blue = lyrics. Better seeing in color. (Color figure online)

Table 2. Main characteristics of the VORAU-253 dataset.

Region	#Regions			#Lines		
Name	Train	Test	Total	Train	Test	Total
Drop-capital	336	232	568	—	—	—
Staff	1194	919	2113	—	—	—
Lyrics	1379	1042	2403	1628	1215	2843

4.2 Evaluation Measures

To the best of our knowledge, there is no common evaluation measure able to assess the results obtained in both tasks jointly, therefore we present a set of metrics for each task.

Task-1 Region segmentation: we report metrics from semantic segmentation and scene parsing evaluations as presented in [10]:

- Pixel accuracy (Pixel$^{acc.}$): $\sum_i \eta_{ii}/\sum_i \tau_i$
- Mean accuracy (Mean$^{acc.}$): $1/K^{t=1}\sum_i \eta_{ii}/\tau_i$
- Mean Jaccard Index (Mean^{IU}): $(1/K^{t=1})\sum_i \eta_{ii}/(\tau_i + \sum_j \eta_{ji} - \eta_{ii})$
- Frequency weighted Jaccard Index (f.w.^{IU}): $(\sum_\kappa \tau_\kappa)^{-1}\sum_i \tau_i\eta_{ii}/(\tau_i + \sum_j \eta_{ji} - \eta_{ii})$.

where η_{ij} is the number of pixels of class i predicted to belong to class j, $K^{t=1}$ is the number of different classes for the task $t = 1$, τ_i the number of pixels of class i, and $\kappa \in \{1,\ldots,K^{t=1}\}$. Even though we present all four metrics, we consider Mean Jaccard Index the most relevant one for the task.

Task-2 Baseline detection: we report precision (P), recall (R) and its harmonic mean (F1) measures as defined specifically for this kind of problem in [8]. Tolerance parameters are set to default values in all experiments (see [8] for details about measure definition, tolerance values and implementation details).

4.3 System Setup

Artificial Neural Network Architecture: like in [11] we define *M-net* as the main network and *A-net* as the adversarial one. Both are trained in parallel, we alternate between one gradient descent step on *M-net* and one step on *A-net* and so on. Both were built with 4×4 convolutional filters with stride of 2. Moreover, the architecture of each network is:

- *A-net*: `C64:C128B:C256B:C512B:C1:Sigmoid`. Activation LeakyReLU.
- *M-net*:
  - Encoder: `C64:C128B:C256B:C512B:C512B:C512B:C512B:C512`. Activation LeakyReLU.
  - Decoder: `C512BD:C512BD:C512BD:C515B:C256B:C128B: C64B:ReLU:C` $K^{t=1} + K^{t=2}$: SoftMax. Activation ReLU.

where `Ck` denotes a convolution layer with `k` filters, `B` a BatchNorm layer and `D` a Dropout layer with a dropout rate of 0.5, and `C`$K^{t=1} + K^{t=2}$ denotes a convolution layer with the number of filters equal to the number of output classes.

Optimization process is performed using minibatch stochastic gradient descent and Adam solver [9], with learning rate of 0.001, and momentum parameters $\beta_1 = 0.5$ and $\beta_2 = 0.999$ during 200 epochs.[3] Due to memory restrictions on the hardware available, minibatch size is set to 8 images of 1024×768 on a single Titan X GPU.

Affine transformations (translation, rotation, shear, scale) and elastic deformations are applied to the input images as a data augmentation technique, where its parameters are selected randomly, and applied on each epoch and image with a probability of 0.5. The source code used to run all experiments is available online at https://github.com/lquirosd/P2PaLA/releases/tag/v0.6.

5 Results

In this section, we evaluate the performance of the DLA proposed approach. Experiments on each corpus were performed from a very low number of training images (16) up to the maximum number of pages available for each corpus.

It is very important to asses the quality of the results even using a few training pages due the high cost of labeling data for training. Even if the results are not perfect, we can use them into an interactive or iterative system to create more training data at a much lower cost.

---

[3] In most of the cases the results can be improved by carefully selecting the hyperparameters of the system based on the data available for each corpus, but in this work we decide to keep hyperparameters fixed across experiments for comparability and homogeneity.

5.1   CAPITÁN

This dataset is very small. We conducted three experiments using 16, 32 and 50 pages for training, which are selected randomly and incrementally added to the training set on the respective experiment.

Task-2 (baseline) results are stable on the first two experiments, but on *Task-1* an increment of around two points in most of the metrics was observed from one experiment to the next one (as expected due to the increment of training data). On the last experiment, a qualitative improvement is observed—especially on baseline detection—, reaching a maximum recall of 97.4% and precision of 85.0%. Errors are mainly related to text lines with a long "white" space between words, as shown in Fig. 3(c, d). However, because of the long space between words, the context an HTR system could use of it is negligible, hence the effect on the transcript.

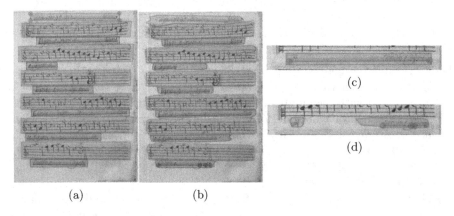

(a)   (b)

Fig. 3. Example of results obtained for the CAPITÁN test set. (a) Ground-truth image. (b) Obtained layout. (c, d) Common error observed, cyan = header, pink = staff, blue = lyrics. Better seeing in color. (Color figure online)

In Table 3, we show the results obtained on each experiment along with the training and inference time (average in seconds per page). Also, we show an example of the extracted layout in Fig. 3.

We also present some results comparing our method with [6]. The definition of layout in both methods is different: on [6] it is defined as a set of vertical aligned image-wide baselines for each region, and in our case we define each region by a polygon, without vertical or horizontal restriction. Consequently, we have to perform some minor adjustments on our output to make it comparable. First, all regions are compressed into the horizontal line which best fits the bottom side of the polygon that defines the region. Then, that line is expanded up to the width of the image.

Table 3. Precision (P), Recall (R), F1, Pixel accuracy (Pixel$^{acc.}$), Mean Pixel accuracy (Mean$^{acc.}$), Mean Jaccard Index (Mean^{IU}) and Frequency weighted Jaccard Index (f.w.^{IU}) results for the CAPITÁN test set. Nonparametric Bootstrapping confidence intervals at 95%, 10 000 repetitions.

	Metric [%]	Number of training pages			CI^a
		16	32	50	
Task-1	Pixel$^{acc.}$	90.1	91.1	92.5	±0.6
	Mean$^{acc.}$	87.7	89.6	91.4	±1.6
	Mean^{IU}	79.3	81.5	84.4	±1.8
	f.w.^{IU}	82.5	84.1	86.3	±1.0
Task-2	P	71.0	71.1	85.0	±5.4
	R	94.9	95.5	97.4	±3.5
	F1	81.2	81.5	90.8	±5.4
Training time [s]		1844.9	2641.5	3517.9	
Inference time [avg. s/page]		0.69	0.69	0.67	

^{a}The confidence intervals of the elements in each row are all within the bounds listed in the corresponding row of the CI column, real values are not present in order to improve the readability of the data. Note that these intervals are not always symmetric.

Using this simple adjustments we obtain a *relative geometric error*, as defined in [6], of 5.2% and a standard deviation of 12.2%, compared to 3.2% and standard deviation of 3.5% in the original paper. Notice that this small difference is mostly related to the slant of the image, since we do not correct the slant, and the ground-truth used on that work is just a horizontal line.

5.2 VORAU-253

The results are shown in Table 4 along with the training and inference time (average in seconds per page). An example of the extracted layout is depicted in Fig. 4.

Although this dataset is far more complex than CAPITÁN, results are satisfactory even with as very little training data as 16 pages, and the quality increases with the number of training data up to 87.0% mean intersection over union and 96.0% F1 measure. These results are enough to be directly processed by most of the state-of-the-art HTR and HMR systems without human intervention.

Table 4. Precision (P), Recall (R), F1, Pixel accuracy (Pixel$^{acc.}$), Mean Pixel accuracy (Mean$^{acc.}$), Mean Jaccard Index (Mean^{IU}) and Frequency weighted Jaccard Index (f.w.^{IU}) results for the VORAU-253 test set. Nonparametric Bootstrapping confidence intervals at 95%, 10 000 repetitions.

	Metric [%]	Number of training pages				CI^a
		16	32	64	128	
Task-1	Pixel$^{acc.}$	92.5	93.7	94.5	94.7	±0.3
	Mean$^{acc.}$	88.6	92.0	93.4	94.2	±0.6
	Mean^{IU}	80.0	84.0	86.6	87.0	±0.7
	f.w.^{IU}	86.2	88.2	89.7	90.1	±0.5
Task-2	P	70.6	84.5	92.2	94.1	±3.0
	R	94.1	97.7	98.0	98.1	±1.2
	F1	80.7	90.6	95.0	96.0	±2.2
Training time [s]		1915.0	2814.4	4361.8	7429.8	
Inference time [avg. s/page]		1.40	1.27	1.22	1.20	

[a]The confidence intervals of the elements in each row are all within the bounds listed in the corresponding row of the CI column, real values are not present in order to improve the readability of the data. Note that these intervals are not always symmetric.

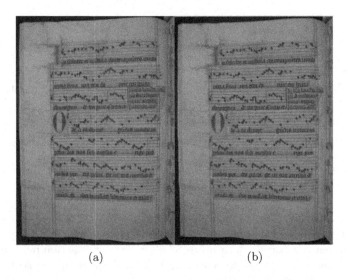

(a)   (b)

Fig. 4. Example of results obtained for the VORAU-253 test set. (a) Ground-truth image. (b) Obtained layout, green = drop-capital, pink = staff, blue = lyrics. Better seeing in color. (Color figure online)

6   Conclusions

On this paper we have proposed a new approach to undertake the DLA problem on music handwritten documents. The input image is segmented into a set of regions of interest and their baselines are detected.

We have demonstrated, via empirical experiments, that the proposed approach is able to obtain an useful and detailed layout for handwritten music documents. It can be used directly by most state-of-the-art HTR and HMR systems.

In addition to the encouraging results obtained, the proposed method's processing time per page is very competitive and does not rely on a detailed labeled data. Both characteristics are necessary for processing large scale archives.

In the future, we plan to extend this method to be able to extract the layout of a document in a more hierarchical way, where the relationship between regions must be extracted along with the layout.

Acknowledgment. This work was partially supported by the Universitat Politècnica de Valècia under grant FPI-420II/899 (PAID-01-18), by the BBVA Foundation trough the 2017–2018 and 2018–2019 Digital Humanities research grants "Carabela" and "Hist-Weather-Dos siglos de Datos Climáticos", the History Of Medieval Europe (HOME) project (Ref.: PCI2018-093122) and through the EU project READ (Horizon-2020 program, grant Ref. 674943). NVIDIA Corporation kindly donated a Titan X GPU used for this research.

References

1. Burgoyne, J.A., Ouyang, Y., Himmelman, T., Devaney, J., Pugin, L., Fujinaga, I.: Lyric extraction and recognition on digital images of early music sources. In: Proceedings of the 10th International Society for Music Information Retrieval Conference, vol. 10, pp. 723–727 (2009)
2. Calvo-Zaragoza, J., Toselli, A.H., Vidal, E.: Probabilistic music-symbol spotting in handwritten scores. In: 16th International Conference on Frontiers in Handwriting Recognition (ICFHR), pp. 558–563, August 2018
3. Calvo-Zaragoza, J., Zhang, K., Saleh, Z., Vigliensoni, G., Fujinaga, I.: Music document layout analysis through machine learning and human feedback. In: 14th IAPR International Conference on Document Analysis and Recognition (ICDAR), vol. 02, pp. 23–24, November 2017
4. Calvo-Zaragoza, J., Castellanos, F.J., Vigliensoni, G., Fujinaga, I.: Deep neural networks for document processing of music score images. Appl. Sci. 8(5), 654 (2018). (2076-3417)
5. Calvo-Zaragoza, J., Toselli, A.H., Vidal, E.: Handwritten music recognition for mensural notation: formulation, data and baseline results. In: 14th IAPR International Conference on Document Analysis and Recognition (ICDAR), vol. 1, pp. 1081–1086. IEEE (2017)
6. Campos, V.B., Calvo-Zaragoza, J., Toselli, A.H., Ruiz, E.V.: Sheet music statistical layout analysis. In: 15th International Conference on Frontiers in Handwriting Recognition (ICFHR), pp. 313–318. IEEE (2016)

7. Castellanos, F.J., Calvo-Zaragoza, J., Vigliensoni, G., Fujinaga, I.: Document analysis of music score images with selectional auto-encoders. In: 19th International Society for Music Information Retrieval Conference, pp. 256–263 (2018)
8. Grüning, T., Labahn, R., Diem, M., Kleber, F., Fiel, S.: READ-BAD: a new dataset and evaluation scheme for baseline detection in archival documents. CoRR abs/1705.03311 (2017). http://arxiv.org/abs/1705.03311
9. Kingma, D.P., Ba, J.: Adam: a method for stochastic optimization. In: 3rd International Conference on Learning Representations (ICLR) (2015)
10. Long, J., Shelhamer, E., Darrell, T.: Fully convolutional networks for semantic segmentation. In: Proceedings of the IEEE Conference on Computer Vision and Pattern Recognition, pp. 3431–3440 (2015)
11. Quirós, L.: Multi-task handwritten document layout analysis. ArXiv e-prints, 1806.08852 (2018). https://arxiv.org/abs/1806.08852
12. Quirós, L., Bosch, V., Serrano, L., Toselli, A.H., Vidal, E.: From HMMs to RNNs: computer-assisted transcription of a handwritten notarial records collection. In: 16th International Conference on Frontiers in Handwriting Recognition (ICFHR), pp. 116–121. IEEE, August 2018
13. Rebelo, A., Fujinaga, I., Paszkiewicz, F., Marcal, A.R., Guedes, C., Cardoso, J.S.: Optical music recognition: state-of-the-art and open issues. Int. J. Multimed. Inf. Retrieval 1(3), 173–190 (2012)
14. Sánchez, J.A., Romero, V., Toselli, A.H., Villegas, M., Vidal, E.: ICDAR2017 competition on handwritten text recognition on the READ dataset. In: 14th IAPR International Conference on Document Analysis and Recognition (ICDAR), vol. 1, pp. 1383–1388. IEEE (2017)
15. Suzuki, S., et al.: Topological structural analysis of digitized binary images by border following. Comput. Vis. Graph. Image Process. 30(1), 32–46 (1985)

Domain Adaptation for Handwritten Symbol Recognition: A Case of Study in Old Music Manuscripts

Tudor N. Mateiu$^{(\boxtimes)}$, Antonio-Javier Gallego, and Jorge Calvo-Zaragoza

Department of Software and Computing Systems, University of Alicante,
Carretera de San Vicente, s/n, 03690 Alicante, Spain
tnm5@alu.ua.es, {jgallego,jcalvo}@dlsi.ua.es

Abstract. The existence of a large amount of untranscribed music manuscripts has caused initiatives that use Machine Learning (ML) for Optical Music Recognition, in order to efficiently transcribe the music sources into a machine-readable format. Although most music manuscript are similar in nature, they inevitably vary from one another. This fact can negatively influence the complexity of the classification task because most ML models fail to transfer their knowledge from one domain to another, thereby requiring learning from scratch on new domains after manually labeling new data. This work studies the ability of a Domain Adversarial Neural Network for domain adaptation in the context of classifying handwritten music symbols. The main idea is to exploit the knowledge of a specific manuscript to classify symbols from different (unlabeled) manuscripts. The reported results are promising, obtaining a substantial improvement over a conventional Convolutional Neural Network approach, which can be used as a basis for future research.

Keywords: Optical Music Recognition · Domain Adaptation · Convolutional Neural Network · Handwritten music symbols

1 Introduction

Optical Music Recognition (OMR) refers to the field of research that studies how to make computers be able to read music notation [1]. OMR promises great benefits to the digital humanities, making accessible and browsable the musical heritage that only exists as written copies distributed all over the world [6].

As in many other fields, modern Machine Learning techniques, such as Deep Neural Networks, have brought significant improvements in OMR [3,8,9]. However, the supervised learning framework assumes that there is an adequate training set from which to learn, and that the system will be used in data that has been generated by the same distribution [5]. Due to the particularities of the musical cultural heritage, this scenario is not that interesting: there are many small-scale musical manuscripts with different graphical characteristics. Given

© Springer Nature Switzerland AG 2019
A. Morales et al. (Eds.): IbPRIA 2019, LNCS 11868, pp. 135–146, 2019.
https://doi.org/10.1007/978-3-030-31321-0_12

the artistic scope of the application domain, it is difficult to indicate exactly what differentiates manuscripts among themselves, but we can roughly add it up to a number of these factors: authors' handwriting style, engraving mechanism, type and color of the paper/parchment, color of the ink, and an amount of possible deterioration, among others. This leads to the need of building a training set from each manuscript to attain reliable results, which ends up in a rather inefficient workflow.

That is why in this work we want to study the use of Domain Adaptation (DA) techniques in the context of music manuscript recognition. DA refers to a scenario in which we want to classify data that come from a distribution (or domain) different from the data used to train, although the set of classification labels is the same [2]. Specifically, we assume the case of semi-supervised DA where data from the new (target) domain is available but not labeled. This represents the typical case of musical manuscripts: it is easy to obtain the images of the manuscripts to be transcribed but costly to annotate them conveniently.

In this first work studying this issue for music manuscripts, we focus on DA for the classification of music-notation symbols. This stage is typically considered within the standard workflow for designing OMR systems [10]. Given a new manuscript, state-of-the-art techniques can be used to detect isolated symbols [4] or assume an interactive environment where the user is responsible for locating the symbols manually with ergonomic interfaces [11].

We conduct comprehensive experiments over five different manuscripts of early music with a state-of-the-art neural architecture for semi-supervised DA. We will evaluate the different parameters for the configuration and training of the neural network, and we will compare the results with those obtained by a model that does not use DA. Our results yield interesting conclusions about the type of adaptation that can be achieved and what are the conditions for it to happen. Eventually, the DA results report a significant improvement over the conventional methods. This outcome can be used as a starting point for future models and a comparative baseline for possible improvements of this type of techniques in the context of document image analysis.

In the rest of the paper, we present the considered methodology (Sect. 2), our experimental setup (Sect. 3), the obtained results and their main outcomes (Sect. 4), and the obtained conclusions along with some avenues for future research (Sect. 5).

2   Methodology

For supervised learning classification algorithms, given X, the input space, and Y, the output space (or label space), we have a source domain D_S over $X \times Y$ from which a labeled set $S = \{(x_i, y_i)\}_{i=1}^{N} \sim (D_S)^N$ is $i.i.d$ drawn, where N is the total number of samples. The objective for these algorithms is to learn a mathematical model, or hypothesis function, $h : X \to Y$, so that the labels of the new samples are predicted with as little error as possible, thus building what is known as a label classifier. Given that the goal of this work is to study the transfer of knowledge from a label classifier to a different domain (or in other words, study the task of DA), we must build a model with certain requirements.

In our scenario, we have two different domains called the *source domain* D_S and the *target domain* D_T, both being distributions over $X \times Y$. The DA learning algorithm will be provided with a labeled source sample S and unlabeled target sample T drawn *i.i.d.* from D_S and from D_T, respectively, $S = \{(x_i, y_i)\}_{i=1}^{n} \sim (D_S)^n$; $T = \{(x_i)\}_{i=1}^{N-n} \sim (D_T)^{n'}$, with $N = n + n'$ being the total number of samples.

Eventually, the goal of the DA algorithm is to build a label classifier, that, just like conventional Convolutional Neural Networks (CNN) classification models, given an input x is able to predict its label y, with the difference being that the input will be drawn from the target domain and the hypothesis must be obtained by applying the previously explained requirements.

2.1 Domain Adversarial Neural Network

The architecture experimented on, the Domain Adversarial Neural Network (DANN), was firstly proposed in the work of [7]. It consists of three parts, two of which are common in any standard feed-forward CNN model: the *feature extractor* and *label classifier* (or *label predictor*), as seen in Fig. 1.

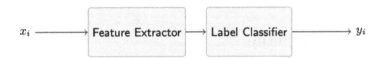

Fig. 1. Basic CNN classification model diagram.

In order for the classification decisions of the label classifier to be made based on features that are both discriminative and invariant to the change of domains, a *domain classifier*, which contains a *Gradient Reversal Layer*, must be added to the model (see Fig. 2).

The Gradient Reversal Layer from the domain classifier multiplies the feature extractor's gradient by a specified negative weight during the back-propagation training. This prevents the training to be performed in a standard way and will carry out what we seek: to maximize the loss of the domain classifier, which in turn ensures domain invariant features to emerge, disabling the model from learning which domain the input belongs to.

These classifiers will work in tandem so that the label classifier is optimized to *minimize* the loss of label predictions, and the domain classifier is optimized to *maximize* the loss of the current domain prediction, assuring that a *domain invariance* exists during the course of learning. Finally, the feature extractor will learn domain invariant features, so we can use the label classifier to classify samples from both the source domain D_S and the target domain D_T.

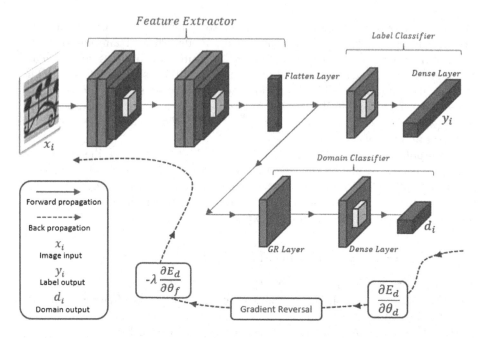

Fig. 2. The considered DANN architecture. The Gradient Reversal Layer (GR Layer), which will maximize domain invariance, is situated in the Domain Classifier structure.

3   Experimental Setup

3.1   Network Configuration

The DANN architecture (see Fig. 2) receives an input image of size 40×40 pixels and transfers it to the Feature Extractor block. The Feature Extractor is composed by two 2D convolutional layers, and ends with a Flatten layer that converts the output of the Feature Extractor into a vector. Then it follows a bifurcation that leads separately to the Label Classifier and the Domain Classifier. Below we will explain the configuration of each of these parts of the network model.

Let $Conv2D(f, (k_1, k_2))$ be a convolutional layer with f filters and kernel size of $k_1 \times k_2$, $MaxPooling((p_1, p_2))$ be a max-pooling operation layer with pool size $p_1 \times p_2$, $Dense(u, a)$ be a dense layer with u units and activation a, and $Dropout(d)$ be a dropout operation layer with a dropout ratio of d. Then the specific configuration of the DANN used in this work is as follows:

– The Feature Extractor considers two blocks, each of which is set as:
$Conv2D(32, (3, 3))$ → $Conv2D(64, (3, 3))$ → $MaxPooling((2, 2))$ → $Dropout(0.25)$ → $Conv2D(64, (3, 3))$ → $Conv2D(64, (3, 3))$ → $MaxPooling((2, 2))$ → $Dropout(0.3)$.

- The Label Classifier is built as: $Dense(128, ReLU) \rightarrow Dropout(0.5) \rightarrow Dense(8, softmax)$, with output being the symbol categories (more details will be given in the experimental section).
- The Domain Classifier consists of: $GRLayer() \rightarrow Dense(128, ReLU) \rightarrow Dropout(0.5) \rightarrow Dense(2, softmax)$, with output being two domains, $d_i \in \{0, 1\}$.

To train the Label Classifier, the standard back-propagation algorithm is used using *Adadelta* as optimizer [12] during 100 epochs. The gradient needed to calculate the weights of the model to classify labels is performed by computing the derivative $\partial E_y / \partial w_y$ of the Label Classifier and transferring it to the Feature Extractor, so the derivative $\partial E_y / \partial w_f$ is calculated in this part of the network, where E_y is the classifier's loss for y_i, and w_y and w_f are the weights of the classifier and feature extractor respectively.

Obtaining the gradient for the model to classify domains is different because of the Gradient Reversal Layer, which will multiply the Domain Classifier's derivative, $\partial E_d / \partial \theta_d$, by a certain negative number, λ, as seen in Fig. 2. By using this negative λ parameter, we force that the features over two different domains are as indistinguishable as possible for inferring the domain of the input, thereby obtaining the desired domain invariance.

3.2 Data and Training

The experimentation was conducted using five sets of music manuscripts from different domains (see Fig. 3). To carry out the experiments, we must first take into account how the data was filtered from these five manuscripts, the amount of data used, the characteristics that describe each domain, and how training has been carried out.

Table 1 shows a summary of the datasets used, as well as their type and number of samples they contain. The initial amount of data (column "Total symbols") contains some labels that are not common in all datasets.

To solve this, the symbol category label sets from each domain have been all intersected in order to obtain a list with only the categories common to all manuscripts, reducing the amount to 15. Additionally, categories which add up to less than 15 elements in all manuscripts have been removed as well, further reducing the amount of categories to 8. This brings us to the final amount of data that will be used for experimentation (see column "Filtered symbols").

As previously mentioned, each of the five manuscripts can vary between each other because of different factors, and, for the most part, these factors are present in all of them creating, like this, five domains. Firstly, data from domains b-3-28 through b-59-850 have been obtained from handwritten manuscripts, while BNE-BDH has been typewritten. Secondly, the manuscripts pose varied characteristics, for example, b-50-747 has a lower resolution so it is more blurry, and b-59-850 has an overexposure or very high intensity of light. Thirdly, manuscripts b-3-28, b-50-747, b-59-850 and BNE-BDH have not been altered by adding any type of additional characteristic, but a distinct type of characteristic was added

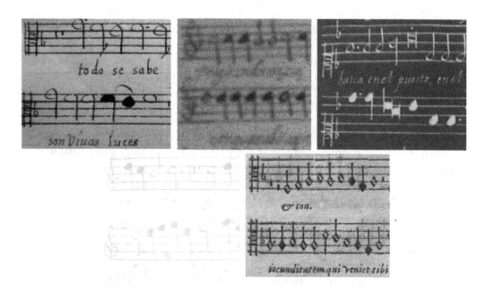

Fig. 3. All five manuscripts used for experimentation. One row at a time, from left to right: b-3-28, b-50-747, b-53-781, b-59-850, BNE-BDH.

Table 1. Description of the five datasets used in the experimentation.

Domain	Type	Total symbols	Filtered symbols
b-3-28	Handwritten	2581	1760
b-50-747	Handwritten	3191	2587
b-53-781	Handwritten	825	570
b-59-850	Handwritten	8467	5225
BNE-BDH	Typewritten	1356	1199

to b-53-781, with the aim of increasing the difficulty and unpredictability of the state in which future manuscripts can be provided, as the DANN model must be robust to these types of alterations and must not require additional manual labor, e.g. image preprocessing, in order to learn and transfer its knowledge. Finally, a synthetic characteristic, which inverts the colors, was manually applied to manuscript b-53-781, which anyway could represent a different scanner mechanism.

Each manuscript's data is split into 80% for training and 20% for validation. However, for training the DANN, two different sets are required. The first, the *training set* T, is the conventional set created from the sample S (source domain), which contains input and label pairs (x_i, y_i). The second, the *domain set* D, is comprised of data from the input samples S and T, along with its domain-label pairs (x_i, d_i), where d_i indicates from which domain it originates (e.g., 0 indicates that it comes from the source domain and 1 indicates that it comes from the target domain).

Given that the DANN model makes use of two specific classifiers which share the weights of a part of the network, a problem arises during its training. If it is trained in a conventional manner just like a CNN model, after fully training one classifier, its knowledge might be lost or become invalid after doing the same to the second classifier. To solve this, the model will make use of a form of pseudo-concurrent training, using small, equally-sized batches obtained from the domain set $\mathcal{D}$ to train one and then the other (e.g., an epoch is comprised of X images, and, using batches of size b, it will be trained X/b times for each epoch). The label classifier will make use of the *training set* $\mathcal{T}$, while the domain classifier will use the *domain set* $\mathcal{D}$.

Experimentation followed a meticulous process with different iterations, where one iteration means all possible permutations by electing each manuscript once as the source domain and the others as the target domain. Additionally, these training iterations were carried out using different values for parameters such as the pseudo-concurrency batch size, values of λ of the gradient reversal layer, the label classifier's learning rate and the domain classifier's learning rate.

4 Results

The experimentation has been carried out in two stages, the first stage studies how the general tendency varies by doing a complete search on the previously mentioned parameters, the batch size, gradient reversal layer's λ, and learning rates, using the values shown in Table 2. The second stage of experimentation carries out tests across all possible manuscript permutations of source-domain using in turn every possible permutation of the parameters. Additionally, we name "D_s Acc." and "D_t Acc." as the average source and target domain accuracy in percentage.

Table 2. Set of values used for the training parameters. Both the CNN and DANN models will use the batch size and the classifier learning rates during the training of the respective models. The λ parameter only affects the DANN model.

Parameter	Values
Batch size	$\{16, 32, 64, 128, 256, 512\}$
λ	$\{0.5, 1.0, 1.5, 2.0\}$
Classifier learning rates	$\{0.5, 1.0, 1.5\}$

Results obtained for the first stage are shown in separated tables, where, for each possible value of the parameter at issue, the average source and target domain accuracies for the CNN and DANN models are shown, when possible. These results are the average obtained as a consequence of carrying out experiments across all possible manuscript permutations of source-target domains, using in turn every possible permutation of the considered parameters. This means that a total of 8640 experiments with different configurations have been carried out: 20 possible manuscript combinations × 2 different models (CNN and DANN) × 6 batch sizes × 4 λ values × 3 label classifier LR × 3 domain classifier LR. These results will allow us to get an idea of how the parameters affect the general trend of the results.

The tendency for the batch size, Table 3, is for results to decrease substantially as the batch size increases. This fulfills what was previously mentioned about the need to use fairly small sizes, as training one classifier with a large batch will undermine the training of the other. The same tendency arises if we use batches that are too small, since in this case the weights are updated using very few samples. The best results tend to originate from a batch size of 64 samples.

Table 3. Influence of batch sizes in the performance of the CNN and DANN architectures.

Batch	CNN		DANN	
	D_s Acc.	D_t Acc.	D_s Acc.	D_t Acc.
16	96.06	18.25	72.20	37.67
32	96.39	18.89	77.60	39.11
64	96.39	18.46	81.30	39.74
128	96.25	18.67	82.10	39.61
256	96.00	19.36	82.78	36.73
512	95.13	19.48	80.15	35.88

The λ parameter (see Table 4) is the variable by which the Gradient Reversal Layer multiplies the derivative of the domain classifier in order for the label classifier to subsequently be trained invariantly to domains. This parameter is equivalent to the learning rate parameter of the back-propagation algorithm, but applied to the learning of the domain invariant characteristics by the Feature Extractor block. Note that this is a particular parameter to the DANN model, and so only its results can be reported. They show a general tendency to obtain better results for small values of λ, so in this case it is better to use a small factor to learn the domain invariant characteristics. This may be due to the fact that if a high value is used, the characteristics shared by the two networks are more adjusted for domain detection, spoiling the result obtained by the label classifier (and undoing what it had already learned). And therefore, it is better to adjust the weights little by little in each iteration, to reach a balance between the two networks.

In Tables 5 and 6, the results obtained by varying the learning rates (LR) of the two networks are analyzed. The general tendency for the best LR values varies between the two classifiers. Table 5 shows the influence of the label classifier's learning rate in the performance of the CNN and DANN architecture.

Table 4. Influence of λ values in the performance of the DANN architecture.

λ	DANN	
	D_s Acc.	D_t Acc.
−0.5	86.77	39.53
−1.0	80.30	38.62
−1.5	76.57	37.24
−2.0	73.83	37.10

In this case, having a LR of 0.5 for the label classifier obtains the best results. Table 6 shows the influence of the domain classifier's learning rate in the performance of the DANN architecture. In this other case, having a LR of 1.0 for the domain classifier obtains the best results. Both classifiers obtain bad results for a high value of their respective LR. As previously argued, this can be motivated by the fact that having a shared weight section, it is better to modify the weights little by little in each iteration. Therefore, it seems that, on average, it is better to use a domain learning rate value slightly higher than that used for the labels.

Table 5. Influence of the label classifier's learning rate in the performance of the CNN and DANN architecture.

LR1	CNN		DANN	
	D_s Acc.	D_t Acc.	D_s Acc.	D_t Acc.
0.5	95.58	20.33	88.36	39.76
1.0	96.21	18.97	79.49	38.89
1.5	96.32	17.27	70.11	35.69

Table 6. Influence of the domain classifier's learning rate in the performance of the DANN architecture.

LR2	DANN	
	D_s Acc.	D_t Acc.
0.5	79.43	37.81
1.0	79.39	38.70
1.5	39.27	37.86

The results for the second stage of the experimentation are shown in Table 7. It reports the results obtained by the CNN and DANN networks for all possible combinations of the five datasets used. In addition, the "target diff" column is added, which shows the difference between the accuracy for the target domain obtained by the DANN and the CNN.

Table 7. Best results obtained for the different combinations of the datasets used as source and target domains. Target Diff column shows DANN's D_t accuracy minus CNN's D_t accuracy.

Source	Target	CNN		DANN		Target diff.
		D_s Acc.	D_t Acc.	D_s Acc.	D_t Acc.	
BNE-BDH	b-3-28	95.83	20.17	88.75	50.28	30.11
	b-50-747	95.83	14.67	92.92	42.08	27.41
	b-53-781	95.83	4.39	90.83	65.79	61.40
	b-59-850	96.25	23.83	90.42	52.82	28.99
b-3-28	BNE-BDH	97.16	37.08	95.17	40.42	3.34
	b-50-747	96.88	73.55	96.02	79.34	5.79
	b-53-781	97.16	16.67	96.02	95.61	78.94
	b-59-850	97.16	53.88	96.02	91.10	37.22
b-50-747	BNE-BDH	97.30	10.42	92.28	32.50	22.08
	b-3-28	97.30	66.48	92.66	59.38	−7.10
	b-53-781	97.68	0.88	94.02	85.09	84.21
	b-59-850	97.49	32.34	95.37	85.84	53.50
b-53-781	BNE-BDH	97.37	35.00	97.37	35.42	0.42
	b-3-28	97.37	8.81	97.37	82.39	73.58
	b-50-747	97.37	4.39	97.37	67.76	60.42
	b-59-850	97.37	23.83	97.37	91.77	64.78
b-59-850	BNE-BDH	99.52	7.50	99.14	29.58	22.08
	b-3-28	99.52	79.55	99.04	85.80	6.25
	b-50-747	99.62	62.17	99.23	80.50	18.34
	b-53-781	99.52	11.40	99.14	95.61	84.21
Average		97.48	29.66	95.33	67.45	37.80

Results for the typewritten manuscript (BNE-BDH) as source and target are promising, as they have good increases in DANN accuracies compared to the CNN ones. An anomaly occurs with b-3-28 and b-53-781 as source domains and BNE-BDH as target, where we obtained results that do not outperform the CNN classifier, given that the target differences are 3.34% and 0.42% respectively. This may be due to the number of samples in the source dataset, since they represent the two datasets with fewer samples.

It is also observed that the pair b-3-28 and b-50-747 obtains poor results in the two possible combinations (source, target), with the following target accuracy difference: 5.79% (b-3-28 as source) and −7.10% (b-50-747 as source). In this case, these worse results are probably caused by the similarity of the domains. In Fig. 3, one can see how these two domains are those that present the most similar writing, only with differences in the overall color of the image. In general, it has been observed that DANN architecture requires the source and target domains to have significant differences, since, if this is not the case, it forces the label classifier to be trained with domain invariance, which results in worse or similar accuracies as the CNN architecture. Note that, if the domains are similar, the DANN is actually modifying the features that are already suitable for both.

Additionally, the opposite happens when there is too much difference between domains, as can be seen whenever b-53-781 is used as the target domain (see Fig. 3). Using b-50-747 and b-59-850 as sources, the DANN architecture obtains the experiments' maximum target accuracy difference: 84.21%.

As a summary, the proposed DANN architecture obtains on average 67.45% target accuracy, which is a 37.80% increase on average than the CNN architecture.

5   Conclusions

This work focuses on the study of the use of DA techniques in the context of musical manuscripts recognition. These techniques dictate that a domain invariance must exist during training so classification decisions can be made based on features that are both discriminative for labels, yet invariant to the change of domains. We implemented an existing DANN architecture, as one of the parts it is comprised of, the domain classifier, includes a gradient reversal layer that will, during back-propagation, ensure these domain invariant features to emerge.

The evaluation of the DANN architecture is carried out by firstly studying how different values of training parameters can affect the general tendency, and then studying the average target domain accuracies using permutations for the five used manuscripts. The parameters evaluated are the training batch size, the gradient reversal layer λ, and the learning rates for both label and domain classifiers.

The classification performance obtained by the proposed architecture for the target domain generally outperforms a CNN approach, as our implementation results in an average of 67.45%, an increase of 37.80% from the CNN's 29.66%. It should be kept in mind that these classification results (of almost 70% on average), are obtained without using any label from target domain; that is, only adapting the knowledge learned from the source domain. It is also worthwhile mentioning the result obtained for the adaptation between typewritten and handwritten domains, reaching in some cases a 66% of accuracy, 61% better than with a CNN.

Future work includes experiments to improve these results with a greater amount of datasets and domains, increasing the number of labels considered. We also intend to evaluate different strategies combining this approach with semi-supervised and incremental methods. Additionally, we would like to extend this research to work with the direct detection of the music symbols in the images [9], instead of assuming a previous segmentation.

Acknowledgements. This work is supported by the Spanish Ministry HISPAMUS project TIN2017-86576-R, partially funded by the EU.

References

1. Bainbridge, D., Bell, T.: The challenge of optical music recognition. Comput. Humanit. **35**(2), 95–121 (2001)
2. Ben-David, S., Blitzer, J., Crammer, K., Kulesza, A., Pereira, F., Vaughan, J.W.: A theory of learning from different domains. Mach. Learn. **79**(1–2), 151–175 (2010)
3. Calvo-Zaragoza, J., Gallego, A.J., Pertusa, A.: Recognition of handwritten music symbols with convolutional neural codes. In: 14th International Conference on Document Analysis and Recognition, Kyoto, Japan, pp. 691–696 (2017)
4. Castellanos, F.J., Calvo-Zaragoza, J., Vigliensoni, G., Fujinaga, I.: Document analysis of music score images with selectional auto-encoders. In: 19th International Society for Music Information Retrieval Conference, Paris, France, pp. 256–263 (2018)
5. Duda, R.O., Hart, P.E., Stork, D.G.: Pattern Classification, 2nd edn. Wiley, New York (2001)
6. Fujinaga, I., Hankinson, A., Cumming, J.E.: Introduction to SIMSSA (single interface for music score searching and analysis). In: 1st International Workshop on Digital Libraries for Musicology, pp. 1–3. ACM (2014)
7. Ganin, Y., Lempitsky, V.S.: Unsupervised domain adaptation by backpropagation. In: Proceedings of the 32nd International Conference on Machine Learning, pp. 1180–1189 (2015)
8. Hajič Jr., J., Dorfer, M., Widmer, G., Pecina, P.: Towards full-pipeline handwritten OMR with musical symbol detection by U-Nets. In: 19th International Society for Music Information Retrieval Conference, Paris, France, pp. 225–232 (2018)
9. Pacha, A., Calvo-Zaragoza, J.: Optical music recognition in mensural notation with region-based convolutional neural networks. In: 19th International Society for Music Information Retrieval Conference, Paris, France, pp. 240–247 (2018)
10. Rebelo, A., Fujinaga, I., Paszkiewicz, F., Marcal, A.R., Guedes, C., Cardoso, J.D.S.: Optical music recognition: state-of-the-art and open issues. Int. J. Multimed. Inf. Retr. **1**(3), 173–190 (2012)
11. Rizo, D., Calvo-Zaragoza, J., Iñesta, J.M.: MuRET: a music recognition, encoding, and transcription tool. In: 5th International Conference on Digital Libraries for Musicology, pp. 52–56. ACM, Paris (2018)
12. Zeiler, M.D.: ADADELTA: an adaptive learning rate method. The Computing Research Repository (CoRR) abs/1212.5701 (2012)

Approaching End-to-End Optical Music Recognition for Homophonic Scores

María Alfaro-Contreras$^{(\boxtimes)}$, Jorge Calvo-Zaragoza, and José M. Iñesta

Software and Computing Systems, University of Alicante, Alicante, Spain
mac77@alu.ua.es, {jcalvo,inesta}@dlsi.ua.es
http://grfia.dlsi.ua.es

Abstract. The recognition of patterns that have a time dependency is common in areas like speech recognition or natural language processing. The equivalent situation in image analysis is present in tasks like text or video recognition. Recently, Recurrent Neural Networks (RNN) have been broadly applied to solve these task with good results in an end-to-end fashion. However, its application to Optical Music Recognition (OMR) is not so straightforward due to the presence of different elements at the same horizontal position, disrupting the linear flow of the time line. In this paper we study the ability of the RNNs to learn codes that represent this disruption in homophonic scores. The results prove that our serialized ways of encoding the music content are appropriate for Deep Learning-based OMR and they deserve further study.

Keywords: Optical Music Recognition · Deep Learning ·
End-to-end recognition · Music encoding

1 Introduction

Optical Music Recognition (OMR) is the field of research that investigates how to computationally decode music notation in document images. It aims at converting the large number of existing written musical sources into a codified format that allows for its computational process [2]. There are many documents in private and public archives, sometimes hidden from public, waiting for being digitized. There are also many digitized funds in specialized portals that are only available as images, without the possibility of study or content search. Since music typesetting is a tedious and expensive process, OMR represents an alternative to efficiently deal with this scenario.

Traditional approaches to OMR [2,14] are based on the usual pipeline of subtasks that characterizes many artificial vision systems, adapted to this particular task: image pre-processing, individual music-object detection, reconstruction of the music semantics by using specific knowledge of the field, and output encoding in a suitable symbolic format.

Recent advances in Machine Learning—namely Deep Learning (DL)—which have achieved great results in similar tasks such as text recognition, allow us

© Springer Nature Switzerland AG 2019
A. Morales et al. (Eds.): IbPRIA 2019, LNCS 11868, pp. 147–158, 2019.
https://doi.org/10.1007/978-3-030-31321-0_13

to be optimistic about developing more accurate OMR systems. The current trends in these fields is the use of end-to-end (or holistic) systems, that face the process in a single stage without explicitly taking into account the necessary sub-steps. To develop these approaches, only training pairs are needed, consisting of problem images, together with their corresponding transcript solutions [3,6].

Due to design reasons, these approaches—typically based on Recurrent Neural Networks or Hidden Markov Models—are only able to formulate the output of the system as one-dimensional sequences. This perfectly fits for natural language tasks (text or speech recognition, or machine translation), since their outputs mostly consists of character (or word) sequences. However, its application to musical notation is not so straightforward due to the presence of different elements sharing horizontal position and skipped relationships. The vertical distribution of these elements disrupts the linear flow of the time line (see Fig. 1). This fact is not trivial to codify and can cause important difficulties in the performance of the recognition systems that make use of the temporal relationships of the recognized elements.

Fig. 1. The recognition process does not follow a linear left-to-right flow.

Although the problem can be drastically simplified by considering that the process will work with each staff independently from the others—a process that could be analogous to the text recognition systems that decompose the document into a series of independent lines—we still have to deal with elements that take place simultaneously in the "time" line, like the notes that make up a chord, irregular groups, or the expression marks, to name a few.

Within the range of music score complexities, one possible simplification of the problem that applies to many sheet music is to assume a *homophonic* music context. In that case, there are multiple parts but they move in the same rhythm. This way, multiple notes can occur simultaneously, but only as a single voice. This way, all the notes starting at the same time last the same, so the score can be segmented in vertical slices that may contain one or more music symbols (see Fig. 2).

Fig. 2. In homophonic music, all the notes starting at the same time last the same.

Even in this simplified context, there is a need of a clear and structured output coding that avoids the ambiguities that the representation of a linear output can show in presence of vertical structures in the data (see Fig. 3).

clefG:L2 digit.4:L2 digit.4:L4 note.half:S1 note.half:S2 note.half:S3 verticalLine:L1

Fig. 3. Ambiguities appear when symbols are stacked. When two notes appear together, that must be played at the same time, a linear symbol sequence without specific marks can be interpreted in more than one way.

There already exist a number of structured formats for music representation and coding, like XML-based music formats [8,11] that are focused on how the score has to be encoded to properly store all its content. That application makes it inappropriate for adopting them as output for an optical recognition system, because the code is plenty of irrelevant marks for the system to generate when it is recognizing the score content (mainly which symbols and where are they in the score).

Due to that, we have designed a specific language to represent an appropriate output for end-to-end OMR, based on serializing the music symbols found in a staff of homophonic music. The sequential nature of music reading must be compatible with the representation of the vertical alignments of some symbols. In addition, this representation has to be easy to generate by the system, which analyzes the input sequentially and produces a linear series of symbols.

The rest of the paper is structured as follows: Sect. 2 describes the recognition framework based on DL, including the ad-hoc serializations for homophonic music; Sect. 3 introduces the experimental setup followed to validate the approach; Sect. 4 shows the results obtained in a controlled scenario, as well as qualitative results with real homophonic sheet music; finally, Sect. 5 concludes the present work, along with some ideas for future research.

2   Recognition Framework

To carry out the OMR task in an end-to-end manner, we use a Convolutional Recurrent Neural Network (CRNN) that permits us to model the posterior probability of generating output symbols, given an input image. Input images are assumed to be single staff-sections, analogously to text recognition that assumes independent lines [15]. This is not a strong assumption, as staves can be easily isolated by means of existing methods [7].

A CRNN consists of one block of *convolutional* layers followed by another block of *recurrent* layers [16]. The convolutional block is responsible for learning how to process the input image, that is, extracting relevant image features for the task at issue, so that the recurrent layers interpret these features in terms of sequences of musical symbols. In our work, the recurrent layers are implemented as Bidirectional Long Short Term Memory (BLSTM) units [9].

The activations of the last convolutional layer can be seen as a sequence of feature vectors representing the input image, $\mathbf{x}$. These features are fed to the first BLSTM layer and the unit activations of the last recurrent layer are considered estimates of the posterior probabilities for each vector:

$$P(\sigma|\mathbf{x}, f), \ 1 \leq f \leq F, \ \sigma \in \Sigma \tag{1}$$

where F is the number of feature vectors of the input sequence and Σ is the set of considered symbols, that must include a *"non-character"* symbol required for images that contain two or more consecutive instances of the same musical symbol [9].

Since both convolutional and recurrent blocks can be trained through gradient descent, using the well-known *Back Propagation* algorithm [17], a CRNN can be jointly trained. However, a conventional end-to-end OMR training set only provides, for each staff image, its corresponding transcription, not giving any type of explicit information about the location of the symbols in the image. It has been shown that the CRNN can be conveniently trained without this information by using the so called "Connectionist Temporal Classification" (CTC) loss function [10]. The resulting CTC training procedure is a form of Expectation-Maximization, similar to the backward-forward algorithm used for training Hidden Markov Models [13]. In other words, CTC provides a means to optimize the CRNN parameters so that is likely to give the correct sequence given an input. The use of the aforementioned *"non-character"* symbol to indicate a separation between symbols is considered essential for adequate CTC training [10].

Once the CRNN has been trained, an input staff image can be decoded into a sequence of music symbols $\hat{\mathbf{s}} \in \Sigma^*$. First, the most probable symbol per frame is computed:

$$\hat{\sigma}_i = \arg \max_{\sigma \in \Sigma} P(\sigma|\mathbf{x}, i), \ 1 \leq i \leq F$$

Then, a pseudo-optimal output sequence is obtained as:

$$\hat{\mathbf{s}} = \arg \max_{s \in \Sigma^*} P(\mathbf{s}|\mathbf{x}) \approx \mathcal{D}(\hat{\sigma}_1, \ldots, \hat{\sigma}_F) \tag{2}$$

where $\mathcal{D}$ is a function that first merges all the consecutive frames with the same symbol, and then deletes the "non-character" symbol [9].

This framework is equal to the one used in text recognition tasks [16], whose expressiveness could be sufficient when working with simples scores where all the symbols have a single left-to-right order. However, as introduced above, we want to extend these approaches so that they are able to model richer scores such as those of homophonic sheet music. In such case, issues like chords may appear, where several symbols share a vertical position. As seen in Fig. 3, a one-dimensional sequence is not expressive enough for this. That is why in the next section we describe our coding proposal to perform end-to-end OMR for homophonic scores.

2.1 Serialization Proposals

Our current research involves the study of four different deterministic, unambiguous and serialized representations to encode the kind of scenarios that happen in homophonic music so that the OMR system becomes more effective when recognizing complex music score images. For that we propose four different types of music representations that differ not in the encoding of the musical symbols themselves but in the way horizontal and vertical distributions of the musical symbols are represented. The grammar for these musical codifications must be deterministic and unambiguous, allowing to analyze a given document in only one way.

Our representation does not make assumptions about the musical meaning of what is represented in the document being analyzed, that is, the elements are identified in a catalog of musical symbols by the shape they have and where they are placed in the score. This has been referred to as "agnostic representation", as opposed to a semantic representation where music symbols are encoded according to their actual music meaning [5].

As aforementioned, the only difference between the four proposed musical codes is how to represent the horizontal and vertical dimensions. Each one of the four codes will have one or two characters that indicate whether, when transcribing the score, the system should move forward, that is, from left to right, or upwards, from bottom to top.

The four different codes proposed are described as follows:

- **Remain-at-position character code:** when transcribing the score, the different musical symbols are assumed to be placed left to right, except when they are in the same horizontal position. In that case they are separated by a slash, "/". This acts as a remain-at-position character, meaning that the system does not advance forward and it has to advance upwards (see Fig. 4a). This behaviour is similar to the backspace of the typewriters. The carriage advances after typing and if we want to align two symbols we need to keep the carriage in a fixed position (by moving it back one position).
- **Advance position character code:** this type of codification uses a "+" sign to force the system to advance forward. This way, when that sign is missing, the output does not move forward and a vertical distribution is being coded (see Fig. 4b).
- **Parenthesized code:** when a vertical distribution appears in the score, the system outputs a parenthesized structure, like **vertical.start** musical_symbol ... musical_symbol **vertical.end** (see Fig. 4c).

– **Verbose code:** this last codification is a combination of the two first ones. It uses the "+" sign as the advance position character to indicate that the system has to move forward, and the "/" sign as the remain-at-position character to indicate that the system has to advance upwards (see Fig. 4d). So, in this codification, every two adjacent symbols are explicitly separated by a symbol indicating whether the system must remain at the same horizontal position or has to advance to the next one.

Note that the four codes are unambiguous representations of the same data, so they are interchangeable and can be translated among them.

clef.G:L2 accidental.flat:L3 digit.4:L2 / digit.2:L4 rest.eighth:L3 dot:S3 slur.start:S2 / note.quarter:S2 slur.end:S2 / note.sixteenth:S2 verticalLine:L1 note.quarter:L1 / note.quarter:L2 note.beamedRight:S2 note.beamedLeft:L2 verticalLine:L1 note.quarter:L1 / bracket.start-S6 note.quarter:S1 / digit.3-S6 note.quarter:L1 / bracket.end-S6 verticalLine:L1

clef.G:L2 + accidental.flat:L3 + digit.4:L2 digit.2:L4 + rest.eighth:L3 + dot:S3 + slur.start:S2 note.quarter:S2 + slur.end:S2 note.sixteenth:S2 + verticalLine:L1 + note.quarter:L1 note.quarter:L2 + note.beamedRight:S2 + note.beamedLeft:L2 + verticalLine:L1 + note.quarter:L1 bracket.start-S6 + note.quarter:S1 digit.3-S6 + note.quarter:L1 bracket.end-S6 + verticalLine:L1

clef.G:L2 accidental.flat:L3 vertical.start digit.4:L2 digit.2:L4 vertical.end rest.eighth:L3 dot:S3 vertical.start slur.start:S2 note.quarter:S2 vertical.end slur.end:S2 note.sixteenth:S2 vertical.end verticalLine:L1 vertical.start note.quarter:L1 note.quarter:L2 vertical.end note.beamedRight:S2 note.beamedLeft:L2 verticalLine:L1 vertical.start note.quarter:L1 bracket.start-S6 vertical.end vertical.start note.quarter:S1 digit.3-S6 vertical.end vertical.start note.quarter:L1 bracket.end-S6 vertical.end verticalLine:L1

clef.G:L2 + accidental.flat:L3 + digit.4:L2 / digit.2:L4 + rest.eighth:L3 + dot:S3 + slur.start:S2 / note.quarter:S2 + slur.end:S2 / note.sixteenth:S2 + verticalLine:L1 + note.quarter:L1 / note.quarter:L2 + note.beamedRight:S2 + note.beamedLeft:L2 + verticalLine:L1 + note.quarter:L1 / bracket.start-S6 + note.quarter:S1 / digit.3-S6 + note.quarter:L1 / bracket.end-S6 + verticalLine:L1

Fig. 4. Musical excerpt presenting a number of different situations where vertical alignments occur and its transcription using the proposed codifications. From top to bottom: (a) remain-at-position character coding, (b) advance-position character coding, (c) parenthesized coding, and (d) verbose coding.

3   Experimental Setup

In this section, the experimental setup, including both the corpora and the evaluation protocol considered, will be described.

3.1   Corpus Generation

As introduced above, the current tendency for the development of OMR systems is to use machine learning techniques which are able to infer the transcription from correct examples of the task, namely, set of pairs (image, transcription). Given the complexity of music notation, for these techniques to produce satisfactory results it is necessary to use a set of sufficient size. To achieve this, a system of automatic generation of labeled data has been developed [1] by using algorithmic composition techniques [12]. The developed system provides two outputs: on the one hand, the expected transcription of the generated score in any of the encodings described above; on the other hand, the score image in PDF format. With both outputs, the necessary pairs for the machine learning algorithm are obtained.

For the generation system, three different methods of algorithmic composition were implemented, allowing to obtain compositions with disparate musical features. The technical details of those composition methods are out of the scope of the present paper (see [1]), but it is important to know that the operation of each one makes them to produce heterogeneous music scores. Therefore, the three of them have been equally used when generating the training set, so that it is not biased in favour of any particular style.

3.2   Evaluation Protocol

A corpus of 8 000 tagged scores, each consisting of a single staff, has been generated using the system for automatic generation of labeled data for OMR research explained in the previous section. Our aim is to evaluate to what extent the CRNN is able to learn the non-linearities in the time line, and which type of encoding yields the best results in the recognition task. This corpus will be used to train the end-to-end neural network described in Sect. 2. Each sample will be a pair composed of the image with a rendered staff and its corresponding representation with the format imposed by one of the four musical encodings proposed, like in the example shown in Fig. 4.

We consider the following evaluation metrics to measure the recognition performance:

- **Sequence Error Rate, Seq-ER** (%): ratio of incorrectly predicted sequences (the sequence of recognized symbols has at least one error).
- **Symbol Error Rate, Sym-ER** (%): computed as the average number of elementary editing operations (insertions, deletions, or substitutions) necessary to match the sequence predicted by the model with the ground truth sequence.

The Seq-ER gives us a fairer evaluation because it does not depend on the amount of symbols needed for coding, that could bias the Sym-ER in favour of the less verbose encodings. On the other hand, the Seq-ER is a much more pessimistic estimate of the performance, because one single error ruins the output. A model can have a very low Sym-ER, e.g. 1%, but if the wrong symbols are equally distributed we can have a very high Seq-ER.

Due to that, in the next section, the Seq-ER will be used for comparing the performance of the neural model using the different encodings (although Sym-ER are also studied), and then the best representation will be used for a qualitative evaluation of the OMR approach on some selected real music fragments.

4   Results

Four different models have been trained, each corresponding to one of the four coding proposals, using the corpus of 8 000 scores generated. From the whole set, 7 000 have been used for training and the remaining 1 000 scores for validation. Note that the main purpose of this experimentation is to compare the different encodings, so we are interested in their error bounds.

First, the convergence of the models learned is shown. That is, how many training epochs the models need for tuning their parameters appropriately. This gives some insights about the complexity for each encoding to be learned by the CRNN. The curves obtained by each type of encoding are shown in Figs. 5 (Seq-ER) and 6 (Sym-ER). From the curves we can observe that the four models converge relatively quickly, needing less than 20 epochs for the *elbow* point.

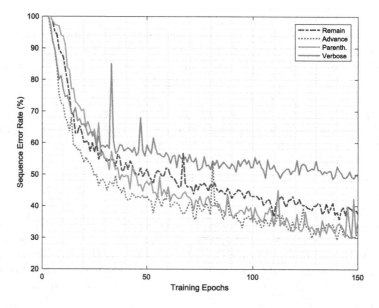

Fig. 5. Convergence analysis: accuracy over the validation set with respect to the training epoch of the deep neural network in terms of Sequence Error Rate. **Remain** stands for *remain-at-position character* coding, **Advance** stands for *advance position character*, **Parenth.** stands for *parenthesized* coding, and **Verbose** stands for *verbose* coding.

Observing Figs. 5 and 6, there is a clear correlation between the Sym-ER and the Seq-ER for the four coding proposals: the encoding with the highest Seq-ER

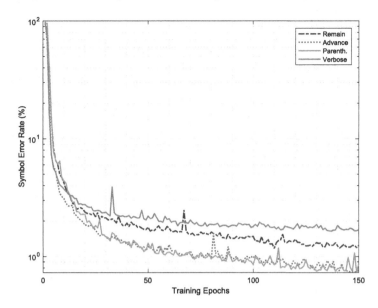

Fig. 6. Accuracy over the validation set with respect to the training epoch of the deep neural network in terms of Symbol Error Rate. Logarithmic units have used of the error representation. **Remain** stands for *remain-at-position character* coding, **Advance** stands for *advance-position character* coding, **Parenth.** stands for *parenthesized* coding, and **Verbose** stands for *verbose* coding.

is also the one with the highest Symb-ER, and vice versa. This way, we can discard that the Sym-ER is biasing the evaluation depending on the verbosity of the used coding, and it can be used for a proper performance evaluation.

It is also noted that the encoding of the output for this OMR task does have an impact on the training, and consequently, on the recognition performance. The advance-position character coding achieved the best results for both metrics: it attains the lowest Sym-ER and Seq-ER. The results were very encouraging, since around 70% of the test scores were error-free recognized, and the symbol recognition error rate is less than 1% of the symbols predicted.

Table 1. Error-bound analysis: best accuracy attained by each coding over the validation set.

	Remain	Advance	Parenthesized	Verbose
Sequence error rate (%)	35.8	29.9	33.1	48.9
Symbol error rate (%)	1.18	0.74	0.77	1.79

4.1   Qualitative Evaluation

All the scores used above are synthetic. An additional experiment has been carried out for transcribing the content of real scores taken from a repository of historical scores (RISM, *Répertoire International des Sources Musicales*[1]). Two musical *incipits* from RISM were presented to the system as independent (unseen) images for qualitative assessment of the proposed approach. The images of this experiment have been distorted to deal with the challenges of a real scenario, as in [4].

Figures 7 and 8 show the images and the predicted sequences using the *Advance* model. From them, we can say that the results obtained (Sym-ER = 5.3% and 9.5%, respectively) are rather accurate even taking into account that the images belong to a different (previously unseen) database than those used for training, and they were rendered using different methods and present distortions and lower quality. These facts explain the higher error values compared to those presented in Table 1, but the performance still shows a high precision in the recognition that can be improved by adding distorted images to the training set.

(a) Input image

```
clef.G:L2 + accidental.flat:L3 + accidental.flat:S4 + accidental.flat:S2 + metersign.C:L3
+ [bracket.start:S0] note.sixteenth:L7 + digit3:S6 note.sixteenth:S4 + note.sixteenth:S4 +
[bracket.end:S-1] note.sixteenth:S3 bracket.start:S6 + note.quarter:L2 + note.sixteenth:L1
bracket.end:S6 + rest.quarter:L3 + rest.eighth:L3 + note.sixteenth:S4 + note.sixteenth:L5
+ note.sixteenth:S5 + note.sixteenth:L5 + note.sixteenth:S5 + note.sixteenth:L6
+ verticalLine:L1 + note.sixteenth:L5 + note.sixteenth:S4 + note.sixteenth:L5 +
note.sixteenth:S5 + note.sixteenth:S4 + note.sixteenth:L4 + note.sixteenth:S4
+ note.sixteenth:L5 + note.eighth:L4 + note.sixteenth:S4 + note.sixteenth:L5 +
note.sixteenth:S5 + note.sixteenth:L5 + note.sixteenth:S5 + note.sixteenth:L6 +
verticalLine:L1
```

(b) Predicted sequence with text in bold typeface indicating mistakes and symbols between brackets denoting missing symbols.

Fig. 7. Qualitative evaluation of the OMR approach for *incipit RISM ID no. 110003911-1_1_1*, yielding a SER of 5.3%.

As it can be observed in Fig. 8, in the last measure there is a three dotted quarter note chord which the model interprets as two clearly differentiated groups: the three quarter notes share the same horizontal space and so do the three corresponding dots, but these are placed at the right of the chord notes,

---

[1] http://www.rism.info/home.html.

(a) Input image

```
clef.G:L2 + accidental.sharp:L5 + accidental.sharp:S3 + accidental.sharp:S5 + digit.8:L2
digit.6:L4 + note.beamedRight2:S2 + note.beamedRight1:S3 + note.beamedLeft1:L4 +
note.quarter:S4 + note.eighth:S3 + verticalLine:L1 + note.quarter:L4 note.quarter:L5 +
dot:S4 [dot:S5] + [note.quarter:S3] note.quarter:S4 + [dot:S3] dot:S4 + verticalLine:L1
+ rest.eighth:L3 + note.beamedRight1:L4 + note.beamedLeft1:S3 + note.quarter:L3 +
note.eighth:S2 + verticalLine:L1 + note.beamedRight1:L2 + note.beamedRight1:S2 +
note.beamedLeft1:L3 + note.quarter:L1 note.quarter:L2 note.quarter:L3 + dot:S1 dot:S2 dot:S3
```

(b) Predicted sequence with text in bold typeface indicating mistakes and symbols
between brackets denoting missing symbols.

Fig. 8. Qualitative evaluation of the OMR approach for *incipit RISM ID no.*
000136642-1_1_1, yielding a SER of 9.5%.

meaning a horizontal advance, which is correctly codified by the model placing
the '+' (advance) character between these two groups. Consequently, this leads
us to the conclusion that the model is able to interpret the vertical and horizontal
relationships in the score and learns how to code them.

5   Conclusions

In this work, we have studied the suitability of the use of the neural network
approach to solve the OMR task in an end-to-end way through a controlled
scenario of homophonic synthetic scores, presenting and analyzing four different
encodings for the OMR output.

As reported in the experiments, our serialized ways of encoding the music
content prove to be appropriate for DL-based OMR, as the learning process is
successful and low Symbol Error Rate figures are eventually attained. In addition, it is shown that the choice of the encoding have some impact on the lower
bound of the Error Rates that can be achieved, which almost directly correlates
with the tendency of the learning curves. These facts reinforce our initial claim
that the encoding of the output for OMR deserves further consideration within
the end-to-end DL paradigm.

Acknowledgements. This work is supported by the Spanish Ministry HISPAMUS
project TIN2017-86576-R, partially funded by the EU.

References

1. Alfaro, M.: Construction of a reference corpus for OMR research (in Spanish). Technical report, University of Alicante (2018)
2. Bainbridge, D., Bell, T.: The challenge of optical music recognition. Comput. Humanit. **35**(2), 95–121 (2001)
3. Baró, A., Riba, P., Calvo-Zaragoza, J., Fornés, A.: From optical music recognition to handwritten music recognition: a baseline. Pattern Recogn. Lett. **123**, 1–8 (2019)
4. Calvo-Zaragoza, J., Rizo, D.: Camera-PrIMuS: neural end-to-end optical music recognition on realistic monophonic scores. In: Proceedings of the 19th International Society for Music Information Retrieval Conference, ISMIR 2018, Paris, France, 23–27 September 2018, pp. 248–255 (2018)
5. Calvo-Zaragoza, J., Rizo, D.: End-to-end neural optical music recognition of monophonic scores. Appl. Sci. **8**(4), 606 (2018)
6. Calvo-Zaragoza, J., Toselli, A.H., Vidal, E.: Early handwritten music recognition with hidden Markov models. In: 15th International Conference on Frontiers in Handwriting Recognition, pp. 319–324. Institute of Electrical and Electronics Engineers Inc. (2017)
7. Campos, V.B., Calvo-Zaragoza, J., Toselli, A.H., Vidal-Ruiz, E.: Sheet music statistical layout analysis. In: 15th International Conference on Frontiers in Handwriting Recognition, pp. 313–318 (2016)
8. Good, M., et al.: MusicXML: an internet-friendly format for sheet music. In: XML Conference and Expo, pp. 03–04 (2001)
9. Graves, A.: Supervised sequence labelling with recurrent neural networks. Ph.D. thesis, Technical University Munich (2008)
10. Graves, A., Fernández, S., Gomez, F., Schmidhuber, J.: Connectionist temporal classification: labelling unsegmented sequence data with recurrent neural networks. In: Proceedings of the 23rd International Conference on Machine Learning, ICML 2006, pp. 369–376. ACM, New York (2006)
11. Hankinson, A., Roland, P., Fujinaga, I.: The music encoding initiative as a document-encoding framework. In: Proceedings of the 12th International Society for Music Information Retrieval Conference, pp. 293–298 (2011)
12. Miranda, E.R.: Composing Music with Computers. Focal Press, Waltham (2001)
13. Rabiner, L., Juang, B.H.: Fundamentals of Speech Recognition. Prentice-Hall Inc., Upper Saddle River (1993)
14. Rebelo, A., Fujinaga, I., Paszkiewicz, F., Marçal, A., Guedes, C., Cardoso, J.: Optical music recognition: state-of-the-art and open issues. Int. J. Multimed. Inf. Retr. **1**(3), 173–190 (2012). https://doi.org/10.1007/s13735-012-0004-6
15. Romero, V., Sanchez, J.A., Bosch, V., Depuydt, K., de Does, J.: Influence of text line segmentation in handwritten text recognition. In: 2015 13th International Conference on Document Analysis and Recognition (ICDAR), pp. 536–540. IEEE (2015)
16. Shi, B., Bai, X., Yao, C.: An end-to-end trainable neural network for image-based sequence recognition and its application to scene text recognition. IEEE Trans. Pattern Anal. Mach. Intell. **39**(11), 2298–2304 (2017)
17. Williams, R.J., Zipser, D.: Gradient-based learning algorithms for recurrent networks and their computational complexity. In: Chauvin, Y., Rumelhart, D.E. (eds.) Back-Propagation: Theory, Architectures and Applications, chap. 13, pp. 433–486. Erlbaum, Hillsdale (1995)

Glyph and Position Classification of Music Symbols in Early Music Manuscripts

Alicia Nuñez-Alcover$^{(\boxtimes)}$, Pedro J. Ponce de León, and Jorge Calvo-Zaragoza

Department of Software and Computing Systems, University of Alicante,
Carretera de San Vicente s/n, 03690 Alicante, Spain
`ana27@alu.ua.es`, {`pierre,jcalvo`}`@dlsi.ua.es`

Abstract. Optical Music Recognition is a field of research that automates the reading of musical scores so as to transcribe their content into a structured digital format. When dealing with music manuscripts, the traditional workflow establishes separate stages of detection and classification of musical symbols. In the latter, most of the research has focused on detecting musical glyphs, ignoring that the meaning of a musical symbol is defined by two components: its glyph and its position within the staff. In this paper we study how to perform both glyph and position classification of handwritten musical symbols in early music manuscripts written in white Mensural notation, a common notation system used for the most part of the XVI and XVII centuries. We make use of Convolutional Neural Networks as the classification method, and we tested several alternatives such as using independent models for each component, combining label spaces, or using both multi-input and multi-output models. Our results on early music manuscripts provide insights about the effectiveness and efficiency of each approach.

Keywords: Optical Music Recognition ·
Handwritten symbol classification · Convolutional Neural Networks ·
Digital preservation

1 Introduction

Music constitutes one of the main vehicles of cultural heritage. A large number of musical manuscripts are preserved in historical archives. Occasionally, these documents are transcribed to a digital format for its easier access and distribution, without compromising their integrity. However, in order to make these heritage really useful, it is necessary to transcribe the sources to a structured format such as MusicXML [3], MEI [11], or MIDI. So far, this has been done manually by experts in early music notation, making the process very slow and costly. Conveniently, Optical Music Recognition (OMR) techniques can help automating the process of reading music notation from scanned music scores [1].

© Springer Nature Switzerland AG 2019
A. Morales et al. (Eds.): IbPRIA 2019, LNCS 11868, pp. 159–168, 2019.
https://doi.org/10.1007/978-3-030-31321-0_14

Typically, the transcription of historical music documents is treated differently with respect to conventional OMR methods due to their particular features; for instance, the use of certain notation systems, or the state of preservation of the original document. Although there exist several works focused on early music documents transcription [4,8], the specificity of each type of manuscript, or its overall writing style makes it difficult to generalize these developments.

In this context, a music transcription system is one that performs the task of obtaining a digital structured representation of the musical content in a scanned music manuscript. The workflow to accomplish such a task could be summarized as follows: First, a document layout analysis step isolates document parts containing music, mostly music staffs. Then, a OMR system detects music symbols contained in these parts, typically producing a sequence or graph of music symbols and their positions with respect to the staff. From this representation, a semantic music analysis step assigns musical meaning to each symbol, as this often depends on the specific location of the symbol in the sequence. Finally, this intermediate music representation is translated by a coding stage into a structured representation in the desired output format.

Unlike other domains, in the particular case of music notation the symbols to be classified have two components: glyph and position in the staff. Traditionally, OMR systems use supervised learning to predict the glyph [2,6,9], whereas the position is determined by heuristic strategies [13]. Since these OMR systems usually perform a pre-process that normalizes the input images (binarization, deskewing, and so on), these heuristic strategies tend to be quite reliable. However, other approaches might use different pre-processing steps, and so traditional heuristics might not work correctly. This is why we propose to deal with the identification of the glyph position by means of supervised learning as well. To this end, we study the best approach to perform classification of a pre-segmented symbol image into its pair of glyph and position. Specifically, we propose different deep learning architectures that make use of Convolutional Neural Networks (CNN) in order to fully classify a given music symbol. We aim to analyze which architecture gives us the best performance in terms of accuracy and efficiency.

The rest of this paper is organized as follows: Sect. 2 presents the methodology considered for the aforementioned task; Sect. 3 describes our experimental setup; Sect. 4 presents and analyzes the results: Sect. 5 concludes the present work and introduces some ideas for future research.

2 Methodology

Our work assumes a segmentation-based approach, in which the locations of symbols that appear in the input music score have already been detected in a previous stage. This can be achieved under an interactive environment where the user manually locates the symbols [10], but can also be automated with object detection techniques [7]. Either way, the next task is to fully classify the symbols in order to retrieve their musical meaning.

We consider that all musical symbols are defined by two components: glyph and position with respect to the staff lines. This is obvious in the case of notes, as these components indicate the duration and the pitch, respectively. We can generalize this to any type, as all symbols are located in a specific position with respect to the lines of the staff. Let $\mathcal{G}$ be the label space for the different glyphs and $\mathcal{P}$ the label space for the different positions. A music symbol is therefore fully defined by a pair $(g, p), s \in \mathcal{G}, p \in \mathcal{P}$. A graphical example is given in Fig. 1.

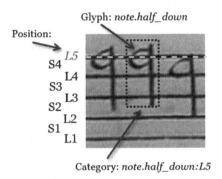

Fig. 1. Example of handwritten music symbols in white Mensural notation, showing its glyph, position, and combined label. Note that position labels refer to the vertical placement of a glyph: Ln and Sn denote symbol positions over or between staff lines, respectively.

Considering the above, the complete classification of a music-notation symbol consists in predicting both its glyph and its position. This opens up several possibilities as regards this dual process, given that the two components are not completely independent. As a base classification algorithm, we resort to Convolutional Neural Networks (CNN), since they represent the state of the art for image classification. These neural networks consist of one or more convolutional layers, which benefits from local connections, tied weights and pooling operations in order to learn a suitable data representation for the task at hand [14]. The convolutional layers are typically followed by one or more fully-connected layers that perform the final prediction stage.

The classical use of CNN is to consider a single image as input, that must be associated with a single class label. However, since we want to know the glyph of a symbol as well as its position within the staff simultaneously, we shall consider different architectures with shared layers, and multi-output and multi-input models, in order to determine which is the best way to obtain the corresponding full symbol classification.

Below we first introduce how to represent the input for classification, after which we propose the neural architectures that allow us to perform the combined glyph and position classification of music symbols.

2.1   Input Scheme and Preprocessing

Two region-based image inputs are used in this approach. Figure 2 shows a part of a music staff from a Mensural notation score. These images are pre-segmented, either manually or automatically, by defining a bounding box around each music symbol in the staff, as shown in Fig. 3 (left). Thus, each bounding box defines an image instance containing a music symbol. These appropriately annotated images can be used to train and evaluate a music glyph classification model. However, in general these instances do not span vertically as to contain all staff lines. Therefore, they do not convey information about the music symbol position. For example, the symbol labeled as 'A' in Fig. 3 (left) is indistinguishable from symbol 'B' in the same image in spite of appearing at different staff positions. This type of images will be referred as *glyph inputs*.

In order to produce models able to correctly classify music symbol positions, a second image set is constructed by enlarging the bounding box frame vertically to a fixed height large enough to contain all staff lines, as shown in Fig. 3 (right). This type of images will be referred as *enlarged inputs* in the following sections. It has been shown that these enlarged images contain enough information for estimating the vertical position of a music symbol within the staff [5].

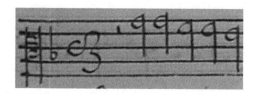

Fig. 2. A sample of a Mensural notation staff.

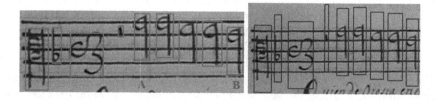

Fig. 3. Left: glyph bounding boxes. Right: enlarged bounding boxes.

2.2   CNN Architectures

Given the aforementioned inputs and targets, we intend to classify every region as one of the available symbols. To accomplish this, we try different approaches that perform the different classifications either simultaneously or independently.

Independent Glyph and Position Models. Our first approach would be to create two different CNN models: one processes a glyph input x_g, and tags it with a glyph label $g \in \mathcal{G}$. The other one processes a enlarged inputs x_p, and tags it with a position label $p \in \mathcal{P}$. These two models are depicted in Fig. 4.

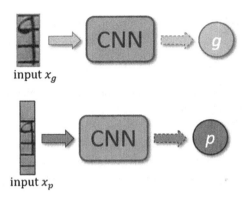

Fig. 4. Top: independent glyph classification model. Bottom: independent position classification model.

Category Output Model. Another approach uses a single enlarged input x_p, and tags it by considering as the label set the Cartesian product of $\mathcal{G}$ and $\mathcal{P}$. We shall refer to the combined label of a symbol as its *category*, denoted by $\mathcal{C}$. Therefore the model tags each input x_p as a pair $c = (g, p)$, such that $g \in \mathcal{G}$ and $p \in \mathcal{P}$. This approach is depicted in Fig. 5.

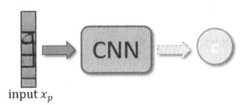

Fig. 5. Category output model: enlarged inputs are provided as input and the model must predict a label from the Cartesian product of glyphs and positions.

Category Output, Multiple Inputs Model. This model uses both glyph and enlarged input images, yet predicting directly the combined category c (Fig. 6).

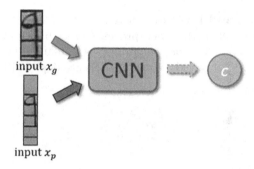

Fig. 6. Category output with multiple inputs: both glyph bounding box and enlarged images are provided as input and the model must predict a label from the Cartesian product of glyphs and positions.

Multiple Outputs Model. This model takes enlarged inputs and predicts the glyph g and position p labels separately, as shown in Fig. 7. The model shares the intermediate data representation layers as input to both final fully-connected classification layers.

Fig. 7. Multiple outputs model: enlarged images are provided as input and the model must predict both the glyph and the position separately.

Multiple Inputs and Outputs Model. Our last model takes both glyph and enlarged inputs x_g and x_p, and predicts glyph g and position p labels as two different outputs, as depicted in Fig. 8.

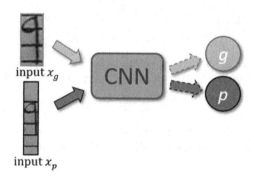

Fig. 8. Multiple inputs and outputs model: both glyph bounding boxes and enlarged images are provided as input and the model must predict both the glyph and the position separately.

3   Experimental Setup

3.1   Network Configuration

Given the proposed general topologies, this section describes the CNN architectures for each model. We have selected a base architecture by means of informal testing. Since this is designed to be used in an interactive scenario, the network must be light to allow for real-time processing.

The input to a convolutional layer is an $m \times n \times r$ image, where m is the height, n is the width of the image and r is the number of channels, being in this case $r = 3$ since we are working with RGB images. Models with one input share the same feature extractor and classification stage. However, they have different inputs and outputs. Their architecture consists of the repeated application of two 3×3 convolutions with 32 filters, each one followed by a rectified linear unit and a 2×2 max-pooling operation with stride 2 for downsampling, then followed by a dropout of 0.25. At the next stack of convolutional layers and max-pooling, we double the number of filters. Finally, after flattening, a fully-connected layer with 256 units followed by a dropout of 0.25 and a *softmax* activation function layer is used for the classification stage.

Models with two inputs have parallel feature extractor for each input, and they are concatenated in their corresponding last stack, after the flattening operation. When using models with two outputs in their classification stages, the network is forked into two different outputs after the last fully-connected layer.

3.2   Dataset

For our experiments, three different manuscripts of handwritten music scores in Mensural notation were available with symbol-level annotation. A summary of the number of samples, and glyph and position classes is shown in Table 1. Combined classes are the result of the Cartesian product of glyph and position classes, as stated above. It is worth noting that most combinations of glyph and position do not appear in our ground-truth, so they have been removed from the set of combined classes.

A major drawback of this ground-truth dataset is the high ratio of label imbalance. Therefore, we took this into account while training our models by weighting the loss for each label according to its relative frequency in the dataset.

Furthermore, since the symbol images can vary drastically in size, and for the sake of creating a dataset suitable for training CNN models, the input images are resized to a fixed size. We resized images to 40×40 pixels for glyph inputs, and to 40×112 pixels for enlarged inputs.

Table 1. Distribution of classes and samples in the Mensural notation symbols dataset.

	Quantity
Glyph classes	37
Position classes	14
Category (combined) classes	157
Total symbols	14373

3.3   Evaluation

In order to evaluate our proposed models, we conducted a 5-fold cross-validation scheme for the six models that were considered. In each fold, 80% of the available data is used for training and 20% for validation. Also, each fold was created by the class imbalance of the original dataset.

Moreover, a grid search was carried out in order to tune hyperparameters. Consequently, we trained every architecture for 15 epochs and a batch size of 32. Additionally, a *RMSprop* optimizer was used [12].

Given that our purpose is to evaluate the correct recognition of music categories, we should consider our models' accuracy taking into account that the glyph and position of a given input are being labeled correctly at once. More precisely, given an input series $x = \{x_1, x_2, \ldots, x_n\}$ we need to calculate the accuracy of the three different outputs of a symbol: position as $p = \{p_1, p_2, \ldots, p_n\}$, glyph as $g = \{g_1, g_2, \ldots, g_n\}$ and category as $c = \{c_1, c_2, \ldots, c_n\}$, for every model.

Another important factor to consider is the complexity of each model since we aforementioned the necessity of good effectiveness and efficiency in the application of these models in a real scenario. In order to provide a value of efficiency that does not depend on the underlying hardware used in the experiments, we consider the number of (trainable) parameters of the neural model as a measure of its complexity.

4   Results

Table 2 presents average and standard deviations achieved by the different classification schemes for the five folds, as well as the complexity (number of trainable parameters) of each model.

The best results in terms of accuracy are obtained by the model with multiple inputs and outputs, then without almost no significant difference, by the model with multiple outputs, followed by the model with independent glyph and position. Moreover, the worst results are obtained by those that consider the full category as output, especially when using multiple inputs.

Another factor for evaluation is the correlation between complexity and accuracy. The models with multiple inputs share the same complexity as well as models with only one input. The best model, with multiple inputs and outputs,

Table 2. Error rate (average ± std. deviation) and complexity with respect to the neural architecture considered for music symbol classification. The complexity of each model is measured as millions of trainable parameters.

Model	Error rate (%)			Complexity (10^6)
	Glyph	Position	Category	
Independent glyph and position	3.0 ± 0.4	5.1 ± 0.6	7.6 ± 0.6	$1.71 + 4.66$
Category output	4.1 ± 0.1	5.9 ± 0.4	8.2 ± 0.3	4.66
Category output, multiple inputs	6.3 ± 0.6	7.5 ± 0.7	11.8 ± 0.9	6.37
Multiple outputs	3.1 ± 0.1	5.2 ± 0.3	7.6 ± 0.3	4.66
Multiple inputs and outputs	2.9 ± 0.1	4.7 ± 0.4	7.1 ± 0.4	6.37

has high complexity compared to the second best one, which is relevant, as both perform comparably well.

Contrary to expectation, it is interesting to point out that models with multiple outputs are performing better. Since glyph and position are dependent features for most instances, we could expect that models predicting combined classes would produce better results. However, we are aware that this outcome might be produced by the possibility of not having enough data to train.

5 Conclusions

In this work, we proposed different segmentation-based approaches to recognize glyph and position of handwritten symbols, since traditional OMR systems mostly focuses on recognizing the glyph of music symbols, but not their position with respect to a staff. Therefore, we propose to classify glyph and position in the same step. Our approach is based on using different CNN architectures where we predict glyph, position, or their combination by training independent models, multi-input and multi-output models.

Experimentation was presented by using a dataset of handwritten music scores in Mensural notation where we evaluated each model by their accuracy on labeling glyph, position, and their combination, as well as their complexity in order to estimate the best model for our purpose.

The results suggest that in order to obtain the best accuracy, models should predict glyph and position labels separately, instead of predicting the combination of both as a single class. We can conclude that interesting insight has been gained with regard to achieving a complete system for extracting the musical content from an image of a score.

As a future work, we plan to consider the use of data augmentation for boosting the performance. However, this has to be designed carefully as traditional data augmentation procedures might not work correctly as regards the positions of the symbols. We also plan to reuse the outcomes of this paper for segmentation-free recognition, which does not need a previous localization of the symbols in the image [10].

As an another step towards the goal of this research, these models must be integrated on a fully-automated transcription system that uses semantic analysis tools to assign actual musical meaning to the output of the considered models.

Acknowledgments. This work is supported by the Spanish Ministry HISPAMUS project TIN2017-86576-R, partially funded by the EU.

References

1. Bainbridge, D., Bell, T.: The challenge of optical music recognition. Comput. Humanit. **35**(2), 95–121 (2001)
2. Calvo-Zaragoza, J., Gallego, A.J., Pertusa, A.: Recognition of handwritten music symbols with convolutional neural codes. In: 14th International Conference on Document Analysis and Recognition, Kyoto, Japan, pp. 691–696 (2017)
3. Good, M., et al.: MusicXML: an internet-friendly format for sheet music. In: XML Conference and Expo, pp. 03–04 (2001)
4. Huang, Y.H., Chen, X., Beck, S., Burn, D., Van Gool, L.: Automatic handwritten mensural notation interpreter: from manuscript to MIDI performance. In: Müller, M., Wiering, F. (eds.) 16th International Society for Music Information Retrieval Conference, Málaga, Spain, pp. 79–85 (2015)
5. Pacha, A., Calvo-Zaragoza, J.: Optical music recognition in mensural notation with region-based convolutional neural networks. In: 19th International Society for Music Information Retrieval Conference, Paris, France, pp. 240–247 (2018)
6. Pacha, A., Eidenberger, H.: Towards a universal music symbol classifier. In: 14th International Conference on Document Analysis and Recognition, IAPR TC10 (Technical Committee on Graphics Recognition), pp. 35–36. IEEE Computer Society, Kyoto (2017)
7. Pacha, A., Hajič Jr., J., Calvo-Zaragoza, J.: A baseline for general music object detection with deep learning. Appl. Sci. **8**(9), 1488–1508 (2018)
8. Ramirez, C., Ohya, J.: Automatic recognition of square notation symbols in western plainchant manuscripts. J. New Music Res. **43**(4), 390–399 (2014)
9. Rebelo, A., Capela, G., Cardoso, J.D.S.: Optical recognition of music symbols. Int. J. Doc. Anal. Recogn. **13**(1), 19–31 (2010)
10. Rizo, D., Calvo-Zaragoza, J., Iñesta, J.M.: MuRET: a music recognition, encoding, and transcription tool. In: 5th International Conference on Digital Libraries for Musicology, pp. 52–56. ACM, Paris (2018)
11. Roland, P.: The music encoding initiative (MEI). In: 1st International Conference on Musical Applications Using XML, pp. 55–59 (2002)
12. Ruder, S.: An overview of gradient descent optimization algorithms. Computer Research Repository abs/1609.04747 (2016)
13. Vigliensoni, G., Burgoyne, J.A., Hankinson, A., Fujinaga, I.: Automatic pitch detection in printed square notation. In: Klapuri, A., Leider, C. (eds.) 12th International Society for Music Information Retrieval Conference, pp. 423–428. University of Miami, Miami (2011)
14. Zeiler, M.D., Fergus, R.: Visualizing and understanding convolutional networks. In: Fleet, D., Pajdla, T., Schiele, B., Tuytelaars, T. (eds.) ECCV 2014. LNCS, vol. 8689, pp. 818–833. Springer, Cham (2014). https://doi.org/10.1007/978-3-319-10590-1_53

Recognition of Arabic Handwritten Literal Amounts Using Deep Convolutional Neural Networks

Moumen El-Melegy[(⊠)], Asmaa Abdelbaset, Alaa Abdel-Hakim, and Gamal El-Sayed

Electrical Engineering Department, Faculty of Engineering, Assiut University, Assiut 71516, Egypt
`moumen@aun.edu.eg`

Abstract. In this paper, we propose using a convolutional neural network (CNN) in the recognition of Arabic handwritten literal amounts. Deep convolutional neural networks have achieved an excellent performance in various computer vision and document recognition tasks, and have received increased attention in the few last years. The domain of handwriting in the Arabic script specially poses a different type of technical challenges. In this work we focus on the recognition of handwritten Arabic literal amount with a limited lexicon. Our experimental results demonstrate the high performance of the proposed CNN recognition system compared to traditional methods.

Keywords: Deep learning · Convolutional neural network · Literal amount recognition · Arabic handwriting recognition · Arabic word recognition

1 Introduction

Machine simulation of human functions has been a very challenging research field. Human reading is one of the most important subjects of this field which has become an intensive research area. Document recognition still presents several challenges to the researchers, especially the recognition of Arabic handwriting. It is important not only for Arabic speaking countries who represent over than seven billion people all over the world, but also for some non-Arabic speaking countries, such as Farsi, Curds, Persians, and Urdu-speakers who use the Arabic characters in writing although the pronunciation is different. Arabic handwriting recognition has an importance in many applications, such as office automation, cheque verification, students' grade reports, and several data entry applications.

There has been less research in the recognition of Arabic handwritten texts compared to the recognition of texts in other scripts, such as Latin, Chinese and Japanese text. The domain of handwriting recognition in the Arabic script presents many technical challenges and obstacles due to its special characteristics, structure and variability in writing styles, where the Arabic script is written from right to left in a cursive way. The shape of an Arabic character depends on its position in the word so it is context

A. Morales et al. (Eds.): IbPRIA 2019, LNCS 11868, pp. 169–176, 2019.
https://doi.org/10.1007/978-3-030-31321-0_15

sensitive and this makes every letter takes up to four different shapes: isolated, at the beginning, in the middle, and at the end. Also some of the Arabic letters have dots associated with them. These diacritical dots can be located above or below the character but not the two simultaneously. Also, the same character may appear differently in its various forms. Moreover, the great similarity between some of the handwritten characters makes the classification of these characters more difficult.

In this work, we focus on Arabic handwritten literal amount recognition. The special and complex nature of the Arabic handwriting adds to the difficulty of this problem. Practically speaking, Arabic literal amount recognition is the least addressed task in Arabic handwritten recognition. It has larger vocabulary than its counterpart problems in other languages, and it has larger amount of variations in writing similar amounts due to complex grammatical rules of Arabic language. Table 1 shows the 50 word classes in the vocabulary of the problem under consideration in this paper. The table gives a real sample of each word class and its closest meaning in English.

The problem of handwritten recognition can be generally solved using one of two different approaches: the first is a traditional handcrafted approach that depends on hand-designed features applied to a classifier or a combination of classifiers (as in the classical approach to machine learning). The second approach is based on deep learning. Deep learning is a new application of machine learning for learning representation of data and models hierarchical abstractions in input data with the help of multiple layers. Deep learning techniques have achieved an excellent performance in computer vision, image recognition, speech recognition and natural language processing, and has acquired a reputation for solving many computer vision problems. Its application to the field of handwriting recognition has been shown to provide significantly better results than traditional methods.

In this work, we propose a *new* solution for Arabic handwritten literal amount recognition using deep learning techniques. *To the best of our knowledge, there are no reported methods in the literature on applying deep learning to Arabic handwritten literal amounts recognition*; all existing methods are based on the classical, handcrafted features approach. As such, our contribution is that the relevant features are learned by a CNN, which leads to significantly better recognition results. In the context of Arabic handwritten character and digit recognition, there are a number of works based on convolutional neural network and deep learning [1, 2, 3]. These works deal with the individual characters not the whole word as we do in our work here, which makes our problem more challenging.

The organization of the paper is as follows. Section 2 presents an overview and related work, In Sect. 3 we present the proposed deep learning approach. Section 4 discusses the experimental results and analysis. Section 5 provides conclusions and future work.

2 Related Work

A number of papers have been done in the field of Arabic handwritten recognition in general and Arabic handwritten literal amount recognition in particular. The notable ones are mentioned below.

Farah et al. [6–8] introduce a number of different works in the field of Arabic handwritten word and literal amount recognition using perceptual high-level features and structural holistic classifier employing holistic features and decision fusion and contextual information. Younis [3] presents a deep neural network for the handwritten Arabic character recognition problem that uses convolutional neural network (CNN) models. They apply the Deep CNN for the AIA9k and the AHDB databases. El-Sawy et al. [2] present a deep learning technique that is applied to recognizing Arabic handwritten digits. Mudhsh and Almodfer [4] present a system for Arabic handwritten alphanumeric character recognition using a very deep neural network. He proposes alphanumeric VGG net for Arabic handwritten alphanumeric character recognition. Alphanumeric VGG net is constructed by thirteen convolutional layers, two max-pooling layers, and three fully-connected layers.

In our previous work [5], we have developed a system for recognition of Arabic handwritten literal amounts depending on the extraction of the structural holistic features from the handwritten literal. In our classification system we used four different classifiers. The results of these classifiers are taken independently. We have tested our system using the standard Arabic Handwritten Database AHDB [9].

Table 1. Classes of Arabic literal amount vocabulary and their meanings in English.

three	ثلاثة	three	ثلاث	two	اثنين	two	اثنان	one	واحد
six	ست	five	خمسة	five	خمس	four	أربعة	four	أربع
eight	ثمانية	eight	ثمان	seven	سبعة	seven	سبع	six	ستة
one	احدى	ten	عشرة	ten	عشر	nine	تسعة	nine	تسع
thirty	ثلاثين	thirty	ثلاثون	twenty	عشرين	twenty	عشرون	two	اثني
sixty	ستون	fifty	خمسين	fifty	خمسون	forty	أربعين	forty	أربعون
eighty	ثمانين	eighty	ثمانون	seventy	سبعين	seventy	سبعون	sixty	ستين
two hundred	مئتين	two hundred	مائتين	hundred	مائة	ninety	تسعين	ninety	تسعون
seven hundred	سبعمائة	six hundred	ستمائة	five hundred	خمسمائة	four hundred	اربعمائة	three hundred	ثلاثمائة
no	لا	other	غير	thousand	ألف	nine hundred	تسعمائة	eight hundred	ثمانمائة

3   Proposed CNN Structure

Our proposed deep learning approach to this problem is based on convolutional neural networks (CNNs). The proposed CNN structure, see Fig. 1, consists of seventeen- layer network, beginning with an image input layer of size 60 by 80 by 1 for height, width, and the channel size, which represents the gray-scale image of the input. After that comes the convolutional layers, and in between exist the batch normalization layer, ReLU layer and max pooling layer, respectively. The first convolutional layer has 8 filters of size 3 × 3. The second consists of 16 filters of size 3 × 3. The third convolutional layer has 32 filters of size 3 × 3. The batch normalization layer is responsible for normalizing the activations and gradients propagating through the network making network training easier. The max pooling layer has a size of 2 × 2 and stride of 2. The last three layers are a fully connected layer, a softmax layer and the classification layer. The fully connected layer combines all the features learned by the previous layers to identify the input patterns. Its output size parameter is set to the number of classes in the target data, which is 50 corresponding to the 50 classes. The softmax layer normalizes the output of the fully connected layer producing positive numbers that sum to one, which can then be used as classification probabilities by the classification layer. The final classification layer uses these probabilities for each input to assign the input to one of the mutually exclusive classes and compute the loss. The proposed CNN is implemented in MatlabR2018a deep learning toolbox.

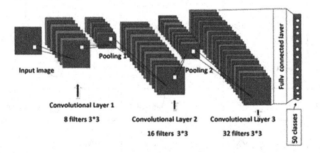

Fig. 1. Proposed structure for our CNN network.

4   Experimental Results and Analysis

In this section, we report the results obtained by applying our proposed recognition system. The AHDB [9] standard dataset is used (see the samples in Fig. 1). It consists of 50 different Arabic literal amounts (classes) which have been written by 100 different writers. It includes 4971 samples with 100 samples for every class, except the class of the word "gher" which has only 71 samples. 80% samples are chosen in random for training, and the remaining 20% are used as an independent test set. This random division of training and test samples is made in equal manner for each class to keep the class balance of the data. Network training is implemented using the Matlab deep learning toolbox. The training function used has several training options. We begin training the network with all default training options, getting an accuracy of 82.7%. We note that there is some significant overfitting, and this was rather expected due to the relatively small number of the used dataset samples. In our trials, we have tuned the training hyperparameters in order to increase the network accuracy and decrease overfitting. For example, when we increase the L2 regularization to 0.1, while fixing the rest of the hyperparameters, the test accuracy increases to 90.04%. Then setting the values for the learning rate drop factor and the learning rate drop period to 0.8 and 20, respectively, an accuracy of 92.56% is now obtained at the same L2 regularization of 0.1. We have also experimented with several values of the L2 regularization parameter: 0.09, 0.06, 0.04, and obtained the respective accuracies of 91.75%, 91.55%, and 93.96%. Then we set the L2 regularization to 0.05, and at this point the accuracy becomes 94.87%. All the previous trials were for 50 epochs of training. When we increase the number of epochs to 100, 300, and 500 (without making any changes to the other parameters), the test accuracies become 95.67%, 95.98%, and 96.58%, respectively. We also note that reducing the number or the size of filters in the network convolutional layers at the same training parameters results in a decrease in the test accuracy. As such, the best result of test accuracy was 96.58% as shown in the curves in Fig. 2.

Through the analysis of the obtained results of our proposed system, we draw several remarks. There are twenty five classes of the total fifty class having a recognition result of 100%. Despite of the difficulty of some of the samples of these classes, they are correctly classified. Also there are 34 samples from different classes of total 994 test samples are misclassified, see the individual recognition rates of these classes in Table 2. See Fig. 3 for samples that are challenging, yet well-classified by our proposed CNN.

There are some factors that have affected the results we have obtained. For example, some samples in the database are badly and poorly written, see the samples in Fig. 4. This may result in errors in the recognition. The results are expected to improve if more examples are used for network training. However, the data we used was the only public-domain dataset that was available to us. So we try to overcome this problem and increase the number of training samples by creating new training data from the original training data using data augmentation. We apply two augmentation methods based on random rotations and translations to our training dataset. Then the augmented data is used to train our CNN at the same hyperparameters we eventually used before. On testing the trained network with the same test data, the accuracy is expectedly increased to 97.8% (see Fig. 5).

We also compare our proposed method to other existing methods based on the classical non-deep learning methods (see Table 3) as applied to the same training and test datasets. All the other classifiers have hand-crafted features, for details please refer to [5]. Clearly the proposed deep CNN has significantly better results.

Table 2. Recognition results for classes that have accuracy less than 100%.

Class	Accuracy	Class	Accuracy
ثلاث	95%	ثلاثة	85%
اربعة	95%	خمسة	95%
سبع	95%	ثمان	95%
ثمانية	95%	تسعة	95%
عشرة	90%	اثنى	95%
ثلاثون	90%	اربعون	95%
خمسين	90%	ستون	95%
ستين	90%	سبعين	95%
ثمانون	95%	تسعون	95%
تسعين	95%	خمسمائة	95%
مائتين	90%	تسعمائة	95%
ستمائة	90%	غير	85.7%
تسعمائة	95%		

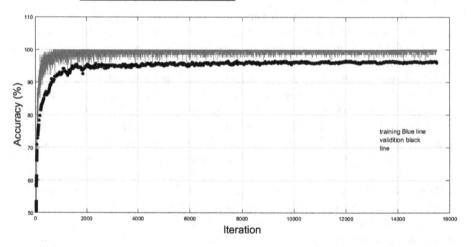

Fig. 2. Training/test curves of the proposed CNN as implemented in Matlab 2018 deep learning toolbox. Blue curve indicates accuracy on training set, while black curve indicates accuracy on the independent test set as training progresses. (Color figure online)

نلوثمانية ثلاثين أربعون ثلاث علاث أربعة أربع اثنا

صشروب انوا عشر احدى ثمان سبعة سبع خمس

سين خمسى اربعين اربعون ثلاثين عشرين ثلاثون

مائه تسعمائة سبعمائة مائتين سبعون ثمانين

Fig. 3. Some examples of challenging test samples, yet well classified by the proposed CNN.

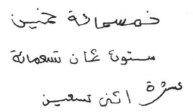

Fig. 4. Some test samples that were misclassified by the proposed CNN.

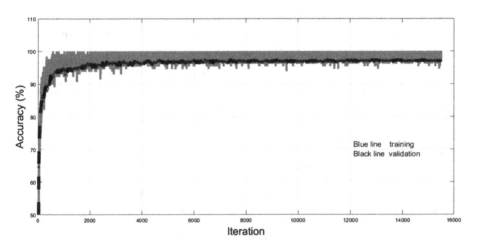

Fig. 5. Training/test curves of the proposed CNN using augmented dataset as implemented in Matlab 2018 deep learning toolbox. Blue curve indicates accuracy on training set, while black curve indicates accuracy on the independent test set as training progresses. (Color figure online)

Table 3. Comparison between different recognition methods.

Classifier	Recognition rate (%)
Bayesian	79.5
k-nearest neighbor	83.5
Decision tree	80.1
Shallow Neural Network	86.5
Our CNN with no augmentation	96.8
Our CNN with augmented data	**97.8**

5   Conclusions

In this paper we have presented a deep CNN network for the recognition of Arabic handwritten literal amounts. In spite of the practical importance of this problem, to the best of our knowledge, *this paper is the first to address this problem using a deep learning approach.* Our experimental results have demonstrated a solid and high performance of 97.8%, when compared to the previous results of classical methods employing hand-crafted features. Our future work is directed to increasing the size of the available data. Since the database we used was the only public-domain dataset that was available to us, we intend to collect our own data to increase the data size for better CNN training and testing.

References

1. Alani, A.A.: Arabic handwritten digit recognition based on restricted Boltzmann machine and convolutional neural networks. MDPI Inf. **8**(4), 142 (2017)
2. El-Sawy, A., Loey, M., El-Bakry, H.: Arabic handwritten characters recognition using convolutional neural network. WSEAS Trans. Comput. Res. **5**(1), 11–19 (2017)
3. Younis, K.S.: Arabic handwritten character recognition based on deep convolutional neural networks. Jordan J. Comput. Inf. Technol. (JJCIT) **3**(3), 186–200 (2017)
4. Mudhsh, M.A., Almodfer, R.: Arabic handwritten alphanumeric character recognition using very deep neural network. MDBI Inf. **8**(3), 105 (2017)
5. El-Melegy, M.T., Abdelbaset, A.A.: Global features for offline recognition of handwritten Arabic literal amounts. In: ITI 5th International Conference on Information and Communications Technology, December 2007
6. Azizi, N., Farah, N., Khadir, M.T., Sellami, M.: Arabic handwritten word recognition using classifiers selection and features extraction/selection. In: Recent Advances in Intelligent Information Systems, pp. 735–742 (2009). ISBN 978-83-60434-59-8
7. Farah, N., Meslati, L.S., Sellami, M.: Classifiers combination and syntax analysis for Arabic literal amount recognition. Eng. Appl. Artif. Intell. **19**(1), 29–39 (2006)
8. Farah, N., Meslati, L.S., Sellami, M.: Decision fusion and contextual information for Arabic words recognition for computing and informatics. Comput. Inform. **24**, 463–479 (2005)
9. Al-Ma'adeed, S., Elliman, D., Higgins, C.A.: A data base for Arabic handwritten text recognition research. Int. Arab J. Inf. Technol. **1**(1), 117–121 (2004)

Offline Signature Verification Using Textural Descriptors

Ismail Hadjadj[1,2(✉)], Abdeljalil Gattal[2], Chawki Djeddi[2], Mouloud Ayad[1], Imran Siddiqi[3], and Faycel Abass[1,2]

[1] Mohand Akli University, Bouira, Algeria
{ismail.hadjadj,faycel.abass}@univ-tebessa.dz,
ayad_moul@yahoo.fr
[2] Larbi Tebessi University, Tebessa, Algeria
abdeljalil.gattal@univ-tebessa.dz,c.djeddi@mail.univ-tebessa.dz
[3] Bahria University, Islamabad, Pakistan
imran.siddiqi@bahria.edu.pk

Abstract. Offline signature verification has been the most commonly employed modality for authentication of an individual and, it enjoys global acceptance in legal, banking and official documents. Verifying the authenticity of a signature (genuine or forged) remains a challenging problem from the perspective of computerized solutions. This paper presents a signature verification technique that exploits the textural information of a signature image to discriminate between genuine and forged signatures. Signature images are characterized using two textural descriptors, the local ternary patterns (LTP) and the oriented basic image features (oBIFs). Signature images are projected in the feature space and the distances between pairs of genuine and forged signatures are used to train SVM classifiers (a separate SVM for each of the two descriptors). When presented with a questioned signature, the decision on its authenticity is made by combining the decisions of the two classifiers. The technique is evaluated on Dutch and Chinese signature images of the ICDAR 2011 benchmark dataset and high accuracies are reported.

Keywords: Offline signature verification ·
Local Ternary Patterns (LTP) ·
oriented Basic Image Features (oBIFs) ·
Support Vector Machine (SVM)

1 Introduction

Biometric authentication [34,36] represents a validation process that relies on the unique biological characteristics of an individual to verify the claimed identity. Typically, biometric systems compare the captured biometric data with the authentic data stored in the database. Among various biometric modalities, signatures represent one of the oldest and most commonly employed traits. Not only acquisition of signatures is simple and does not require any specialized hardware,

© Springer Nature Switzerland AG 2019
A. Morales et al. (Eds.): IbPRIA 2019, LNCS 11868, pp. 177–188, 2019.
https://doi.org/10.1007/978-3-030-31321-0_16

they enjoy widespread social acceptability for authentication purposes in banking, legal and official documents. Signatures are produced as a result of complex elementary movements, or strokes, which are concatenated in such a way that their execution produces the desired trajectory with the minimal effort [3].

Two types of signatures are employed in the authentication systems, offline (static) and online (dynamic). Offline signatures are images of signatures digitized from paper versions using a camera or a scanner. Online signatures, on the other hand, are acquired on specialized devices which are capable of recording the signature trajectories (and other useful information like pressure etc.). While online signatures carry more information as compared to their offline counterparts [2,3,9,15], a major factor limiting their widespread acceptability is the requirement of special hardware for acquisition purposes. This study focuses on the former of the two techniques, that is, offline signature verification. From the view point of technical contribution, the signature verification techniques reported in the literature either target the feature extraction [27,30,31] or the classification part of the system [17,28,32].

A number of International competitions have also been organized on offline signature verification [4,20–22] in conjunction with the various editions of International Conference on Document Analysis and Recognition (ICDAR). Such competitions not only serve to provide an idea on the state-of-the-art performance on this problem but also allow an objective comparison of various techniques under the same experimental settings. The increasing number of participants in these competitions speaks off the research attention this problem has been attracting over the years.

In this study, we investigate the effectiveness of two textural measures in characterizing signatures, the oriented Basic Image Features (oBIFs) and Local Ternary Patterns (LTP). For classification, we employ the Support Vector Machine (SVM) classifier. Two separate classifiers are trained (using dissimilarity measures computed from each of the features) and each classifier aims to discriminate between genuine and forged signatures. The (partial) decisions of the two SVMs are then combined using the sum rule. Experiments on the benchmark ICDAR 2011 signature verification dataset report low error rates. More details are presented in the subsequent sections of the paper.

2   Related Works

Signature verification has been researched for many decades and the contributions have been summarized in a number of reviews on this problem [18,19,37]. As discussed earlier, the techniques reported in the literature focus either on enhancement of feature extraction or on proposition of classifiers to effectively discriminate between genuine and forged signatures. Among relatively recent contributions to verification of signatures, Guerbai et al. [14] propose a writer-independent framework that employs curvelet transform with the One-Class Support Vector Machine (OC-SVM) using only genuine signatures in the training. Experiments on CEDAR and GPDS signature datasets report low error rates.

Likewise, Zois et al. [38] propose a grid-based template matching scheme with SVM classifier for verification of signatures and evaluate the technique on four different signature databases. In another work, Soleimani et al. [32] propose Histogram of Oriented Gradients (HOG) and Discrete Radon Transform (DRT) with Deep Multitask Metric Learning (DMML).

Among other recent works, Hafemann et al. [16] present an interesting technique where features are extracted using deep convolutional neural networks in a writer-independent mode while classifiers are trained in a writer-dependent mode. Experimental study of the system is carried out on GPDS-960 and Brazilian PUC-PR datasets reporting promising results. A writer-independent approach based on deep metric learning is presented in [29]. The model learns signature embeddings in a high dimensional space. The technique relies on comparing triplets of two genuine and one forged signature for performance enhancement. Das et al. [7] propose to build multi-script signatures aggregating many single-script signatures. An analysis on nine different signature databases in five scripts conclude that Bhattacharyya distance can be employed to analyze multi-script against single-script scenarios. Diaz et al. [8] propose a set of linear and non-linear transformations which simulate the signing process. This allows duplicating the signatures. The duplicator is evaluated using four existing signature verification techniques on two public datasets resulting in an enhancement in the overall performance. In other recent studies, Bouamra et al. [5] exploit run-length features with a One-Class Support Vector Machine (OC-SVM) while Zois et al. [39] propose to compute transitions between asymmetrical arrangements of pixel structures.

We presented an overview of the recent works on offline signature verification. The discussion by no means is exhaustive and serves to provide an idea of the recent trends in this domain. Comprehensive surveys on verification of signatures can be found in [18,19,37].

3   Methods

This section presents the details of the proposed technique to validate the authenticity of a signature. An overview of the key steps involved in the methodology is presented in Fig. 1. Like any pattern classification system, the technique comprises two key phases, training and evaluation. Training involves extracting features from images of signatures and training two separate SVM models using dissimilarity measures computed from the two features (LTP and oBIFs histograms). During classification, a questioned signature image is fed to the system, features are extracted and decisions of the two SVMs are combined to arrive at a final decision on the genuineness of the signature. Details on feature extraction, dissimilarity measure and classification are presented in the following.

3.1   Local Ternary Patterns

Local binary patterns (LBP) [26] and many of its variants [6] have been widely employed to characterize the textural information in an image and have reported

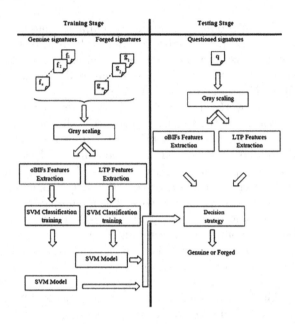

Fig. 1. An overview of the key steps in the proposed technique

promising performance on diverse problems. Among different variants of original LBP descriptor, Tan and Tiggs [33] proposed an extended version called Local Ternary Pattern (LTP). The key idea of the LTP descriptor is to extend the two valued $(0, 1)$ LBP codes to three values $(-1, 0, 1)$. LTP compute a representation based on the distribution of neighboring pixels into three values instead of thresholding values to 0 and 1, as summarized in Eq. 1.

$$LTP_{P,R} = \sum_{p=0}^{P-1} 2^p (i_p - i_c); s(x, t) = \begin{cases} 1, x \geq t \\ 0, -t < x < t \\ -1, x \leq -t \end{cases} \tag{1}$$

Considering t to be the threshold value and x be the value of the central pixel, the upper and lower threshold values are set as $x+t$ and $x-t$ respectively. Neighboring pixels (with reference to the central pixel) taking values between these thresholds are assigned 0. A value 1 is assigned to the pixels with value greater than the upper threshold while the value -1 is assigned to the pixels with value less than the lower threshold. The generated ternary code is divided into two new codes; the upper pattern and the lower pattern. Finally, an LTP histogram is computed that is employed as feature to characterize the signature. Figure 2 illustrates the computation of LTP on a 3×3 block of an image.

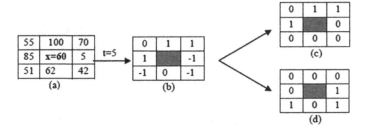

Fig. 2. Computation of LTP code (a): 3 × 3 Image Block (b): Ternary Pattern (c): Upper Pattern (d): Lower Pattern

3.2   OBIFs Histogram

Since its first appearance in [12,13], the oriented Basic Image Features (oBIFs) have been employed as an effective textural descriptor for applications like digit recognition [10,11], character recognition [23], classification of texture [25] and identification of writers from handwritten documents [1,24]. The oriented Basic Image Features (oBIFs), an extending the Basic Image Features (BIFS) [13], combine local orientation with the local symmetry information. The computation of oBIFs involves classifying each pixel in the image into one of the local symmetry classes based on the response of a bank of six Derivative-of-Gaussian (DoG) filters. The computation of features is controlled by the scale parameter σ and the supplementary parameter ϵ. The defined classes include dark line on light, light line on dark, dark rotational, light rotational, slope, saddle-like or flat. In addition to the symmetry class, an orientation is also assigned to each pixel, orientations being quantized into n possible values. No orientation is assigned if the pixel is attributed to the dark rotational, light rotational or the flat class, n possible orientations can be assigned to the dark line on light, light line on dark and the saddle-like classes while $2n$ possible orientations can be assigned for the slope class. This gives a feature vector of dimension $5n + 3$ for each image.

From the view point of signature verification problem, each signature image can be viewed as a unique texture that can be exploited to characterize the corresponding individuals. In our implementation, we fix the orientation quantization parameter to $n = 4$ resulting in a total of $(5 \times 4 + 3)$ 23 entries in the oBIFs dictionary. The number of pixels in each of the 23 classes is counted, the resulting histogram is normalized and is employed as the signature descriptor. Figure 3 illustrates a sample handwritten signature encoded using oBIFs.

3.3   Dissimilarity Measure

Unlike the typical classification framework where features extracted from classes under study are directly fed to the classifier, signature verification requires modeling of intra and inter class distances to authenticate the validity of a questioned signature. While the signatures of different individuals occupy different regions

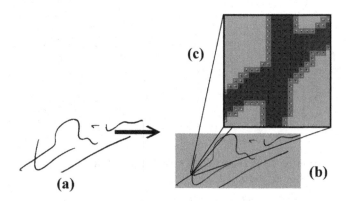

Fig. 3. Computation of oBIFs code (a): Original signature image (b): Signature encoded using the oBIFs (c): A segment of the encoded signature

in the feature space, the dissimilarities between the signatures (feature vectors) of the same individuals are likely to be low. On the other hand, the inter-writer dissimilarities are likely to be high. Exploiting the same idea, we employ the L1 norm to compute the dissimilarity between two signatures, $Z = |V - Q|$, where V and Q represent the feature vectors of two signatures being compared. The dissimilarities are computed using pairs of signature images (genuine and forged pairs) and are fed to the training model as discussed in the following.

3.4 Decision Strategy

For classification, we train separate Support Vector Machine (SVM) classifiers [35] for LTP and oBIF features. It is important to mention that we employ a write-independent approach where a single global classifier is designed rather than training separate models for each of the individuals (writer-dependent approach). For a questioned signature image, the dissimilarity measures computed from LTP and oBIF features are fed to the respective classifiers. Each classifier outputs the scores (probabilities) of the signature being genuine or forged. The scores of the two classifiers are added to arrive at the final decision about the genuineness of a given signature as summarized in Eq. 2.

$$f = max(f_F(x_{LTP}) + f_F(x_{oBIF}), f_G(x_{LTP}) + f_G(x_{oBIF})) \qquad (2)$$

f_F and f_G refer to genuine and forged scores respectively while x_{LTP} and x_{oBIF} refer to the dissimilarities computed using LTP and oBIF features. f is the maximum value selected from sum of genuine and forged scores provided by the two SVM classifiers. From the view point of implementation details, the training of the SVM requires selecting two parameters, the regularization parameter (C) and the Radial Basis Function (RBF) kernel parameter (σ). In our experiments, we investigated different values of σ and the soft margin parameter C in the interval $[1, 50]$ to empirically select the optimal combination.

4   Experiments and Results

This section presents the details of the database employed in our study along with the experimental protocol and the realized results. To evaluate the system performance standard metrics including accuracy, False Acceptance Rate (FAR) and False Rejection Rate (FRR) are used. To objectively compare the performance of our system with other techniques, we have employed the dataset from the International Competition on Signature Verification (ICDAR Sig-Comp2011) [20] that was held in conjunction with the International Conference on Document Analysis and Recognition (ICDAR 2011). The competition included both online and offline tasks and the offline datasets contained signatures of Chinese and Dutch signers. We employ the same dataset in our experimental study.

The Dutch training set includes signature images of 10 signers and for each contributor, there are, 24 genuine signatures and, on the average, 12 skilled forgeries. Likewise, for the Chinese signatures, there are 10 signers and for each signer there are 24 genuine samples and on the average 34 skilled forgeries. The test sets of both the subsets include a *'reference'* and a *'questioned'* set. The reference signatures are the known genuine signatures while the questioned signatures are either genuine or forged. The Dutch test set contains 648 reference signatures and 1286 questioned signature for 54 authors while the Chinese test set includes 116 reference and 487 questioned signatures provided by 10 authors. A summary of the these statistics is presented in Table 1.

Table 1. Number of Authors (A) and number of Genuine (G) (Reference (GR) and Questioned (GQ)) and Forged (F) (Questioned Forged (FQ)) Signatures in the ICDAR Sig-Comp2011 Dataset

Dataset	Training set			Test set			
	A	G	F	A	GR	GQ	FQ
Dutch	10	240	123	54	648	648	638
Chinese	10	235	340	10	116	120	367

We carried out a comprehensive series of experiments on the Sig-Comp2011 dataset. In case of LTP features, we investigated different combinations of the radius r and the number of neighbors p. The threshold t in computation of LTP features was fixed to $t = 0.3$ after empirical study in the range $t \in [0.1, 0.9]$ with steps of 0.1. The accuracies on the Dutch and Chinese test sets using different setting of LTP computation are summarized in Table 2.

Similar to LTP features, we also study the impact of scale parameter σ ($\sigma = 1, 2, 4, 8, 16$) while computing the oBIFs histogram on the overall accuracy. The accuracy values on Dutch and Chinese signatures on various values of σ are summarized in Table 3.

It can be seen that LTP features with $r = 1$ and $n = 16$ outperform other configurations reporting the highest accuracies on both Dutch (92.31%) and

Table 2. Accuracy on the Dutch Chinese Test Set with LTP Features

Descriptor	LTP parameters			Accuracy	
	r	p	*Dim.*	Dutch	Chinese
LTP ($t = 0.3$)	1	8	118	91.53	74.74
	2	8	118	78.55	74.74
	4	8	118	76.38	74.54
	8	8	118	65.11	74.95
	1	16	486	**92.31**	**75.36**
	2	16	486	84.23	75.36
	4	16	486	75.91	75.36
	8	16	486	71.09	75.36
	16	16	486	72.03	75.36

Table 3. Accuracy on the Chinese and Dutch Test Sets with oBIF histograms

Descriptor	Parameters		Accuracy	
	σ	*Dim.*	Dutch	Chinese
oBIFs	2	23	**96.19**	66.94
	4	23	78.79	72.69
	8	23	50.35	73.72
	16	23	50.27	**75.98**

Chinese (75.36%) signatures. In case of oBIF histograms accuracies of 95.19% and 75.98% are reported on the Dutch and Chinese signatures respectively. To study the effectiveness of combining the decisions of classifiers trained on LTP and oBIFs individually, we chose the best set of parameters for each of the features. The decisions are combined according to the fusion rule presented in Eq. 2. Table 4 summarizes the combined accuracies on the two datasets where it can be seen that overall accuracies of 97.74% and 75.98% are reported on the Dutch and the Chinese signatures respectively.

Table 4. Accuracy on the Chinese and Dutch Test Sets by combining the decisions of individual classifiers

Descriptor	Parameters	Dim.	Dutch	Chinese
LTP	$r = 16$, $p = 1$, $t = 0.3$	486	92.31	75.37
oBIFs	$\sigma = 16$ $\epsilon = 0.001$	23	96.19	75.97
Combined	–	–	97.74	75.98

Table 5. Comparison of proposed technique with participants of ICDAR 2011 competition [20] - Dutch Signatures

Rank	ID	Accuracy	FRR	FAR
1	**Proposed method**	**97.74**	**2.16**	**2.36**
2	Qatar-I	97.67	2.47	2.19
3	Qatar-II	95.57	4.48	4.38
4	HDU	87.80	12.35	12.05
5	Sabanci	82.91	17.93	16.41
6	Anonymous-I	77.99	22.22	21.75
7	DFKI	75.84	23.77	24.57
8	Anonymous-II	71.02	29.17	28.79

Table 6. Comparison of proposed technique with participants of ICDAR 2011 competition [20] - Chinese Signature

Rank	ID	Accuracy	FRR	FAR
1	Sabanci	80.04	21.01	19.62
2	**Proposed method**	**75.98**	**24.07**	**20.00**
3	Anonymous-I	73.10	27.50	26.70
4	HDU	72.90	27.50	26.98
5	DFKI	62.01	37.50	38.15
6	Anonymous-II	61.81	38.33	38.15
7	Qatar-I	56.06	45.00	43.60
8	Qatar-II	51.95	50.00	47.41

We also compare the performance of the proposed technique with those of the systems submitted to the ICDAR 2011 competition [20]. A total of seven systems were submitted to the competition. The evaluation protocol in our experiments was kept similar to that of the competition to allow a meaningful comparison. The comparative results are summarized in Tables 5 and 6 for Dutch and Chinese signatures respectively. It can be observed from the two tables that the proposed technique as well as the participating systems report high accuracies on the Dutch signatures as compared to the Chinese signatures. This may be attributed to the challenging images in the Chinese dataset as well as presence of noise in the signature images. Comparing the performance of our technique with the competition participants, it can be seen that the proposed technique outperforms the submitted systems on Dutch signatures (Table 5) while it is ranked second on the Chinese signatures (Table 6). These high accuracies validate the effectiveness of LTP and oBIFs in discriminating between genuine and forged signatures.

5 Conclusion

We presented an effective technique to characterize signature using textural descriptors. The textural information in a signature image is captured using two descriptors, the local ternary pattern and the oriented basic image features. Different configurations are investigated during the feature extraction step to find the optimal set of parameters for each of the features. Distances between the feature vectors (of genuine and forged signatures) are employed to train a separate SVM classifier for each of the two descriptors. During verification, the decisions of the two SVMs are combined to come to a final conclusion about the authenticity of a signature. The technique is evaluated on the ICDAR 2011 benchmark dataset containing Dutch and Chinese signatures and high accuracies comparable/superior to the state-of-the-art are reported.

In our further study on this problem, we intend to investigate other textural measures for signature verification and incorporate feature selection techniques to identify the most appropriate textural descriptors for this problem. With multiple features, sophisticated feature as well as classifier combination techniques can also be investigated such as the new approaches based on deep. Another interesting direction could be to carry out a comprehensive series of experiments by varying the number of signatures in the training set to identify the minimum number of samples required for acceptable performance for example using a more popular database.

References

1. Abdeljalil, G., Djeddi, C., Siddiqi, I., Al-Maadeed, S.: Writer identification on historical documents using oriented basic image features. In: 2018 16th International Conference on Frontiers in Handwriting Recognition (ICFHR), pp. 369–373. IEEE (2018)
2. Al-Omari, Y.M., Abdullah, S.N.H.S., Omar, K.: State-of-the-art in offline signature verification system. In: 2011 International Conference on Pattern Analysis and Intelligence Robotics, vol. 1, pp. 59–64. IEEE (2011)
3. Bhattacharyya, D., Ranjan, R., Alisherov, F., Choi, M., et al.: Biometric authentication: a review. Int. J. u- and e-Serv. Sci. Technol. **2**(3), 13–28 (2009)
4. Blankers, V.L., van den Heuvel, C.E., Franke, K.Y., Vuurpijl, L.G.: ICDAR 2009 signature verification competition. In: 2009 10th International Conference on Document Analysis and Recognition, pp. 1403–1407. IEEE (2009)
5. Bouamra, W., Djeddi, C., Nini, B., Diaz, M., Siddiqi, I.: Towards the design of an offline signature verifier based on a small number of genuine samples for training. Expert Syst. Appl. **107**, 182–195 (2018)
6. Brahnam, S., Jain, L.C., Nanni, L., Lumini, A., et al.: Local Binary Patterns: New Variants and Applications. Springer, Heidelberg (2014). https://doi.org/10.1007/978-3-642-39289-4
7. Das, A., Ferrer, M.A., Pal, U., Pal, S., Diaz, M., Blumenstein, M.: Multi-script versus single-script scenarios in automatic off-line signature verification. IET Biom. **5**(4), 305–313 (2016)

8. Diaz, M., Ferrer, M.A., Eskander, G.S., Sabourin, R.: Generation of duplicated off-line signature images for verification systems. IEEE Trans. Pattern Anal. Mach. Intell. **39**(5), 951–964 (2017)

9. Diaz, M., Ferrer, M.A., Impedovo, D., Malik, M.I., Pirlo, G., Plamondon, R.: A perspective analysis of handwritten signature technology. ACM Comput. Surv. **51**(6), 117:1–117:39 (2019). https://doi.org/10.1145/3274658

10. Gattal, A., Djeddi, C., Chibani, Y., Siddiqi, I.: Isolated handwritten digit recognition using oBIFs and background features. In: 2016 12th IAPR Workshop on Document Analysis Systems (DAS), pp. 305–310. IEEE (2016)

11. Gattal, A., Djeddi, C., Chibani, Y., Siddiqi, I.: Oriented basic image features column for isolated handwritten digit. In: Proceedings of the International Conference on Computing for Engineering and Sciences, pp. 13–18. ACM (2017)

12. Griffin, L.D., Lillholm, M.: Symmetry sensitivities of derivative-of-Gaussian filters. IEEE Trans. Pattern Anal. Mach. Intell. **32**(6), 1072–1083 (2010)

13. Griffin, L.D., Lillholm, M., Crosier, M., van Sande, J.: Basic Image Features (BIFs) arising from approximate symmetry type. In: Tai, X.-C., Mørken, K., Lysaker, M., Lie, K.-A. (eds.) SSVM 2009. LNCS, vol. 5567, pp. 343–355. Springer, Heidelberg (2009). https://doi.org/10.1007/978-3-642-02256-2_29

14. Guerbai, Y., Chibani, Y., Hadjadji, B.: The effective use of the one-class SVM classifier for handwritten signature verification based on writer-independent parameters. Pattern Recogn. **48**(1), 103–113 (2015)

15. Hafemann, L.G., Sabourin, R., Oliveira, L.S.: Offline handwritten signature verification - literature review. CoRR abs/1507.07909 (2015). http://arxiv.org/abs/1507.07909

16. Hafemann, L.G., Sabourin, R., Oliveira, L.S.: Writer-independent feature learning for offline signature verification using deep convolutional neural networks. In: 2016 International Joint Conference on Neural Networks (IJCNN), pp. 2576–2583. IEEE (2016)

17. Hafemann, L.G., Sabourin, R., Oliveira, L.S.: Learning features for offline handwritten signature verification using deep convolutional neural networks. Pattern Recogn. **70**, 163–176 (2017)

18. Hou, W., Ye, X., Wang, K.: A survey of off-line signature verification. In: Proceedings of 2004 International Conference on Intelligent Mechatronics and Automation, pp. 536–541. IEEE (2004)

19. Impedovo, D., Pirlo, G.: Automatic signature verification: the state of the art. IEEE Trans. Syst. Man Cybern. Part C (Appl. Rev.) **38**(5), 609–635 (2008)

20. Liwicki, M., et al.: Signature verification competition for online and offline skilled forgeries (SigComp 2011). In: 2011 International Conference on Document Analysis and Recognition, pp. 1480–1484. IEEE (2011)

21. Malik, M.I., et al.: ICDAR 2015 competition on signature verification and writer identification for on- and off-line skilled forgeries (SigWiComp 2015). In: 2015 13th International Conference on Document Analysis and Recognition (ICDAR), pp. 1186–1190. IEEE (2015)

22. Malik, M.I., Liwicki, M., Alewijnse, L., Ohyama, W., Blumenstein, M., Found, B.: ICDAR 2013 competitions on signature verification and writer identification for on- and offline skilled forgeries (SigWiComp 2013). In: 2013 12th International Conference on Document Analysis and Recognition, pp. 1477–1483. IEEE (2013)

23. Newell, A.J., Griffin, L.D.: Natural image character recognition using oriented basic image features. In: 2011 International Conference on Digital Image Computing: Techniques and Applications, pp. 191–196. IEEE (2011)

24. Newell, A.J., Griffin, L.D.: Writer identification using oriented basic image features and the delta encoding. Pattern Recogn, **47**(6), 2255–2265 (2014)
25. Newell, A.J., Griffin, L.D., Morgan, R.M., Bull, P.A.: Texture-based estimation of physical characteristics of sand grains. In: 2010 International Conference on Digital Image Computing: Techniques and Applications, pp. 504–509. IEEE (2010)
26. Ojala, T., Pietikäinen, M., Mäenpää, T.: Multiresolution gray-scale and rotation invariant texture classification with local binary patterns. IEEE Trans. Pattern Anal. Mach. Intell. **7**, 971–987 (2002)
27. Okawa, M.: From BoVW to VLAD with KAZE features: offline signature verification considering cognitive processes of forensic experts. Pattern Recogn. Lett. **113**, 75–82 (2018)
28. Parziale, A., Diaz, M., Ferrer, M.A., Marcelli, A.: SM-DTW: stability modulated dynamic time warping for signature verification. Pattern Recogn. Lett. **121**, 113–122 (2019)
29. Rantzsch, H., Yang, H., Meinel, C.: Signature embedding: writer independent offline signature verification with deep metric learning. In: Bebis, G., et al. (eds.) ISVC 2016. LNCS, vol. 10073, pp. 616–625. Springer, Cham (2016). https://doi.org/10.1007/978-3-319-50832-0_60
30. Serdouk, Y., Nemmour, H., Chibani, Y.: New off-line handwritten signature verification method based on artificial immune recognition system. Expert Syst. Appl. **51**, 186–194 (2016)
31. Sharif, M., Khan, M.A., Faisal, M., Yasmin, M., Fernandes, S.L.: A framework for offline signature verification system: best features selection approach. Pattern Recogn. Lett. (2018)
32. Soleimani, A., Araabi, B.N., Fouladi, K.: Deep multitask metric learning for offline signature verification. Pattern Recogn. Lett. **80**, 84–90 (2016)
33. Tan, X., Triggs, W.: Enhanced local texture feature sets for face recognition under difficult lighting conditions. IEEE Trans. Image Process. **19**(6), 1635–1650 (2010)
34. Tuyls, P., Akkermans, A.H.M., Kevenaar, T.A.M., Schrijen, G.-J., Bazen, A.M., Veldhuis, R.N.J.: Practical biometric authentication with template protection. In: Kanade, T., Jain, A., Ratha, N.K. (eds.) AVBPA 2005. LNCS, vol. 3546, pp. 436–446. Springer, Heidelberg (2005). https://doi.org/10.1007/11527923_45
35. Vapnik, V., Golowich, S.E., Smola, A.J.: Support vector method for function approximation, regression estimation and signal processing. In: Advances in Neural Information Processing Systems, pp. 281–287 (1997)
36. Wayman, J., Jain, A., Maltoni, D., Maio, D.: An introduction to biometric authentication systems. In: Wayman, J., Jain, A., Maltoni, D., Maio, D. (eds.) Biometric Systems, pp. 1–20. Springer, London (2005). https://doi.org/10.1007/1-84628-064-8_1
37. Zhang, Z., Wang, K., Wang, Y.: A survey of on-line signature verification. In: Sun, Z., Lai, J., Chen, X., Tan, T. (eds.) CCBR 2011. LNCS, vol. 7098, pp. 141–149. Springer, Heidelberg (2011). https://doi.org/10.1007/978-3-642-25449-9_18
38. Zois, E.N., Alewijnse, L., Economou, G.: Offline signature verification and quality characterization using poset-oriented grid features. Pattern Recogn. **54**, 162–177 (2016)
39. Zois, E.N., Alexandridis, A., Economou, G.: Writer independent offline signature verification based on asymmetric pixel relations and unrelated training-testing datasets. Expert Syst. Appl. **125**, 14–32 (2019)

Pencil Drawing of Microscopic Images Through Edge Preserving Filtering

Harbinder Singh[1], Carlos Sánchez[2(✉)], Gabriel Cristóbal[2], and Gloria Bueno[3]

[1] Chandigarh Engineering College, Landran, Mohali, India
`harbinder.ece@cgc.edu.in`
[2] Instituto de Óptica (CSIC), Serrano 121, 28006 Madrid, Spain
`carlos.sanchez@io.cfmac.csic.es`
[3] VISILAB, Univ. Castilla la Mancha, Cuidad Real, Spain
`https://www.cgc.edu.in/`

Abstract. Automatic diatom identification approaches have revealed remarkable abilities to tackle the challenges of water quality assessment and other environmental issues. Scientists often analyze the taxonomic characters of the target taxa for automatic identification. In this process the digital photographs, sketches or drawings are recorded to analyze the shape and size of the frustule, the arrangement of striae, the raphe endings, and the striae density. In this paper, we describe two new methods for producing drawings of different diatom species at any stage of their life cycle development that can also be useful for future reference and comparisons. We attempt to produce drawings of diatom species using Edge-preserving Multi-scale Decomposition (EMD). The edge preserving smoothing property of Weighted Least Squares (WLS) optimization framework is used to extract high-frequency details. The details extracted from two-scale decomposition are transformed to drawings which help in identifying possible striae patterns from diatom images. To analyze the salient local features preserved in the drawings, the Scale Invariant Feature Transform (SIFT) model is adopted for feature extraction. The generated drawings help to identify certain unique taxonomic and morphological features that are necessary for the identification of the diatoms. The new methods have been compared with two alternative pencil drawing techniques showing better performance for details preservation.

Keywords: Edge-preserving filters · Diatom identification · Taxonomic · Drawings · Feature analysis

1 Introduction

Diatoms are microscopic microalgae found in aquatic ecosystems that are used as bio-indicators and contribute to the primary production in the aquatic ecosystems. Diatoms have unique taxonomic features called frustule. The main parts of the diatom consist of the nucleus, cell wall, cytoplasm, and plasma membrane.

© Springer Nature Switzerland AG 2019
A. Morales et al. (Eds.): IbPRIA 2019, LNCS 11868, pp. 189–200, 2019.
https://doi.org/10.1007/978-3-030-31321-0_17

To date, different Light Microscopy (LM) and Scanning Electron Microscopy (SEM) techniques have been developed to visualize the microstructures of the frustule. These taxonomic features are important determinants to classify and identify the diatoms. The automation of classification and identification process plays an important role in a wide range of applications that include forensic examination, ecology, and palaeo-ecology. Nowadays most diatom identification methods demand automatic feature extraction models, which can analyze the photographs and drawings of the diatom species. It is therefore imperative that a method for preparing drawings is needed to record the salient diatom features.

In ADIAC project [5], several attempts to diatom identification and classification were reported. For geometric and texture analysis, different descriptors were utilized that include Scale-invariant Feature Transform (SIFT), Gray-Level Occurrence Matrix (GLCM), Fourier Transform (FT) and Gabor wavelets. It was reported that a 97.97% accuracy can be achieved in diatom identification and classification process by using FT and SIFT for the classification of 38 species. Another attempt to explain the diatom classification mechanism is described in DIADIST project [1], in which study on visual indexing of photographs and drawings of microscopic species had been conducted for taxonomic purposes. Extended Depth-of-Focus (DOF) and image-to-drawing conversion methods were applied for taxonomic classification automatically. In a recent study on morphological observation of common pennate diatoms represents that unique features associated with the surface of the frustule are useful for the classification of the diatom [7]. Diatom species were identified based on specific features including the structure of raphe, fibulae and striae, and pore arrangement. Fareha et al. have identified 25 diatom species from the 16 genera from the two estuaries [6].

Drawings have not only morphological or aesthetic interest. They serve to highlight the differential details allowing species discrimination. New diatom taxa need to be accompanied by an illustration showing the morphological differentiation. In recent years there have been some attempts to develop a model for producing drawings of diatom species. McLaughlin has described various methods of drawing generation [14]. The simplest method of creating a drawing is freehand design in which no special apparatus is needed. This process requires observing and recording of fine structures of the frustule manually, which is very time-consuming. Moreover, highly professional and experienced diatomist with a highly developed degree of illustrative ability is required for accurate illustrations. Hicks et al. [12] have proposed an automatic drawing generation model based on frequency and orientation of silica shell patterns, and sternum shapes. Frequency descriptors [20] were utilized to detect the frequency and orientation of silica shell patterns. This model was restricted to the analysis of pennate microscopic species having simple striae patterns on their shells and tested successfully on 12 species. The model was not suitable for species with non-striae patterns, and it seems therefore important to develop automatic drawing generation model that would be adopted for analysis of species with complex and non-striae patterns.

The rest of this paper is organized as follows. The proposed drawing generation models are described in Sect. 2. The experimental results, quantitative and qualitative evaluation, and comparison to other state-of-the-art methods are presented in Sect. 3, and finally, the paper is concluded in Sect. 4.

2   The Proposed Method Using Edge Preserving Filtering

We propose a model for producing drawings of different diatom species based on edge-preserving filter (EPF). We advocate the use of base-detail decomposition technique [8] for multi-scale detail extraction. In many computer graphics and image processing technique, such as HDR tone mapping, detail enhancement [8], image denoising [17] and image fusion [18], it is paramount to operate on images at multiple scales. A number of EPF filter have been proposed [8,11,16,17, 19]. We refer the interested reader to [21] for more comprehensive performance analysis of EPF filters. Among these EPF, the WLS has recently emerged as an excellent tool for multi-scale decomposition, which is suited for single image detail enhancement [8].

WLS Filtering and Base-Detail Layer Decomposition. Prior to generating the drawings of microscopy images of diatom species, we apply multi-scale decomposition, which gives an efficient solution to enhance salient details, while avoiding objectionable artifacts. We would demonstrate that it helps in detecting possible striae patterns from the photographs of diatom species that plays a vital role in producing drawings automatically. In our implementation, let v denote the input microscopy image for which we seek to produce a new image w, which preserve salient edges and contours while smoothing the details between such salient features. For a pixel P, the WLS can be formalized as seeking the minimum of

$$\sum_p \left((w_p - v_p)^2 + \gamma \left(q_{x,p}(v) \left(\frac{\partial w}{\partial x} \right)_p^2 + q_{y,p}(v) \left(\frac{\partial w}{\partial y} \right)_p^2 \right) \right) \qquad (1)$$

where γ is a parameter that controls the influence of smoothness weights: increasing the value of γ yields progressively smoother results w. The goal of the expression term $(w_p - v_p)^2$ is to minimize the distance between w and v, while the second (regularization) term strives to achieve smoothness by minimizing the partial derivatives of w. The smoothness requirement is enforced in a spatially varying manner via the smoothness weights q_x and q_y, which depend on v:

$$q_{x,p}(v) = \left(\left| \frac{\partial l}{\partial x}(p) \right|^\alpha + \epsilon \right)^{-1}, q_{y,p}(v) = \left(\left| \frac{\partial l}{\partial y}(p) \right|^\alpha + \epsilon \right)^{-1} \qquad (2)$$

where l is the log-luminance channel of the input image v, the exponent α determines the sensitivity to the gradients of v, while ϵ is a small constant (default value is 0.0001) that avoids division by zero in regions where v is constant.

Using the WLS filter described above, we seek to extract details at multiple scales, and we rely on decomposition similar to Farbman et al. [8]. More specifically, we extract the base layer b and detail layer d from a given input image v. Let $w^1, \ldots, w^k$ represent progressively coarser versions of v. A set of detail layers d are computed as differences between successive coarser levels, and is defined as follows:

$$di = w^{i-1} - w^i \quad \text{where } i = 1, \ldots, k \text{ and } w^0 = v \tag{3}$$

In our model the input image is repeatedly smoothed, each time increasing the value of parameter α by using a fixed value of γ (typically 1.2). Thus, we define:

$$w^{i+1} = F_{c^i \alpha}(v) \tag{4}$$

We have experimented with a six-layer decomposition, in which the initial value of $\alpha = 0.1$ and $c = 2$. Therefore, α is increased by a factor of 2 at each iteration (i.e. 0.1, 0.2, 0.4, 0.8, 1.6, 3.2). We found that this strategy is suitable to migrate salient features to detail layers for most of the cases while avoiding the effect of noise present in the input images. Several experiments were conducted to decide the free parameters used in WLS filtering. Finally, the detail enhanced image is constructed from the coarsest base layer (i.e. b^6) and six detail layers $d^1, d^2, d^3, d^4, d^5, d^6$, respectively. To avoid resharpening, sigmoid curve $S(a, x) = 1/(1 + exp(-ax))$ is adopted for detail layer manipulation during reconstruction process. More precisely, our detail enhanced image is computed as:

$$\hat{w} = b^6 + \sum_{i=1}^{6} S\left(a_i, d_P^i\right) \tag{5}$$

where a_i is the constant depending on subscript $i = 1, \ldots, k$. Therefore, $a_1, a_2, a_3, a_4, a_5, a_6$ are acting as boosting factors for the manipulation of corresponding detail layers $(d^1, d^2, d^3, d^4, d^5, d^6)$. In our experiments, we have found that $a_1 = 15$, $a_2 = 20$, $a_3 = 80$, $a_4 = 80$, $a_5 = 80$, $a_6 = 80$ are good default setting for most of the cases.

In this paper, we propose two models for generating pencil sketch drawings automatically. The workflow of the first approach called EMD1 is summarized as follows:

1. Load the microscopy image of diatom species.
2. Region of Interest (ROI) selection from input images of diatom species.
3. Construct the detail-enhanced image using Eq. 5.
4. Compute the gradient magnitudes of the image obtained in step 2, using Eq. 6.
5. Construct the negative image of output obtained in step 3, using Eq. 8.
6. Construct the binary image of output obtained in step 4, using Otsu's method [15].

Region of Selection (ROI) Selection. To select a portion of an image that we want to consider for pencil sketch drawing generation, a binary mask is created. In this binary mask image, pixels that belong to the region of interest (ROI) are set to 1 and all other pixels are set to 0. To select the ROI from microscopy images of diatom species, edge detection and morphological filtering based technique [9] is adopted in this paper. In future work, more sophisticated segmentation methods will be adopted for handling multiple touching objects in the microscopy images of diatom species.

Gradient Computation. In order to detect the striae pattern in an image constructed using Eq. 5, we compute the gradient magnitudes at each point, which is formulated as follows:

$$G_{\hat{w}} = \sqrt{Gx^2 + Gy^2} \tag{6}$$

where Gx and Gy are the Sobel convolution kernels:

$$Gx = \begin{bmatrix} -1 & 0 & +1 \\ -2 & 0 & +2 \\ -1 & 0 & +1 \end{bmatrix}, Gy = \begin{bmatrix} +1 & +2 & +1 \\ 0 & 0 & 0 \\ -1 & -1 & -1 \end{bmatrix} \tag{7}$$

We found that this simple tool is very effective for detecting striae patterns in microscopy images of diatom species.

Inversion and Binarization. To obtain the photographic inverted image of image computed from Eq. 6, we subtract the gradient magnitudes of every pixel from 255:

$$I_{neg}(x, y) = 255 - G_{\hat{w}}(x, y) \tag{8}$$

In the last step, an optimum threshold is computed from gray-level histogram [15] for binarization of drawings. Because of simplicity and suitability, we rely on Otsu's method that is based on the minimization of inter-class variance. In this approach, the weighted sum of variances of two classes is utilized for computing the optimal threshold T. The final binary image is obtained by applying thresholding operation on image computed from Eq. 8, which is defined as follows:

$$I_O = \begin{cases} 1 & \text{if} \quad I_{neg} > T, \\ 0 & \text{if} \quad I_{neg} \leq T \end{cases} \tag{9}$$

In the second model of generating pencil sketch drawing called EMD2, all the steps are same except step 3. Instead of computing gradient magnitudes, to detect the image block having salient features such as striae patterns, we apply the filter on $\hat{w}$ having 6×6 kernel:

$$h_e = \begin{bmatrix} 1 & 1 & 1 & 1 & 1 & 1 \\ 1 & 1 & 1 & 1 & 1 & 1 \\ 1 & 1 & -8 & -8 & 1 & 1 \\ 1 & 1 & -8 & -8 & 1 & 1 \\ 1 & 1 & 1 & 1 & 1 & 1 \\ 1 & 1 & 1 & 1 & 1 & 1 \end{bmatrix}, \tag{10}$$

In the forthcoming section, we will compare the results of proposed models with the existing drawing generation models.

3   Results and Analysis

In this section, we first describe two alternative pencil sketch generation methods available in the literature and then show how proposed multi-scale decomposition helps to produce better results. We also compare our method with recently published work by Hicks et al.

Pencil line drawing is a popular technique to record something that can be utilized as a quick way of graphically signifying a photograph. A pencil drawing also allows an artist to record something to create uncomplicated sketches for later use. In DODGE method [2], the variance of Gaussian was used to blur the inverted grayscale input image. The blend mode technique [3] was used to highlights the salient details such as boldest edges. In another method (2D-CONV) [4], 2D convolution filters were used to transform input photographs into pencil line drawings. Different filter banks were suggested to produce appealing pencil line drawings. Inversion and thresholding operations were used to enhance salient details with some precision to reduce noise. At this point, it is interesting to compare our models to alternative methods. Recall that our objective is to be able to extract shape, size, and pattern of striae on silica shells, so we must keep this objective in mind when analyzing and comparing the alternative methods. Figure 1 demonstrates the results generated from EMD1. The drawings generated from EMD1 preserve the shape, size, and pattern of striae on silica shells. As we mentioned earlier in Sect. 1 that Hicks et al. [12] attempted to address the problem of automatic drawing generation of pennate diatom species having striae pattern. For comparison purpose, we illustrate the drawings of four diatom species in Fig. 2, which are generated automatically through a model proposed by Hicks et al. They also seek objective detecting of striae pattern of the input images, whereas as demonstrated in Fig. 1 we explore the possibilities of extracting an original pattern with shape and size of striae on frustule. The proposed work opens several areas of future research to identify and classify the diatom species from drawings.

Figure 3 shows a comparison of our drawings (see Fig. 3(c, f, i, l)) with the results generated from [4] (see Fig. 3(a, d, g, j)) and [2] (see Fig. 3(b, e, h, k)). A comparison with previous drawing generation models indicates that EMD2 model preserves more salient details from out-of-focus and low contrast regions, as demonstrated in Fig. 3(c, f, i, l). Moreover, EMD2 model provides the control needed to generate more compelling drawings from noisy and low contrast microscopy images of diatom species. Adjusting the free parameter associated with EPF filter in Eq. 5 provides an excellent tool for further interactive adjustments, but for a fair comparison, all the free parameters are kept fixed in this paper.

To measure the accuracy of proposed EMD2 model, the Scale Invariant Feature Transform (SIFT) model [13] is adopted for feature extraction. The local

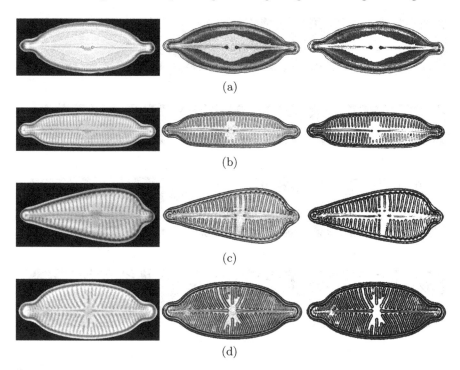

Fig. 1. Results of proposed EMD1 model: input images (left column), pencil sketches (middle column), and binary images (right column) generated automatically from 4 diatom species having striae patterns. (a) *Caloneis amphisbaena*, (b) *Cymbella hybrida*, (c) *Gomphonema augur*, and (d) *Navicula constans*.

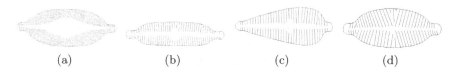

Fig. 2. Drawings of Hicks et al. [12], generated automatically from 4 diatom species having striae patterns. (a) *Caloneis amphisbaena*, (b) *Cymbella hybrida*, (c) *Gomphonema augur*, and (d) *Navicula constans*.

features extracted from SIFT are highly distinctive and invariant to scale change, affine distortion, change in illumination and addition of noise. [13] demonstrated that invariant local feature matching could be extended to image and pattern recognition problems in which local features were matched against a large number of image data sets. Therefore, to analyze the performance of drawing generation models, we must consider a highly distinctive feature transform such as SIFT. The keypoints extracted using SIFT are shown in Fig. 4. The keypoints extracted from the drawings of [4] are illustrated in Fig. 4(a, d, g, j), and Fig. 4(b, e, h, k) demonstrate the keypoints extracted from the drawings of [2].

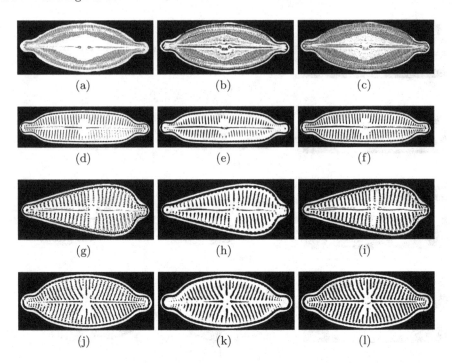

Fig. 3. Comparison of drawings generated from the microscopy images of 4 diatom species: *Caloneis amphisbaena* (top row, left-to-right), *Cymbella hybrida* (second row, left-to-right), *Gomphonema augur* (third row, left-to-right), and *Navicula constans* (bottom row, left-to-right). (a, d, g, j) 2D-CONV [4], (b, e, h, k) DODGE [2], and (c, f, i, l) Proposed EMD2 model.

In the rightmost drawings (see Fig. 4(c, f, i, l)) of proposed EMD2 model, the goal to preserve more distinctive local features has been achieved. A comparison with previous drawing generation models indicates that EMD2 model preserves more keypoints in the generated drawings.

The objective performance analysis of two state-of-the-art pencil line drawing generation methods and proposed models on 4 microscopy images are tabulated in Table 1. In this table, better values are shown in bold (i.e. higher the percentage match of keypoints between input images and drawings is, better the quality of drawings generated). It can be observed from the table that EMD1 has provided better performance for *Gomphonema augur* and *Placoneis constans* image data sets. It can be seen that the DODGE outperforms for *Caloneis amph.* and *Cymbella hybrida* image data sets but EMD2 models yield the second highest value for *Caloneis amph.* data set. Similarly, EMD1 models yield the second highest value for *Cymbella hybrida* data set. Therefore, from the simulation results shown in Table 1, it is clear that among the considered state-of-the-art models, the proposed models have shown better performance.

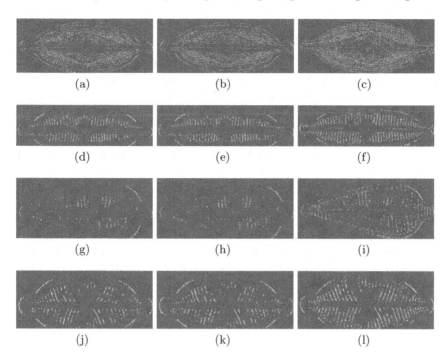

Fig. 4. Visualization of keypoints extracted from drawings of 4 diatom species using SIFT [13]: *Caloneis amphisbaena* (top row, left-to-right), *Cymbella hybrida* (second row, left-to-right), *Gomphonema augur* (third row, left-to-right), and *Navicula constans* (bottom row, left-to-right). (a, d, g, j) 2D-CONV [4], (b, e, h, k) Dodge [2], and (c, f, i, l) Proposed EMD2 model. The keypoint are extracted from drawings shown in Fig. 3. In order to better visualize the keypoints, a constant value is multiplied to each pixel e.g. 200.

It is well known that textural feature analysis measures the visual quality of images from different aspects and also useful in image classification problems [10]. In this paper, in addition to visual inspection, to assess the performance of proposed models and other schemes, we consider four textural features computed from grayscale spatial dependencies [10]. In particular, we consider the following Haralick's co-ocurrence features,

- Angular Second Moment (ASM), which is a measure of homogeneity of the image.
- Contrast (C), which is a measure of local intensity variations.
- Difference Variance (DV).
- Difference Entropy (DE), which is a measure of information.

For the 4 images of diatom species, comprehensive texture analysis comparisons of proposed models with other schemes are given in Table 2. From Table 2, we can observe that the C, DV, and DE values of EMD1 are always

Table 1. Quantitative analysis of keypoints extracted from microscopy input images and corresponding drawings using SIFT.

Image Name	Input	DODGE	2D-CONV	EMD1	EMD2
Caloneis amph.	1322	1472	1503	1261	2255
	Feature Matched	**155 (11.7%)**	79(5.97%)	31(2.34%)	126(9.53%)
Cymbella hybrida	786	573	1503	1120	705
	Feature Matched	**35(4.45%)**	15(1.90%)	24(3.05%)	17(2.16%)
Gomphonema augur	616	186	489	1253	471
	Feature Matched	20(3.25%)	14(2.27%)	**25(4.06%)**	17(2.76%)
Placoneis constans	889	452	356	2061	700
	Feature Matched	40(4.50%)	25(2.81%)	**77(8.66%)**	48(5.40%)

the largest, and the ASM of *Caloneis amph.* and *Cymbella hybrida* images are close the highest value. Therefore, the proposed drawing generation models can more effectively extract prominent salient details from input images of diatom species.

Table 2. Comprehensive texture analysis of drawings of microscopy images.

Image	Metric	DODGE	2D-CONV	Hicks	EMD1	EMD2
Caloneis amph.	ASM	0.432	0.434	**0.495**	0.465	0.435
	C	4.967	3.561	0.594	**6.621**	4.906
	DV	4.805	3.450	0.588	**6.400**	4.746
	DE	0.328	0.261	0.078	**0.396**	0.325
Cymbella hybrida	ASM	0.464	0.457	**0.518**	0.516	0.458
	C	1.984	2.203	0.297	**4.871**	2.576
	DV	1.929	2.14	0.302	**4.712**	2.500
	DE	0.169	0.183	0.040	**0.324**	0.206
Gomphonema augur	ASM	0.488	0.465	0.499	**0.550**	0.492
	C	1.942	2.613	0.223	**3.994**	2.490
	DV	1.888	2.535	0.230	**3.867**	2.417
	DE	0.167	0.208	0.030	**0.282**	0.201
Placoneis constans	ASM	0.452	0.440	0.528	**0.558**	0.455
	C	2.908	3.370	0.483	**6.558**	3.469
	DV	2.820	3.265	0.481	**6.340**	3.361
	DE	0.225	0.250	0.058	**0.394**	0.256

4   Conclusions

In this paper, we proposed two models for generating pencil sketch drawings from the images of diatom species. Our proposed models are based on edge-preserving smoothing property of WLS optimization framework. We exhibit the potential of WLS filter for base-detail decomposition, which greatly improves the performance of the proposed models. The experimental results clearly demonstrate that the proposed models can extract the original pattern of stria with shape and size better than other drawing generation models. It should be noted that the WLS fixed parameter set used in multi-scale decomposition can generally obtain good results for most of the images of diatom species. Despite this, we also hypothesize that better results can be generated by choosing different parameters settings according to different types of source images of diatom species, which should be investigated in our future work. In future work, authors would also like to investigate the applicability of proposed models to images of diatom species having non-stria patterns and for other micro-algae organisms.

Acknowledgements. The authors acknowledge financial support of the Spanish Government under the Aqualitas-retos project (Ref. CTM2014-51907-C2-R-MINECO). The authors thank Saúl Blanco for providing subjective evaluation of the results.

References

1. British Diatomists of the 19th Century Database. http://rbg-web2.rbge.org.uk/DIADIST/dia_intro.htm
2. How to turn any image into a pencil sketch with 10 lines of code. https://bit.ly/2FgkzCV. Accessed 28 Apr 2019
3. Photoshop Blend Modes Explained. https://bit.ly/2RtJU1z. Accessed 28 Apr 2019
4. Simple filters and pencil line drawing effect. https://bit.ly/2Vzn1Pg. Accessed 28 Apr 2019
5. du Buf, H., Bayer, M.M.: Automatic Diatom Identification. Series in Machine Perception and Artificial Intelligence, vol. 51, pp. 1–316. World Scientific, Singapore (2002)
6. Hilaluddin, F., Leaw, C.P., Lim, P.: Fine structure of the diatoms Thalassiosira and Coscinodiscus (Bacillariophyceae): light and electron microscopy observation. Ann. Microsc. **10**, 28–35 (2010)
7. Hilaluddin, F., Leaw, C.P., Lim, P.: Morphological observation of common pennate diatoms (Bacillariophyceae) from Sarawak estuarine waters. Ann. Microsc. **11**, 12–23 (2011)
8. Farbman, Z., Fattal, R., Lischinski, D., Szeliski, R.: Edge-preserving decompositions for multi-scale tone and detail manipulation. ACM Trans. Graph. **27**(3), 67:1–67:10 (2008)
9. Gonzalez, R.C., Woods, R.E.: Digital Image Processing. Prentice Hall, Upper Saddle River (2008)
10. Haralick, R.M., Shanmugam, K., Dinstein, I.: Textural features for image classification. IEEE Trans. Syst. Man Cybern. **SMC-3**(6), 610–621 (1973). https://doi.org/10.1109/TSMC.1973.4309314
11. He, K., Sun, J., Tang, X.: Guided image filtering. IEEE Trans. Pattern Anal. Mach. Intell. **35**(6), 1397–1409 (2013)
12. Hicks, Y., Marshall, D., Rosin, P.L., Martin, R.R., Mann, D.G., Droop, S.: A model of diatom shape and texture for analysis, synthesis and identification. Mach. Vis. Appl. **17**(5), 297–307 (2006)
13. Lowe, D.G.: Distinctive image features from scale-invariant keypoints. Int. J. Comput. Vis. **60**(2), 91–110 (2004)

14. McLaughlin, R.B.: An Introduction to the Microscopical Study of Diatoms. Hooke College of Applied Sciences, Westmont (2012)
15. Otsu, N.: A threshold selection method from gray-level histograms. IEEE Trans. Syst. Man Cybern. **9**(1), 62–66 (1979). https://doi.org/10.1109/TSMC.1979.4310076
16. Paris, S., Hasinoff, S.W., Kautz, J.: Local Laplacian filters: edge-aware image processing with a Laplacian pyramid. ACM Trans. Graph. **30**(4), 68:1–68:12 (2011)
17. Perona, P., Malik, J.: Scale-space and edge detection using anisotropic diffusion. IEEE Trans. Pattern Anal. Mach. Intell. **12**(7), 629–639 (1990)
18. Singh, H., Kumar, V., Bhooshan, S.: Weighted least squares based detail enhanced exposure fusion. ISRN Sig. Process. **2014**, 18 (2014)
19. Tomasi, C., Manduchi, R.: Bilateral filtering for gray and color images. In: Proceedings of the Sixth International Conference on Computer Vision, ICCV 1998. IEEE Computer Society (1998)
20. Zahn, C.T., Roskies, R.Z.: Fourier descriptors for plane closed curves. IEEE Trans. Comput. **C-21**(3), 269–281 (1972)
21. Zhu, F., Liang, Z., Jia, X., Zhang, L., Yu, Y.: A benchmark for edge-preserving image smoothing. IEEE Trans. Image Process. **28**, 3556–3570 (2019). A publication of the IEEE Signal Processing Society

Line Segmentation Free Probabilistic Keyword Spotting and Indexing

Killian Barrere[1], Alejandro H. Toselli[2(✉)], and Enrique Vidal[2]

[1] Univ Rennes, Rennes, France
[2] Universitat Politècnica de València, Valencia, Spain
alto@upv.es

Abstract. Probabilistic Keyword Spotting and Indexing (PKWSI) allows effective search through untranscribed large collections of images. However, when text-line detection fails to detect foreground text, the PKWSI techniques also fail dramatically. In this paper, we develop a new line segmentation-free approach using a uniform line-sized image slicing instead of previous text-line detection. As a result, new issues arise due to overlapping slices, leading to several spot hypotheses for the same word. We develop solutions to take advantage of multiple spots and to consolidate them into single hypotheses. We test our approach on a difficult historical handwritten dataset and it yields promising results.

Keywords: Keyword spotting · Probabilistic indexing ·
Handwritten text recognition · Segmentation free

1 Introduction

In recent works, Probabilistic Keyword Spotting and Indexing (PKWSI) has proven to be a promising approach to make the textual contents of untranscribed handwritten text images accessible [1,4,8,12]. This applies to search and retrieval, as well as to other highly demanded information extraction tasks [4,8]. PKWSI has achieved in recent years a great level of maturity, allowing effective and accurate access to textual contents of large collections of handwritten text images.

One remaining bottleneck of this technology is the need for previous detection and extraction of the relevant text lines. While this can be very reliably done for clean, well-written handwritten documents, it becomes a problematic bottleneck for many historic manuscript collections. Clearly, when the text-line detection fails to detect foreground text, the PKWSI techniques also fail dramatically, often leading to useless probabilistic indices of many images of the collection.

In this paper, we develop and test a new PKWSI approach which does not rely on previous line detection. Instead, each image is uniformly scanned vertically into line-sized rectangular image slices. In this new approach, a new issue arises due to the fact that several line-shaped slices often become all relevant, leading to several spot hypotheses for the same word. We study this problem formally

© Springer Nature Switzerland AG 2019
A. Morales et al. (Eds.): IbPRIA 2019, LNCS 11868, pp. 201–213, 2019.
https://doi.org/10.1007/978-3-030-31321-0_18

and develop solutions to consolidate multiple spots of the same word into a single spot with its corresponding relevance probability. The new approach is tested on a difficult dataset of historical handwritten images and the experiments yield promising results.

In this paper, Sect. 2 starts by describing current keyword spotting approaches. Then we introduce and explain formally our approach in Sect. 3. Following that, we present the different experiments and results in Sect. 4.

2   Probabilistic Keyword Spotting and Indexing

Keyword spotting can be seen as a binary classification problem to decide whether a particular image region x is relevant for a given query word v, i.e. try to answer the following question: "Is v actually written in x?". As in [8], we aim to compute the image region word relevance probability $P(R = 1 \mid X = x, V = v)$. From now on, for the sake of clarity, we will omit the random variable names, and for $R = 1$, we will simply write R.

Image region word relevance probabilities are computed without taking into account where the considered word may appear in the region x. Nevertheless, it should be pointed out that the precise positions of the words within x are easily obtained as a byproduct. The relevance probability of an image region x for a keyword v, $P(R \mid x, v)$, is approximately computed as:

$$P(R \mid x, v) \approx \max_{b \sqsubseteq x} P(R \mid x, b, v) \qquad (1)$$

where b is a word-sized image region or Bounding Box (BB), and with $b \sqsubseteq x$ we mean the set of all BBs contained in x [8,12–14].

Since b is assumed to (tightly) contain only one word (hopefully v), it is straightforward to see that $P(R \mid x, b, v) = P(v \mid x, b)$. This is just the posterior probability needed to "recognize" the BB image (x, b), or more formally speaking, to classify the BB (x, b) into one of the possible words of some vocabulary. The approximate computation of this probability is carried out using processing, training, optical and language models steps, similar to those employed in handwritten text recognition, even though no actual text transcripts are produced in PKWSI. Instead, for each image x, the distribution $P(R \mid x, b, v)$ itself is obtained and adequately indexed to allow efficient textual search and information retrieval [8,12–14].

It is important to remark that the PKWSI framework is not limited to just a single word query; it can also accommodate sequences of words or characters [8]. This raises a distinction into the two approaches referred to as Lexicon-Based (LB) and Lexicon-Free (LF). In general, LB methods are known to be faster and more accurate than LF ones. However, since LB PKWSI relies on a predefined lexicon, fixed in the training phase, it does not support queries involving out-of-vocabulary keywords.

On the other hand, the LF PKWSI approach works usually at character level, but it attempts to keep the good performance of LB indexing by actually producing relevance probabilities for what character sequences called *pseudo-words*,

which are automatically "discovered" in the very test images being indexed
[8,11]. This approach has proved to be very robust, and it has in actually been
used to very successfully index two iconic large collections: The French Chancery
Collection [1],[1] and BENTHAM PAPERS,[2] containing 90 000 manuscript images
written in old English.

Target Image Regions
Up to this point we have not clearly specified what the image regions x are.
Depending on the size and shape of x, Eq. (1) may become more or less difficult
to compute. In the traditional keyword spotting literature, word-sized regions
have often been considered. This is reminiscent of segmentation-based methods
which required previously cropped accurate word BBs. However, as discussed
in Sect. 1, this is not realistic for large image collections. More importantly, by
considering isolated words, it is difficult for the underlying word recognizer to
take advantage of word linguistic contexts to achieve good spotting precision.

On the other hand, we may consider whole page images, or relevant *text
blocks* thereof, as the search target image regions. While it can be sufficiently
adequate for many textual content retrieval applications, a page may typically
contain many instances of the word searched for and, on the other hand, users
generally like to get narrower responses to their queries. A particularly interesting intermediate search target level consists of *line-shaped regions*. Lines are
useful targets for indexing and search in practice and, in contrast with word-
sized image regions, lines generally provide sufficient linguistic context to allow
to compute accurate word classification probabilities. Moreover, as discussed in
[8,12,14], line region relevance probabilities can be very efficiently computed.

3   Full Segmentation-Free PKWSI

So far, to use line-shaped image regions in PKWSI, it is assumed that text lines
have previously been detected. As discussed in Sect. 1, it constitutes a serious
bottleneck. Here we propose a new approach which does not rely on previous
line detection. Instead, each image is scanned vertically and is uniformly sliced
into line-shaped, rectangular image regions, where the methods discussed in
previous sections are applied. Thanks to the robustness of the relevance probability estimates, those image regions where no text is actually present generally
get low probability for any word, while in regions which actually contain text,
word relevance probabilities become high as in the previous line-based approach.
According to the usual terminology in the field of keyword spotting [3], we will
refer to our new approach as Line Segmentation-Free PKWSI (LSF-PKWSI).

3.1   Vertical Sampling of Line-Shaped Image Regions

The principle of vertical sampling is to extract consecutive line-shaped rectangular images using a page-wide sliding window of fixed height. This window

---

[1] http://prhlt-kws.prhlt.upv.es/himanis.
[2] http://prhlt-carabela.prhlt.upv.es/bentham.

determines the region to extract and is shifted by a fixed number of pixels at each step. While in line-based PKWSI there is one line-image per line of text, in LSF-PKWSI there are typically several overlapping rectangular windows, possibly containing the same (parts of a) text line. We also expect a lot of regions without text; mainly in the borders of pages and between columns, if they have more than one.

Figure 1 shows several consecutive windows resulting from the proposed vertical sampling process. It showcases that the method can extract relevant image regions even when text lines are significantly slanted. In the example, we expect to retrieve all the existing words, but on separate sliding windows.

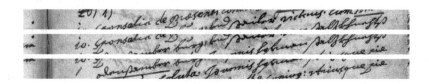

Fig. 1. Consecutive windows extracted with a Vertical Sampling Rate of 1.5.

By extracting more windows than those strictly required, LSF-PKWSI aims to avoid the need for text-line detection, thereby circumventing problems related with no keywords being spotted in image regions where lines are poorly detected.

3.2 Estimating Vertical Sampling Parameters

To adapt the vertical sampling to the writing density of each text image, an estimate of the height of the text lines composing each page is needed.

We are aware that there are already very good text-line detection methods [2]. Obviously, such systems could be used to estimate line heights, but we believe that a much simpler method should be sufficient for our purposes. In comparison, state-of-the-art line detection methods require the usage of Artificial Neural Networks and may require training. In contrast, our method is mainly based on applying Fourier transform on a signal representing the amount of ink in each row of the image. Figure 2 illustrates the whole process.

To obtain our estimates, we first pre-process the page images. We apply a Gaussian Blur to remove the high frequencies and the noise. Then, we binarize the image at a local level, using the Sauvola algorithm [10]. By applying a horizontal Run Length Smoothing Algorithm (RLSA) [15], we aim to highlight the text-line-like regions. After that, we compute the average number of black pixels in each row to obtain a signal which approximately represents line vertical positions. Finally, we apply a Fourier transform to estimate several line heights, $\hat{h}$, as the highest values of the amplitude spectrum.

Once line heights are estimated, two important parameters remain to be determined. The actual height of the sampling window, which is proportional to $\hat{h}$. We are referring to this proportional factor as the *Height Factor*. Then,

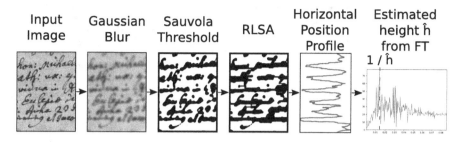

Fig. 2. Steps of the process to estimate the height of line-shaped windows.

the *Vertical Sampling Rate* which represents the number of extracted window images a pixel belongs to. It impacts the vertical overlapping of the sampling windows and the distance between each consecutive window.

Both parameters affect the chance that words are spotted correctly, hopefully leading to better relevance probabilities. However increasing that chance might also affect the speed and memory consumption of the whole process.

From what we have tested so far, this approach seems to be good for the dataset we are using. However, in datasets where the text orientation highly differs from the horizontal, this process would not be suitable.

3.3 Consolidating Multiple Spots of the Same Word Region

Figure 3 shows different kinds of BBs obtained with LSF-PKWSI. In a typical case, we obtain many high relevance-probability BBs which significantly overlap with each other around a true-positive spot (Fig. 3a). Conversely, false-positive spots often correspond to "lone" BBs with no other BBs for the same word in their neighborhood (Fig. 3b). We also observe very elongated BBs in areas of the image where there is no text, generally with low relevance-probability (Fig. 3d). However, elongated BBs containing a wide area without text, followed by a word on their right side may also appear (Fig. 3c). They appear to have high relevance-probability for the true-positive word. In future works we will try to avoid this kind of spots, which we believe are mostly due to a mismatch between the shapes and sizes of the line (shaped) image regions considered for training and testing.

Users generally want a single spot for each keyword. It is also expected to have a high difference in the relevance probability between true-positives and false-positives. Based on the previous observations (Fig. 3), we should group all the BBs that are overlapping into a single spot with a high relevance-probability, while either discarding or lowering the relevance-probability of lone BBs.

Let x be again a full image and b the true BB of a word-sized image region for the keyword v. In LSF-PKWSI, b is unknown, but there are several word-sized regions β, associated with b, where v is likely to be written. $P(R \mid x, b, v)$ can be computed by considering all possible BBs, β, for which v may be relevant:

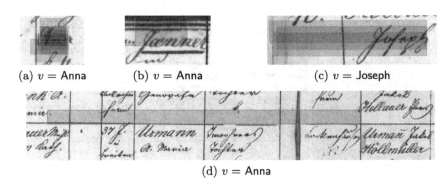

(a) $v = $ Anna   (b) $v = $ Anna   (c) $v = $ Joseph

(d) $v = $ Anna

Fig. 3. Different situations arising with our method: (a) a true-positive with many overlapping BBs, (b) a lone false-positive, (c) a true-positive with elongated BBs, and (d) a very elongated false-positive.

$$P(R \mid x, b, v) \equiv P(v \mid x, b) = \sum_{\beta} P(v, \beta \mid x, b) = \sum_{\beta} P(\beta \mid x, b) P(v \mid x, b, \beta)$$
$$\approx \sum_{\beta \sqcap b} P(\beta \mid x, b) P(v \mid x, \beta) \tag{2}$$

We say that two BBs β_1 and β_2 (viewed as sets of pixels of x) θ-*significantly overlap* (written as $\beta_1 \sqcap \beta_2$) if $|\beta_1 \cap \beta_2| / |\beta_1 \cup \beta_2| \geq \theta$, where θ, $0 < \theta \leq 1$, is a fixed parameter for the minimum fraction of pixels that β_1 and β_2 must share (a typical value for θ is 0.5). In this case we assume that v is conditionally independent of b given β, otherwise (i.e. $\beta \not\sqcap b$), $P(\beta \mid x, b) = 0$.

In plain words, the relevance probability of b for v is computed as a weighted average of the relevance probabilities of all the BBs associated with v (obtained as explained in Sect. 2). Therefore, a best BB $\hat{b}$, for some image region where v is written should be one that maximizes $P(v \mid x, b)$. According to Eq. (2):

$$\hat{b} = \arg\max_{b} \sum_{\beta \sqcap b} P(\beta \mid x, b) \, P(v \mid x, \beta) \tag{3}$$

An algorithmic solution to this optimization problem does not seem easy. Thus, we leave it for future studies. Here, we instead adopt a simple but effective heuristic approach, as discussed later in this section.

The weights, $P(\beta \mid x, b)$, of Eqs. (2) and (3) can be considered as prior probabilities of the different BBs relevant for v, conditioned by the position and shape of the unknown true BB, b. For a BB β, this probability should be high if it shares significant parts with b (and also with other nearby BBs), and low for lone BBs. In addition, this prior should be high if the shape (size) of β is adequate to hold v, and should be low if it is too large or too small for v. Based on that we built Algorithm 1.

Algorithm 1 has two parameters: θ (explained above) and τ. τ is the minimum overlapping fraction between all the BBs $\in g$ and a new BB β in order to be

---

Algorithm 1. Create groups of Bounding Boxes of a given word v

---
Start with no groups; i.e. $G = \emptyset$
for all BBs β of v **do**
    for all $g \in G$ **do**
        if β overlaps with more than τ of the BBs $\in g$ **then** Insert β in g
    if β was not inserted in any group **then** Create $g = \{\beta\}$; $G = G \cup \{g\}$

---

added into g. Algorithm 1 returns groups of overlapping BBs. Each BB β in a group g share at least a fraction θ of its pixels with at least a fraction τ of the other BBs in the same group g.

For each relevant BB group g produced by Algorithm 1, a single merged BB $\hat{b}$ (as in Eq. (3)) has to be obtained, for which an estimate of $P(\beta \mid x, \hat{b})$ is initially needed. We explored several approximations to this single-BB prior probability. Based on these, the simple heuristic presented in Eq. (4) (which ignores the dependence on b) gave the best empirical results:

$$P(\beta \mid x, \hat{b}) \approx P(\beta) = F\left(\sum_{\beta' \in g: \beta' \neq \beta} \frac{|\beta \cap \beta'|}{|\beta \cup \beta'|} \right) \qquad (4)$$

For each BB β, we sum the relative overlap for each other BB β' and apply a customized sigmoid function, $F(x) = 1 / (1 + e^{-ax+b})$, with parameters tuned.

Then, we compute the merged BB (our heuristic approximation to $\hat{b}$) as a weighted sum of the coordinates of all overlapping β's in g; that is:

$$\hat{b} \equiv \hat{\boldsymbol{b}} = \sum_{\beta \in g} P(\beta) P(v \mid x, \beta) \, \boldsymbol{\beta} \qquad (5)$$

with $\hat{\boldsymbol{b}}$ and $\boldsymbol{\beta}$ the 4-dimensional vectors of the coordinates (center and size) of $\hat{b}$ and β, respectively.

Finally, to obtain $P(v \mid x, \hat{b})$ following Eq. (2), we should do a weighted average, as in Eq. (5). However, additional experiments suggest that somewhat better results are obtained by using the simpler maximum approximation:

$$P(v \mid x, \hat{b}) \approx \max_{\beta \in g} P(\beta) P(v \mid x, \beta) \qquad (6)$$

Examples of merged BBs could be seen in Fig. 4b and c or by using both Raw and Consolidated demonstrators which are explained in Sect. 5.

4 Experiments and Results

Assessment measures, data set and partitions, query sets, experimental setup and results are presented in this section.

4.1   Assessment

PKWSI performance is assessed in terms of standard *recall* vs. *interpolated precision* curves [5], from which the *average precision* (AP) [6] and the mean AP (mAP) [9,16] are obtained. While the AP is computed from a *global* ranked list containing all the results from all queries, the mAP is the mean of the APs of the individual queries. For the results presented in this section, AP and mAP have been computed using a publicly available tool called KwsEvalTool.[3]

In line-based PKWSI, these scores are computed at line level. Yet the transcripts of the image windows extracted by the LSF-PKWSI approach are not (precisely) known. To compare with previous results, we then decided to use a fair evaluation that could be used in both line-based PKWSI and LSF-PKWSI, without incurring the high cost of creating a detailed BB-based ground truth.

To this end, a simple idea is to evaluate the performance at the page level. This amounts to ask whether each keyword is written on a page or not. We compute the relevance probability at the page level for each keyword v, by taking the maximum of the $P(v \mid x, \hat{b})$ from Eq. (6) according to Eq. (1). It should be noted that, however with this approximation, we ignore repeated occurrences of keywords in each given page, which may be rather likely for some keywords.

4.2   Dataset and Experimental Partition

The 289 images of the dataset used in this work were selected from a subset of 57 222 scans of more than 800 000 sacramental register images belonging to the Passau Diocesan Archives[4]. The images show a great variety in the evolution of handwriting, record keeping and more and more standardized table forms introduced in the early 19th century. For more details about this collection and dataset, refer to [4].

Table 1 shows relevant details of the dataset used in this work. 179 images were selected for training, 21 for validation and the remaining 89 for testing. The large number of test-set image windows needed in our approach typically requires relatively important amounts of computing time and space. Hence, we decided to select a small subset of 10 test-set images from which we obtained the first encouraging results. This subset was later used as a further validation set to tune the parameters of the methods discussed in Sect. 3. It includes 4 images with large tables and 6 pages without. We will refer to this set as TestVal.

4.3   Query Set

The query set used in this work was adopted according to the most common criteria, where most of the words seen in the test set are chosen as keywords. Besides being a meaningful choice from an application point of view, it ensures that all the keywords are relevant (appear in one or more test images), thereby

---

[3] https://github.com/PRHLT/KwsEvalTool.git.
[4] Openly available at http://data.matricula-online.eu/de/deutschland/passau.

Table 1. The Passau experimental dataset. All the figures correspond to a transliterated version where all letters were case and diacritics folded.

Number of:	Train+Val	TestVal	Test	Total
Pages	200	10	89	289
Lines	29 314	2 053	16 376	45 690
Running words	72 848	5 204	37 354	110 207
Running words excluding punctuation	–	3 712	26 709	15 141
Different words	11 160	1 191	5 801	16 169
Different characters	99	48	87	102
Query words	–	1 043	5 725	–

allowing mAP to be properly computed. All the test-set words longer than 1 character are used, making a total of 1 043 and 5 725 transliterated keywords for the TestVal and the Test partitions respectively (see Table 1).

4.4 Experimental Setup

As discussed in Sects. 2 and 3, a primary step in the LSF-PKWSI approach is to obtain the $P(v \mid x, b)$ (as a byproduct of computing $P(R \mid x, v)$), see Eq. (1)). As shown in [8,12–14], a very appropriate way to obtain this relevance probability is by using previously trained optical and language models, similar to those employed in handwritten text recognition. In this work, as in [4], we use a Convolutional Recurrent Neural Network (CRNN) [7] for character optical modeling and a 6-gram character language model. Details about the (meta-)parameter settings employed for training/decoding with these models and producing the required relevance probabilities can be seen in [4,7].

Most of the experiments have been carried out with the TestVal set and in all of them, performances are measured at the page level. For the 89 pages of the entire Test set, the process requires 72h on a GeForce GTX 1080 and a 2 core Intel Core i3-6100 CPU. This time can be drastically reduced in several ways, but we believe that this is a secondary goal, since our main aim is to prove that LSF-PKWSI can bring competitive results.

Regarding the pre-processing of the page images (required to obtain an estimate of the line height $\hat{h}$) as described in Sect. 3.2, we used a kernel size of 25 for the Gaussian blur, a local thresholding covering areas of 201×201 pixels and a value of 20 for the RLSA parameter. We obtained the two best line-height estimates using a fast Fourier transform.

Then, we optimized the vertical sampling parameters (Height Factor and Vertical Sampling Rate). Table 2 shows the results we obtained on the TestVal set by using LSF-PKWSI without any consolidation of the BBs. Despite the fact that after a value of 2, the Vertical Sampling Rate does not seem to impact a lot the AP, we believe that it matters more after the consolidation process. Based on the above reason, the AP obtained and the resources usage, we extracted lines of height $1.25 \cdot \hat{h}$, with a Vertical Sampling Rate of 3 windows per unit of height $\hat{h}$. We leave as future works the continuation of these experiments, as it might have an important impact.

Table 2. AP measured when changing the parameters of the line extraction process. Both Height Factor and Vertical Sampling Rate depend of the estimated line height $\hat{h}$. The results are obtained on the TestVal set at the page level.

Height factor	Vertical sampling rate			
	1	2	3	4
1.0	–	–	0.658	–
1.25	0.641	0.712	**0.714**	0.722
1.5	–	–	0.660	–

Then, we tuned the vertical sampling parameters. First, we dealt with the parameters for Algorithm 1. For θ, which is the minimum fraction of pixels that two BBs have to share to be considered as overlapping BBs, we used a value of 0.5. Concerning the parameter τ, we selected a value of 0.2. It is used in Algorithm 1 as the minimum fraction of BBs in a group g a new BB β has to overlap with to be inserted in g.

Finally, for the value of the customized sigmoid, $F(x) = 1 / (1 + e^{-ax+b})$ we used $a = 8$ and $b = 2.75$. It allows keeping most of the BBs overlapping with each other, while lowering the probability of lone BBs.

Table 3. Comparison between the results obtained with line-based PKWSI method (Baseline), and our method before (LSF-PKWSI Raw) and after being consolidated (LSF-PKWSI Consolidated) as described in Sect. 3.3.

Experiment	TestVal		Test	
	mAP	AP	mAP	AP
Baseline	0.90	0.84	0.73	0.73
LSF-PKWSI Raw	0.88	0.72	0.62	0.54
LSF-PKWSI Consolidated	**0.89**	**0.76**	**0.64**	**0.60**

4.5   Results

Table 3 shows a comparison of our new approach and the previous results obtained with manually extracted lines. We compare the line-based PKWSI (Baseline) with our approach before consolidation (i.e. with overlapping BBs, referred to as LSF-PKWSI Raw) and after the consolidation, as described in Sect. 3.3 (LSF-PKWSI Consolidated). The consolidated LSF-PKWSI test set results are 0.13 points behind the baseline in terms of AP and 0.09 points in terms of mAP.

The LSF-PKWSI approach obtained three times more spots on the test set than the baseline method, with many of them being false-positives. We believe that it could be explained by the number of lines extracted and also their width.

Using a relevance probability threshold for which precision is equal to recall, our new approach obtained 17% less true-positives (*hits*) and 44% more false-positives than the baseline. Therefore, to further improve the LSF-PKWSI results, we should focus on reducing the number of false-positives. However, at this point, the number of spots detected by the baseline is similar to those detected by LSF-PKWSI (less than 0.1% difference). This significant improvement with respect to the previous three-fold difference shows that, although we do obtain many raw spots, most of them are false-positives with a very low probability. They can be easily filtered out for queries.

It is worth noting that our approach is often capable to outperform the baseline. In cases where the provided text-lines were not correct (e.g. because of mistakes in the ground truth), the baseline either fails or obtains low probabilities (Fig. 4a), whereas LSF-PKWSI is able to obtain a correct spot with a high probability (Fig. 4b and c).

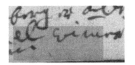 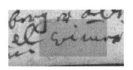 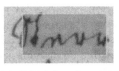

(a) $P(R \mid \mathsf{Wimer}) = 0.027$ (b) $P(R \mid \mathsf{Wimer}) = 0.99$ (c) $P(R \mid \mathsf{Sterr}) = 0.94$
        (Baseline)   (LSF-PKWSI)   (LSF-PKWSI)

Fig. 4. Two examples where LSF-PKWSI performs better than the baseline. In (b) LSF-PKWSI leads to better score and BB compared to the baseline (a), where the provided text line was wrong. (c) is an example where LSF-PKWSI spotted the correct keyword (**Sterr**) while the baseline failed.

Moreover, with our hands-on experience using the demonstrators (links in Sect. 5), we feel that the practical performance is really better than the results of Table 3 suggest. Hence, we believe that improving the results by a fair amount should not be difficult.

5   Conclusions

A new, line segmentation-free approach to probabilistic keyword spotting and indexing has been introduced and tested in a series of experiments with a difficult dataset of historical handwritten images. Despite having results a bit lower than the baseline, they are still promising and we are confident about their possible improvements. To allow practical testing of three different approaches (baseline,

raw and consolidated spot BBs) three demonstrators have been implemented for
the TestVal set and are publicly available: baseline,[5] where text lines were manually detected; line segmentation-free, raw spot BBs,[6] without post-processing;
and line segmentation-free, consolidated by merging overlapping BBs[7]. A demon-
strator for the full Test set is also available.[8]

The similar idea can be used for other applications such as probabilistically
indexing text in natural scene images which may include text regions, or even
to probabilistically index other objects of interest.

Future works will be devoted to improve the consolidation of word BBs. Especially, we expect to improve the results by sticking to the formal development
and avoiding heuristics and tunable parameters as much as possible. In addition,
we also plan to carry out experiments to measure precision-recall performance at
the word BB level, rather than the rough, page-image level assessment reported
in this paper. Moreover, we want to compare our approach with state-of-the-art
methods using automatic text-line extraction, and also with full segmentation
free approaches instead of manually extracted text lines as in the current baseline results. Lastly, we might also consider improving the computing time and
memory taken by our approach.

Acknowledgements. This work was partially supported by the BBVA Foundation
through the 2017–2018 and 2018–2019 Digital Humanities research grants "Carabela"
and "HistWeather – Dos Siglos de Datos Climáticos", and by EU JPICH project
"HOME – History Of Medieval Europe" (Spanish PEICTI Ref. PCI2018-093122). This
is also partially supported by the École Normale Supérieure de Rennes and Univ Rennes
through a funded internship grant.

References

1. Bluche, T., et al.: Preparatory KWS experiments for large-scale indexing of a vast medieval manuscript collection in the HIMANIS project. In: 14th ICDAR (2017)
2. Diem, M., Kleber, F., Fiel, S., Grüning, T., Gatos, B.: CBAD: ICADR 2017 com-petition on baseline detection. In: 2017 14th IAPR International Conference on Document Analysis and Recognition (ICDAR), vol. 1, pp. 1355–1360. IEEE (2017)
3. Giotis, A.P., Sfikas, G., Gatos, B., Nikou, C.: A survey of document image word spotting techniques. Pattern Recogn. **68**, 310–332 (2017)
4. Lang, E., Puigcerver, J., Toselli, A.H., Vidal, E.: Probabilistic indexing and search for information extraction on handwritten German parish records. In: 2018 16th International Conference on Frontiers in Handwriting Recognition (ICFHR), pp. 44–49, August 2018
5. Manning, C.D., Raghavan, P., Schtze, H.: Introduction to Information Retrieval. Cambridge University Press, Cambridge (2008)

[5] http://prhlt-carabela.prhlt.upv.es/passau-LSF-TestVal/Baseline.
[6] http://prhlt-carabela.prhlt.upv.es/passau-LSF-TestVal/Raw.
[7] http://prhlt-carabela.prhlt.upv.es/passau-LSF-TestVal/Consolidated.
[8] http://prhlt-carabela.prhlt.upv.es/passau-LSF/.

6. Perronnin, F., Liu, Y., Renders, J.M.: A family of contextual measures of similarity between distributions with application to image retrieval. In: 2009 IEEE Conference on Computer Vision and Pattern Recognition, pp. 2358–2365, June 2009
7. Puigcerver, J.: Are multidimensional recurrent layers really necessary for handwritten text recognition? In: Proceedings of 14th ICDAR (2017)
8. Puigcerver, J.: A probabilistic formulation of keyword spotting. Ph.D. thesis, Universitat Politècnica de València (2018)
9. Robertson, S.: A new interpretation of average precision. In: Proceedings of the International Conference on research and Development in Information Retrieval, SIGIR 2008, pp. 689–690 (2008)
10. Sauvola, J., Pietikäinen, M.: Adaptive document image binarization. Pattern Recogn. **33**(2), 225–236 (2000)
11. Toselli, A.H., Puigcerver, J., Vidal, E.: Two methods to improve confidence scores for lexicon-free word spotting in handwritten text. In: Proceedings 15th ICFHR, pp. 349–354 (2016)
12. Toselli, A.H., Vidal, E., Romero, V., Frinken, V.: HMM word graph based keyword spotting in handwritten document images. Inf. Sci. **370–371**, 497–518 (2016)
13. Toselli, A.H., Vidal, E., Puigcerver, J., Noya-García, E.: Probabilistic multi-word spotting in handwritten text images. Pattern Anal. Appl. **22**, 23–32 (2018)
14. Vidal, E., Toselli, A.H., Puigcerver, J.: A probabilistic framework for lexicon-based keyword spotting in handwritten text images. Technical report, arXiv (2017)
15. Wahl, F.M., Wong, K.Y., Casey, R.G.: Block segmentation and text extraction in mixed text/image documents. Comput. Graph. Image Process. **20**(4), 375–390 (1982)
16. Zhu, M.: Recall, Precision and Average Precision. Working Paper 2004-09 Department of Statistics & Actuarial Science - University of Waterloo (2004)

Other Applications

Incremental Learning for Football Match Outcomes Prediction

José Domingues[1], Bernardo Lopes[1], Petya Mihaylova[2],
and Petia Georgieva[1(✉)] (iD)

[1] Department of Electronics, Telecommunications and Informatics,
University of Aveiro, Aveiro, Portugal
petia@ua.pt,petia.georgiev@gmail.com
[2] Technical University of Sofia, Sofia, Bulgaria

Abstract. Generating predictions for football match results is an expanding research area due to the commercial assets involved in the betting process. Traditionally, the results of the matches are predicted using statistical models verified by domain experts. Nowadays, this approach is challenged by the increasing amount of diverse football related information that need to be processed. In this paper, we propose an incremental learning method to predict the football match outcome category (home win, draw, away win) based on prior to the game publicly available information. The proposed framework is illustrated with data for the Portuguese first division football teams for 2017/2018 season. Factor Analysis was applied to extract most discriminating features which allowed gradual convergence of the prediction error to 32.4% after accumulation of about one third of the season games. Our approach outperforms traditional models in the gambling industry today and implies potential financial opportunities. The proposed prediction model is useful for researchers, football betting crowd, bookmakers, sport managers.

Keywords: Football match outcome prediction · Betting market ·
Sport data mining · Factor analysis · Machine learning

1 Introduction

More than four out of ten people consider themselves football fans which makes this sport the most popular in the world, [1]. This is specially noticed in Europe, since in terms of revenue and overall level of competition, it is the most prominent football market in the world [2]. The statistics reveals that the European football leagues are on the top of average public attendance as shown in Fig. 1(a) for the season 2016/17. For the same year the European professional football market revenue achieved the impressive 25.5 Billion U.S dollars as seen in Fig. 1(b). Due to this large market size, there is an increasing number of betting companies

This Research work was funded by National Funds through the FCT - Foundation for Science and Technology, in the context of the project UID /CEC/00127/2013.

© Springer Nature Switzerland AG 2019
A. Morales et al. (Eds.): IbPRIA 2019, LNCS 11868, pp. 217–228, 2019.
https://doi.org/10.1007/978-3-030-31321-0_19

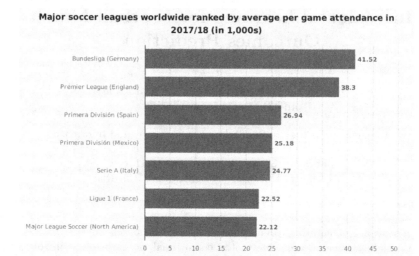

(a) Major soccer leagues attendance in 2016/17

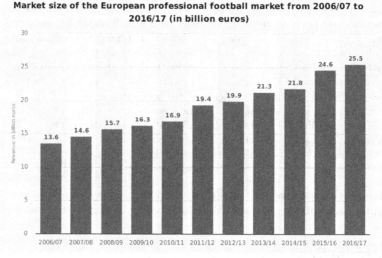

(b) European professional football market size in 2016/17

Fig. 1. Football industry statistics (source: www.statista.com)

and stakeholders seeking game prediction knowledge to make monetary gains. In fact, sport betting is expected to grow at double-digit rates over 2015–2020, (Fig. 2). The growth of the betting market is further boosted by the availability of betting websites and online services.

The traditional prediction approaches based on domain experts forecasting and statistical methods are challenged by the increasing amount of diverse

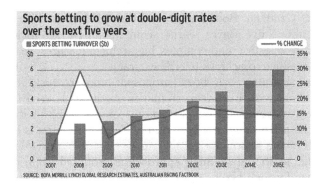

Fig. 2. Sports betting growth 2007–2015 (source: bofAML data analytics)

football related information that need to be processed [3]. The search for intelligent prediction models that help discover interesting knowledge and make more reliable predictions has placed the machine learning at the forefront of sport results forecasting. However, the leading football data providers do not make their complete information publicly available, such as player performance indicators, opposition information, or not directly match-related features. The betting stakeholders need to scour various public football databases to find suitable data or to use web scraping techniques to create their own database. To address this problem, in this paper we propose a softmax regression model to predict the football match outcomes based on free accessible web data (previous match results and betting odds data). The prediction is formulated as a classification problem with three classes, Home Team wins, Draw or Away Team wins and is illustrated on data for the Portuguese first division football teams.

2 Related Works

Football betting is a multi-billion dollar market and according to the research carried out by [4] it is a must to have an intelligent system to enable forecasting of results. Previous studies in this area differ largely in the game information considered (prior to the game or in-game data, multiple seasons or single season data) and the prediction methods. The bookmakers forecasting in the recent past relied basically on highly qualified domain experts and complex statistical methods. In recent times, majority of bookmakers are investing in intelligent prediction systems hoping to make profits and avoid financial risks through improved prediction rates. For example, in 2014 the Microsoft company Bing correctly predicted the final result of all the games in the knockout round of the World Cup in that year and became the top betting house.

Bayesian models dominate the supervised machine learning approach in the business of betting. For example in [5] Bayesian learning principle was used in predicting results of football matches in England. In [6] various classification methods (Naive Bayes, Logistic regression + Principal Component Analysis,

Random Forest, Gradient Boosting) have been tried to predict the outcome of cricket matches. Naive Bayes learner combined with significant data pre-processing, feature selection and complex hierarchical features was found as the most effective classifier to achieve 62.4% accuracy.

Unsupervised learning has been proposed more recently in [7], where k-Nearest Neighbor (with Euclidean similarity measure) is used to predict the winners of Barclays Premier League based on seasonal results between 2004/05 and 2013/14. The prediction for the second half of each season is based on data for the first half of the season.

An interesting sport data mining approach is proposed in [8] to predict the winners of college football bowl games. Based on historical results of the games and in-game events (e.g. passing attempts, rushing attempts, turnovers for and against), four engineered measures (Rating Percentage Index, Pythagorean wins, offensive strategy, turnover differential) are computed and used in a statistical model to output the team winning percentage.

Similarly, in [9] different machine learning techniques have been explored to predict the score (a regression problem) and the outcome (classification problem) of football matches based on both prior to the match data and in-game events (shots, passes, corners, penalties). Expected goals model is combined with offensive (attacking) and defense team ratings to achieve a similar accuracy to bookmakers models (around 52%).

Regarding the game information considered there are two major approaches - multiple seasons or single season of data. In the multiple seasons approach data from earlier seasons is used to train the model and then applied to predict the current season results, [10]. In the single season approach data from the current (target) season is somehow split into training and testing sub sets. Multiple seasons approach is more suitable for sports with less changes from year to year. Since at the end of the football season the two teams at the bottom of the ranking list are relegated to the second league and the top two teams from the second league are promoted to the first one, prediction models within each season are more relevant. The football competitions are organized in rounds, with teams usually playing one match in each round. In the case where one season of data is on hand, the number of rounds that will be used for training the model, and the number of rounds that will be used for testing the model needs to be determined.

Due to the variety of game information used to build predictions, the results of different studies cannot be compared directly, however, one thing is clear, more accurate models are needed.

3   The Proposed Framework

Our study is inspired by the machine learning (ML) framework presented in [11]. This paper provides a critical analysis of the literature in ML, focusing on the application of artificial neural networks (ANNs) to sport results prediction. The authors propose round-by-round training/test spits where round 1 to $n-1$ are used to train the model and round n is the test data set, for each round n in N, where N is the total number of rounds in the competition. In the present paper, we extend this idea, however instead of round-by-round prediction, the first n rounds are used to train the model and then applied to predict the future $N - n$ rounds in the same season. After each round of games, the training data set is incremented and the model is retrained starting from the optimal parameters found at the previous round of training. Instead of ANN model we propose the softmax regression as the prediction model.

Softmax regression is a generalization of logistic regression more relevant for mutually exclusive multi-class problems (which is the case of football outcome prediction) than the multi-class logistic regression. Given an input vector of features $x^{(i)}$ with a label $y^{(i)}$, the softmax function estimates the probability that this example belongs to each of the class labels $j = 1, 2, ..c$

$$p(y^{(i)} = j | x^{(i)}; \theta) = \frac{e^{\theta_j^T x^{(i)}}}{\sum_{j=1}^{c} e^{\theta_j^T x^{(i)}}} \tag{1}$$

The network outputs c dimensional vector of the estimated probabilities, where θ is the matrix of the model parameters.

$$\hat{y}(x^{(i)}) = \frac{1}{\sum_{j=1}^{c} e^{\theta_j^T x^{(i)}}} \begin{bmatrix} e^{\theta_1^T x^{(i)}} \\ e^{\theta_2^T x^{(i)}} \\ ... \\ e^{\theta_c^T x^{(i)}} \end{bmatrix} \tag{2}$$

The denominator in Eq. (2) normalizes the distribution to sum to one. Given a batch of m training examples, the softmax cost function with L2 regularization term (parametrized by λ) is computed as

$$J_{reg}(\theta) = -\frac{1}{m} [\sum_{i=1}^{m} \sum_{j=1}^{c} 1\{y^{(i)} = j\} \log \frac{e^{\theta_j^T x^{(i)}}}{\sum_{j=1}^{c} e^{\theta_j^T x^{(i)}}}] + \frac{\lambda}{2} \sum_{j=1}^{c} \sum_{l=1}^{n} \theta_{jl}. \tag{3}$$

One of the arguments for choosing the softmax classifier is the lack of hyper-parameters (with exception to the regularization parameter λ) to be set up in advance. For example the Neural Network-based classifiers need to specify at least the number and type of nodes (for a standard 3 layers NN architecture) or the number of layers (for deeper architectures). For Support Vector Machines the hyper-parameters are C gain, the kernel function and its parameters. The systematic approach for the choice of the best hyper-parameters imposes an exhaustive search over intervals of potential values, and this is often overlooked by the proposed prediction schemes. Further to that, the probabilistic interpretation of softmax regression provides a convenient measure to account for the classifier confidence. Marginal outputs (close to 0 or 1) are a good start in choosing a betting strategy. In contrast, outputs around 0.5 should be considered as less reliable and therefore further information is required before defining the betting strategy.

Data management in the single season approach is not a trivial task, there are as many division policies as publications on the topic. Our approach considers the games in one season as a time series where training data increases incrementally after each round of matches to predict the future results in the same season. The train-test data split is shown in Fig. 3, where X_{train} is the train data over a sequence of the first p rounds of games and X_{test} is the test data over the rest of the game rounds in the same season. The model training is divided into epochs, reflecting the number of rounds. At the first training epoch, X_{train} contains data from the first round of games to predict the next games of the season. At the second epoch, X_{train} contains data from the first and the second rounds to predict the future game outcomes. Thus at each next epoch X_{train} batch increases, accumulating data for the past round of games to predict the next games of the same season. At each training epoch the model parameters are updated over a number of iterations until the cost function $J_{reg}(\theta)$ converges. At each next training epoch θ initial values are set to the optimal values computed at the previous epoch. Once the internal iterative process in one epoch is over, the model with frozen parameters is used to predict all future games in that season X_{test}. This framework is denoted as an incremental learning where after each round of games the model is retrained starting with parameters computed at the previous round.

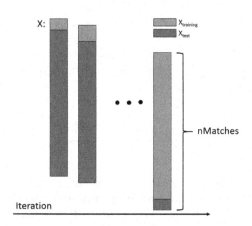

Fig. 3. Incremental train-test data split

4   Experiments and Results

4.1   Data Set

The Portuguese football first league was used to illustrate the prediction framework proposed in Sect. 3. Publicly available data for 306 football matches during the season 2017/18 were retrieved from [12]. The league consists of 18 teams, it has 34 rounds with 9 matches per round. The database consists of 62 features listed in the Appendix (A). Since, the purpose of the model is to guide gamblers and promote their betting activity before the game start, only available information prior to the game is used. Therefore the *Match Statistics* features (number of goals and the result in the half time of the game, number of given red or yellow cards, number of shots, target shots, corners and fouls) were excluded. 46 features including prior game statistics and match betting odds of various houses (Betbrain, Pinnacle, Bet365, etc.) were the inputs to the model. The histogram of data class distribution, depicted in Fig. 4, shows that the chances of the home team to win are much higher, 51.6% of the matches played at home have been a victory. This is expected taking into account the familiarity of the home team with the local environment and the support of fans and cheering crowd in general.

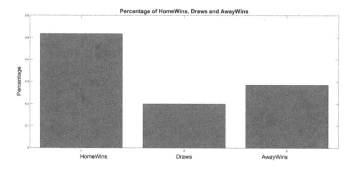

Fig. 4. Data class distribution

4.2 Performance Evaluation

The regularization term in Eq. (3) prevents the model from overfitting which is expected due to the high number of features (46). The performance of the model for different values of the regularization parameter λ in the interval [0:0.1:2.3] was evaluated as shown in Fig. 5. The average prediction error over all predicted rounds of games achieved a minimum of 31.75% for $\lambda = 0.1$.

The evolution of the prediction error as a function of the accumulating training data (games) with the complete set of 46 features and $\lambda = 0.1$ is shown in Fig. 6. As seen from the figure, the error does not converge, it oscillates around 45% and therefore the learning to predict the game outcomes does not improve in the course of the competition season.

As an attempt to improve the prediction performance we apply factor analysis to discover the most discriminative features that define the match outcome categories.

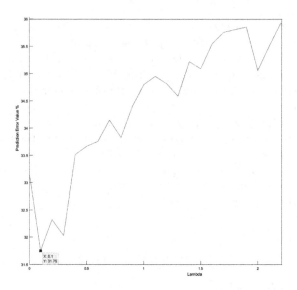

Fig. 5. Average prediction error versus λ

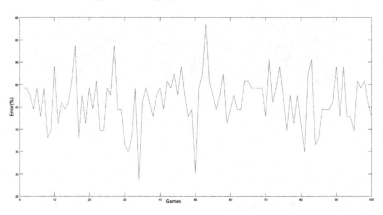

Fig. 6. Prediction error evolution (complete set of 46 features)

4.3   Factor Analysis for Feature Dimensionality Reduction

Factor Analysis (FA) is an explorative method similar to Principal Components Analysis (PCA), [13]. Both methods try to reduce the dimensionality of the dataset based on the concept of common variance. Common variance is the amount of variance that is shared among a set of features. Features that are highly correlated will share a lot of variance. PCA assumes that the total variance is build on all common variances, while the FA assumes that total variance can be partitioned into common and unique variance. Unique variance is any portion of variance that is not common. Thus, in the FA framework, the total variance is sum of the common variance and unique variance.

FA identifies the latent (unobserved) factors which represent the hidden organization of the observable measures. Factor scores (aka factor loadings) indicate how each latent factor is associated with the observable variables used in the analysis. If the goal is to simply reduce the variable list down into a linear combination of smaller number of components then PCA is the way to go. However, if the goal is to determine what are the features that describe a latent factor then the FA is more appropriate, [13].

Table 1. Factor loading of 4-factor FA model explaining 69% of total data variance (presented in bold if absolute values are greater that 0.7).

Features	F1	F2	F3	F4
HT (HomeTeam)	**0.90**	−0.10	0.43	0.01
AT (AwayTeam)	**0.87**	0.13	−0.8	0.03
HTP (Home Team Points)	−0.14	**0.88**	0.12	0.06
ATP (Away Team Points)	−0.02	**0.83**	−0.04	0.08
FTHG (Full Time Home Team Goals)	−0.23	−0.11	**0.77**	−0.14
FTAG (Full Time Away Team Goals)	−0.34	−0.02	**−0.71**	−0.15
PSCH (Pinnacle closing home odds)	0.34	0.2	−0.24	−0.65
PSCD (Pinnacle closing draw odds)	0.14	0.01	−0.37	0.56
PSCA (Pinnacle closing away win odds)	0.15	0.05	0.25	0.39
Variance (%)	26	17	14	12

In Table 1 are presented the FA results of a 4-factor model performed with software R libraries. In the first column are represented the dominant features that identify the four hidden factors. The other columns are the factor loadings on these factors. For example the first factor loadings in the second column (0.90 and 0.87) indicate that the observable measures HomeTeam (HT) and AwayTeam (AT) can be used to "describe" Factor 1 (F1); in other words, F1 has characteristics very similar to features HT and AT. The other features (HTP, ATP,...PSCA) are not useful to describe F1 because their factor loadings on F1 are small (less than 0.70). The same interpretation holds true for the factor loadings on factors F2, F3, F4. The negative factor loading of the feature Full Time Away Team Goals (FTAG) on hidden Factor F3 (−0.71) means that F3 has the characteristic "opposite" of FTAG.

Based on FA, a reduced feature set was constructed accounting for features with factor loadings with absolute values greater that 0.7:
Reduced feature vector = [HT, AT, HTP, ATP, FTHG, FTAG], HT and AT are binary features (0,1), HTP and ATP are the points of each team before the game starts, FTHG and FTAG correspond to the number of goals scored so far during the current season.

The evolution of the prediction error with the new feature set is shown in Fig. 7. The incremental learning is clearly observable: at the beginning of the season, the prediction success rate is slightly better than a chance rate (55%); at about one third of the season games (around 100 games) the prediction error gradually decreases to 32.4%.

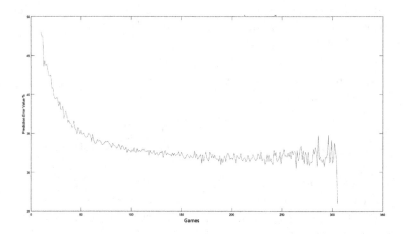

Fig. 7. Prediction error evolution (FA-based features)

5   Conclusions and Future Work

The traditional bookmakers approaches include domain experts forecasting and statistical methods. In this paper, a machine learning model (namely softmax regression) is proposed to predict the football match outcome category (home win, draw, away win) based on prior to the game data. The model exploits incremental learning by subsequent accumulation of more match data in the course of the competition season. The model is illustrated by analyzing match statistics and bookmakers odds for the Portuguese first division football teams for season 2017/2018, retreated from publicly available web data. Factor Analysis was applied to extract most discriminating features which allowed gradual convergence of the prediction error to 32.4% after accumulation of about one third of the season games. This is an improvement upon levels present in the gambling industry today and implies potential financial opportunities. Therefore the proposed model can be useful not only for academics, but also advantageous for betting crowd, bookmakers, club managers.

With regards to future work, there is most certainly room for improvement. Alternative avenues could be explored either in the form of new algorithms or in the form of new features. Further to that, the efficacy of more sophisticated prediction strategies with both machine and human intelligence need to be assessed. It is very important for betting houses and bookies to count on human control and guidance, beyond the autonomous machine learning models, in order to ensure consistent results.

A   Appendix - Football Data

(retrieved from [12])
FTR = Full Time Result (H = Home Win, D = Draw, A = Away Win)
HT, AT = Home Team/Away Team
FTHG, FTAG = Full Time Home Team Goals/Full Time Away Team Goals
HTP, ATP= Home Team Points/Away Team Points

Match Statistics
HTHG, HTAG = Half Time Home Team Goals/Half Time Away Team Goals
HTR = Half Time Result (H = Home Win, D = Draw, A = Away Win)
HS, AS = Home Team Shots/Away Team Shots
HST, AST = Home Team Shots on Target/Away Team Shots on Target
HC, AC = Home Team Corners/Away Team Corners
HF, AF = Home Team Fouls Committed/Away Team Fouls Committed
HY, AY = Home Team Yellow Cards/Away Team Yellow Cards
HR, AR = Home Team Red Cards/Away Team Red Cards

Match betting odds data
B365H, B365D, B365A = Bet365 home win odds /draw odds/ away win odds
BWH, BWD, BDA = Bet&Win home win odds /draw odds/ away win odds
IWH. IWD, IWA = Interwetten home win odds /draw odds/ away win odds
LBH, LBD, LBA = Ladbrokes home win odds /draw odds/ away win odds
PSH, PSD, PSA = Pinnacle home win odds /draw odds/ away win odds
VCH, VCD, VCA = VC Bet home win odds /draw odds/ away win odds
WHH, WHD, WHA = William Hill home win odds /draw odds/ away win odds

Bb1X2 = BetBrain bookmakers to compute match odds avrg& max
BbMxH, BbAvH = Betbrain maximum home win odds/average home win odds
BbMxD, BbAvD = Betbrain maximum draw odds/average home win odds
BbMxA, BbAvA = Betbrain maximum away win odds/average home win odds

Total goals betting odds
BbOU = BetBrain bookmakers for over/under 2.5 total goals avrg & max
BbMx>2.5, BbAv>2.5 = Betbrain max over 2.5 goals/average over 2.5 goals
BbMx<2.5, BbAv<2.5 = Betbrain max under 2.5 goals/average under 2.5 goals

Asian handicap betting odds
BbAH = N° of BetBrain bookmakers used to Asian handicap avrg & max
BbAHh = Betbrain size of handicap (home team)
BbMxAHH = Betbrain maximum Asian handicap home team odds
BbAvAHH = Betbrain average Asian handicap home team odds
BbMxAHA = Betbrain maximum Asian handicap away team odds
BbAvAHA = Betbrain average Asian handicap away team odds

Closing odds (last odds before match starts)
PSCH, PSCD, PSCA = Pinnacle closing home/draw/away win odds

References

1. Bloomberg, Soccer is the world's most popular sport and still growing - Bloomberg. https://www.bloomberg.com/news/articles/2018-06-12/soccer-is-the-world-s-most-popular-sport-and-still-growing. Accessed 13 Nov 2018
2. Statista, Soccer - Statistics & Facts — Statista. https://www.statista.com/topics/1595/soccer/. Accessed 13 Nov 2018
3. Buchdahl, J.: Fixed Odds Sports Betting: Statistical Forecasting and Risk Management. High Stakes Publisher, London (2003)
4. Smith, M., Paton, D., Williams, L.: Do bookmakers possess superior skills to bettors in predicting outcomes? J. Econ. Behav. Organ. **71**(2), 539–549 (2009)
5. Joseph, A., Fenton, N., Neil, M.: Predicting football results using Bayesian nets and other machine learning techniques. Knowl.-Based Syst. **19**(7), 544–553 (2006)
6. Kampakis, S., Thomas, W.: Using machine learning to predict the outcome of English County twenty over cricket matches, Cornell University. https://arxiv.org/abs/1511.05837 (2015)
7. Esumeh, E.O.: Using machine learning to predict winners of football league for bookies. Int. J. Artif. Intell. **5**, 22 (2015)
8. Leung, C.K., Joseph, K.W.: Sports data mining: predicting results for the college football games. Procedia Comput. Sci. **35**, 710–719 (2014)
9. Herbinet, C.: Predicting football results using machine learning techniques. MEng thesis, Imperial College London (2018)
10. Cao, C.: Sports data mining technology used in basketball outcome prediction. Master's Thesis, Dublin Institute of Technology, Ireland (2012)
11. Bunker, R.P., Fadi Thabtah, F.: A machine learning framework for sport result prediction. Appl. Comput. Inf. **15**, 27–33 (2019)
12. www.football-data.co.uk/portugalm.php.
13. Tucker, L., MacCallum, R.: Exploratory factor analysis. University of Illinois and Ohio State University (1997)

Frame by Frame Pain Estimation Using Locally Spatial Attention Learning

Jun Yu[1(⊠)], Toru Kurihara[1], and Shu Zhan[2]

[1] School of Information, Kochi University of Technology, Kochi, Japan
`218005n@gs.kochi-tech.ac.jp`
[2] School of Computer and Information, Hefei University of Technology, Anhui, China

Abstract. Estimating pain intensity for patient is a challenging area in clinic treatment and medical diagnosis. Painful facial expression only relates to some areas of face. Inspired by this fact, we introduce end-to-end locally spatial attention learning for pain estimation. By focusing on important region in the face with 1×1 locally convolutional layer, the local features related to pain intensity can be captured. Furthermore, facial expression is the dynamic deformation of face in the time domain. In order to model the information, the long short-term memory network (LSTM) is incorporated into our architecture. The feature extracted by the convolutional neural network (CNN) with the locally spatial attention learning is fed to the LSTM network. The results show that our locally spatial attention learning can provide the fine-grained variation on the face region for pain intensity assessment.

Keywords: Pain intensity · Spatial attention · CNN · LSTM

1 Introduction

Pain is an unpleasant feeling which is related to tissue damage and unhealthy condition. Accurate pain intensity estimation is a central problem in mental health and clinical treatment. Traditionally, pain intensity is evaluated by the observation of expert and self-reported data, such as Observer Pain Intensity (OPI), Visual Analog Scale (VAS). However, for elderly people with dementia who lack the ability to express pain intensity, evaluating pain intensity becomes a basic issue in some medical diagnosis applications. In addition, manual pain estimation is time-consuming, inaccurate without professional training and not available for real-time pain assessment. In such situations, accurate pain intensity evaluation plays an important role in medical treatment and health care. Hence, there is a large demand to build automatic assessment system for pain intensity estimation.

To solve the great demand, a large majority of research has focused on automatic pain intensity assessment. In the initial stage, detecting pain in video by facial action units has been proposed by Lucey et al. [7]. Later, some methods

© Springer Nature Switzerland AG 2019
A. Morales et al. (Eds.): IbPRIA 2019, LNCS 11868, pp. 229–238, 2019.
https://doi.org/10.1007/978-3-030-31321-0_20

have been proposed to evaluate pain intensity using multimodal data, such as thermal and depth data from camera [3], biomedical signals from the electrocardiogram signals (ECG) and the electromyogram signals (EMG) [17]. Recently, deep convolutional neural networks have achieved great successful results in face recognition, face detection and so on. Therefore, deep neural networks are attracting widespread interest in the fields of facial expression recognition, especially pain intensity estimation. Recurrent Convolutional Neural Network used for object detection was utilized by Zhou et al. [20] for pain intensity estimation. Another method was developed by fine-tuning deep face verification net with regularized regression loss [15]. Pau et al. [11] proposed a method by combining deep convolutional neural networks with long short-term memory networks for pain intensity estimation. Their study suggested extracting features from both the spatial space and the temporal space can obtain good performance for frame-level pain intensity estimation. Tavakolian et al. [14] developed a method by using binary coding of discriminative statistical feature representation from the convolutional neural network. Hamming distance is applied in the new loss function and benefits the training of the whole framework.

Attention mechanism is one of key properties of human being's visual system which can selectively concentrate on the important areas in an image or a scene for better understanding. Inspired by that, there are several attempts to utilize attention mechanism to improve the performance in Image captioning [19] and other applications. More recently, a concise attention module was proposed by Hu et al. [2] to build the relationship between different channels inside the neural network. The global average pooling was used for estimating the channel-wise attention. Later, Woo et al. [18] proposed new attention model called Convolutional Block Attention Module (CBAM). In the CBAM module, the attention mechanism was applied not only in the channel space, but also in the spatial space. Extensive experimental results have shown that CBAM module can achieve the best performance in both the image classification task and the object detection task.

Until now, a few research in the field of pain intensity estimation has attempted to utilize the attention mechanism in their research work. The purpose of this study is to propose and examine end-to-end locally spatial attention learning architecture for pain intensity assessment. The overview of the pipeline of our approach is illustrated in Fig. 1. The approach we have applied in this work aims to exploit "where" is important in spatial space for pain intensity estimation. Besides, our architecture also exploits the relationship between different frames in the video sequence. The proposed attention-based architecture is validated in the widely used benchmark database [8]. The paper is organized as follows: In the method part, we describe the details of our approach. In the experiments part, we investigate and analyze our proposed method in the database. Finally, the conclusion is given for our research.

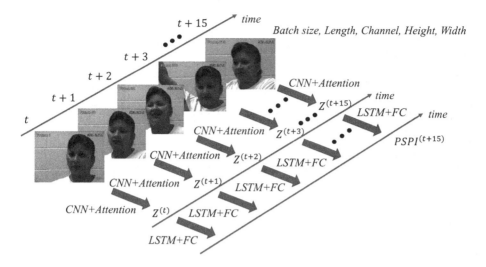

Fig. 1. The illustration of the whole pipeline of our architecture. The input of the architecture is a five-dimensional tensor, including batch size, length of the sequence, channel, height, width.

2 Proposed Method

Feeling pain always generates painful facial expression which is the structural and geometric change in the face. The purpose of this study is to estimate the pain intensity directly from the patient's face in a recorded video or in the real-time surveillance system. The motivation of our method is that each region of the face is not equally contributed to the painful expression. In order to capture the local detailed variation of face, we propose the locally spatial attention learning architecture for pain assessment. Our structure is based on the VGG network [12] and previous face recognition work [13]. The input tensor is rescaled by the locally spatial attention model, which can enhance the ability of our network for extracting features from images. Some previous behavioral and emotion study suggests the dynamic information of facial expression is useful and efficient for emotional assessment [5]. In our architecture, the LSTM network is adopted for capturing the dynamic information in the temporal domain. By combining the spatial variation and the temporal variation in the video sequence of patient's face, we are able to estimate the pain intensity robustly. Our architecture consists of CNN with spatial attention model and LSTM. Each frame in the video sequence is fed into the whole architecture. The CNN block extracts features from single frame, then puts the feature vector into the LSTM block to estimate the pain intensity. The details of our architecture is shown in Fig. 2. In the next section, we elaborate on the details of our approach.

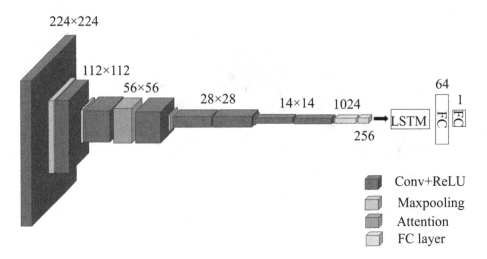

Fig. 2. In our study, we utilize CNN network and LSTM network as the backbone of our architecture. We incorporate the locally spatial attention learning model into the convolutional network as illustrated in this figure.

2.1 Locally Spatial Attention Learning

Figure 3 shows the overview of our locally spatial attention learning model. For extracting the static features from the patient face, we utilize the convolutional neural network which is based on VGG11 network (configuration A) [12] for pain intensity estimation. The motivation of our study is that each region of the face is not equally contributed to the painful expression. In an attempt to design the spatial attention model, our intention is to provide a way of detecting the important region for pain intensity estimation. The proposed model which consists of two layers is inserted inside the third block in the convolutional neural network to capture the detailed information from the previous layers.

The aim of the locally spatial attention learning is to capture more detailed information from the face region. A major problem with the previous attention model based on the conventional convolutional kernel is that the generated attention map is translation invariant so that the local details of the image are hard to capture. The geometry attribute of the face image is symmetrical and structural. Inspired by previous face recognition study [13] and other research field based on the facial expression [9], we propose the locally spatial attention learning model which is incorporated into the convolutional neural network. Given the output tensor T from the previous building block in the convolutional neural network, the shape of the tensor is $C \times H \times W$, which C is the channel number of the tensor while H and W represent height and width respectively. The spatial attention model takes the input tensor T and generates a 2D spatial attention map A_s, with size $H \times W$. To generate the spatial attention map, the locally convolutional layer is adopted in the first layer of the spatial attention model. For each location in the spatial dimension of the input tensor, the first layer of our locally spatial attention learning model uses different convolutional kernel for extracting discriminative appearance feature. The P_{ij} denotes the different weights for the input tensor T of different location T_{ij}. Each T_{ij} has its own receptive field of the face image. In hence, the more the spatial attention model is behind, the larger the receptive field of the attention model becomes. The kernel size of the locally convolutional layer is 1×1, so shape of the output tensor of the first layer is $R \times H \times W$, $R = 16$. The tanh function is used as activate function in the first layer. In the second layer of spatial attention model, we apply the conventional 1×1 convolutional kernel to generate the spatial attention map, which describes the informative parts of the face region. The sigmoid function is applied on the top of the spatial attention model to let the attention weight lie from zero to one. The shape of the attention map A_s is $1 \times H \times W$. In short, the spatial attention model is calculated as:

$$T_{res} = T \otimes A_s \tag{1}$$

where, the operation $\otimes$ denotes the element-wise multiplication. The output of the attention map rescales the input tensor T. In our implementation, the attention map A_s is broadcasted in the channel dimension of the input tensor T. Then, the rescaled tensor T_{res} is fed to the latter convolutional layer in the convolutional neural network. We utilized dropout strategy to avoid the overfitting problems. The dropout ratio of the fully connected layer was set to 0.3. The arrangement of the two layers inside the attention model is a key problem for pain intensity estimation. We compare different spatial attention models in the experiments section, and the results demonstrate that locally convolutional layer in the first order is better than other structures.

2.2 Temporal Learning

The structure we have used in this study aims not only to extract features from the single frame, but also to build the dynamic temporal relationship among different frames. For this purpose, the LSTM network for learning the long-term relationship among sequence data is adopted in our temporal model. The output feature vector $Z^{(t)}$ of convolutional neural network is fed into the LSTM, so the input node of the LSTM is 256-dimensional. $Z^{(t)}$ denotes the feature vector of tth frame in the video sequence. In our temporal model, we only use one LSTM layer. The hidden node of our LSTM is 128-dimensional. The dropout ratio of our temporal model is 0.3. The LSTM network has three gates to control the information from the previous sequence data and the existing sequence data. The video sequence of the patient is divided into small groups to train our architecture. For instance, the first sequence $s_1 = \{f_1, f_2, \ldots, f_{16}\}$, is 16 consecutive frames from one video, the next sequence is $s_2 = \{f_2, f_3, \ldots, f_{17}\}$, until the last sequence of this video. Here, f_i denotes the ith frame in the video sequence. The frame label of our training database is Prkachin and Solomon's Pain Intensity metric (PSPI) [10]. The label of each sequence is the PSPI label of the last frame in this sequence. Therefore, the PSPI label is predicted by considering all the 16 frames. Finally, two fully connected layers are used for predicting the PSPI value based on the output of the LSTM network. The dimension of the first layer is 64 and the dimension of the second layer is only 1 for estimating pain intensity. Considering that estimating the pain intensity is the regression task, we chose the mean squared error loss for training our architecture.

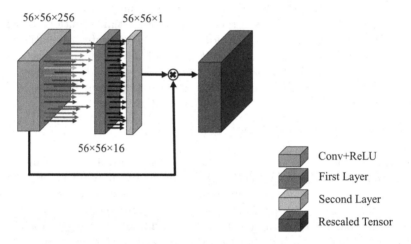

Fig. 3. The overview of the locally spatial attention learning model. As illustrated in Fig. 2, the spatial attention model is inserted inside the third block of convolutional neural network. The first layer is 1×1 locally convolutional layer. The second layer is the conventional 1×1 convolutional layer for dimension reduction. The input of the attention model is the orange block which is the output of first layer in the third block of convolutional network. The spatial attention map is used for rescaling the input tensor. (Color figure online)

3   Experiment

In this section, we will discuss the details of our experiments and results for our proposed architecture.

3.1   Database and Preprocessing Details

We trained and validated our spatial attention architecture in the UNBC-McMaster shoulder pain database [8], which consists of 25 subjects with 200 videos. All the participants in this database have got shoulder pain. In the recording stage, they did a list of active and passive range-of-motion tests with their limbs under the professional guidelines. The database provides three types of labels for calculating pain intensity, including VAS, OPI and PSPI. As reported in the previous research [16], the PSPI label in the database is not always reliable. In their finding, both VAS and OPI labels for some subjects are not zeros which means the subject feels pain. However, the PSPI label for that subject is zero. It is known from the literatures [6], that some action units related to the pain expression is not calculated in the original PSPI equation. To investigate the ability of our spatial attention model, we did the data cleaning process for reliable PSPI labels in the database. Therefore, we excluded one subject which has no obvious pain (101-mg101 subjects, including 9 video sequences) and some video sequences without reliable PSPI labels (bg096t2afaff, ib109t2aeaff). Totally, 24 subjects with 189 video sequences were used in our experiments. We followed the previous research [15] to preprocess the PSPI label by transforming the range of value from 0–15 to 0–5. Data preprocessing is a major part in training deep neural networks. As demonstrated in Fig. 4, the OpenFace2.0 toolkit [1] was utilized in our experiments for face alignment and cropping.

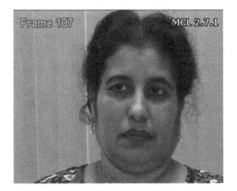 

Fig. 4. Original image and processed image by OpenFace 2.0 toolkit. All the images in the database are resized to 224 × 224.

3.2   Implementation and Analysis

The convolutional neural network in our architecture was trained from scratch and the whole structure was trained in the end-to-end manner. As mentioned before, we only used 24 subjects with 189 video sequences to train our network. Therefore, we evaluated our network in 24 subjects leave-one-subject-out cross validation. The learning rate of our network was set to 0.0001. Our network was trained in 20 epochs. Furthermore, we utilized Adam [4] optimizer with weight decay 0.001 to train our architecture. The whole architecture was implemented by PyTorch framework with batch size 32 in 4 GPUs. The order of the locally convolutional layer and the conventional convolutional layer in the spatial attention model is an essential issue in our study. Here, we compare two different spatial attention models. As illustrated in Fig. 5, the left attention map is derived from the locally spatial attention learning model used in our architecture, while the right attention map is from the different spatial attention model which the first layer is the convolutional layer and the second layer is the locally convolutional layer. As can be seen from Fig. 5, our proposed locally spatial attention learning model indicates that the cheek of the face and the region between eyebrows are important for pain intensity estimation. The right attention map shows nearly same importance in the whole face region which indicates this architecture cannot detect significant region for pain intensity estimation. Comparison of two different attention maps shows our proposed model can capture the important region of face more effectively than another model with different order. We also compare the performance of our method with the general architecture and the previous research. Here, the general architecture only contains CNN network without locally spatial attention learning and LSTM network. The mean absolute error (MAE), mean squared error (MSE) and Pearson Correlation Coefficient (PCC) are reported in Table 1. As listed in Table 1, it is obvious that the locally spatial attention model can achieve an improvement on both MAE and MSE with comparison of general architecture. Comparison between our method and previous research shows the performance of our architecture is not perfect. It should, however, be noted that we use smaller training database for our neural network. Accurate and reliable PSPI label is important for training deep neural networks. Estimating the Pain intensity should be accurately related to the painful expression and feeling which are crucial issues for some medical diagnosis.

Table 1. Comparison of different methods

Methods	MAE	MSE	PCC
Zhou et al. [20]	N/A	1.54	0.65
Wang et al. [15]	0.456	0.804	0.651
Rodriguez et al. [11]	0.5	0.74	0.78
Tavakolian et al. [14]	N/A	0.69	0.81
Spatial attention	0.58	1.22	0.4
General	0.61	1.29	0.4

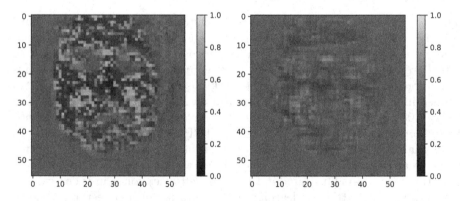

Fig. 5. Example of two different attention maps for image is displayed in Fig. 4. The left attention map is derived from our proposed attention model. The right is comparison model which exchanges the order of 1×1 locally convolutional layer and 1×1 common convolutional layer.

4   Conclusion

Automatic pain intensity estimation is a key technique in some medical applications. In this paper, we propose locally spatial attention learning method to find important region on the face and to enhance the performance of the whole architecture. The results indicate that our proposed method can capture the important area of face for pain intensity estimation. Our current study expands the prior work in this research area and provides a new method for future study on painful expression analysis. We conducted our experiments in the shoulder pain database. At present, the results show the performance of our architecture is better than the general structure without locally spatial attention learning and is not outstanding compared with the state-of-the-art methods. We will improve our spatial attention architecture for better results by effective network engineering in the future.

References

1. Baltrusaitis, T., Zadeh, A., Lim, Y.C., Morency, L.P.: Openface 2.0: facial behavior analysis toolkit. In: 2018 13th IEEE International Conference on Automatic Face & Gesture Recognition (FG 2018), pp. 59–66. IEEE (2018)
2. Hu, J., Shen, L., Sun, G.: Squeeze-and-excitation networks. In: Proceedings of the IEEE Conference on Computer Vision and Pattern Recognition, pp. 7132–7141 (2018)
3. Irani, R., et al.: Spatiotemporal analysis of RGB-DT facial images for multimodal pain level recognition. In: Proceedings of the IEEE Conference on Computer Vision and Pattern Recognition Workshops, pp. 88–95 (2015)
4. Kingma, D.P., Ba, J.: Adam: a method for stochastic optimization. arXiv preprint arXiv:1412.6980 (2014)
5. Krumhuber, E.G., Kappas, A., Manstead, A.S.: Effects of dynamic aspects of facial expressions: a review. Emot. Rev. **5**(1), 41–46 (2013)
6. Kunz, M., Lautenbacher, S.: The faces of pain: a cluster analysis of individual differences in facial activity patterns of pain. Eur. J. Pain **18**(6), 813–823 (2014)
7. Lucey, P., et al.: Automatically detecting pain in video through facial action units. IEEE Trans. Syst. Man Cybern Part B (Cybern.) **41**(3), 664–674 (2011)
8. Lucey, P., Cohn, J.F., Prkachin, K.M., Solomon, P.E., Matthews, I.: Painful data: the UNBC-McMaster shoulder pain expression archive database. In: 2011 IEEE International Conference on Automatic Face & Gesture Recognition and Workshops (FG 2011), pp. 57–64. IEEE (2011)
9. Pei, W., Dibeklioğlu, H., Baltrušaitis, T., Tax, D.M.: Attended end-to-end architecture for age estimation from facial expression videos. arXiv preprint arXiv:1711.08690 (2017)
10. Prkachin, K.M., Solomon, P.E.: The structure, reliability and validity of pain expression: evidence from patients with shoulder pain. Pain **139**(2), 267–274 (2008)
11. Rodriguez, P., et al.: Deep pain: exploiting long short-term memory networks for facial expression classification. IEEE Trans. Cybern. **99**, 1–11 (2017)
12. Simonyan, K., Zisserman, A.: Very deep convolutional networks for large-scale image recognition. arXiv preprint arXiv:1409.1556 (2014)

13. Taigman, Y., Yang, M., Ranzato, M., Wolf, L.: Deepface: closing the gap to human-level performance in face verification. In: Proceedings of the IEEE Conference on Computer Vision and Pattern Recognition, pp. 1701–1708 (2014)
14. Tavakolian, M., Hadid, A.: Deep binary representation of facial expressions: a novel framework for automatic pain intensity recognition. In: 2018 25th IEEE International Conference on Image Processing (ICIP), pp. 1952–1956. IEEE (2018)
15. Wang, F., et al.: Regularizing face verification nets for pain intensity regression. In: 2017 IEEE International Conference on Image Processing (ICIP), pp. 1087–1091. IEEE (2017)
16. Werner, P., Al-Hamadi, A., Limbrecht-Ecklundt, K., Walter, S., Gruss, S., Traue, H.C.: Automatic pain assessment with facial activity descriptors. IEEE Trans. Affect. Comput. 8(3), 286–299 (2017)
17. Werner, P., Al-Hamadi, A., Niese, R., Walter, S., Gruss, S., Traue, H.C.: Automatic pain recognition from video and biomedical signals. In: 2014 22nd International Conference on Pattern Recognition (ICPR), pp. 4582–4587. IEEE (2014)
18. Woo, S., Park, J., Lee, J.-Y., Kweon, I.S.: CBAM: convolutional block attention module. In: Ferrari, V., Hebert, M., Sminchisescu, C., Weiss, Y. (eds.) ECCV 2018. LNCS, vol. 11211, pp. 3–19. Springer, Cham (2018). https://doi.org/10.1007/978-3-030-01234-2_1
19. Xu, K., et al.: Show, attend and tell: neural image caption generation with visual attention. In: International Conference on Machine Learning, pp. 2048–2057 (2015)
20. Zhou, J., Hong, X., Su, F., Zhao, G.: Recurrent convolutional neural network regression for continuous pain intensity estimation in video. In: Proceedings of the IEEE Conference on Computer Vision and Pattern Recognition Workshops, pp. 84–92 (2016)

Mosquito Larvae Image Classification Based on DenseNet and Guided Grad-CAM

Zaira García[1] , Keiji Yanai[2] , Mariko Nakano[1(✉)] ,
Antonio Arista[1] , Laura Cleofas Sanchez[1] , and Hector Perez[1]

[1] Instituto Politecnico Nacional, Mexico City, Mexico
zgn_1607@hotmail.com, mnakano@ipn.mx
[2] The University-of-Electrocommunications, Tokyo, Japan

Abstract. The surveillance of Aedes aegypti and Aedes albopictus mosquito to avoid the spreading of arboviruses that cause Dengue, Zika and Chikungunya becomes more important, because these diseases have greatest repercussions in public health in the significant extension of the world. Mosquito larvae identification methods require special equipment, skillful entomologists and tedious work with considerable consuming time. In comparison with the short mosquito lifecycle, which is less than 2 weeks, the time required for all surveillance process is too long. In this paper, we proposed a novel technological approach based on Deep Neural Networks (DNNs) and visualization techniques to classify mosquito larvae images using the comb-like figure appeared in the eighth segment of the larva's abdomen. We present the DNN and the visualization technique employed in this work, and the results achieved after training the DNN to classify an input image into two classes: Aedes and Non-Aedes mosquito. Based on the proposed scheme, we obtain the accuracy, sensitivity and specificity, and compare this performance with existing technological approaches to demonstrate that the automatic identification process offered by the proposed scheme provides a better identification performance.

Keywords: Mosquito larvae · Classification · Deep Neural Network ·
Mosquito control · Mosquito surveillance

1 Introduction

Worldwide viral diseases transmitted by mosquitoes have the most significant repercussions in public health. The principal viruses spread by the bite of the female Aedes mosquitoes (Aedes aegypti and Aedes albopictus) are arboviruses, which cause Dengue, Zika, Chikungunya and Yellow fever. When an Aedes mosquito bites a person, who has been infected previously with an arbovirus, the mosquito can become a carrier of the virus. If this mosquito bites another person, then the person also can be infected with arbovirus [1]. In the Dengue fever, the World Health Organization (WHO) estimates that the four serotypes of this virus are a menace for 40% of the world population that lives in subtropical and tropical areas [2]. In 2016, the WHO declared Zika virus a public health emergency and advised that pregnant women need to be protected from mosquitoes, because Zika fever can cause serious birth defects in her unborn baby [3].

A. Morales et al. (Eds.): IbPRIA 2019, LNCS 11868, pp. 239–246, 2019.
https://doi.org/10.1007/978-3-030-31321-0_21

Aedes aegypti and Aedes albopictus use natural and artificial water-holding containers to lay their eggs. After hatching, larvae grow and develop into pupae and subsequently into a terrestrial flying adult mosquito. Mosquito surveillance is a key component of any local integrated vector control program. Collecting data of the mosquito population in many geographic areas to identify the presence or absence of the Aedes aegypti or Aedes albopictus, support the creation of maps to determine the specific areas to fumigate. The specimen collection can be divided into three main surveys: ovitraps, immature stage surveys and adult mosquito trapping. Nevertheless, the transportation of specimens to laboratories and the analysis performed by an entomologist with special equipment are very time-consuming tasks in comparison with the mosquito's short lifecycle which takes about ten days from eggs to adult mosquitos under a favorable condition [4].

Nowadays, exist some technological approaches to support the identification of Aedes mosquitoes at the larvae stage focused on the eighth segment of its abdomen, denominated comb-like figure region. The first attempt presented in [5] used texture descriptors, such as Co-Occurrence matrix, Local Binary Pattern (LBP) and 2D Gabor filters, to extract characteristics and classify them using a Support Vector Machine (SVM). The better results obtained by 2D Gabor filters together with the SVM achieve an accuracy of the 79%. In [6], AlexNet DNN architecture with transfer learning technique is used, showing high sensitivity but low specificity. The work [7] proposed using VGG-16 in combination with VGG-19 to enhance the accuracy obtained during the classification. However, all before mentioned works use larvae images in grayscale and the comb-like figure area is cropped manually before the classification, so these methods need human intervention to carry out the recognition.

In this paper, we propose Aedes mosquito's identification algorithm using Dense-Net 121 [8] and Guided Grad-CAM [9] to visualize the area of interest on which the trained DenseNet focuses. The area of interest determined by Dense-Net coincides with the region of the comb-like figure of larva's abdomen, which is efficiently used to distinguish larvae of Aedes mosquitos from other genera. The experimental results of the proposed algorithm show higher accuracy, sensitivity and specificity of classification, comparing with the before mentioned previous methods [5–7].

The rest of the paper is organized as follows: Sect. 2 describes the DenseNet 121 that is applied in the proposed method to achieve the classification of raw mosquito larvae images. Section 3 presents the obtained results applying Guided Grad-CAM to identify the area of interest in which the DNN focus to achieve the classification. Finally, in Sect. 4 the conclusions are presented.

2   Proposed Methods

The proposed method uses a database of raw mosquito larvae images, in other words we use larvae images without any manual interventions. As the comb-like figure area is relatively small in the input raw image, we decided to use Dense Convolutional Network (DenseNet) [8], in which a different connectivity of pattern between layers is processed in feed-forward way. Whereas other DNNs have L connections if the number of layers is L, DenseNet has N direct connections, which is given by

$$N = \frac{L(L+1)}{2} \tag{1}$$

In Fig. 1 an example of DenseNet architecture is presented. Dense Blocks are made up of 1×1 convolution and 3×3 convolution layers.

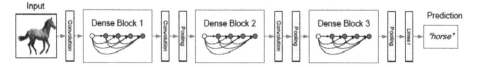

Fig. 1. DenseNet architecture example.

This configuration gives DenseNet many advantages that let the DNN improve the information flow between layers, strengthen feature propagation and feature reuse although the region of interest is relatively small. Also, this configuration reduces overfitting on tasks with smaller training set sizes [8]. Thus, the feature maps of preceding layers to be used as inputs into all subsequent layers are concatenated, supporting the final classifier decides based on all of them.

$$x_\ell = H_\ell([x_0, x_1, \ldots, x_{\ell-1}]) \tag{2}$$

In (2), x_ℓ denotes the output of the ℓth layer, $H_\ell(\cdot)$ a non-linear transformation and $[x_0, x_1, \ldots, x_{\ell-1}]$ is the feature maps concatenation of the layers $0, \ldots, \ell-1$.

The Guided Grad-CAM is a visual explanation technique that allows us visualizing the regions learned by the DNNs. The Guided Grad-CAM does not require the architectural changes or re-training, also it can highlight fine-grained details in the image, and it provides class discriminative capability [9]. This technique can detect the importance of the neurons into the decisions of interest, using the gradient information into the last convolutional layer. The neuron importance of the feature map k for the target class c, α_k^c is given by

$$\alpha_k^c = \frac{1}{Z} \overbrace{\sum_i \sum_j}^{\text{global average pooling}} \underbrace{\frac{\partial y^c}{\partial A_{ij}^k}}_{\text{gradients via backprop}} \tag{3}$$

First the gradient of the score for class c, y^c (before softmax) are computed with respect to A^k feature maps of a convolutional layer $\frac{\partial y^c}{\partial A^k}$. The gradients flow back and average-pooled to obtain (3). Then, ReLU is applied to the linear combination of forward activation maps (4).

$$L^c_{Grad-CAM} = ReLU \overbrace{\left(\sum_k \alpha_k^c A^k \right)}^{\text{Linear combination}} \tag{4}$$

The reason of the use of ReLU in (4) is because only the features that have positive influence on the area of interest are maintained. Figure 2 shows how Grad-CAM and Guided Backpropagation are obtained and the combination of them using point wise multiplication to generate Guided Grad-CAM.

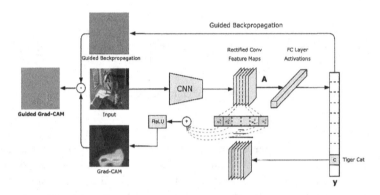

Fig. 2. Obtaining guided Grad-CAM.

3 Experimental Results

In this section, we show and explain the results achieved after training DenseNet 121. We modify the classification layer of the network to be able to discriminate into two classes: Aedes and Non-Aedes mosquito. Figure 3 illustrates an example of the images used during training, validation and test phases. The principal feature of larvae of Aedes

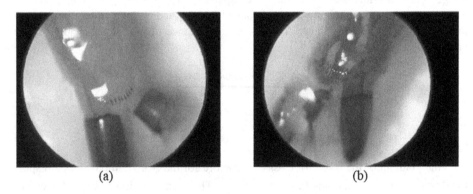

Fig. 3. 8th segment of larva's abdomen. (a) Aedes sample. (b) Non-Aedes sample.

mosquito is single line well-defined comb-like figure as shown by Fig. 3(a), while other genera of mosquitoes shows disordered comb-like figure as shown by Fig. 3(b).

Our mosquito database is made up of 760 images divided in three sets. 600 images as training set, 60 as validation set and 100 for test. We use data augmentation technique to generate more data with random transformation such as rotation, width and height shift, flip, shear and zoom. It is important to mention that this procedure is only applicable to the training data and the final size of this data is 4,200 images. We use different values of hyperparameters to get different models to obtain an optimum model. Finally, we employed the Grid Search parameter tuning technique to determine one from the accuracy point of view, which is given by Table 1.

Table 1. Hyperparameters used to train DenseNet 121.

Hyperparameter	Values
Epoch	150
Batch size	10
Learning rate	0.01
Momentum	0.2
Decay	$10e^{-5}$

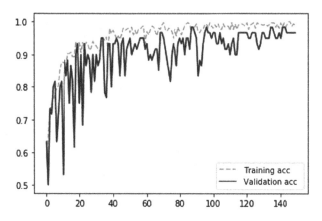

Fig. 4. Training and validation accuracy.

Although we determine the best hyperparameters, the several models obtained before were also used as inputs for Guided Grad-CAM to determine the area of the image that the network is focused on, because we are interested in making the DenseNet 121 identify the area of the comb-like figure. The model obtained with the hyperparameters•in Table 1 achieves 97% accuracy in the classification of the samples as it shows in Fig. 4.

Table 2 presents the confusion matrix obtained using the test set composed by 50 larvae of Aedes mosquitoes and 50 of other genera, after the DNN with optimum hyper-parameter is trained.

Table 2. Confusion matrix results with 97% of accuracy.

	Aedes	Non-Aedes
Aedes	48	2
Non-Aedes	1	49

Table 3 shows a comparison of the accuracy, sensitivity and specificity values between the before mentioned methods and our method.

Table 3. Comparison of our method with previous methods.

Methods	Accuracy (%)	Sensitivity (%)	Specificity (%)
Co-Occurrence Matrix and SVM [5]	67.00	64.51	58.06
LBP and SVM [5]	72.00	67.74	77.41
2D Gabor filters and SVM [5]	79.00	77.41	80.64
AlexNet DNN [6]	80.92	100	70
VGG-16 with VGG-19 [7]	91.28	94.18	88.37
Proposed scheme			
DenseNet 121 and Guided Grad-CAM	**97.00**	**97.00**	**96.00**

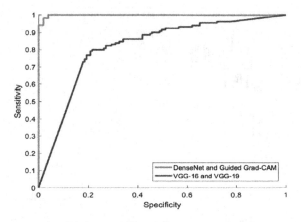

Fig. 5. Comparison of ROC curves.

Figure 5 shows the receiver operation characteristics (ROC) curve of the proposed method using DenseNet and Guided Grad-CAM and the method which employs VGG-16 and VGG-19 [7]. The ROC curves demonstrate that our proposed method draws a curve nearer to the optimum value (that is one) than the VGG-16 and VGG-19 method [7] which is the recent one with high values of accuracy, sensitivity and specificity in the state of arts.

As a verification phase, the Guided Grad-CAM is applied to the obtained model to visualize the region of interest. In Fig. 6 the area of interest determined by the network for each class: Aedes and Non-Aedes mosquitoes are shown for one training data and one test data. From Fig. 6, we can observe clearly that the interest region of the Aedes larva indicated by the DenseNet is the comb-like figure, and although this region for Non-Aedes sample is not as clear as the Aedes one in both training and testing stages, these regions are highlighted as we expected. The visualization of the interest region where the trained DNN focused on allows us an efficient verification of the training process of the DNN. Additionally, the visualization of interest region helps us to select not only an appropriate architecture of the DNN but also hyper-parameters used in it.

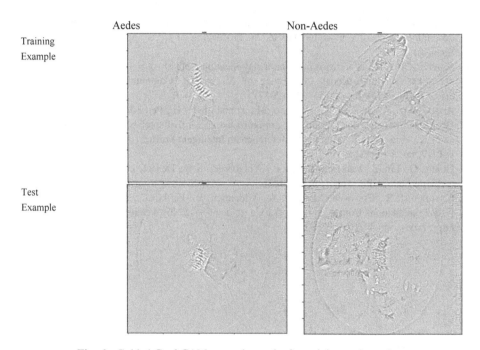

Fig. 6. Guided Grad-CAM example results for training and test data.

4 Conclusions

In this work we proposed a novel method to classify mosquito genera from the mosquito larvae images using the DenseNet 121 as the DNN architecture and the Guided Grad-CAM as a visualization technique of the interest region on where the trained DenseNet focused. The experimental results show that the proposed method provides the accuracy of 97%, which is superior to the performance of the previously reported methods [5–7]. The proposed scheme does not need any special pre-processing for the images such as manual segmentation of the area where the comb-like figure is presented or convert a color image to a grayscale one. This characteristic allows automatic recognition of the larvae without any human intervention, once larva's image is captured and introduced into the system. The future work includes building a bigger dataset in order to reinforce the training and obtain a higher classification accuracy.

References

1. Salud Documentos Técnicos. https://drive.google.com/file/d/0B0K9c-Z-JA2nSWRrYWRf VzhXa0E/view. Accessed 31 Jan 2019
2. KidsHeatlh. https://kidshealth.org/en/parents/dengue.html?WT.ac=pairedLink. Accessed 05 Feb 2019
3. CDC. https://www.cdc.gov/zika/pdfs/Draft-Environmental-Assessment-Mosquito-Control-for-publication.pdf. Accessed 05 Feb 2019
4. CDC. https://www.cdc.gov/chikungunya/pdfs/Surveillance-and-Control-of-Aedes-aegypti-and-Aedes-albopictus-US.pdf. Accessed 07 Feb 2019
5. Garcia-Nonoal, Z., Sanchez-Ortiz, A., Arista-Jalife, A., Nakano, M.: Comparison of image descriptors to classify mosquito larvae. In: Proceeding of CAIP, pp. 271–278 (2017). (in Spanish)
6. Sanchez-Ortiz, A., et al.: Mosquito larva classification method based on convolutional neural networks. In: International Conference of Electronics, Communications and Computers CONIELECOMP, pp. 155–160 (2017)
7. Arista-Jalife, A., Sanchez, A., Nakano, M., Tunnermann, H., Perez-Meana, H., Shouno, H.: Deep Learning employed in the recognition of the vector that spreads Dengue, Chikungunya and Zika viruses. In: International Conference on Intelligent Software Methodologies SoMeT, vol. 17, no. 1 (2018)
8. Huang, G., Liu, Z., Van Der Maaten, L., Weinberger, K.Q.: Densely connected convolutional networks. In: CVPR, vol. 1, p. 3 (2017)
9. Selvaraju, R.R., Cogswell, M., Das, A., Vedantam, R., Parikh, D., Batra, D.: Grad-CAM: visual explanations from deep networks via gradient-based localization, vol. 7, no. 8 (2016). https://arxiv.org/abs/1610.02391v3

Towards Automatic Rat's Gait Analysis Under Suboptimal Illumination Conditions

Ana F. Adonias[1,2]([✉]), Joana Ferreira-Gomes[3,4,5], Raquel Alonso[3,4,5], Fani Neto[3,4,5], and Jaime S. Cardoso[1,2]

[1] Faculdade de Engenharia, Universidade do Porto, Porto, Portugal
anafilipaadonias@gmail.com
[2] Instituto de Engenharia de Sistemas e Computadores,
Tecnologia e Ciência, Porto, Portugal
[3] Instituto de Investigação e Inovação em Saúde,
Universidade do Porto, Porto, Portugal
[4] IBMC - Instituto de Biologia Molecular e Celular,
Universidade do Porto, Porto, Portugal
[5] Departamento de Biomedicina, Unidade de Biologia Experimental,
Faculdade de Medicina, Universidade do Porto, Porto, Portugal

Abstract. Rat's gait analysis plays an important role in the assessment of the impact of certain drugs on the treatment of osteoarthritis. Since movement-evoked pain is an early characteristic of this degenerative joint disease, the affected animal modifies its behavior to protect the injured joint from load while walking, altering its gait's parameters, which can be detected through a video analysis. Because commercially available video-based gait systems still present many limitations, researchers often choose to develop a customized system for the acquisition of the videos and analyze them manually, a laborious and time-consuming task prone to high user variability. Therefore, and bearing in mind the recent advances in machine learning and computer vision fields, as well as their presence in many tracking and recognition applications, this work is driven by the need to find a solution to automate the detection and quantification of the animal's gait changes making it an easier, faster, simpler and more robust task. Thus, a comparison between different methodologies to detect and segment the animal under degraded luminance conditions is presented in this paper as well as an algorithm to detect, segment and classify the animal's paws.

Keywords: Computer vision · Gait analysis · Rat · Osteoarthritis

1 Introduction

Osteoarthritis is a common degenerative disease that has a significant incidence worldwide, resulting in a considerable economic and social impact [1].

© Springer Nature Switzerland AG 2019
A. Morales et al. (Eds.): IbPRIA 2019, LNCS 11868, pp. 247–259, 2019.
https://doi.org/10.1007/978-3-030-31321-0_22

This disorder damages the joints, structures responsible for the movement of the body, causing pain, stiffness and functional impairment which as a tremendous impact on the patients' routines. Although there are some treatments that aim to relieve osteoarthritis' symptoms, these therapeutic strategies remain unsatisfactory once they fail to relieve pain or they trigger undesirable side-effects. Consequently, several researchers have attempted to develop more efficient drugs for the treatment of osteoarthritis by relying on the use of animal experimentation [2]. Among the animal species used in research, mice and rats are the most widely used models to study the pathophysiological mechanisms underlying osteoarthritis and to test the therapeutic efficacy of targeted drugs [3]. One of the experimental approaches consists on inducing osteoarthritis in the animal, and afterwards administering a drug and quantifying the changes in the behavioral signs of the animal over several time points of the treatment and/or disease progression. Since movement-evoked pain is a common characteristic of osteoarthritis, the affected animal usually modifies its behaviour after the induction of the model as an attempt to protect the injured joint from load while walking, and therefore decrease pain [4]. As consequence, the gait's parameters will vary over time and will differ between a normal (not ill) and an arthritic rodent. The analysis of these differences can be an useful tool to evaluate, through a continuous monitoring, the therapeutic impact of distinct approaches, namely drugs, on osteoarthritis treatment.

The gait changes can be assessed through the use of videos, which can be analyzed manually or automatically. The manual observation and quantification of the gait changes is a laborious and time-consuming intensive task that is prone to high user variability [5]. Currently, there are some commercially available video-based gait systems capable of detecting and identifying the animal's limbs and automatically compute spatial, temporal, kinematic and dynamic gait parameters. However, they are often expensive closed solutions and some even induce stress to the animal by forcing it to walk on a treadmill at a certain speed in a certain trajectory, which might affect the results [6]. Thus, researchers often select different gait pattern descriptors to report in the literature and some opt for the development of a customized system for the acquisition of the videos and their manual analysis. One approach to do so is to carefully examine videos and choose a set of frames with the animal walking (two paws on the platform) and with the animal standing still (four paws on the platform). After selecting the frames, a software like ImageJ[1] may be used to extract the metrics of interest: the footprints are manually selected and a threshold value is defined by the user, above which the number and intensity of pixels are quantified, allowing the comparison of the area and mean intensity applied by each paw [7].

In order to automate this task and make the system more robust, flexible and able to solve problems related not only with the time and cost but also with the reproducibility inherent to the assessment process conducted by the researchers, a demand for an automated system of gait analysis in laboratory animals like rats and mice arises naturally. Moreover, the availability of such

---

[1] imagej.nih.gov/ij/.

systems opens the possibility to rethink gait analysis itself, since the typical testing time scale can be easily extended, thus diversifying the gait parameters extracted and the context of evaluation. In this way, this paper contributes with the development of an automated, quantitative tool for rat gait assessment using recent computer vision techniques. A comparison between different methodologies to detect and segment the animal under degraded luminance conditions is made and an algorithm to detect, segment and classify the animal's paws is proposed. Furthermore, in order to evaluate the proposed methodologies, it was created a database with the rats' gait and its manual annotation which is made available to the community.

2 Related Work

Advances in the fields of computer vision and machine learning have been providing the detection, recognition and tracking of objects. Since mice and rats are the most used animal models in the studies of the mechanisms related with human diseases, some sophisticated algorithms have been proposed for mice and rats detection and tracking to ease and optimize time-consuming and laborious tasks usually done manually. The majority of these studies are, however, related with mouse behaviour analysis having a different goal and a different setup from the one addressed in this work [8–11]. On those studies, the camera records the animal from above in normal light conditions.

As previously mentioned, there are some commercially available video-based gait systems which combine video-tracking technology with image analysis methodologies to quantify and characterize mice and rats gait [12–16]. Besides all the commercial systems quoted, it is also worth to mention the work of Mendes *et al.* who developed the MouseWalker, an open source software which aims to evaluate mice gait [17]. This work has the same acquisition procedure as the one used in this paper but it has a much more controlled environment since there are not background movements and the trajectory of the animal is restricted to a walkway, being the mouse always within the frame and walking in the same direction. Another work found, authored by Leroy *et al.*, proposes an automated gait analysis for laboratory mice with the same setup but different lighting conditions [18]. In this study, a background subtraction method is applied to segment the mouse and a motion analysis is performed. Unlike the setting of our work, their hardware allows them to detect the footprints by resorting to a color filter, enhancing the pixels that correspond to the region of the paws with a strong red color component.

Nonetheless, as much as is possible to ascertain, there is no work that meets the same degraded luminosity conditions and works robustly in open field platforms.

3 Methodology

3.1 Data Acquisition

Animal Model. The data used for this study was acquired at Faculdade de Medicina da Universidade do Porto, Portugal. All experimental procedures were performed in compliance with all ethical norms required [19] and appropriate measures were taken to minimize pain or discomfort of the animals. All the conditions were assured in order to maintain reliable results without external interference. Under brief anaesthesia, the osteoarthritis model was induced on the right hind knee of adult male rats [7]. In order to have a control of the disease, a similar number of animals were injected with saline, under the same conditions.

CatWalk Test. This paper focuses on data acquired by a custom-made gait analysis system that allows the animal to walk freely on an open field platform located in a darkened room, with no stressors or rewards. The hardware configuration used is based on the CatWalk system, illustrated in Fig. 1 (adapted from [20]). A fluorescent light beam is sent through a glass platform being reflected internally in all the plane with the exception of the areas where the paws are placed, where light is refracted (A); images are recorded by a video camera placed under the glass platform in which animals are allowed to walk freely. The video camera is connected to a computer equipped with video acquisition software (Ulead Video Studio, Freemont, CA) (B); the signal intensity depends on the contact area of the paw with the surface and increases with the pressure applied by the paw (C); an example of the obtained images is represented in (D).

After recording the animal's gait, the analysis of the CatWalk behavior consists on measuring the mean intensity and the contact area of each paw with the glass platform, to evaluate the disability induced by the model over time [7].

3.2 Dataset

The dataset is composed by 15 videos of 4 different animals in different time points of the experience, each one with different duration, averaging about 5 min each. The videos have a resolution of 640 horizontal × 480 vertical pixels and a frequency of 25 frames per second. All these videos had already been analyzed by an expert researcher, who had already selected the frames of interest (approximately 9 frames per video) and the associated threshold used to compute the area and the mean intensity of the paws in each frame.

In order to evaluate the algorithm's performance, 130 frames, selected by experts, were annotated using LabelMe, an open source graphical image annotation tool. Six classes were used to annotate all the body parts of interest of the rat: the body, the tail and right-hind, left-hind, right-fore and left-fore paws. All these classes were annotated using polygons. The detailed masks of the paws (Fig. 2(C)) were obtained by applying the threshold defined by the experts. Rat and tail annotations were thereafter confirmed by an expert researcher.

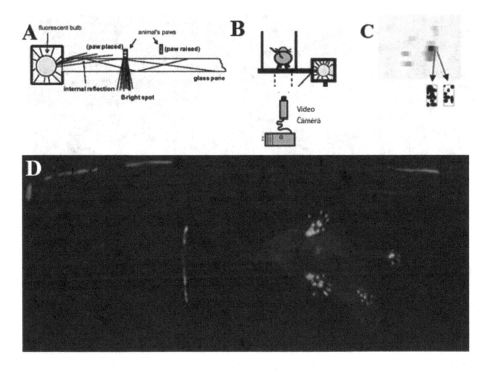

Fig. 1. Principle of the CatWalk setup.

The dataset is available for the community.[2]

Dataset Heterogeneity. Some artifacts and variations from video to video, and even within each video, hinder the automation process. Since the videos were acquired in the dark, degraded luminosity conditions is a feature of the dataset. Furthermore, although the data was acquired using a static camera in a slightly controlled environment, it is clear the presence of noise in the background, whether due to the interference of researchers with the animal or simply due to contamination of the platform by the animal. Since the platform area is larger than the area captured by the camera, periods when the animal leaves completely the field of view are also present in the dataset. Moreover, as the rat is a non-rigid object, its shape varies during the video sequence, making it more difficult to track. Some natural behaviors such as sitting, lying, scratching, getting up, among others, hinder the detection and segmentation of the paws since these behaviors promote the contact of the body with the platform, enhancing some non-paws derived pixels.

---

[2] To access the videos and the annotations, please contact the author Joana Ferreira-Gomes through the email jogomes@med.up.pt.

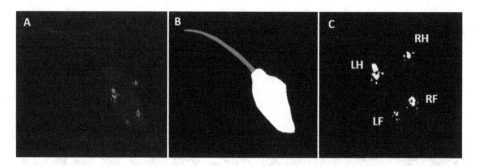

Fig. 2. A. input frame; **B.** *White:* ground truth rat's body; *Gray:* ground truth tail; **C.** ground truth paws: *LH:* left-hind; *RH:* right-hind; *LF:* left-fore; *RF:* right-fore.

3.3 Methodologies

Based on the current state of the art, automated recognition of the animal's gait pattern can be done by using different computer vision techniques. The generic video-based gait framework is presented in Fig. 3 and it may be briefly described as follows: rat detection and segmentation, paws detection, segmentation and classification and gait features extraction. We did not adopt the current trend of deep learning approaches since the dataset is small, not favouring deep learning solutions, and the strong field knowledge about the setting facilitates the feature-engineering approach.

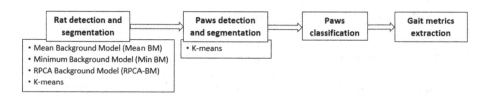

Fig. 3. The generic video-based gait framework.

Rat Detection and Segmentation. The goal of this first step is to robustly detect and segment the body of the animal. This operation is key in the adopted methodology since its success will impact the following steps. Several segmentation techniques were performed, evaluated and compared. Before the segmentation, a pre-processing step was applied to all frames in order to remove noise. This operation results in a smoother image and is performed by computing a simple convolution between each frame (f) and a 3-by-3 box linear kernel (k).

Since one of the dataset characteristics is the presence of background fluctuations, three different approaches were tested to model the background. The first consisted on averaging the whole video sequence (v), which was previously pre-processed (Eq. 1). In the second, because the animal is ideally the only bright object in motion, the minimum value of each pixel (p_ij) in the entire video sequence was computed (Eq. 2).

$$BG(v) = \frac{1}{|v|} \sum_{f \in v} K \otimes f \tag{1}$$

$$BG_{ij}(v) = min_{f \in v}(p_{ijf}) \tag{2}$$

Both of these offline approaches result in a static background image which is subtracted to each frame to obtain the foreground, the animal in motion (Eq. 3). A threshold operation is performed to obtain the mask of the rat and morphological operations are applied to remove some background noise.

$$FG(v, f) = |f - BG(v)| \tag{3}$$

The third tested approach to model the background was presented by Candès *et al.* [21] and it considers the data matrix (M) a sum of a low-rank component (L) with a sparse component (S). Each column in M corresponds to a frame of the video, properly vectorized. The intuition is that the observed data is the sum of the background information and the objects' data. Since the background is essentially static, it should lead to a low-rank matrix; since the objects tend to be small, they should correspond to a sparse matrix. The low-rank matrix can be then representative of the background, being able of comprising small fluctuations on it and on the other hand, the sparse matrix can formulate the objects in motion, in this case the rat, which can be seen as an outlier. In this method, the principal component pursuit is used to solve the robust principal component analysis (RPCA) following the Eq. 4, where $\| A \|_1$ denotes the vector l_1 norm of the matrix A and $\| A \|_*$ denotes the nuclear norm, the sum of the singular values of A. Morphological operations were applied to the sparse matrix to remove small blobs of the background by filtering it.

$$min(\| L \|_* + \lambda \| S \|_1) \quad \text{subject to} \quad M = L + S \tag{4}$$

K-means was another of the methodologies used to segment both rat body and paws. This clustering algorithm aims the partitioning of the image into homogeneous regions containing pixels with similar characteristics. The number of clusters was set to three, where one represents the background, other the body of the rat and the third one the paws. This technique was applied to the foreground after a pre-processing algorithm that aims to improve the frames' quality, facilitating the segmentation process. This algorithm was divided in two main steps: a contrast enhancement, achieved by normalizing the frame histogram between zero and its maximum value, followed by a convolution between the enhanced image and a sharpening filter (S) (Eq. 5) which resulted in a well-defined image.

$$S = \begin{bmatrix} -1 & -1 & -1 & -1 & -1 \\ -1 & +1 & +1 & +1 & -1 \\ -1 & +1 & +9 & +1 & -1 \\ -1 & +1 & +1 & +1 & -1 \\ -1 & -1 & -1 & -1 & -1 \end{bmatrix} \qquad (5)$$

In all the mentioned methodologies the tail was removed by resorting to an opening morphological operation with a square structuring element of size 13×13. The convex hull of the result of the opening operation was computed, obtaining thus a mask of the body of the rat. This was done to standardize the output of this step since the tail is not always detected and segmented together with the body of the animal.

Paws Detection, Segmentation and Classification. As mentioned before, K-means was the algorithm used to segment the paws. This methodology was performed to the foreground with and without the pre-processing algorithm mentioned above in order to test its efficacy. After having the paws segmented, their classification as left-hind (LH), right-hind (RH), left-fore (LF) and right-fore (RF) must occur. For this purpose, the animal's orientation must be first determined. As the body of the animal can be represented by an ellipse, this shape was used to fit the rat's body mask, the output of the rat segmentation step. Two approaches were performed to obtain the tail point, the extreme point of the ellipse nearest to the tail. The first consisted on a simple subtraction method between the mask obtained before the tail removal in the rat segmentation stage and after. By this way, when the tail is segmented simultaneously with the body there will exist a considerable difference in one of the ellipse's sides. However, as already referred, sometimes the tail is not visible on the frame, not being detected in the first place. In these cases, the difference between both images is not substantial and the tail point is computed by resorting to the second methodology. This second approach computes the minimum difference between the mean of the rat (region inside the ellipse) and its neighborhood in an extension of 250 pixels. A spatial distribution of the minimum points is obtained and each one of the points is associated to the nearest extreme point of the ellipse. The tail point is computed based on the median distance and on the standard deviation between the grouped points and its associated extreme. After determining the tail point, the ellipse is splitted by quadrants which are classified as LH, RH, LF and RF and each paw is assigned to the respective quadrant. This association is made based on euclidean distances.

4   Results

The algorithm was implemented in Python with the exception of the RPCA algorithm, which was implemented in MATLAB being available via the GitHub library of Sobral [22].

Rat Detection and Segmentation. In order to evaluate the different segmentation methods, the intersection level A_f between the ground truth bounding box of the animal's body GT and the tracked bounding box T, was computed to each frame (f) according to the expression $A_f = \mathrm{Area}(GT_f \cap T_f)/\mathrm{Area}(GT_f \cup T_f)$. A known measure to evaluate the performance on a video sequence is to count the number of successful frames whose A_f is larger than a defined threshold. According to the PASCAL criterion [23], this threshold value is 0.5, however, as the use of an unique threshold value may not be representative enough of this measure, the success plot showing the ratios of successful frames at thresholds ranging from 0 to 1 was computed for each of the methodologies (Fig. 4).

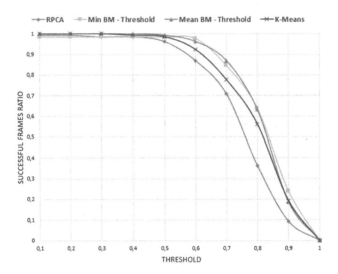

Fig. 4. Rat segmentation successful plot for the 130 frames manually labelled.

In general, all the methodologies are capable of detecting correctly the animal in the majority of the frames. The Mean BM methodology demonstrated to be the most robust, presenting the most similar results with the ground truth. The RPCA methodology showed to be the less vigorous due to the presence of the animal in the low-rank matrix. Since the algorithm was applied to the input video, without any previous processing and computed at intervals of 2000 frames due to computational costs, if the animal stays in the same place for a while, it will be assumed as belonging to the background, hindering its segmentation. The Min BM method showed also good results, being less able, however, of handling large background fluctuations and failing completely the detection of the rat in frames where the researcher interferes with the animal, for example, or when there is a contamination of the platform. Figure 5 shows two examples of the obtained bounding boxes for each methodology.

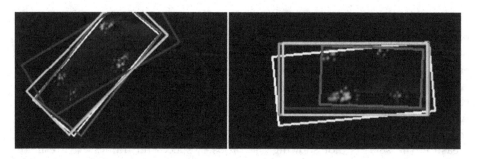

Fig. 5. Rat segmentation examples for the proposed methodologies: **White:** Ground truth; **Blue:** K-means; **Yellow:** Min BM; **Green:** Mean BM; **Red:** RPCA; (Color figure online)

Paws Segmentation. As the ultimate goal of this work is to quantify the rat's gait metrics, such as the area and the mean intensity of the paws, it is important to evaluate paws segmentation by using pixel-based metrics since the algorithm must be meticulous enough to return the referred gait metrics efficiently. While the true positives (TP) return the number of correctly detected paws pixels, the true negatives (TN) give the number of background pixels correctly detected. In contrast, the false positives (FP) and the false positives (FN) are pixels that are falsely recognized as foreground and background, respectively. From these metrics, the true positive rate (TPR), false positive rate (FPR) and F-score were computed. The TPR reports how frequently the algorithm correctly detects the paws pixels, being given by $TPR = TP/(TP + FN)$. The FPR refers to the number of times the paws are falsely detected and it is given by $FPR = FP/(FP + TN)$. The F-score measure combines TPR and precision (P), given by $P = TP/(TP+FP)$, by their harmonic mean being a measure of the algorithm's accuracy.

The results of paws' segmentation are present in Table 1. Comparing both non pre-processed and pre-processed data, it can be concluded that this step introduces an improvement in the algorithm's performance. A visual example of this pre-processing step can be found in Fig. 6 and two examples of the K-means output applied to the pre-processed frames are present in Fig. 7.

Table 1. Results of paws' segmentation.

Method	TRP	FPR	F-score
K-means (with pre-processing)	0,85 ± 0,06	0,00 ± 0,00	0,77 ± 0,09
K-means (without pre-processing)	0,61 ± 0,15	0,00 ± 0,00	0,73 ± 0,10

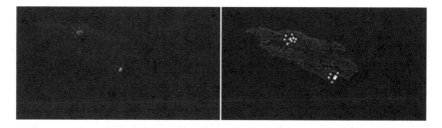

Fig. 6. Right: input frame; **Left:** output from the pre-processing step.

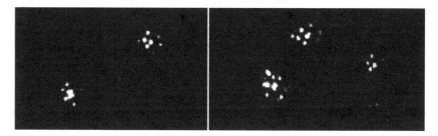

Fig. 7. Paws' segmentation examples: **White:** TP pixels; **Black:** TN pixels; **Blue:** FP pixels; **Red:** FN pixels (Color figure online)

Paws' Classification. The results obtained for the paws' classification are presented in Table 2. In the same way that the pre-processing improves paws' segmentation, it also improves their classification. A frame is only considered well classified when it detects exactly the same paws as the ones present in the GT image in the same region of the image. From the 32% of frames that the classification of frames fails, about one-third is because the algorithm fails in detecting the animal's orientation. This happens when the rat is occluded, when the rat segmentation is not meticulous, when the tail is not well removed or when the animal is near a bright object. The remaining frames' classification fails when the rat has its belly or other part of the body in contact with the platform, enhancing some non-paws derived pixels, or when the two fore paws are very close to each other and the algorithm assumes that it is only one footprint. When comparing the obtained results for each paw, the RH and LH paws' classification is correct in 82% and 85% of the frames, respectively, while RF and LF are successfully classified in 78% and 76% of the frames, respectively.

Table 2. Results of paws' classification.

Method	Frames well classified	Paws well classified
K-means (with pre processing)	68%	81%
K-means (without pre processing)	48%	75%

5 Conclusions

In this paper, an automated method for the assessment of the rat's gait is presented, with the aim of not only providing data that is less prone to user variability but also alleviating the work of scientists who resort to the analysis of the gait of animal models to evaluate the effectiveness of drugs used in the treatment of osteoarthritis. The contributions include an annotated dataset publicly available to the scientific community and a framework capable of detecting the animal, segmenting its body and detecting, segmenting and classifying its paws under degraded lighting conditions.

In between the proposed methodologies to segment the rat, modeling the background and subtract it from each frame gave the best results. Averaging the entire video sequence demonstrated to be the best approach to model the background. Quantitative and visual inspections of the results of paws' segmentation demonstrate a good performance of the algorithm. The proposed algorithm to classify the paws showed promising results.

Future work will focus on the development of an algorithm to select the frames of interest, and on the extraction of more complex gait metrics. We aim to keep extending our dataset and evaluating deep learning methodologies as an alternative to the adopted framework. We also plan the development of a web platform deploying the proposed framework as a free service available to researchers.

References

1. Arthritis Foundation: Arthritis by the numbers. Arthritis Foundation (2018)
2. Benson, R.A., McInnes, I.B., Garside, P., Brewer, J.M.: Model answers: rational application of murine models in arthritis research. Eur. J. Immunol. **48**(1), 32–38 (2018)
3. Hedrich, H., Bullock, G.: The Laboratory Mouse: Handbook of Experimental Animals. Elseviner Limited, London (2004)
4. Jacobs, B.Y., Kloefkorn, H.E., Allen, K.D.: Gait analysis methods for rodent models of osteoarthritis. Current Pain Headache Rep. **18**(10), 456 (2014)
5. Pathak, S.D., et al.: Quantitative image analysis: software systems in drug development trials. Drug Disc. Today **8**(10), 451–458 (2003)
6. Lakes, E.H., Allen, K.D.: Gait analysis methods for rodent models of arthritic disorders: reviews and recommendations. Osteoarthritis Cartilage **24**(11), 1837–1849 (2016)
7. Ferreira-Gomes, J., Adães, S., Castro-Lopes, J.M.: Assessment of movement-evoked pain in osteoarthritis by the knee-bend and CatWalk tests: a clinically relevant study. J. Pain. **9**(10), 945–954 (2008)
8. da Silva Monteiro, J.P., et al.: Automatic behavior recognition in laboratory animals using kinect (2012)
9. De Chaumont, F., et al.: Computerized video analysis of social interactions in mice. Nat. Methods **9**(4), 410 (2012)
10. De Chaumont, F., et al.: Live Mouse Tracker: real-time behavioral analysis of groups of mice. BioRxiv **16**(2), 345132 (2018)

11. Noldus, L.P.J.J., Spink, A.J., Tegelenbosch, Ruud A.J.: EthoVision: a versatile video tracking system for automation of behavioral experiments. Behav. Res. Methods Instrum. Comput. **33**(3), 398–414 (2001)
12. Digigait. http://www.mousespecifics.com/digigait/. Accessed 15 Apr 2019
13. CatWalk. http://www.noldus.com/animal-behavior-research/products/catwalk. Accessed 15 Apr 2019
14. TreadScan. http://www.cleversysinc.com/CleverSysInc/wp-content/uploads/2013/02/Datasheet-TreadScan-2017.pdf. Accessed 15 Apr 2019
15. RunwayScan. http://www.cleversysinc.com/docs/software-RunwayScan.pdf. Accessed 15 Apr 2019
16. GaitScan. http://www.cleversysinc.com/CleverSysInc/wp-content/uploads/2013/02/Datasheet-GaitScan-2010.pdf. Accessed 15 Apr 2019
17. Mendes, C.S., Bartos, I., Márka, Z., Akay, T., Márka, S., Mann, R.S.: Quantification of gait parameters in freely walking rodents. BMC Biol. **13**(1), 50 (2015)
18. Leroy, T., Silva, M., D'Hooge, R., Aerts, J.-M., Berckmans, D.: Automated gait analysis in the open-field test for laboratory mice. Behav. Res. Methods **41**(1), 148–153 (2009)
19. Zimmermann, M.: Ethical guidelines for investigations of experimental pain in conscious animals. Pain **16**(2), 109–110 (1983)
20. Vrinten, D.H., Hamers, F.F.T.: 'CatWalk' automated quantitative gait analysis as a novel method to assess mechanical allodynia in the rat; a comparison with von Frey testing. Pain **102**(1–2), 203–209 (2003)
21. Candès, E.J., Li, X., Ma, Y., Wright, J.: Robust principal component analysis? J. ACM (JACM) **58**(3), 11 (2011)
22. Sobral, A., Bouwmans, T., Zahzah, E.: LRSLibrary: Low-Rank and Sparse Tools for Background Modeling and Subtraction in Videos. Robust Low-Rank and Sparse Matrix Decomposition: Applications in Image and Video Processing. CRC Press, Boca Raton (2015)
23. Everingham, M., Van Gool, L., Williams, C.K.I., Winn, J., Zisserman, A.: The pascal visual object classes (VOC) challenge. Int. J. Comput. Vis. **88**(2), 303–338 (2010)

Impact of Enhancement for Coronary Artery Segmentation Based on Deep Learning Neural Network

Ahmed Ghazi Blaiech[1,2(✉)], Asma Mansour[1,3], Asma Kerkeni[1,3], Mohamed Hédi Bedoui[1], and Asma Ben Abdallah[1,3]

[1] Laboratoire de Technologie et Imagerie Médicale,
Faculté de Médecine de Monastir, Université de Monastir,
5019 Monastir, Tunisia
ahmedghaziblaiech@yahoo.fr

[2] Institut Supérieur des Sciences Appliquées et de Technologie de Sousse,
Université de Sousse, 4003 Sousse, Tunisia

[3] Institut supérieur d'informatique et de Mathématiques, Université de Monastir,
5019 Monastir, Tunisia

Abstract. X-ray Coronary angiograms are intended to specify the global state of the artery system and therefore to detect and locate the zones of narrowing. Accurate coronary artery segmentation is a fundamental step in computer aided diagnosis of many diseases. In this paper, deep neural network based on U-Net architecture is proposed in order to improve the segmentation task for coronary images. In this context, various enhancement methods, like the adaptive histogram equalization and the multiscale technique with a Frangi filter are tested not only in normal conditions but also in the presence of noise to improve the system performance and to ensure its robustness against real conditions. Promising result are obtained and discussed for different performance criteria. This work will serve as a reference and motivation for researchers interested in the field of blood vessel segmentation by deep learning neural networks.

Keywords: Coronary artery segmentation · Enhancement · Deep learning

1 Introduction

Despite the progress of non-invasive exams, X-ray Coronary Angiography (XCA) remains the best reliable way to assess clinically significant Coronary Artery Diseases (CADs) [1]. The angiograms obtained by XCA reveal the initial CAD symptoms by perfectly visualizing all coronary arteries placed in the crown around the heart, which will bring the blood necessary for the proper functioning of the organ. However, the complex vessel structure, image noise, poor contrast and non-uniform illumination make vessel tracking a tedious task. For accurate coronary vessel detection in various medical imaging applications, like stenosis detection [2], 3D reconstruction [3] and cardiac dynamics assessing [4], two steps are necessary: enhancement and segmentation.

The enhancement step is necessary to improve the vessel delineation while reducing background artifacts. It aims to ameliorate the image quality to make it more

© Springer Nature Switzerland AG 2019
A. Morales et al. (Eds.): IbPRIA 2019, LNCS 11868, pp. 260–272, 2019.
https://doi.org/10.1007/978-3-030-31321-0_23

appropriate for further applications like segmentation. In this context, several methods have been put forward in the literature to improve X-ray angiographic images, such as adaptive histograms with limited contrast (CLAHE) [5], multiscale-based Hessian filtering [6, 7] and Gabor filtering [8]. The subsequent segmentation step allows a better analysis of the vessels for diagnosis, treatment planning and execution and enables the evaluation of clinical outcomes. Many studies have been proposed for coronary vessel segmentation such as thresholding [3, 9, 10], mathematical morphology [11–13], deformable models [14] and machine learning algorithms like decision trees, random forests, genetic algorithms and Artificial Neural Networks (ANNs). ANNs have gained researchers' interest, in the recent years, for its good performance to become the best solution for classification tasks. Particularly, Deep Neural Network (DNN) architecture is widely used for successful classification, where the output to an image is a single class label, such as lung diseases [15] and skin cancer classification [16]. Other tasks can be well-achieved with DNN architecture, like semantic segmentation, where the desired output should include a class label supposed to be assigned to each pixel. Some of these segmentation tasks have focused on blood vessels [17, 18]. In [17], the authors presented an automatic segmentation of blood vessels in fundus images. A DNN with 10-layer architecture composed of Convolutional Neural Networks (CNNs) and a max-pooling layer were used in this work and implemented to segment blood vessels. This implementation was tested on a publicly available DRIVE dataset, and our results demonstrate the high effectiveness of the deep learning approach which can achieve 0.9466 and 0.9749 average accuracy and area under the curve, respectively. In [18], the authors depicted a Fully Convolutional Network (FCN) for blood vessel segmentation. The FCN architecture consisted of convolution and deconvolution layers. The encoder phase extracted high-level abstract patterns. From these underlying structures, the decoder phase would try to recover the segmentation of an input. FCN training went from data augmentation and transfer learning to patch-based segmentation. In the latter, the segmentation of an image involved patch extraction, segmentation, and aggregation. The experimental results were presented on the task of retinal blood vessel segmentation using the well-known publicly available DRIVE database.

However, little work has suggested blood vessel segmentation on the coronary by DNNs. We cite the work of [19], where the authors presented a method for detecting vessel regions in angiography images based on the deep learning approach using CNNs. Firstly, the intended angiogram was processed to enhance the image quality. In this phase, a multi-scale top-hat transform [20] was applied to enhance the contrast of input angiogram images. Secondly, a patch around each pixel was fed into a trained CNN to determine whether the pixel was of vessel or background regions. The experiments showed that the proposed algorithm had a dice score of 81.51 and accuracy of 97.93.

To improve the performance segmentation on coronary, we will give great importance to the enhancement step, which influences the segmentation result. This step is even more important than the profile of the noise rate on the image.

The aim in this paper is to investigate the impact of enhancement methods in the presence of noise to get accurate coronary segmentation based on DNNs. Segmentation is established by U-Net deep learning network architecture, which has been successfully used on other vascular images of the retina [21]. For enhancement, CLAHE and multiscale-based Hessian filtering methods, namely Frangi [6], and Ranking the Orientation Responses of Path Operators (RORPO) [22] are used, thus increasing the system performance. Our system must withstand noise to ensure robustness under real conditions.

This paper is organized as follows: Sect. 2 presents an overview of some enhancement techniques that are used afterwards. In the same section, the U-Net deep learning architecture is detailed for the coronary artery segmentation step. Section 3 describes the database used for the experiment and the implementation software environment and it discusses the obtained implementation results. The impact of enhancement is highlighted as regards the noise rate available in the input image.

2 Materials and Methods

Figure 1 presents the implementation process. The first step is the pre-processing of coronary images, which consists in utilizing the enhancement methods like CLAHE, Frangi and RORPO filters. The second step is the segmentation of blood vessels from enhanced images using DNN architecture.

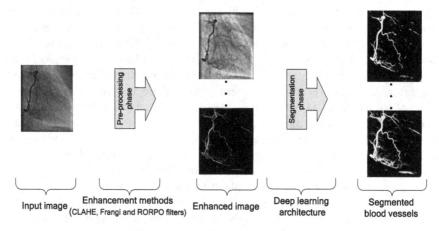

| Input image | Enhancement methods (CLAHE, Frangi and RORPO filters) | Enhanced image | Deep learning architecture | Segmented blood vessels |

Fig. 1. Implementation process

2.1 Image Pre-processing

Various types of images, like medical images, suffer from different problems like poor contrast, non-uniform illumination and noise. Therefore, it is essential to enhance the quality of images for further processing. A lot of techniques have already been proposed by different researchers [22]. We present in this work various enhancement techniques applied to vascular structures to prepare them for the segmentation step.

CLAHE

It is an enhancement method given by Zuiderveld [23] to improve the image contrast for medical imaging applications so as to overcome the amplification of the noise problem. This method produces the optimal equalization in terms of maximum entropy, and it also limits the contrast of an image.

A lot of work has used this method for image pre-processing. For example, in [24], the authors adopted the CLAHE method to ameliorate the image quality. Hitam et al. presented in [25] a mixture of CHAHE in RGB and HSV (Hue Saturation Value) color spaces to enhance the underwater images. Buzuloiu et al. [26] suggested a variant of the CLAHE method of color-image enhancement using the adaptive neighborhood method.

Frangi Filter

It was proposed by Frangi et al. [6]. This filter is often considered as the current gold-standard due to its simplicity, intuitive formulation and good vascular structure enhancement of medical angiographic images [27, 28]. The thought behind this filter is that the Hessian eigenvalues are geometrically interpreted as principal vessel curvatures. Generally, vessels have a strong variation perpendicular to the vessel centerline (high λ_2) and a weak variation along it (weak λ_1). On the basis of these observations, Frangi et al. introduced two measures to describe structures in images: RB = λ_1/λ_2 is the blob like structure measure, and $S = \|H_F\| = \sqrt{\lambda_1^2 + \lambda_2^2}$. The Frobenius norm of the Hessian matrix is the second-order structureness measure. These measures are combined in a vesselness function as given by the following equation:

$$V_\sigma(p) = \begin{cases} 0 & \text{if } \lambda_2 < 0 \\ \exp\left(\frac{-R_b^2}{2\beta^2}\right)\left(1 - \exp\left(\frac{-S^2}{2c^2}\right)\right) & \text{Otherwise} \end{cases} \tag{1}$$

where parameters β and c are thresholds that control the filter's sensitivity to R_b and S, respectively.

According to the scale space theory, $V_\sigma(p)$ will be the maximum only when the width of the vessel in pixel p matches a suitable scale factor σ. As a consequence, in the multi-scale vessel enhancement algorithm, V_σ is computed for multiple scales. Then the largest one is taken as the final filter output. In order to normalize the intensity values across the scale space, parameter σ^γ is introduced [6]. Parameter γ can be fixed to determine the preference for particular scales. It is equal to 1 if no scale is preferred. Thus, for $\gamma < 1$, small scales are favored, and for $\gamma > 1$, the large ones are favored.

RORPO Filter

It is a new filter proposed by Odyssée [29]. This morphological filter is used for thin structures in nD images characterized by a lower size in one of their n dimension compared with others without requiring post-processing.

The patterns present in a linear structure are one of the most difficult to handle. This has motivated introducing this type of filter in the field of medical imaging [29]. This filter is based on path operators, which include a path opening and a path closing. Usually, two types of structures can be found: curvilinear structures, which are directionally elongated, and blob-like structures, which are approximately isotropic. A path opening in a given orientation that preserves structures within at least one path. RORPO based on a path opening aims to detect curvilinear structures. Along with the intensity feature, RORPO can provide for each point of a curvilinear structure, an estimation of its orientation.

2.2   U-Net DNN Architecture

The U-Net DNN architecture is illustrated in Fig. 2. It was used with success in retinal blood vessels in [21]. Such architecture is composed of a contracting path (left side) and an expansive one (right side). The former follows the typical architecture of a convolutional network. It consists in a repeated application of two 3 × 3 convolutions, each followed by a Rectified Linear Unit (ReLU) and a 2 × 2 max pooling operation with stride 2 for down-sampling. At every down-sampling step, we double the number of feature channels.

The expansive path consists in up-sampling the feature map followed by a 2 × 2 convolution (up-convolution) which halves the number of feature channels:

- A concatenation with the correspondingly cropped feature map from the contracting path
- Two 3 × 3 convolutions, each followed by a ReLU.

Cropping is necessary due to the loss of border pixels in every convolution. At the final layer, a 1 × 1 convolution is utilized to map each 64-component feature vector to the desired number of classes.

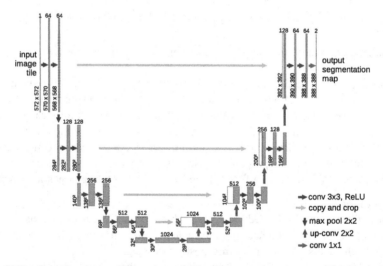

Fig. 2. U-Net deep learning architecture: each blue box corresponds to a multi-channel feature map. The number of channels is depicted on top of the box. The x-y size is provided at the lower left edge of the box. White boxes represent copied feature maps. The arrows represent the different operations [21]. (Color figure online)

3   Experiments and Results

In this section, we first introduce the dataset and the enhancement steps. Then we describe the hardware and software environment for our work. We finish this section by presenting the results provided for the implementation of U-Net architecture with various enhancement methods on four dataset types.

3.1   Dataset

A total of 10 obtained images (dataset DS1) are selected for coronary artery segmentation in 2D X-ray angiograms. Each image has a size of 512×512 pixels and 256 gray levels with a resolution of 0.3×0.3 mm. This dataset is acquired during routine cardiac catheter examinations using a single plane Siemens Artis Zee angiography system available at the Cardiology Department of The University Hospital Fattouma Bourguiba, Monastir, Tunisia [30].

Difficult situations that may occur in X-ray coronary angiograms can happen, like the low signal-to-noise ratio, the severe stenosis and the overlapping of many vessels. Our objective is to ensure the robustness and reliability of our system. For this, we add noise in the original images. Thus, we vary the Gaussian noise for DS1 in the order of 0.001, 0.01 and 0.1, which results in three other DS2, DS3 and DS4 datasets. Each one is split into two datasets: 2/3 and 1/3 for training and validation, respectively.

3.2   Experiment Setup

To evaluate the different enhancement methods on four datasets, the manual segmentation of images has been done and verified by clinicians. These annotated images are used as ground truth. Each image is divided into 57,000 patches of size 48*48 for better memory allocation. Hence, a significant number of patches allows increasing the performance. The ROI (Region of Interest) detection is considered on the whole image given that the probability of the presence of blood vessels is the same for each pixel. The implementation of deep learning architecture is carried out with Keras tools by using GeForce GTX 950 graphics card. The model is trained for a total of 100 epochs utilizing a stochastic gradient descent optimizer. This number of epochs is chosen to ensure the convergence of the network, i.e. seek the minimum of validation errors while avoiding overfitting. Each epoch represents the training of the model throughout all data. The input data are divided into batches of 64.

3.3   Performance Metrics

The quantitative analysis is used to evaluate the segmentation results in terms of their performance compared to that given by the expert. We opt for accuracy, sensitivity, precision, specificity and F-score in this situation. An optimal segmentation is the one that reaches up to 100% for all metrics that provide information about both the ground truth image and how it is segmented by the U-Net architecture.

There are four parameters described with the following definitions:

- TP (True Positive): pixels representing the blood vessel (white color) correctly predicted
- TN (True Negative): pixels not representing the blood vessel (black color) correctly predicted
- FP (False Positive): pixels not representing the blood vessel incorrectly predicted
- FN (False Negative): pixels representing the blood vessel incorrectly predicted

Global accuracy is to simply report the percentage of pixels in the image, which are correctly classified. Pixel accuracy is commonly reported for each class separately and globally across all classes. This metric provides misleading results because the class representation (blood vessel in white) is small within the image. This metric is given by Eq. (2) as follows:

$$Accuracy = (VP + VN)/(VP + VN + FP + FN) \tag{2}$$

Indeed, sensitivity, called also recall, reports the percentage of pixels representing the blood vessel (white color), which are correctly classified. This metric is very important in our case. This metric is given by Eq. (3) as follows:

$$Sensitivity = VP/(VP + FN) \tag{3}$$

Moreover, precision is an interesting metric given that it presents the percentage of pixels not representing the blood vessel but classified as a white color. This metric is given by Eq. (4) as follows:

$$Precision = VP/(VP + FP) \tag{4}$$

However, specificity is not an interesting metric as it reflects only the percentage of pixels, which are correctly classified on the outside blood vessel. This metric is given by Eq. (5) as follows:

$$Specificity = VN/(VN + FP) \tag{5}$$

The last used metric is the F-score, which is calculated by Eq. (6) as follows:

$$F\text{-Score} = 2 * (precision * sensitivity)/(precision + recall) \tag{6}$$

3.4 Results and Discussion

Figure 3 shows a ground truth image from the original database (DS1) for validation. Figures 4, 5, 6 and 7 respectively depict DS1, DS2, DS3 and DS4 datasets as follows:

- Input original image (a)
- Input images with CLAHE (b), Frangi (c) and RORPO (d) enhancements
- Output original image (e)
- Output images with CLAHE (f), Frangi (g) and RORPO (h) enhancements

Fig. 3. Ground truth image from DS1 for validation

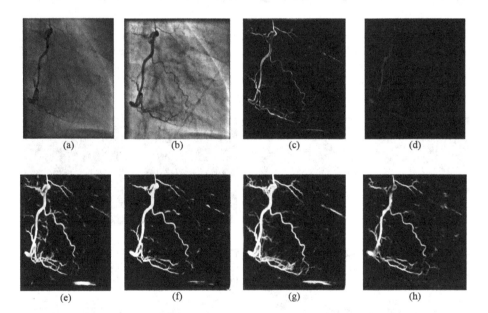

Fig. 4. Implementation images for DS 1: input original image (a), input image with CLAHE method (b), input image with Frangi filter (c), input image with RORPO filter (d), output original image (e), output image with CLAHE method (f), output image with Frangi filter (g), output image with RORPO filter (h)

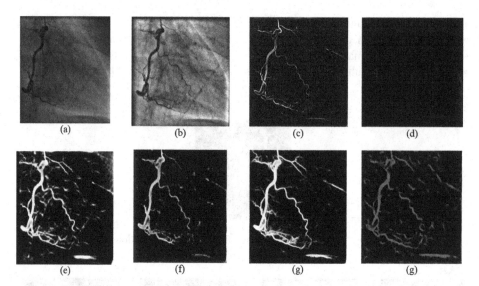

Fig. 5. Implementation images for DS 2: input original image (a), input image with CLAHE method (b), input image with Frangi filter (c), input image with RORPO filter (d), output original image (e), output image with CLAHE method (f), output image with Frangi filter (g), output image with RORPO filter (h)

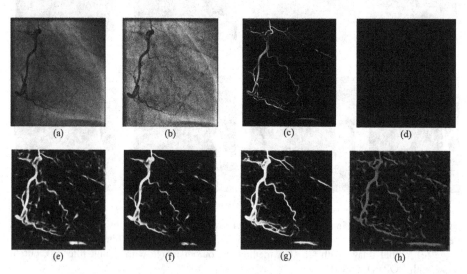

Fig. 6. Implementation images for DS 3: input original image (a), input image with CLAHE method (b), input image with Frangi filter (c), input image with RORPO filter (d), output original image (e), output image with CLAHE method (f), output image with Frangi filter (g), output image with RORPO filter (h)

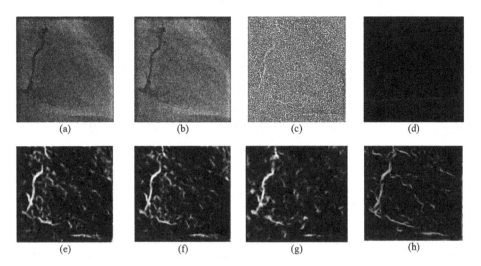

Fig. 7. Implementation images for DS 4: input original image (a), input image with CLAHE method (b), input image with Frangi filter (c), input image with RORPO filter (d), output original image (e), output image with CLAHE method (f), output image with Frangi filter (g), output image with RORPO filter (h)

Tables 1, 2, 3 and 4 provide the implementation results for the four datasets. Each dataset is tested with three enhancement methods: CLAHE, Frangi and RORPO.

Table 1. Performance metrics of implementation results for DS 1

Performance metrics	Without pretreatment	CLAHE method	Frangi filter	RORPO filter
Global accuracy	0.948	**0.951**	0.930	0.912
Specificity	0.959	0.969	0.932	**0.970**
Sensitivity	0.817	0.752	**0.909**	0.269
Precision	0.643	**0.685**	0.543	0.445
F-score	0.719	**0.951**	0.680	0.335

Table 2. Performance metrics of implementation results for DS 2

Performance metrics	Without pretreatment	CLAHE method	Frangi filter	RORPO filter
Global accuracy	0.928	**0.948**	0.938	0.912
Specificity	0.936	0.973	0.942	**0.990**
Sensitivity	0.836	0.658	**0.896**	0.033
Precision	0.538	**0.690**	0.579	0.238
F-score	0.654	0.673	**0.704**	0.058

Table 3. Performance metrics of implementation results for DS 3

Performance metrics	Without pretreatment	CLAHE method	Frangi filter	RORPO filter
Global accuracy	0.937	**0.942**	0.930	0.911
Specificity	0.953	0.961	0.932	**0.990**
Sensitivity	0.753	0.730	**0.909**	0.030
Precision	0.591	**0.627**	0.543	0.216
F-score	0.662	0.675	**0.680**	0.053

Table 4. Performance metrics of implementation results for DS 4

Performance metrics	Without pretreatment	CLAHE method	Frangi filter	RORPO filter
Global accuracy	0.920	0.925	**0.943**	0.923
Specificity	0.956	0.962	**0.985**	0.976
Sensitivity	**0.519**	0.509	0.476	0.322
Precision	0.514	0.543	**0.739**	0.548
F-score	0.516	0.525	**0.579**	0.406

From the four tables, we have obtained good performance results, especially for the global accuracy varying from 0.951 to 0.942 for DS1–DS4 datasets. Similarly, the sensitivity metric is high in particular for DS1, DS2 and DS3 datasets, which is between 0.909 and 0.896.

Indeed, Table 1 presents an image without noise. We note that the best metric is selected with the CLAHE method, especially for global accuracy, precision and F-Score. However, the Frangi filter provides good sensitivity.

Tables 2 and 3, with the presence of noise of around 0.001 and 0.01 respectively, the Frangi filter keeps the performance of sensitivity and improves its performance in F-Score. The CLAHE method preserves almost the same global accuracy and precision.

In Table 4, where noise in around 0.1, the Frangi filter has the best metric of segmentation. This confirms that the filter is robust against noise since it helps to extract thin and peripheral vessels and avoid the blockage caused by low vesselness values in vascular regions.

Actually, we note that the RORPO filter is not a good method for segmentation, with and without noise, compared to the CLAHE method and the Frangi filter.

We note as well that CLAHE is a good method for an image without noise, but it does not keep the same performance for a noisy one. Finally, we conclude that all the results show that the quality of segmentation depends on the enhancement method chosen.

4   Conclusion and Perspectives

In this paper, we have presented U-Net DNN architecture used to improve the segmentation result of coronary images. In addition, we have estimated the impact of enhancement methods as a function of the presence of noise to have accurate coronary segmentation and ensure its robustness against real conditions. The segmentation quantitative evaluation is performed on real angiography images without noise or with increased noise. The obtained results have given a good idea about the choice of the enhancement method in blood vessel segmentation according to noise.

In the future, we will automatically choose the best enhancement method according to the image quality. To attain this aim, we will realize this procedure on a larger database for coronary images or other blood vessel organs. The objective is to ensure a performed image segmentation based on deep learning coupled with the adequate enhancement method.

References

1. Li, Z., Zhang, Y., Liu, G., Shao, H., Li, W., Tang, X.: A robust coronary artery identification and centerline extraction method in angiographies. Biomed. Sig. Process. Control **16**, 1–8 (2015)
2. Sato, Y., Araki, T., Hanayama, M., Naito, H., Tamura, S.: A viewpoint determination system for stenosis diagnosis and quantification in coronary angiographic image acquisition. IEEE Trans. Med. Imaging **17**(1), 121–137 (1998)
3. Liao, R., Luc, D., Sun, Y., Kirchberg, K.: 3-D reconstruction of the coronary artery tree from multiple views of a rotational X-ray angiography. Int. J. Cardiovasc. Imaging **26**, 733–749 (2010)
4. Zheng, S., Zhou, Y.: Assessing cardiac dynamics based on X-ray coronary angiograms. J Multimed. **8**(1), 48–55 (2013)
5. Elloumi, Y., Akil, M., Kehtarnavaz, N.: A mobile computer aided system for optic nerve head detection. Comput. Methods Programs Biomed. **162**, 139–148 (2018)
6. Frangi, A.F., Niessen, W.J., Vincken, K.L., Viergever, M.A.: Multiscale vessel enhancement filtering. In: Wells, W.M., Colchester, A., Delp, S. (eds.) MICCAI 1998. LNCS, vol. 1496, pp. 130–137. Springer, Heidelberg (1998). https://doi.org/10.1007/BFb0056195
7. Fazlali, H.: Vessel region detection in coronary X-ray angiograms. In: IEEE International Conference on Image Processing (ICIP), pp. 1493–1497 (2015)
8. Fernando, C.S., Ivan, C.A.: Coronary artery segmentation in X-ray angiograms using Gabor filters and differential evolution. Appl. Radiat. Isot. **138**, 18–24 (2018)
9. Condurache, A., Aach, T.: Vessel segmentation in angiograms using hysteresis thresholding. In: MVA, pp. 269–272 (2005)
10. Andriotis, A., Zifan, A., Gavaises, M.: A new method of three-dimensional coronary artery reconstruction from X-ray angiography: validation against a virtual phantom and multislice computed tomography. Cathet. Cardiovasc. Interv. **71**(1), 28–43 (2008)
11. Ko, C.-C., Mao, C.-W., Sun, Y.-N., Chang, S.-H.: A fully automated identification of coronary borders from the tree structure of coronary angiograms. Int. J. Biomed. Comput. **39**(2), 193–208 (1995)
12. Sun, K., Sang, N., Zhao, E.: Extraction of vascular tree on angiogram with fuzzy morphological method. Int. J. Inf. Technol. **11**(9), 119–127 (2005)
13. Dufour, A., Tankyevych, O., Naegel, B., Talbot, H., Ronse, C., Baruthio, J., et al.: Filtering and segmentation of 3D angiographic data: advances based on mathematical morphology. Med. Image Anal. **17**(2), 147–164 (2013)
14. Brieva, J., Gonzalez, E., Gonzalez, F., Bousse, A., Bellanger, J.: A level set method for vessel segmentation in coronary angiography. In: IEEE Engineering in Medicine and Biology (2005)

15. Anthimopoulos, M., Christodoulidis, S., Ebner, L., Christe, A.: Lung pattern classification for interstitial lung diseases using a deep convolutional neural network. IEEE Trans. Med. Imaging **35**, 1207–1216 (2016)
16. Dorj, U.O., Lee, K.K., Choi, J.Y., Lee, M.: The skin cancer classification using deep convolutional neural network. Multimed. Tools Appl. **77**(8), 9909–9924 (2018)
17. Melinscak, M., Prentasic, P.: Retinal vessel segmentation using deep neural networks. In: International Conference on Computer Vision Theory and Applications, vol. 57, pp. 577–582 (2015)
18. Birgui Sekou, T., Hidane, M., Olivier, J., Cardot, H.: Retinal Blood Vessel Segmentation Using a Fully Convolutional Network – Transfer Learning from Patch- to Image-Level. In: Shi, Y., Suk, H.-I., Liu, M. (eds.) MLMI 2018. LNCS, vol. 11046, pp. 170–178. Springer, Cham (2018). https://doi.org/10.1007/978-3-030-00919-9_20
19. Nasr-Esfahani, E., Karimi, N., Jafari, M.H.: Segmentation of vessels in angiograms using convolutional neural networks. Biomed. Sig. Process. Control. **40**, 240–251 (2018)
20. Bai, F., Zhou, B.: Image enhancement using multi scale image features extracted by top-hat transform. Opt. Laser Technol. **44**, 328–336 (2012)
21. Ronneberger, O., Philipp, F., Thomas, B.: U-Net: convolutional networks for biomedical image segmentation. https://arxiv.org/abs/1505.04597
22. Krig, S.: Image Pre-Processing. In: Krig, S. (ed.) Computer Vision Metrics, pp. 39–83. Springer, Berkeley (2014). https://doi.org/10.1007/978-1-4302-5930-5_2
23. Zuiderveld, K.: Contrast limited adaptive histogram equalization. In: Graphics Gems IV, pp. 474–485 (1994)
24. Pisano, E.D., et al.: Contrast limited adaptive histogram equalization image processing to improve the detection of simulated spiculations in Dense mammograms. J. Digit. Imaging **11**(4), 193–200 (1998)
25. Hitam, M.Z., Awalludin, E.A., Yussof, W., Awalludin, E.A., Bachok, Z.: Mixture contrast limited adaptive histogram equalization for underwater image enhancement. In: IEEE International Conference (2013)
26. Buzuloiu, V., Ciuc, M., Rangayyan, R.M., Kij, L., Constantin, V.: Histogram equalization of colour images using the adaptive neighborhood approach. In: Proceedings of the SPIE, Nonlinear Image Processing X, vol. 3646, p. 330 (1999)
27. Greenberg, S., Aladjem, M., Kogan, D.: Fingerprint image enhancement using filtering techniques. Real Time Imaging **8**(3), 227–236 (2002)
28. Cabrera, F.D., Salinas, H.M., Puliafito, C.A.: Automated detection of retinal Layer structures on optical coherence tomography images. Opt. Express **13**(25), 10200–10216 (2005)
29. Odyssée, M., Benoit, N., Hugues, T., Laurent, N, Nicolas, P.: 2D filtering of curvilinear structures by ranking the orientation responses of path operators (RORPO). In: Image Processing On Line, 01 October 2017 (2017)
30. Kerkeni, A., Ben, A.A., Manzanera, A., Bedoui, M.H.: A coronary artery segmentation method based on multiscale analysis and region growing. Comput. Med. Imaging Graph. **48**, 49–61 (2016)

Real-Time Traffic Monitoring with Occlusion Handling

Mauro Fernández-Sanjurjo$^{(\boxtimes)}$, Manuel Mucientes, and Víctor M. Brea

Centro Singular de Investigación en Tecnoloxías Intelixentes (CiTIUS),
Universidade de Santiago de Compostela, Santiago de Compostela, Spain
{mauro.fernandez,manuel.mucientes,victor.brea}@usc.es

Abstract. Traffic surveillance through vision systems is a highly demanded task. To solve it, it is necessary to combine detection and tracking in a way that meets the requirements of operating in real time while being robust against occlusions. This paper proposes a traffic monitoring system that meets these requirements. It is formed by a deep learning-based detector, tracking through a combination of Discriminative Correlation Filter and a Kalman Filter, and data association based on the Hungarian method. The viability of the system has been proved for roundabout input/output analysis with near 1,000 vehicles in real-life scenarios.

Keywords: Multiple object tracking · Traffic monitoring · Roundabout analysis

1 Introduction

Detection and tracking of vehicles in a video allows to estimate every vehicle trajectory while they remain in the scene. This has applications in a wide range of tasks: vehicle counting, accident detection, roundabout entry/exit analysis or assisted traffic surveillance. In a real-life scenario speed and robustness are a must, which translate to the requisites of real-time performance and occlusion handling.

In terms of the current tracking solutions we can distinguish two types: *low-level* and *high-level* trackers. The former exploits the visual information in the current frame to find the object of interest while the latter can use more complex information to estimate the new object position (probabilistic models,

---

This research was partially funded by the Spanish Ministry of Economy and Competitiveness under grants TIN2017-84796-C2-1-R and RTI2018-097088-B-C32 (MICINN/FEDER), and the Galician Ministry of Education, Culture and Universities under grant ED431G/08. Mauro Fernández is supported by the Spanish Ministry of Economy and Competitiveness under grant BES-2015-071889. These grants are co-funded by the European Regional Development Fund (ERDF/FEDER program). We thank Aplygenia S.L. for their collaboration.

A. Morales et al. (Eds.): IbPRIA 2019, LNCS 11868, pp. 273–284, 2019.
https://doi.org/10.1007/978-3-030-31321-0_24

environment maps, etc.). Current *low-level* trackers [2,5,7] cannot handle total occlusions and do not provide a framework for multiple object tracking. In addition, the best current solutions require a high end GPU or do not operate in real time with multiple objects on CPU [7,20].

In recent years, the high-level tracking problem has been focused as a tracking-by-detection approach [1]. This framework considers the tracking task as a data association problem between detections and trackers over time. This assumes the existence of reliable detections in every frame of a video, something that in a real-life scenario is not a valid option as current state-of-the-art deep-learning based detectors operate above 75 ms per frame [17].

In this paper, we present a traffic monitoring system that performs multiple object detection and tracking in a video in real-time handling total occlusions. The system is composed of a deep-learning based detector, a low-level Discriminative Correlation Filter (DCF) based tracker, a high-level Kalman Filter based tracker and data association based on the Hungarian algorithm. The contributions of our proposal are:

- A traffic monitoring system that can process **more than 400 vehicles simultaneously** in videos with HD resolution in real-time.
- The system also **handles occlusions** by detecting the upcoming occlusion and searching the occluded vehicle in a zone called *ROI (Region-Of-Interest)* that is proportional to the error degree in the tracking process. We provide a metric for **on-line tracking failure detection** by estimating the distance between two independent tracking methods allowing us to update the system's tracking error accordingly.
- We extend our system for solving a **real-life traffic application**: roundabout I/O (Input/Output) with near 1,000 vehicles.

The rest of this paper is structured as follows. Section 2 gives an overview of closely related work. In Sect. 3 we explain the details of our approach. In Sect. 4 we discuss the implementation details of our system and introduce the traffic application developed. Finally, conclusions are given in Sect. 5.

2   Related Work

Traffic monitoring systems detect and track all the vehicles in a video sequence. This task presents two main challenges: to manage total occlusions and to operate in real-time with multiple vehicles.

The work in the field of object detection is mainly based on deep convolutional neural networks (ConvNets). One of the first works in this area was R-CNN [12] which uses a region proposal algorithm (such as selective search [23] or edge boxes [25]) and applies a classification network to each of them. Improving the previous approach, Fast-RCNN [11] introduces the regions in an intermediate stage of the network, thus, saving a lot of computing time. Finally, becoming the milestone in the object detection field, Faster-RCNN [22] introduces a region proposal algorithm based entirely on a neural network called the

Region Proposal Network (RPN). The RPN uses the information from intermediate layers of a standard classification network to provide different locations in which an object may appear.

To improve the performance of the proposal of regions in all possible scales, Lin et al. [18] replicate the RPN from Faster-RCNN in several layers of the network in which deeper feature maps are combined with shallower ones. The shallower the layer the smaller the object it will locate. This approach, called Feature Pyramid Network (FPN) obtains outstanding results as shown in the COCO detection challenge 2016 [19]. All these approaches present a high level of performance but, their main limitation is their computational cost, which makes them harder to use in applications that demand real-time performance.

In the last years, top trackers from the Visual Object Tracking (VOT) challenge [15] are based on two approaches: Discriminative Correlation Filters (DCF) based trackers, and deep-learning based trackers. On the one hand, DCF based trackers predict the target position training a correlation filter that can differentiate between the object of interest and the background [5,6,13]. On the other hand, deep-learning based trackers use ConvNets. SiamFC [2] is one of the first approaches of this kind. This tracker consists of two branches that apply an identical transformation—deep features extractor—to two inputs: the search image and the exemplar. Then, both representations are combined through cross-correlation, generating a score map that indicates the most probable position of the object.

Due to the increase in performance of deep learning detectors in recent years, the task of tracking is increasingly being seen as a data association problem, i.e. tracking-by-detection. In this approach, the primary concern is to assign detections to trackers over time. Some international challenges [1] have emerged to rank solutions to this problem, evaluating precision, robustness and speed among other performance metrics. In the past few years, complex solutions to this tracking approach that obtain outstanding results have appeared. Some of them focus on extending traditional high-level tracking approaches. As an example, Kim et al. [14] and Chen et al. [4] propose extensions to the classical multiple hypotheses tracking (MHT) [21]. The former introduces on-line appearance representations while the latter enhances the classical MHT by incorporating a detection model that includes detection-scene and detection-detection analysis.

All these approaches have demonstrated good performance in classic multiple object tracking metrics as commented before. Their fundamental limitation is the speed, as none of the work discussed in this section shows performance metrics above 2.6 Hz even without accounting for the detection time. Also, they assume the existence of detections in every frame of a video without taking into account high performance object detectors inference time.

Some work in the traffic monitoring field has been done in the recent years [8]. In [10], vehicle counting is performed employing an environment segmentation strategy. In [9] a tracking approach using background subtraction and Kalman filter tracking to tackle the data collection in roundabouts is proposed. These approaches usually run at real-time speed due to the use of background subtraction for detecting mobile objects. These object identification methods could represent a limitation in scenarios that present camera movement (on-board cameras), shadows, image artifacts, or objects that appear very close to each other since they usually are identified as only one by the background subtraction algorithm.

3   Video Traffic Monitoring

We propose a complete traffic monitoring system that combines tracking and detection and can operate as a baseline for multiple applications.

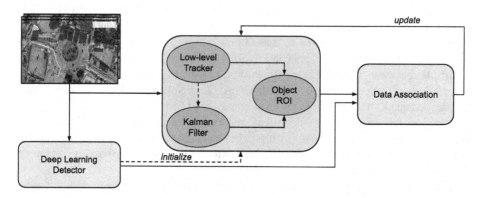

Fig. 1. Architecture of our traffic monitoring system. It is formed by three modules: detection (yellow), tracking (red) and data association (blue). (Color figure online)

Our system is made up of three blocks (Fig. 1): detection, tracking and data association. To detect vehicles in an image, we use a deep learning based detector. For tracking, we combine a DCF-based tracker with a Kalman-based one, which enables to calculate a failure detection metric to identify occluded vehicles. Finally, in the data association module, we assign each detection with its correspondent tracker through the Hungarian method [16,24] and perform an update of the trackers.

Algorithm 1 presents the main steps of the system. The inputs to the system at every time instant t are the new frame (Im_t) of the video, and the set of trackers in the previous time instant (Φ_{t-1}). First, the trackers positions in the new image (Im_t) are estimated. We start calculating the new position of the object with a DCF tracker (Algorithm 1, line 3—Algorithm 1:3—). Tracking based just on DCF trackers has two limitations: (i) we cannot handle occlusions (Fig. 3); (ii) it does not provide a robust tracking failure detection (i.e. knowing when the tracking fails) as the PSR (Peak to Sidelobe Ratio) value [3], which measures the spread degree of the convolution operation of the correlation filter, is not a reliable measure. As shown in Fig. 4, the PSR takes different threshold values for different videos and scenarios, which makes difficult to identify when a tracker is lost.

Algorithm 1: Traffic Monitoring System

Require:
(a) Im_t: Image frame at current time t
(b) $\Phi_{t-1} = \{\varphi_{t-1}^1, \varphi_{t-1}^2, \ldots, \varphi_{t-1}^n\}$

1 **Function** Main(Im_t, Φ_{t-1}):
2   **for** $i=1$ to n **do**
3      $dcf\_roi_t^i$ =DCF_Track (φ_{t-1}^i)
4      $kf\_roi_t^i$ = Kalman_Predict (φ_{t-1}^i)
5      $\overline{\varphi}_t^i \leftarrow < roi_t^i >$=Estimate_ROI $(dcf\_roi_t^i, kf\_roi_t^i)$
6   **if** $time\_elapsed > \tau$ **then**
7      $\Psi_t \leftarrow \{\psi_t^1, \psi_t^2, \ldots, \psi_t^m\}$ =ConvNet_Detect()
8   **for** $i=1$ to n **do**
9   **for** $j=1$ to m **do**
10          $IOU_t^{i,j} = \frac{\overline{\varphi}_t^i \cap \psi_t^j}{\overline{\varphi}_t^i \cup \psi_t^j}$
11      $\{<\overline{\varphi}_t^\alpha, \psi_t^\beta>\}$ =Hungarian (IOU_t)
12   **for** every α, β in $\{<\overline{\varphi}_t^\alpha, \psi_t^\beta>\}$ **do**
13   update_tracker $(\overline{\varphi}_t^\alpha, \psi_t^\beta)$
14   **for** $i=1$ to n **do**
15   **if** not_updated $(\overline{\varphi}_t^i)$ **then**
16   check_tracker_deletion $(\overline{\varphi}_t^i)$
17   **for** $j=1$ to m **do**
18   **if** not_updated (ψ_t^j) **then**
19   new_tracker (ψ_t^j)
20    $\Phi_t = \overline{\Phi}_t$
21   **return** (Φ_t)

To provide a solution to both problems, we introduce a Kalman Filter (KF) tracker that, by modeling the movement of the object can handle occlusions and, in combination with the DCF tracker, can estimate the error in the tracking process. So, once the vehicle's new position is calculated by the DCF tracker, we estimate the position using the Kalman filter. We use a linear constant velocity model in the KF, so the state of each vehicle is modeled as:

$$\mu := [x, y, v_x, v_y] \tag{1}$$

Here x and y are the position of the object, and v_x and v_y represent the linear velocity in both axes. We perform Kalman prediction in Algorithm 1:4. With the bounding boxes proposed by both methods, we estimate the region of interest (ROI) in which the object might be located (Algorithm 1:5). The larger the difference between the two trackers, the larger the ROI. Occlusions can be determined in cases where both predictors propose very different bounding boxes, since the bounding boxes provided by DCF will remain static, while those

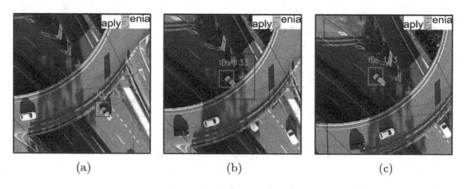

(a)   (b)   (c)

Fig. 2. Creation of a search ROI for occlusion handling. (a) Both tracking methods agree on the object position. (b) As the DCF fails to track the occluded object, the distance between both estimations increases and so it does the search ROI. Finally in (c), when the detector finds the vehicle at the other side of the road and the tracker recovers. Images courtesy of Aplygenia S.L.

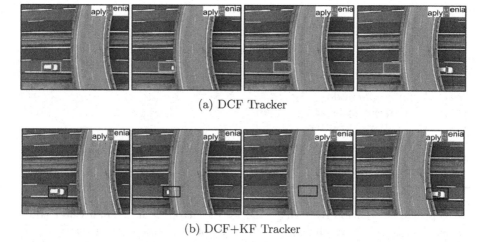

(a) DCF Tracker

(b) DCF+KF Tracker

Fig. 3. (a) The low-level DCF tracker (in green) cannot recover the identity of the object once occluded as it only relies on appearance. (b) The combination of a DCF and a KF manages occlusions, as it also takes into account the object motion model. Images courtesy of Aplygenia S.L. (Color figure online)

from the Kalman filter will follow the previous movement pattern of the object (Fig. 2).

Our system is robust enough so we do not need to call the detector in every frame. The aim of the detection component is twofold. First, it initializes every tracker or object of interest in the scene. Second, it refines the location and size of the bounding boxes of the trackers along their trajectories through the data association component (see Fig. 1), improving tracking performance metrics. If the time elapsed since the previous detection is greater than or equal to τ,

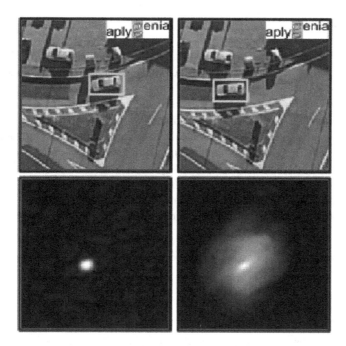

Fig. 4. PSR values are poor predictors of tracking failures for the DCF tracker. The image shows a case in which the object is being tracked successfully but the PSR value changes and shows a high degree of dispersion. The opposite case, a tracking failure not detected by the PSR values, is also frequent. Images courtesy of Aplygenia S.L.

detection is performed using a convolutional neural network (Algorithm 1:6–7), which returns a set of detections Ψ_t. In practice, this is performed with a fully convolutional network called FPN [18], which uses feature maps information at different scales to locate from small to large objects, through a pyramidal architecture with lateral connections between them. The FPN provides high precision at a high computational cost, taking about 130 ms to perform a full detection in an HD image. If no detection is performed at current time t, tracking prediction alone ($\overline{\Phi}_t$) determines the current trackers state (Φ_t, Algorithm 1:20).

The data association block aims to assign each detection to its corresponding tracker and to identify objects that enter or leave the scene. In so doing, we build up the cost matrix IOU_t (see Algorithm 1:8–10), where every entry is the Intersection Over Union (IOU) between a tracker $\overline{\varphi}_t^i$ and a detection ψ_t^j. That association is solved by the Hungarian Method (Algorithm 1:11). For every successful assignation ($<\varphi_t^\alpha, \psi_t^\beta>$), tracker φ_t^α is updated with detection ψ_t^β (Algorithm 1:13). Finally, trackers not updated in the data association phase are candidates for being deleted, and detections not assigned are initialized as new trackers (Algorithm 1:14–19).

4   Results

The proposed system (Fig. 1) runs on a server with an Intel Xeon E52623v4 2.60 GHz CPU, 128 GB RAM and an Nvidia GP102GL 24 GB [Tesla P40] as GPU. Table 1 shows the times of the two most computational expensive operations of our system: detection and tracking—computing times of other tasks are negligible. In a 30 fps video, we have 0.03 seconds per frame for the tracking task. Using 15 threads for parallelization, theoretically, the system is able to process up to 148 objects in the image while maintaining real-time performance, i.e. 30 fps. As mentioned before, detection is the slowest part of our system, taking an average 0.135 s in an HD image and 0.075 s in VGA resolution. These values are below the 0.2 threshold required by the system for the detection module, as we only perform detection 5 times every 30 frames.

Table 1. Computational times for the detection and tracking modules of the traffic monitoring system.

Tracking	
Frames processed by second	30 frames of 30
Total max. time with parallel computing	0.0121 s (60 objects, 15 threads)
Max. number of objects in 0.03 s	148 objects
Detection	
Frames processed by second	5 frames of 30
Average time per HD image	0.135 s

4.1   Roundabout Monitoring

In this section, we analyze our complete system (Fig. 1) for roundabout monitoring. The objective of the system is to identify the entry and the exit a vehicle takes, maintaining its identity while it remains in the roundabout. The final goal is to provide the I/O matrix R, in which every element $(R(i,j))$ represents the number of vehicles that joined the roundabout taking entry i and exit j. If a vehicle enters the roundabout and exits it with the same ID we count that as a tracking success. On the contrary, if the identity changes along the video, then we count that vehicle as a tracking failure.

For performing the metrics, we use a video dataset which consists of five videos of roundabouts recorded from an Unnamed Aerial Vehicle (UAV) at 30 fps with HD resolution[1]. The videos have different conditions that are challenging for traffic monitoring: shadows, total occlusions (two level roads), camera movement, etc. Figure 5 shows a snapshot of some of these videos[2].

---

[1] These videos where recorded by the company Apligenia S.L.
[2] A demonstration video can be downloaded from: http://bit.ly/roundabout_samp le_video.

Table 2. Computational times for the fast version of our traffic monitoring system.

Tracking	
Frames processed by second	10 frames of 30
Total max. time with parallel computing	0.0121 s (60 objects, 15 threads)
Max. number of objects in 0.1 s	492 objects
Detection	
Frames processed by second	5 frames of 30
Average time per HD image	0.135 s

As explained before, the robustness of our system allows us to avoid calling the detector at every frame. This led us to develop a fast version that performs tracking in one of every 3 frames and detection in one of every 6 frames, without degrading the performance metrics for roundabout monitoring. Table 2 shows the times for this version fast version.

Table 3. Results in the video dataset for roundabout monitoring. The columns are: video, number of occlusions (#*occ*), number of vehicles occluded (#*vocc*), duration of video, total number of vehicles (#*vehicles*) and success rate obtained by our tracking system.

Video	# occ	#vocc	Time (min:sec)	#vehicles	Success
usc_vr_1	308	160	05:11	320	86,50%
usc_pl_1	–	–	11:12	138	88%
usc_rb_1	–	–	11:48	230	95%
usc_sx_1	–	–	09:26	255	91%
usc_ou_1	22	11	02:49	52	96%
Total	*330*	*171*	*40:26*	*995*	*91,30%*

Table 3 shows the results obtained from processing the I/O matrix of five videos with 995 vehicles in total. We have used the fast version of our traffic monitoring system to highlight the robustness of the proposal even when processing just 10 of each 30 frames. Theoretically, the system can track up to 492 objects, although in these videos the maximum number of concurrent objects was 60. An average success rate of 91% is obtained. Results also show our system's ability to handle occlusions as two of the videos are scenarios with a high rate of total occlusions: in one of them the 50% of the vehicles are totally occluded nearly twice on average.

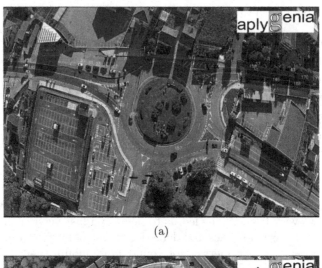

(a)

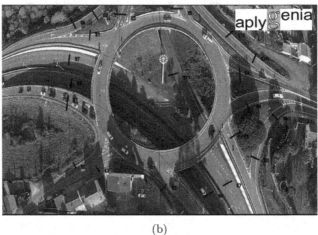

(b)

Fig. 5. Example frames of some videos of the roundabout monitoring dataset. These videos are recorded from an UAV flying over a roundabout. Images courtesy of Aplygenia S.L., their distribution is restricted.

5   Conclusions

We have presented a traffic monitoring system that combines a convolutional neural network detection, DCF and Kalman trackers, and a Hungarian data association. The system is able to track hundreds of objects in real-time while being robust to occlusions. The combination of the DCF and Kalman filters allows to estimate the error of each tracker, thus increasing the robustness and reliability of the system. We have applied the traffic monitoring system to the problem of roundabout monitoring. Our system achieves a 91% success rate for the I/O matrix, even in cases with high occlusion rates, shadows and movement of the UAV onboard camera.

References

1. MOTChallenge the multiple object tracking benchmark. https://motchallenge. net/. Accessed 18 Dec 2018
2. Bertinetto, L., Valmadre, J., Henriques, J.F., Vedaldi, A., Torr, P.H.S.: Fully-convolutional siamese networks for object tracking. In: Hua, G., Jégou, H. (eds.) ECCV 2016. LNCS, vol. 9914, pp. 850–865. Springer, Cham (2016). https://doi. org/10.1007/978-3-319-48881-3_56
3. Bolme, D.S., Beveridge, J.R., Draper, B.A., Lui, Y.M.: Visual object tracking using adaptive correlation filters. In: IEEE Conference on Computer Vision and Pattern Recognition (CVPR) (2010)
4. Chen, J., Sheng, H., Zhang, Y., Xiong, Z.: Enhancing detection model for multiple hypothesis tracking. In: IEEE Conference on Computer Vision and Pattern Recognition Workshop (2017)
5. Danelljan, M., Bhat, G., Khan, F.S., Felsberg, M.: ECO: efficient convolution operators for tracking. In: IEEE Conference on Computer Vision and Pattern Recognition (CVPR) (2017)
6. Danelljan, M., Häger, G., Khan, F.S., Felsberg, M.: Discriminative scale space tracking. IEEE Trans. Pattern Anal. Mach. Intell. **39**(8), 1561–1575 (2017)
7. Danelljan, M., Robinson, A., Shahbaz Khan, F., Felsberg, M.: Beyond correlation filters: learning continuous convolution operators for visual tracking. In: Leibe, B., Matas, J., Sebe, N., Welling, M. (eds.) ECCV 2016. LNCS, vol. 9909, pp. 472–488. Springer, Cham (2016). https://doi.org/10.1007/978-3-319-46454-1_29
8. Datondji, S.R.E., Dupuis, Y., Subirats, P., Vasseur, P.: A survey of vision-based traffic monitoring of road intersections. IEEE Trans. Intell. Transp. Syst. **17**(10), 2681–2698 (2016)
9. Dinh, H., Tang, H.: Development of a tracking-based system for automated traffic data collection for roundabouts. J. Mod. Transp. **25**(1), 12–23 (2017)
10. Engel, J.I., Martín, J., Barco, R.: A low-complexity vision-based system for real-time traffic monitoring. IEEE Trans. Intell. Transp. Syst. **18**(5), 1279–1288 (2017)
11. Girshick, R.: Fast R-CNN. In: IEEE International Conference on Computer Vision (ICCV) (2015)
12. Girshick, R., Donahue, J., Darrell, T., Malik, J.: Rich feature hierarchies for accurate object detection and semantic segmentation. In: IEEE Conference on Computer Vision and Pattern Recognition (CVPR) (2014)
13. Henriques, J.F., Caseiro, R., Martins, P., Batista, J.: High-speed tracking with kernelized correlation filters. IEEE Trans. Pattern Anal. Mach. Intell. **37**(3), 583–596 (2015)
14. Kim, C., Li, F., Ciptadi, A., Rehg, J.M.: Multiple hypothesis tracking revisited. In: IEEE International Conference on Computer Vision (ICCV) (2015)
15. Kristan, M., et al.: The sixth visual object tracking VOT2018 challenge results, pp. 3–53, January 2019
16. Kuhn, H.W.: The hungarian method for the assignment problem. Naval Res. Logistics Q. **2**(1–2), 83–97 (1955)
17. Li, Z., Peng, C., Yu, G., Zhang, X., Deng, Y., Sun, J.: Light-head R-CNN: in defense of two-stage object detector. arXiv preprint arXiv:1711.07264 (2017)

18. Lin, T.Y., Dollár, P., Girshick, R., He, K., Hariharan, B., Belongie, S.: Feature pyramid networks for object detection. In: IEEE Conference on Computer Vision and Pattern Recognition (CVPR) (2017)
19. Lin, T.-Y., et al.: Microsoft COCO: common objects in context. In: Fleet, D., Pajdla, T., Schiele, B., Tuytelaars, T. (eds.) ECCV 2014. LNCS, vol. 8693, pp. 740–755. Springer, Cham (2014). https://doi.org/10.1007/978-3-319-10602-1_48
20. Nam, H., Han, B.: Learning multi-domain convolutional neural networks for visual tracking. In: IEEE Conference on Computer Vision and Pattern Recognition (CVPR) (2016)
21. Reid, D., et al.: An algorithm for tracking multiple targets. IEEE Trans. Autom. Control **24**(6), 843–854 (1979)
22. Ren, S., He, K., Girshick, R., Sun, J.: Faster R-CNN: towards real-time object detection with region proposal networks. In: Advances in Neural Information Processing Systems (NIPS) (2015)
23. Uijlings, J.R., Van De Sande, K.E., Gevers, T., Smeulders, A.W.: Selective search for object recognition. Int. J. Comput. Vis. (IJCV) **104**, 154–171 (2013)
24. Wu, B., Nevatia, R.: Detection and tracking of multiple, partially occluded humans by bayesian combination of edgelet based part detectors. Int. J. Comput. Vis. **75**(2), 247–266 (2007)
25. Zitnick, C.L., Dollár, P.: Edge boxes: locating object proposals from edges. In: Fleet, D., Pajdla, T., Schiele, B., Tuytelaars, T. (eds.) ECCV 2014. LNCS, vol. 8693, pp. 391–405. Springer, Cham (2014). https://doi.org/10.1007/978-3-319-10602-1_26

Image Based Estimation of Fruit Phytopathogenic Lesions Area

André R. S. Marcal[1,2(✉)] ⓘ, Elisabete M. D. S. Santos[2],
and Fernando Tavares[1,3] ⓘ

[1] Faculdade de Ciências, Universidade do Porto, Porto, Portugal
andre.marcal@fc.up.pt
[2] INESC TEC, Porto, Portugal
[3] CIBIO - Centro de Investigação em Biodiversidade e Recursos Genéticos, InBIO,
Laboratório Associado, Universidade do Porto, Vairão, Portugal

Abstract. A method was developed to measure the surface area of walnut fruit phytopathogenic lesions from images acquired with a basic calibration target. The fruit is modelled by a spheroid, established from the 2D view ellipse using an iterative process. The method was tested with images of colour circular marks placed on a wooden spheroid. It proved effective in the estimation of the spheroid semi-diameters (average relative errors of 0.8% and 1.0%), spheroid surface (1.77%) and volume (2.71%). The computation of the colour mark surface area was within the expected error, considering the image resolution (up to about 4%), for 22 out of 28 images tested.

Keywords: Fruit phytopathogenic lesion · 3D modeling · Spheroid · Volume · Surface area

1 Introduction

The early detection of plant diseases caused by phytopathogenic bacteria is of upmost importance to assist the phytosanitary regulatory bodies in the implementation of timely and efficient control measures [2].

Xanthomonas arboricola pv. juglandis (X_{aj}) is the etiological agent of the "walnut (*Juglandis regia* L.) bacteria blight" (WBB) affecting leaves, and the "brown apical necrosis" (BAN) resulting in premature fruit drop. The presumptive diagnosis of these walnut diseases, which affect severely the walnut production worldwide [7], is made by the observation of symptomatic lesions characterized by necrotic spots or patches on walnut leaves and fruits. The presence of dark brown spot lesions is an initial sign of a possible infection by X_{aj}, which can be further confirmed by their development over time, both in terms of average area covered and incidence within a walnut symptomatic tree or an orchard. Therefore, the identification of these lesions is an immediate indication for both phytosanitary inspectors and producers, to implement suitable phytosanitary

© Springer Nature Switzerland AG 2019
A. Morales et al. (Eds.): IbPRIA 2019, LNCS 11868, pp. 285–295, 2019.
https://doi.org/10.1007/978-3-030-31321-0_25

practices and to make comparative assessments between samples at different time points to infer the aggressiveness of the disease, and gather informative data for epidemiological records. However, in order to analyze the temporal development of the lesions, that may help to determine the virulence of a particular X_{aj} strain or the resistance of a specific walnut host cultivar, quantitative measurements are required.

There are a number of methods for the computation of fruit area, but most of these methods are not suitable to quantitatively evaluate the extension of the lesions and consequently the disease aggressiveness. Several papers present methodologies for the evaluation of fruit shape parameters, such as surface area or volume [1,3,8,9]. However, these approaches compute the full surface area, and not a specific sub-region. Furthermore, most of these methods involve stereoscopic cameras [3] and well controlled illumination [6,8], which is not adequate for in field measurements.

This work is part of the EVOXANT project - Evolution of Xanthomonas arboricola species complex beyond culturability. One of its objectives is to develop a low cost, easy to use system for the acquisition of lesion area data from fruits in the field by farmers or phytosanitary technicians. This paper presents the system that is being developed for that purpose, focusing on the geometric aspects of the problem. The data used is collected by a simple camera (such as those available in smartphones and tablets) and a basic calibration target. The paper includes three additional chapters to this introduction. Section 2 presents the methodology, including imaging geometry, 3D fruit modeling and error estimation; Sect. 3 presents an experimental evaluation using color marks of fixed size over a wooden spheroid; and Sect. 4 highlights the main conclusions.

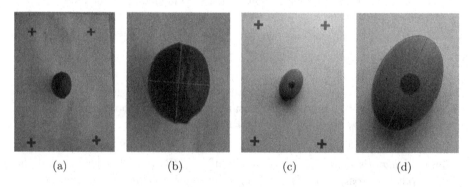

(a)   (b)   (c)   (d)

Fig. 1. (a) Walnut fruit acquired with A4 calibration target; (b) 2D ellipse modelled for (a); (c) wooden spheroid in A4 target; (d) 2D ellipse modelled for (c). (Color figure online)

2   Methodology

2.1   Image Acquisition and Pre-processing

The images are acquired with the object of interest placed over a calibration target, which was designed to assist in the spacial referencing. The calibration target consists of a set of 4 marks (red crosses) on a white paper, forming a rectangle, with the distance between their centres known. It can be created with a standard printer, with matte paper of any format. The only requirements are that the marks are red, they form a rectangle, and the rectangle length and width are known. For in-field use, it is preferable to use a rigid material, for example wood or plastic. The images can be acquired with any standard camera, preferably with the rectangle formed by the red marks roughly aligned with the image grid and covering most of the image. As an example, Fig. 1(a) shows an image of a walnut fruit and Fig. 1(c) a wooden spheroid, both acquired with an A4 calibration target. The test images used in this work were obtained with an Apple iPAD mini 5MP camera (2592 × 1936 pixels).

A number of pre-processing tasks are performed in order to identify and evaluate the object or sub-region of interest. The initial step is to identify the calibration marks, which is done by thresholding in the HSV colour domain, followed by a sequence of morphological operations [5]. The image coordinates of the mark centres are used to compute 6 estimates of the average image resolution at the calibration plane (R_0 in pixel/mm). The region of interest is established as the polygon defined by the 4 marks, removing a margin so that the red marks are not included. A global thresholding is applied to the HSV image, with further morphological operations performed to segment the single largest object present in the region of interest [5].

Considering that the object of interest is adequately segmented, the next step is to identify a sub-region within the object. It is a 2-class classification problem, which can also be considered as a binary segmentation of an RGB image. In the case of a walnut fruit, the background is green and the foreground brown (lesion). An alternative case is considered here, with a spheroid with light brown background and a blue circular mark, as illustrated in Fig. 1(c). The choice of colours was made to simplify the segmentation task, as the paper focus on the geometric aspects of the problem, particularly the measurement of a 3D object size using a single image.

2.2   Imaging Geometry

The reconstruction of the position and shape of objects from photographs requires a pair of stereoscopic images [4]. However, a single photograph might be enough for a plane object, provided that the distance between the projection centre and the plane along the principal axis (D_0) is known. As the photogrammetric equations used for the central projection of a plane requires several inner orientation (camera) parameters [4], not available for standard cameras, a simplified approach is considered for near-vertical images. It requires only 1 camera

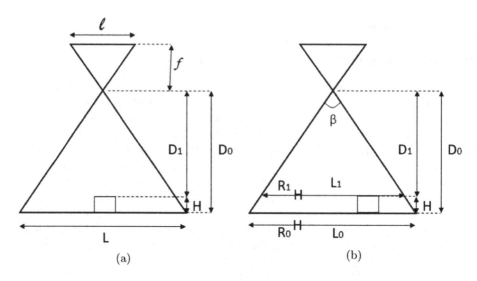

Fig. 2. Schematic representation of near vertical imaging geometry.

parameter - the principal distance, also referred to as or focal length (f), or the camera field of view (β). A schematic representation of the imaging geometry is presented in Fig. 2(a). The image scale in the reference plane (S_0) is computed as $S_0 = f/D_0$, or alternatively as $S_0 = l/L$. For an object of height H, placed over the reference plane, a new scale factor (S_1) needs to be computed, using (1). It requires the knowledge of H, S_0, and D_0 or f.

$$S_1 = \frac{f}{D_1} = \frac{f}{D_0 - H} = \frac{S_0 D_0}{D_0 - H} \tag{1}$$

$$R_1 = \frac{L_0 R_0}{L_1} = \frac{D_0 R_0}{D_1} = \frac{D_0 R_0}{D_0 - H} = \frac{R_0 L_0}{L_0 - 2H tan(\beta/2)} \tag{2}$$

A perhaps more relevant parameter than S (dimensionless) is the image resolution (R, in pixel/mm). An estimate for the calibration plane (R_0) is computed from the reference marks. For an object of some height (H), the image resolution (R_1) is computed by (2), illustrated in Fig. 2(b). The parameter β is assumed constant for a given camera. If unknown, it can be estimated by acquiring an image and measuring D_0. The only unknown is thus the object height (H), considering that both R_0 and L_0 are computed directly from the identification of the calibration marks.

2.3   Object Modeling

The object of interest (fruit) is modeled by a spheroid. Initially, an ellipse is adjusted to the 2D object view, obtained from segmentation, as illustrated in Fig. 1(b, d). This ellipse is used to establish a new coordinate system with X along its major axis and Y along the minor axis. The origin of the XYZ coordinate system is the centre of the spheroid (3D object) and the third axis (Z) is perpendicular to the reference plane.

The 2D ellipse is also used to make an initial estimate for the spheroid minor (m_0) and major semi-diameters (M_0), computed at the reference plane level (with R_0). The spheroid centre height is initially estimated to be m_0. An iterative process is then performed by computing new estimates for the image resolution at a height m_{i-1} (R_i), and with this resolution a new value for m (m_i). The process is repeated until the absolute change in m_i is <0.01 mm. The process converges fast, with usually only 4 or 5 iterations required. As an example, for the image presented in Fig. 1(d), the final values were obtained after 5 iterations: $m_0 = 21.62$, $m = m_5 = 19.95$ (the real value of m is 20.0 mm); $M_0 = 32.59$, $M = 30.08$ (the real value of M is 30.0 mm).

The intersection of the spheroid with a plane (P) perpendicular to the major axis results in a circle with radius r_x, which can be computed as a function of the distance between the plane and the centroid (x) using (3). On that circle, the value of z can be computed as a function of x, y by (4), and the object height (H) can be thus estimated as $H(x,y) = z(x,y) + m$.

$$\frac{x^2}{M^2} + \frac{r_x^2}{m^2} = 1 \Leftrightarrow r_x = \frac{m}{M}\sqrt{M^2 - x^2} \tag{3}$$

$$y^2 + z^2 = r_x^2 \Leftrightarrow z = m\sqrt{1 - \frac{x^2}{M^2} - \frac{y^2}{m^2}} \tag{4}$$

Another important parameter is the spheroid surface area imaged by a pixel. The exact calculation is mathematically and computationally hard, thus a simplified approach is used instead. The surface area (A) for the pixel centered at x, y, z is considered to be approximately the product of two arc lengths: on the ellipse of plane X0Z (l_e) and on a circle on plane P (l_c). It is computed by (5–9), where Δ is the pixel size (in mm) estimated for height $H(x,y)$.

$$A = l_c.l_e \tag{5}$$

$$l_c = r_x|\theta_+ - \theta_-| \tag{6}$$

$$\theta_\pm = \cos^{-1}\left(\frac{y \pm \Delta/2}{r_x}\right) \tag{7}$$

$$l_e = \sqrt{\Delta^2 + (y_+ - y_-)^2} \tag{8}$$

$$y_\pm = \frac{m}{M}\sqrt{M^2 - (x \pm \Delta/2)^2} \tag{9}$$

Once the spheroid is defined, the x, y coordinates of each pixel in the 2D object are used to compute z and $H(x,y)$ (4), the image resolution $R(x,y)$ (2), $\Delta(x,y) = 1/R(x,y)$, and the surface area $A(x,y)$ (5–9). A total of 3 auxiliary

(a) $H(x,y)$ - Height   (b) $R(x,y)$ - Resolution   (c) $A(x,y)$ - Surface Area

Fig. 3. Auxiliary images modeling the spheroid height (a), resolution (b) and surface area (c), for each pixel (x, y), created for the test image of Fig. 1(c).

images are thus created - $H(x,y)$, $R(x,y)$ and $A(x,y)$, such as the example presented in Fig. 3, produced for the test image of Fig. 1(c). In this example, the positive values of $H(x,y)$ are between 19.97 and 39.90 mm and the Resolution ($R(x,y)$) between 10.22 and 11.15 pixels/mm, with linear transformations applied to create the grayscale images presented in Fig. 3(a, b). The Surface Area ($A(x,y)$) of the spheroid pixels varies smoothly, except near the edges where very high values occur. The linear transformation used to produce the greyscale image of Fig. 3(c) only covers values up to $60\,(\times 10^{-3}\,\mathrm{mm}^2)$, but the range of values present in this case is from 8.0 to $1609.2\,(\times 10^{-3}\mathrm{mm}^2)$.

2.4 Error Estimation

A number of factors influence the accuracy of the measurement of a sub-region area on the object surface. Some are related to the imaging geometry and object shape, which are addressed in the methodology proposed, but others, such as the image resolution and the segmentation process, can also have a substantial contribution. A rough estimate of the error due to these factors is performed in this section, in order to establish what can be a reference baseline.

It is reasonable to consider that the maximum error in the segmentation process along the sub-region contour is 1 pixel (if successful), which will result in a maximum error (δ) of ± 2 pixels/R_0 in the estimation of the circle diameter. The use of the observed diameter (d_{obs}) instead of the reference diameter (d_{ref}) will result in an error in the area computed (A_{obs}), in regards to the reference area (A_{ref}), that can be estimated using (10). An estimate of the relative error (ε_0) for the area is obtained by (11). The last term in (10) is considerably smaller that the error term, by a factor of $\delta/2d_{ref}$ (or approximately $1/d_{obs}$, with d_{obs} in pixel units), thus neglected. As an example, the test image presented in Fig. 1(c) has a value of $R_0 = 9.43$ pixels/mm and, considering that $d_{ref} = 12.0$ mm, an estimated relative error $\varepsilon_0 = 0.035$ (or 3.5%). In this case the last term in (10) would contribute with only 0.03% to ε_0, 2 orders of magnitude smaller than ε_0.

$$A_{obs} = \frac{\pi}{4}(d_{obs})^2 \quad = \frac{\pi}{4}(d_{ref} + \delta)^2 \simeq A_{ref} + \underbrace{\frac{\pi}{2}d_{ref}\delta}_{error} + \underbrace{\frac{\pi}{4}\delta^2}_{\approx 0} \tag{10}$$

$$\varepsilon_0 = \frac{|error|}{A_{ref}} = \frac{\frac{\pi}{2}d_{ref}\delta}{\frac{\pi}{4}(d_{ref})^2} = \frac{2\delta}{d_{ref}} = \frac{4}{R_0 d_{ref}} \tag{11}$$

3 Results

The method was evaluated using a wooden prolate spheroid of $40 \times 40 \times 60$ mm (diameters), and blue circular marks of two sizes: I - 4.0× mm radius ($50.3 \times$ mm^2 area), II - 6.0 mm radius ($113.1 \times$ mm^2 area). Two calibration targets were used - of A4 and A5 size. The spheroid volume and surface area [10] are $50265 \times$ mm^3 and $6767 \times$ mm^2, respectively.

3.1 Marks on Plane

A preliminary test was carried out, placing the colour marks on the reference plane itself. A total of 20 images were tested - 5 images for each mark (I and II) and calibration target (A4 and A5). The areas computed (A_{obs}) were compared with the reference (ideal) values (A_{ref}), with the results presented in Table 1. In the worst case scenario (mark I / A4), the average relative error (ε_R) is 3.32%, below the maximum expected error (ε_0) due to limited image resolution (about 5%). A slightly better case is for mark II (with A4 target), with an average ε_R of 1.15% and ε_0 of about 3%. Considering the small size of the marks, it is much better to use an A5 calibration target. The average relative errors with this target are 0.67% (mark I) and 0.72% (mark II), well below ε_0 (around 2.5–3.5%).

This test provides an estimate of the error that can be expected due to the segmentation process and limited image resolution. This baseline error can thus be expected to be roughly between 2.5% (II/A5) and 5.5% (I/A4) or, more generally, about 4%.

3.2 Marks on 3D Object

Images of the colour mark on the spheroid were acquired in 7 positions, presented schematically in Fig. 4. The figure shows the location of the mark in the 2D view ellipse. In positions #1 to #6 the spheroid is roughly aligned with the calibration target grid, whereas in position #7 the XY axes of the spheroid are at about 45° with the target grid. A total of 28 test images were thus used (7 positions × 2 marks × 2 targets). The results are presented in Table 2 for mark I (4 mm radius) and in Table 3 for mark II (6 mm radius). For each image tested, the following values are presented: computed/observed area of the mark (A_{obs})

Table 1. Results for images with marks I and II on the reference plane: area computed (A_{obs}) in mm^2, maximum expected error (ε_0) and relative error (ε_R).

Target	#	A_{obs} (I)	ε_0 (I)	ε_R (I)	A_{obs} (II)	ε_0 (II)	ε_R (II)
A_4	1	48.33	5.36%	3.84%	111.45	3.69%	1.46%
	2	48.63	5.48%	3.25%	110.19	3.84%	2.57%
	3	48.74	5.37%	3.04%	112.47	3.39%	0.56%
	4	48.92	5.12%	2.67%	113.18	3.39%	0.07%
	5	48.35	5.73%	3.80%	114.33	3.55%	1.09%
A_5	1	49.86	3.52%	0.82%	113.63	2.24%	0.47%
	2	49.94	3.67%	0.65%	114.00	2.46%	0.80%
	3	50.04	3.71%	0.44%	113.74	2.54%	0.57%
	4	50.72	3.42%	0.91%	114.07	2.48%	0.86%
	5	50.53	3.30%	0.53%	114.11	2.61%	0.90%

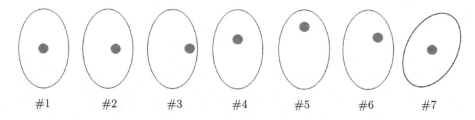

#1   #2   #3   #4   #5   #6   #7

Fig. 4. Test position 2D views of the spheroid and mark location. In position #7 the XY axes are aligned at about 45° in regards to the calibration target grid.

in mm^2, relative error (ε_R) in %, ellipsoid semi-diameters (m and M) in mm, relative errors (ε_R) for the computation of the spheroid surface area and volume.

The iterative process proved to be effective in establishing a 3D spheroid out of the observed 2D ellipse. The values computed for M and m are generally very close to the real spheroid size ($M = 30$ and $m = 20$ mm), with average absolute errors of only 0.25 mm for M and 0.20 mm for m (corresponding to relative errors of 0.8% and 1.0%, respectively). For the 28 images tested, the computation of the prolate spheroid volume and surface area [10], using the values M and m obtained from the iterative process, result in average relative errors of 1.77% (for surface area) and 2.71% (for volume).

An initial evaluation of the colour mark area computation is done for the test image of Fig. 1(c) (mark II, A4 target, position #7). The maximum expected error due to the segmentation process and limited image resolution (baseline) was estimated to be about 3.5% (Sect. 2.4). For this image, the computed mark area was 110.18 mm^2 (instead of 113.1 mm^2), corresponding to a relative error (ε_R) of 2.6% (Table 3). It is worth mentioning that if the mark area was computed for the calibration plane, ignoring the fact that the mark is over a 3D object, the result would be 150.4 mm^2 ($\varepsilon_R = 33.0\%$).

For mark I (Table 2) the relative error in the colour mark area computation is low (up to about 4%) in all 7 positions for the A4 calibration target and 5 positions for the A5 target. For the A5 target, there are 2 observation (positions #2 and #3) where the error is slightly above the expected (5.1% and 6.5%). For mark II (Table 3) the relative error in the colour mark area computation is low (up to about 4%) in 5 out of the 7 positions tested, for both calibration targets. The observation with high relative errors are for positions #4 and #6, for both targets, with relative errors of up to 12.6% (A5 target, position #4).

Table 2. Results for mark I ($50.3\,\mathrm{mm}^2$) on spheroid - area computed (A_{obs}) [mm^2] and relative error (ϵ_R) [%], spheroid semi-diameters m and M [mm], relative errors for spheroid surface and volume estimation [%].

Target	#	A_{obs}	ϵ_R (Area)	m	M	ϵ_R (Surface)	ϵ_R (Volume)
A_4	1	51.07	1.6%	20.02	29.92	0.11%	0.09%
	2	51.16	1.8%	20.32	30.34	2.82%	4.35%
	3	52.17	3.8%	20.14	30.27	1.54%	2.27%
	4	51.37	2.2%	20.04	29.89	0.02%	0.09%
	5	51.29	2.0%	20.27	29.77	1.01%	1.92%
	6	50.53	0.5%	20.11	29.82	0.19%	0.50%
	7	51.21	1.9%	20.34	30.52	3.43%	5.18%
A_5	1	51.42	2.3%	20.17	30.31	1.86%	2.76%
	2	52.82	5.1%	20.26	30.22	2.15%	3.35%
	3	53.52	6.5%	20.20	30.03	1.30%	2.13%
	4	51.14	1.8%	20.14	30.19	1.35%	2.06%
	5	51.31	2.1%	20.10	30.01	0.65%	1.06%
	6	51.70	2.9%	20.25	30.13	1.86%	2.95%
	7	51.20	1.9%	20.20	30.31	2.03%	3.06%

Possible reasons to explain the high relative errors observed in the colour mark area computation observed for some cases, include: (i) the simplification in the computation of the surface area imaged by a pixel; (ii) the fact that the images acquired are not exactly vertical (the principal axis is not perpendicular to the surface observed); (iii) the planar shape of the colour marks. As it happens, the colour marks are plane stickers, thus having a tendency to be slightly detached from the spheroid, particularly in areas of strong curvature, as those close to the edge of the spheroid along its major axis. Furthermore, the spheroid surface is not perfect, thus limiting the ability of the sticker to properly adhere to the surface. But the most likely reason is (i), as the arc length along the ellipse of plane X0Z tends to be underestimated. It is a matter for further investigation, preferably with an experimental evaluation with spheroids of various sizes.

Overall, the results are in line with the expected errors (baseline), except for 6 out of 28 images tested. The relative error in the computation of the mark

Table 3. Results for mark II (113.1 mm^2) on spheroid - area computed (A_{obs}) [mm^2] and relative error (ϵ_R) [%], spheroid semi-diameters m and M [mm], relative errors for spheroid surface and volume estimation [%].

Target	#	A_{obs}	ϵ_R (Area)	m	M	ϵ_R (Surface)	ϵ_R (Volume)
A_4	1	110.17	2.6%	19.99	29.97	0.13%	0.18%
	2	110.27	2.5%	20.20	30.14	1.56%	2.46%
	3	112.38	0.6%	20.18	30.05	1.21%	1.96%
	4	102.89	9.0%	19.77	29.37	3.06%	4.37%
	5	111.45	1.5%	20.24	29.91	1.21%	2.10%
	6	104.44	7.7%	19.67	28.88	4.92%	6.91%
	7	110.18	2.6%	19.95	30.08	0.11%	0.26%
A_5	1	108.39	4.2%	20.26	30.17	1.99%	3.15%
	2	108.44	4.1%	20.21	30.07	1.50%	2.41%
	3	111.23	1.7%	20.25	30.08	1.74%	2.81%
	4	98.87	12.6%	19.62	29.41	3.81%	5.65%
	5	109.56	3.1%	20.31	30.25	2.58%	4.04%
	6	100.20	11.4%	19.56	29.26	4.59%	6.76%
	7	108.68	3.9%	20.06	30.16	0.77%	1.11%

area of those 6 cases is nevertheless acceptable. In fact, even in the worst cases (up to 12.6% error) the results are much better than what would be obtained if the 3D shape of the object was ignored (which would result in relative errors up to 80% for the images tested).

The errors for position #1 and #7 are comparable, which indicates that the orientation of the spheroid XY axes in regard to the target grid is not critical. Generally it is preferable to have the colour mark as close to the centre of the imaged ellipse as possible (such as positions #1 and #7). Given the rather small size of the colour marks tested, the results could be improved by using a higher resolution camera. It is worth mentioning that the 5MP camera used can be considered to be close to the bottom level of current devices (smartphones and tablets).

4   Conclusion

The method proposed in this paper can be useful for the computation of sub-region surface area of a 3D object, modeled by a spheroid, using a single image acquired with a standard camera and a simple calibration target. The experimental evaluation indicates that the proposed iterative process is effective in establishing a 3D spheroid out of the observed 2D ellipse, with average relative errors of 0.8% and 1.0% in the estimation its semi-diameters, which result in low errors in the computation of both the spheroid surface area (1.77%) and volume

(2.71%). The computation of the colour mark surface area proved to be effective regardless of the mark location in the 2D ellipse, with errors up to about 4% in 22 out of 28 images tested.

The application of the proposed methodology to fruits lesions poses potentially new challenges. A major one is the segmentation, both of the fruit and the lesion, which is a critical issue for the subsequent processing. Of particular concern is the lesion, which might be difficult to properly distinguish. Other issues include the fruit shape (it might not be similar to a spheroid), fruit surface roughness, and the actual image acquisition geometry, which can be difficult to control in field. There is nevertheless a great potential in having low cost, easy to use systems to acquire and process in field data for fruit lesion area estimation. The methods presented in this paper provide a set of tools that can be a basis for such system.

Acknowledgments. This work was co-financed by the European Structural & Investment Funds (ESIFs) through the Operational Competitiveness and Internationalization Programme – COMPETE 2020, and by National Funds through FCT – Fundação para a Ciência e a Tecnologia, within the framework of the project EVOXANT (PTDC/BIA-EVF/3635/2014, POCI-01-0145-FEDER-016600).

References

1. Clayton, M., Amos, N.D., Banks, N.H., Morton, R.H.: Estimation of apple fruit surface area. NZ J. Crop Hortic. Sci. **23**, 345–349 (1995)
2. EUDirective - European, Union: Council Directive 2000/29/EC of 8 May 2000 on protective measures against the introduction into the Community of organisms harmful to plants or plant products and against their spread within the Community. European Union Official Journal L 169 of 10 July 2000, pp. 1–112 (2000)
3. Khojastehnazhand, M., Omid, M., Tabatabaeefar, A.: Determination of orange volume and surface area using image processing technique. Int. Agrophysics **23**, 237–242 (2009)
4. Kraus, K.: Photogrammetry Volume 1 - Fundamentals and Standard Processes, 4th edn. Ferdinand Dummlers Verlag, Bonn (1993)
5. MATLAB and Image Processing Toolbox Release 2017a, The MathWorks Inc, Natick, Massachusetts, United States (2017)
6. Momin, M.A., Rahman, M.T., Sultana, M.S., Ziauddin, A.T.M., Igathinathane, C., Grift, T.E.: Geometry-based mass grading of mango fruits using image processing. Inf. Process. Agric. **4**, 150–160 (2017)
7. Moragrega, C., Matias, J., Aletá, N., Montesinos, E., Rovina, M.: Apical necrosis and premature drop of Persian (English) walnut fruit caused by Xanthomonas arboricola pv. juglandis. Plant Dis. **95**, 1565–1570 (2011)
8. Sabliov, C., Boldor, D., Keener, K., Farkas, B.: Image processing method to determine surface area and volume of axi-symmetric agricultural products. Int. J. Food Prop. **5**(3), 641–653 (2002)
9. Sadrnia, H., Rajabipour, A., Jafary, A., Javadi, A., Mostofi, Y.: Classification and analysis of fruit shapes in long type watermelon using image processing. Int. J. Agric. Biol. **1**(9), 68–70 (2007)
10. Wolfram MathWorld, Prolate Spheroid http://mathworld.wolfram.com/ProlateSpheroid.html. Accessed 26 Feb 2019

A Weakly-Supervised Approach for Discovering Common Objects in Airport Video Surveillance Footage

Francisco Manuel Castro[1]([✉])[iD], Rubén Delgado-Escaño[1][iD], Nicolás Guil[1][iD], and Manuel Jesús Marín-Jiménez[2][iD]

[1] University of Málaga, Málaga, Spain
{fcastro,rubende,nguil}@uma.es
[2] University of Córdoba, Córdoba, Spain
mjmarin@uco.es

Abstract. Object detection in video is a relevant task in computer vision. Standard and current detectors are typically trained in a strongly supervised way, what requires a huge amount of labelled data. In contrast, in this paper we focus on object discovery in video sequences by using sets of unlabelled data. Thus, we present an approach based on the use of two region proposal algorithms (a pretrained Region Proposal Network and an Optical Flow Proposal) to produce regions of interest that will be grouped using a clustering algorithm. Therefore, our system does not require the collaboration of a human except for assigning human understandable labels to the discovered clusters. We evaluate our approach in a set of videos recorded at *apron area*, where the aeroplanes park to load passengers and luggage. Our experimental results suggest that the use of an unsupervised approach is valid for automatic object discovery in video sequences, obtaining a CorLoc of 86.8 and a mAP of 0.374 compared to a CorLoc of 70.4 and mAP of 0.683 achieved by a supervised Faster R-CNN trained and tested on the same dataset.

Keywords: Object discovery · Weakly-supervised learning · Region proposal · Deep neural networks

1 Introduction

The goal of *object detection* is to define the spatial extent and the kind of objects present in an image or video sequence. The object detection problem has been studied for a long time from multiple points of view [1,7,20,23]. Traditional approaches are based on manually designed descriptors that are computed, and then classified, along the image in a sliding window setup [7,23]. With the

Supported by project TIC-1692 (Junta de Andalucía) and TIN2016-80920R (Spanish Ministry of Science and Tech.). We gratefully acknowledge the support of NVIDIA Corporation with the donation of the Titan X Pascal GPU used for this research.

© Springer Nature Switzerland AG 2019
A. Morales et al. (Eds.): IbPRIA 2019, LNCS 11868, pp. 296–308, 2019.
https://doi.org/10.1007/978-3-030-31321-0_26

advent of deep learning techniques, the features are self-learnt by the model, what greatly boosted the performance of the proposed approaches [1,20].

However, all those approaches required to be trained in a strongly supervised way, that is, using manually labelled data for training. This requirement, coupled with the large amount of data needed for training a deep learning approach, makes costly the use of deep learning models with new classes that are not present in public datasets. Therefore, in order to add new classes to a dataset, it is necessary to label thousands of images in order to perform a good training process. Thus, there is a bottleneck that hampers the scaling of the detection models to larger number of classes: the lack of annotated images, as the annotation process is tedious and time-consuming.

The ideal solution to this problem would be to train the models using the huge amount of unlabelled data available in many online media sharing pages, such as Flickr. However, only few works apply fully unsupervised solutions, and most of them have serious limitations, like having only one object in the scene [3,12] or being focused on a specific kind of object (*e.g.* humans) [29]. Moreover, the unsupervised solutions produce lower results than using a supervised one. Borrowing ideas from supervised and unsupervised learning, we found the weakly-supervised learning that uses unlabelled data combined with either some labelled samples or coarse grained information about some samples. The key idea is to combine the huge amount of unlabelled information available with some labelled data in order to facilitate the training process. In this category of learning approaches we find works that use image-level labels (*i.e.* the label of the visually dominant object) to learn to localise the object [8,25], or, on the other hand, they use some labelled samples like in [4,31].

In contrast to all those previous works, we propose a weakly-supervised approach for object discovery in videos. In Fig. 1 we show a sketch of our pipeline. Firstly, we automatically find regions of candidate objects in a sequence of frames by using the Region Proposal Network (RPN) of a Convolutional Neural Network (CNN) previously trained for object detection. In addition, we compute the optical flow maps between consecutive frames to discover areas with moving objects. The obtained regions are described by a feature vector obtained from a pretrained CNN. After that, the descriptors are grouped using a clustering algorithm in order to find similar objects. Finally, to assign the labels to the detected objects, a human may optionally revise some samples from each cluster. This labelling process will be performed only when new clusters are obtained, what means that there are new classes present in the video. Thus, instead of labelling hundreds of thousands images, it is only necessary to label a set of clusters.

Therefore, the main contributions of this work are: *(i)* a novel pipeline for weakly-supervised object discovery in videos; *(ii)* a combination of two complementary region proposals specially designed for video sequences; *(iii)* a fast weakly-supervised labelling process based on a clustering process; and, *(iv)* a thorough experimental study to validate the proposed framework.

In our experiments, we use videos obtained from a RGB camera that is continuously recording the *apron area* (area where the aeroplanes park to load

passengers and luggage) of the Gdansk Airport, although it can be used in other
scenarios with static cameras, like bus or train stations, ports, etc. According to
the results, our region proposal approach is more robust than a fully-supervised
approach (fine-tuned model for this specific domain), specially with small regions
or classes with changes in the shape such as persons. However, during the automatic labelling of discovered regions, our approach obtains worse results than
the fully-supervised approach, mainly for classes with few samples.

The rest of the paper is organised as follows. We start by reviewing related
work in Sect. 2. Then, Sect. 3 explains the proposed approach. Sect. 4 contains
the experiments and results. Finally, we present the conclusions in Sect. 5.

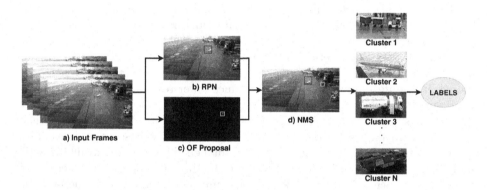

Fig. 1. Pipeline for weakly-supervised object discovery. (a) The input is a
sequence of RGB video frames. (b) The RPN process a set of detections for each input
frame. (c) The OFRP obtains another set of detections. (d) Non-maximum suppression
is applied to combine overlapped detections. A final clustering step is performed to
obtain a label per detection.

2   Related Work

Weakly-Supervised Approaches. Weakly-supervised learning has gained
importance in the last years [5,17,26–28] due to the huge amount of publicly
available unlabelled data. By contrast, the amount of labelled data is very
reduced, what penalises supervised approaches. Focusing on the problems of
object detection/localisation, many researches use the key idea of having an
image level label that only contains the object that appears in the image, without bounding-boxes (like in an image classification problem). In [8], the authors
propose a multiple-instance learning approach that iteratively trains a detector and infers the bounding-boxes. Kantorov *et al.* [10] use two context aware
models to improve the localisation taking into account the surrounding context
region of the bounding-box. In [13], Li *et al.* propose the use of an object detector trained only with positive samples of a class to produce a set of heat maps

that are refined and segmented to localise the trained class. Wang *et al.* [25] propose a different approach using a spatio-temporal minimisation process for video object discovery and segmentation across videos with irrelevant frames. In contrast to those previous approaches, some authors propose the use of some labelled samples together with a big amount of unlabelled samples. An example of this kind of work is the proposed in [4] where the authors use conditional random fields initialised from generic knowledge and, iteratively, they adapt to new classes. Shi *et al.* [21] propose a different approach using a Bayesian joint topic modelling that uses a single generative model for all objects improving learning and localisation. A more complex approach is presented in [31] where the authors introduce a deep model that uses a joint learning to localise and segment objects. A completely different approach is presented in [30] where the weak information is provided as a sentence that describes the image.

Unsupervised Approaches. Unlike weakly-supervised learning, unsupervised approaches do not use any kind of labelled data. Due to the difficulty of this type of approach, there are very few works that applies unsupervised learning to the object detection/localisation task. In [3,12], the authors propose an approach for dominant object discovery and tracking in videos using region proposals and a matching scheme in order to produce spatio-temporal tubes of detections in videos. A more recent point of view of this work is presented in [24] where the authors reformulate the approach as an optimization process improving previous results by a wide margin. Koh *et al.* [11] propose an approach for primary object discovery in videos exploiting the recurrence of a primary object in a video using a modelling scheme from motion and colour proposals. A specific approach for pedestrian detection in proposed in [29] where an iterative process of object discovery, object enforcement and label propagation is performed for training a progressive latent model for pedestrian detection. Ošep *et al.* [16] propose an automatic approach for object discovery in stereo videos using a generic tracker to find the objects, and a clustering process to group similar tracks. Note that, although this approach is somehow similar to ours, they use stereo cameras together with a tracker, while we use a single camera as input. Moreover, we rely on the optical flow maps to find out moving objects, while they use a pretrained object tracker that could tend to detect only objects that were seen during training. A similar approach is presented in [18] where they also use stereo cameras and depth information to perform object discovering and a clustering process to group similar detections.

In this work, we explore the weakly-supervised object discovery problem in videos obtained from common RGB cameras. Our approach is composed of two main parts: a region proposal step to produce interest areas and a clustering step to group similar areas. Finally, the obtained set of clusters can be manually labelled in order to use human understandable labels. Note that our approach can be applied in a fully unsupervised way using as labels the cluster indices.

3   Proposed Approach

In this section we describe our proposed framework to address the problem of weakly-supervised object discovery in videos using CNNs. The pipeline proposed is represented in Fig. 1. Using a sequence of consecutive frames as input, the following steps are performed: *(i)* region proposal using the RPN branch of a Faster R-CNN [20]; *(ii)* region proposal based on the optical flow maps obtained from consecutive frames; *(iii)* non-maximum suppression to combine and remove overlapped regions from both proposals; *(iv)* clustering process to group similar regions; and, *(v)* manual labelling of the obtained clusters.

3.1   Region Proposal

We describe here the two region proposal strategies used in our approach. In particular, we use the RPN branch of a pretrained Faster R-CNN [20] and the optical flow maps obtained from a pretrained Spatial Pyramid Network (SpyNet) [19]. Our intuition is that the RPN will be able to detect big objects present in the foreground, which are the most common kind of detections used for training this type of networks. Similarly, the region proposal based on the optical flow maps will be able to find out subtle and small objects.

Region Proposal Network. In our approach, we use the Region Proposal Network (RPN) of a pretrained Faster R-CNN [20] model. A RPN uses an image as input to produce a set of object proposals. To generate the regions, the input sample is fed into a set of convolutional layers in order to produce a feature map. After that, a small network is slided over that feature map with a window size of $n \times n$. This small network is composed of two sibling fully-connected layers, a box regression layer (*reg*) to obtain the bounding box coordinates and a box-classification layer (*cls*) to obtain the class of the detected object. Note that the sliding window is implemented in a direct way by using a $n \times n$ convolutional layer followed by two sibling 1×1 convolutional layers for *reg* and *cls*. Note that in this step the pretrained model applies a suppression of detections whose score is lower than a threshold T_S. The regions kept after the filtering process are fed into the next step of our pipeline.

Optical Flow Proposal. The optical flow proposal (OFRP) is based on the optical flow maps obtained from a pretrained Spatial Pyramid Network (SpyNet) [19] model. Two consecutive frames are fed into the network to produce an optical flow map F_t. To remove noise from the optical flow map produced by changes in the illumination or the conditions of the scenario, all positions whose optical flow components (x and y) are smaller than a threshold T_F are set to 0. After that, we binarize the optical flow map and find the contours to obtain the regions of the objects present in the map. In order to do this, we use the well known algorithm proposed by Suzuki *et al.* [22]. To avoid intermittent detections, we track each region in F_t to the next optical flow map F_{t+1} using the mean optical flow components of that region. If the region is missed in F_{t+1}, we remove the original detection from F_t. Finally, in order to prevent insignificant regions, we remove those proposed regions whose area is smaller than a threshold T_A. The regions kept after the filtering process are fed into the next step of our pipeline.

3.2 Non-maxima Suppression

Since there are two region proposal algorithms running at the same time, when a big object is moving in consecutive frames, it is probable that the RPN proposal and the OFRP produce detections of the same object. Therefore, it is necessary to combine both detections into a single one to avoid overlapping regions. To combine both detections, we compute the Intersection over Union (IoU) metric between both detections. If the IoU is bigger than a threshold T_I and the aspect ratio of both regions is similar (*i.e.* the ratio of the biggest one between the smallest one must be bigger than a threshold T_{AR}), we keep the region proposed by the RPN algorithm, as it is more accurate than the optical flow one. Moreover, we apply non-maxima suppression to each individual proposal algorithm to remove overlapped regions whose IoU is greater than the same threshold T_I. After this step, we have the final set of regions used by the clustering algorithm.

3.3 Clustering

The objective of this step is grouping similar regions into clusters. By this way, instead of learning a classifier that assigns a label according to the features of a region, we just find the closest cluster to a region. In order to do this, we first have to describe the detected regions to fed that information into the clustering algorithm. To describe the regions we employ a pretrained ResNet-50 [9] model as feature extractor, where descriptors are given by the activations of the average pooling before the classification layer. Once the features of each region have been extracted, we apply a L2-normalisation to the features and we reduce their dimensionality to 128 with the UMAP algorithm [15]. This method is specially useful as it is able to reduce the dimensionality keeping the global structure of the data but preserving local neighbours relations. Finally we use them as input to the clustering algorithm. For this purpose, we use the HDBSCAN clustering algorithm [2] which is able to deal with different cluster shapes and densities with a good performance. These capabilities are really important in our problem since there could exist objects with many different number of samples or even objects with many different shapes, and the algorithm should deal with those situations.

3.4 Labelling Process

The last step of our pipeline is completely optional, since it is only necessary in order to assign a human understandable label to each cluster obtained in the previous step. To do this, we show the set of N samples with the highest scores, obtained by HDBSCAN, for each cluster to a human who establishes the labels of the clusters. Note that a higher score indicates a better membership of a sample to a given cluster. In order to assign a label to a cluster, at least, half of the showed cluster samples must belong to one of the considered classes. By this way, the labelling is robust against outliers. Note that this is the only step where the intervention of a human is necessary. Once clusters are labelled, that information can be used in order to remove false positives detections if they are grouped into different clusters. Thus, if a cluster only contains false positives, we can ignore that cluster to produce better detections.

Fig. 2. Dataset. Different frames obtained from a RGB camera recording the *apron area* of the Gdansk Airport. Top row shows the ground-truth labels obtained manually. Bottom row shows the output of our weakly-supervised approach.

4   Experiments and Results

4.1   Dataset

In our experiments, we are going to use a video dataset obtained from a RGB camera that is continuously recording the *apron area* (area where the aeroplanes park to load passengers and luggage) of the Gdansk Airport. Those cameras are publicly available online[1]. The dataset consists of 96 video-clips of one minute length recorded by a FullHD camera which provides a video stream with a resolution of 1920 × 1080 pixels and a frame rate of approximately 15 fps. Approximately, 60% of the videos are recorded during the morning and the other 40% are recorded during the afternoon/evening in order to deal with different illumination conditions. Some examples can be seen in Fig. 2. Note that in our experiments we are going to focus on the closest apron area since the other areas are excessively far to identify the appearing objects. Thus, in our experiments

---

[1] Live cameras: http://www.airport.gdansk.pl/airport/kamery-internetowe.

we are going to consider the following categories of objects: *car ('car'), fire-truck ('ft'), fuel-truck ('fuel'), luggagetrain-manual ('lgm'), luggagetrain ('lg'), mobile-belt ('mb'), person ('pe'), plane ('pl'), pushback-truck ('pb'), stairs ('st')* and *van ('van')*. Note that the abbreviated name of each class used in the tables is included in parenthesis. For training the clustering algorithm and our dimensionality reduction, we use odd id video clips and, for testing, the even id video clips, *i.e.* half for training and half for testing.

Finally, in order to obtain test metrics and compare our approach with a fully-supervised method, we have manually labelled all videos. Then, the training labels are used for training a supervised approach and the test labels are used to compute the CorLoc and mAP metrics. Roughly, we have labelled a total of 32238 objects where the less frequent class is 'van' with 30 samples and the most common is 'mobile-belt' with 6082.

4.2   Implementation Details

We ran our experiments on a computer with 32 cores at 2.3 GHz, 256 GB of RAM and a GPU NVidia Titan X Pascal running with Python 3.6 and Ubuntu 18.04. Faster R-CNN is implemented in TensorFlow 1.13 and we use the pretrained weights on the Open Images dataset provided in the TensorFlow detection model zoo[2]. SpyNet is implemented in PyTorch 1.0 and we use the pretrained weights available in the project repository[3]. Finally, the we use a ResNet-50 model implemented in TensorFlow 1.3 with the pretrained weights available in the samples repository of TensorFlow[4]. For the clustering process, we use the implementation of HDBSCAN available in pip repository. Regarding the parameters commented in Sect. 3, after a cross-validation process on a subset of the training data, we have selected the following values: $T_S = 0.3, T_{F_x} = 0.3, T_{F_y} = 0.003, T_A = 200, T_{AR} = 0.5, T_I = 0.75$.

4.3   Performance Evaluation

We use two metrics to evaluate the performance of our approach. On the one hand, for region localisation we use the *Correct Localisation* metric (CorLoc), adopted as well in [12,24]. This metric is defined as the percentage of objects correctly localised according to the Pascal criterion: the IoU between the predicted region and the ground-truth region is bigger than 0.5 for the RPN and bigger than 0.3 for the OFRP. Note that we use a smaller threshold for the optical flow case because the bounding-boxes tend to include the shadows, *i.e.* making the bounding-boxes wider or higher. On the other hand, for the object classification task, we use the Mean Average Precision (mAP) [6], which is the mean of the average precision (AP) across all classes.

---

[2] We use the model `faster_rcnn_inception_resnet_v2_atrous_oid_2018_01_28.tar.gz`.

[3] SpyNet: https://github.com/sniklaus/pytorch-spynet.

[4] ResNet-50: https://github.com/tensorflow/models/tree/master/official/resnet.

4.4   Experimental Results

We first examine the impact of our two region proposal algorithms (*i.e.* RPN and OFRP) described in Sect. 3 according to their individual CorLoc metrics. Secondly, we evaluate the accuracy of the clustering algorithm compared with other traditional clustering algorithms. Finally, we compare the performance of our proposed approach with a pretrained CNN for object detection and the same CNN but fine-tuned for our dataset.

Region Proposal Comparative. In this experiment, we evaluate the performance of each region proposal algorithm (*i.e.* RPN and OFRP), in terms of the CorLoc metric, by comparing the proposals produced by each approach with the annotated ground-truth. Moreover, we compare the performance of each algorithm depending on the category of object to be detected.

Table 1 summarizes the CorLoc results (higher is better) for our two proposal algorithms ('RPN' and 'OFRP'), together with the combination of both ('RPN+OFRP') and our final approach ('Ours') considering the clustering labels to filter detections (more details in Sect. 3.4). Moreover, we also include the results of a fine-tuned Faster R-CNN with manually labelled training data ('Supervised-Faster') as explained in Sect. 4.1. Note that each row contains results using a different IoU threshold during the computation of the metric. In the first row, we use the standard value of 0.5. However, as explained in the previous section, the OFRP requires a smaller threshold. Thus, in the second row we use two different thresholds, one for the 'RPN' (0.5) and a second one for the optical flow (0.3). According to the results, it is clear that the 'OFRP' produces more and more accurate regions than the 'RPN', mainly due to the huge scale differences between objects, as we pointed out in Sect. 3. If we focus on the effect of the threshold, we can see that the performance increases clearly when the threshold is softer because it allows the metric to take into account bounding-boxes that contain objects and their shadows. Finally, if we apply the clustering step and the manual label of clusters, we are able to remove detections grouped into useless clusters (*i.e.* containing only false positives) improving the results as shown in column 'Ours'. Comparing our final results with the fully-supervised network, our approach is able to produce better proposals even using the more restrictive metric with an IoU of 0.5.

In order to clarify the contribution of each proposal algorithm, we measure the CorLoc metric over the true positive set of each class. By this way, we will see the detection capabilities per class of the two algorithms. Table 2 summarizes the results for this experiment. As we can see, the 'OFRP' produces better regions for most of the classes and only for the 'plane' class obtains worse results. This is because the planes appear in a static situation in most of the frames, thus, there is no optical flow in that situation. Focusing on the 'RPN' results, we can see that it only obtains better results for bigger object classes (*i.e.* planes, cars, vans, and different types of tracks), what validates our intuition and makes necessary the use of the 'OFRP' for small objects (*e.g.* persons) since the RPN has never seen objects with such an small area.

Table 1. CorLoc results for our two region proposal algorithms. Each row shows the results for a different IoU threshold used to compute the metric. Each column represents a different approach. Best result is marked in bold. More details in the text.

	RPN	OFRP	RPN+OFRP	Ours	Supervised-faster
CorLoc (0.5)	12.3	24.8	37.1	**86.8**	70.4
CorLoc (RPN@0.5, OFRP@0.3)	12.3	33.5	45.8	**96.2**	-

Table 2. CorLoc results per class using only true positives during the computation. Each row represents a different proposal algorithm and each column represents a different class. Best results are marked in bold. More details in the text.

	Car	ft	Fuel	lgm	lg	mb	pe	pl	pb	st	Van
RPN	8.3	37.5	41.4	3.7	11.3	7.1	1.3	**95.2**	7.7	3	25
OFRP	**78.4**	**60.7**	**57.8**	**74.2**	**88.2**	**72.8**	**79.1**	4.8	**64.7**	**84.8**	**37.5**

Class Prediction Comparative. In this experiment we focus on the class prediction part of the detection problem. Therefore, we try to predict the class appearing in the bounding boxes obtained from the previous step. Firstly, we compare two clustering algorithms: k-Means [14] and HDBSCAN (*i.e.* the one selected). Moreover, we compare two algorithms for dimensionality reduction: UMAP (see Sect. 3) and PCA. To measure the performance of the different approaches, we compare the AP per class and the mAP using the labels obtained after the clustering process of the detected objects. Table 3 summarises the results for this experiment. Each row represents a different algorithm, where 'kNC' means k-Means with NC clusters and 'HSC' means HDBSCAN with SC samples per cluster. Moreover, each row includes the dimensionality reduction algorithm used (*i.e.* PCA or UMAP). On the other hand, each column represents a different class. 'mAP' column represents the mean AP for all classes and

Table 3. AP results per class. Each row represents a different algorithm and each column represents a different class. 'mAP' column represents the mean AP for all classes and 'mAPv' is the mean of valid AP values. Only classes marked with '*' are considered for mAPv metric. Best results are marked in bold. More details in the text.

Algorithm	car	ft*	fuel*	lgm	lg	mb*	pe	pl*	pb	st	van	mAP	mAPv
k50+PCA	-	0.718	0.532	-	-	0.249	0.515	0.303	-	-	-	0.210	0.450
H30+PCA	-	0.714	0.471	0.159	-	0.341	0.367	0.364	-	-	-	0.219	0.472
H30+UMAP	-	0.718	0.746	0.264	-	0.487	0.451	0.419	0.001	-	-	0.280	0.592
H50+PCA	-	0.383	0.439	-	-	0.347	0.436	0.404	-	-	-	0.182	0.393
H50+UMAP	-	0.730	0.970	-	-	0.513	0.542	0.454	-	-	-	0.291	0.666
Baseline	-	0.561	0.640	-	-	0.102	-	0.831	-	-	-	0.194	0.533
Supervised-Faster	0.899	0.782	0.980	0.873	0.970	0.969	0.105	0.969	0.949	0.04	0	**0.685**	**0.925**

'mAPv' is the mean of the common classes to all rows (*i.e.* 'ft', 'fuel', 'mb' and 'pl'). Note that '-' means that there is no cluster that predicts that specific class, so there are no valid AP results and we ignore them during the computation of 'mAPv'. Focusing on the k-Means results, we can see that the performance is worse than using HDBSCAN with a cluster size of 30 samples with PCA. Therefore for the next experiments we focus on HDBSCAN. Comparing PCA with UMAP, we can see that in all cases, UMAP clearly boosts the results demonstrating their better dimensionality reduction performance. Then, if we compare the different cluster size values for HDBSCAN, it is clear that bigger clusters benefit the performance of our approach, specially if we only take into account the valid classes. We want to clarify that non-valid classes are due to the low number of samples (*i.e.* around one hundred compared to thousands of images for the other classes) available during training for those classes. Thus, the clustering algorithms tend to assign them to bigger clusters. For example, 'car' and 'push-back' objects only appear in the scene during the arrival and departure of the planes, respectively. Therefore, the number of samples is very limited compared with other classes that appear more frequently in the scene. Finally, the last two rows summarise the results for the baseline using a pretrained Faster R-CNN and a fine-tuned model using our trained data, respectively. Comparing both results with our best approach (H50+UMAP), we can see that our approach improves the results obtained by the baseline. However, the fine-tuned model obtains the best results on average, as it has been trained in a supervised way. Comparing the class-specific results with our best approach, we can see that the model also suffers with classes that have few samples (*i.e.* stairs or van). Moreover, focusing on *person* class – which appears in the videos with low contrast, very noisy frames, many shape changes and small bounding-boxes – our approach overcomes the results obtained by the fine-tuned model. Mainly, because our OF-based region proposal is more robust in those situations, as the fine-tuned model is not able to obtain a good representation for *persons*.

5   Conclusions

We have presented a weakly-supervised approach for automatic object discovery in videos. Our method consists of two main components: a region proposal, which produces the bounding-boxes, and a clustering algorithm, which groups similar detections to assign them an unsupervised label. We have tested our approach on video sequences of the *apron area* of an airport showing that it is able to detect and classify automatically objects appearing on those videos. Moreover, the collaboration of the human is only necessary in order to assign human understandable labels. Therefore, our approach is able to run automatically without the collaboration of the human.

Regarding the region proposal algorithms (RPN or OFRP), we have demonstrated that the combination of a pretrained RPN together with an pretrained OFRP, is able to improve the results obtained by a fine-tuned model for the specific problem. Moreover, our approach is especially robust dealing with small regions and classes with changes in the shape (*e.g.* persons).

Regarding the clustering algorithm, our results show that HDBSCAN combined with UMAP improves traditional approaches such as k-Means and PCA. In this case, the fully-supervised approach (fine-tuned Faster R-CNN) obtains the best results, but our weakly-supervised approach is able to obtain better results than the evaluated baseline. As future work, we plan to use the detections produced by our approach to retrain iteratively the CNN models in order to obtain better results in each iterative step with a minimum labelling process. By this way, the gap between our weakly-supervised approach and the fully-supervised network should decrease in each iteration.

References

1. Cai, Z., Vasconcelos, N.: Cascade R-CNN: delving into high quality object detection. In: CVPR, pp. 6154–6162 (2018)
2. Campello, R.J., Moulavi, D., Zimek, A., Sander, J.: Hierarchical density estimates for data clustering, visualization, and outlier detection. TKDD 10(1), 5 (2015)
3. Cho, M., Kwak, S., Schmid, C., Ponce, J.: Unsupervised object discovery and localization in the wild: Part-based matching with bottom-up region proposals. In: CVPR, pp. 1201–1210 (2015)
4. Deselaers, T., Alexe, B., Ferrari, V.: Localizing objects while learning their appearance. In: ECCV, pp. 452–466 (2010)
5. Durand, T., Mordan, T., Thome, N., Cord, M.: WILDCAT: weakly supervised learning of deep convnets for image classification, pointwise localization and segmentation. In: CVPR, pp. 642–651 (2017)
6. Everingham, M., Van Gool, L., Williams, C.K.I., Winn, J., Zisserman, A.: The PASCAL visual object classes challenge 2012 (VOC2012) results. http://www.pascal-network.org/challenges/VOC/voc2012/workshop/index.html
7. Felzenszwalb, P.F., Girshick, R.B., McAllester, D., Ramanan, D.: Object detection with discriminatively trained part-based models. IEEE PAMI 32(9), 1627–1645 (2010)
8. Gokberk Cinbis, R., Verbeek, J., Schmid, C.: Multi-fold mil training for weakly supervised object localization. In: CVPR, pp. 2409–2416 (2014)
9. He, K., Zhang, X., Ren, S., Sun, J.: Deep residual learning for image recognition. In: CVPR, pp. 770–778 (2016)
10. Kantorov, V., Oquab, M., Cho, M., Laptev, I.: ContextLocNet: context-aware deep network models for weakly supervised localization. In: Leibe, B., Matas, J., Sebe, N., Welling, M. (eds.) ECCV 2016. LNCS, vol. 9909, pp. 350–365. Springer, Cham (2016). https://doi.org/10.1007/978-3-319-46454-1_22
11. Koh, Y.J., Kim, C.S.: Unsupervised primary object discovery in videos based on evolutionary primary object modeling with reliable object proposals. IEEE Transact. Image Process. 26(11), 5203–5216 (2017)
12. Kwak, S., Cho, M., Laptev, I., Ponce, J., Schmid, C.: Unsupervised object discovery and tracking in video collections. In: ICCV, pp. 3173–3181 (2015)
13. Li, Y., Liu, L., Shen, C., van den Hengel, A.: Image co-localization by mimicking a good detector's confidence score distribution. In: Leibe, B., Matas, J., Sebe, N., Welling, M. (eds.) ECCV 2016. LNCS, vol. 9906, pp. 19–34. Springer, Cham (2016). https://doi.org/10.1007/978-3-319-46475-6_2
14. Lloyd, S.: Least squares quantization in PCM. IEEE Transact. Inf. Theor. 28(2), 129–137 (1982)

15. McInnes, L., Healy, J., Melville, J.: Umap: Uniform manifold approximation and projection for dimension reduction. arXiv preprint arXiv:1802.03426 (2018)
16. Ošep, A., Voigtlaender, P., Luiten, J., Breuers, S., Leibe, B.: Large-scale object discovery and detector adaptation from unlabeled video. arXiv preprint arXiv:1712.08832 (2017)
17. Peyre, J., Sivic, J., Laptev, I., Schmid, C.: Weakly-supervised learning of visual relations. In: ICCV, pp. 5179–5188 (2017)
18. Pot, E., Toshev, A., Kosecka, J.: Self-supervisory signals for object discovery and detection. arXiv preprint arXiv:1806.03370 (2018)
19. Ranjan, A., Black, M.J.: Optical flow estimation using a spatial pyramid network. In: CVPR, pp. 4161–4170 (2017)
20. Ren, S., He, K., Girshick, R., Sun, J.: Faster R-CNN: towards real-time object detection with region proposal networks. In: NIPS, pp. 91–99 (2015)
21. Shi, Z., Hospedales, T.M., Xiang, T.: Bayesian joint topic modelling for weakly supervised object localisation. In: ICCV, pp. 2984–2991 (2013)
22. Suzuki, S., et al.: Topological structural analysis of digitized binary images by border following. Comput. Vis. Graph. Image Process. **30**(1), 32–46 (1985)
23. Viola, P., Jones, M., et al.: Rapid object detection using a boosted cascade of simple features. CVPR **1**, 511–518 (2001)
24. Vo, H.V., Bach, F., Cho, M., Han, K., LeCun, Y., Perez, P., Ponce, J.: Unsupervised image matching and object discovery as optimization. arXiv preprint arXiv:1904.03148 (2019)
25. Wang, L., Hua, G., Sukthankar, R., Xue, J., Niu, Z., Zheng, N.: Video object discovery and co-segmentation with extremely weak supervision. IEEE PAMI **39**(10), 2074–2088 (2017)
26. Wang, X., Peng, Y., Lu, L., Lu, Z., Bagheri, M., Summers, R.M.: ChestX-ray8: hospital-scale chest x-ray database and benchmarks on weakly-supervised classification and localization of common thorax diseases. In: CVPR, pp. 2097–2106 (2017)
27. Wei, Y., et al.: STC: a simple to complex framework for weakly-supervised semantic segmentation. IEEE PAMI **39**(11), 2314–2320 (2017)
28. Xu, Y., Kong, Q., Wang, W., Plumbley, M.D.: Large-scale weakly supervised audio classification using gated convolutional neural network. In: Proceedings of ICASSP, pp. 121–125 (2018)
29. Ye, Q., Zhang, T., Ke, W., Qiu, Q., Chen, J., Sapiro, G., Zhang, B.: Self-learning scene-specific pedestrian detectors using a progressive latent model. In: CVPR, pp. 509–518 (2017)
30. Yu, H., Siskind, J.M.: Sentence directed video object codiscovery. IJCV **124**, 312–334 (2017)
31. Zhang, D., Han, J., Yang, L., Xu, D.: SPFTN: a joint learning framework for localizing and segmenting objects in weakly labeled videos. In: IEEE PAMI (2018)

Standard Plenoptic Camera Calibration for a Range of Zoom and Focus Levels

Nuno Barroso Monteiro[1,2(✉)] and José António Gaspar[1]

[1] Institute for Systems and Robotics, University of Lisbon, Lisbon, Portugal
{nmonteiro,jag}@isr.tecnico.ulisboa.pt
[2] Institute for Systems and Robotics, University of Coimbra, Coimbra, Portugal

Abstract. Plenoptic cameras have a complex optical geometry combining a main lens, a microlens array and an image sensor to capture the radiance of the light rays in the scene in its spatial and directional dimensions. As conventional cameras, changing the zoom and focus settings originate different parameters to describe the cameras, and consequently a new calibration is needed. Current calibration procedures for these cameras require the acquisition of a dataset with a calibration pattern for the specific zoom and focus settings. Complementarily, standard plenoptic cameras (SPCs) provide metadata parameters with the acquired images that are not considered on the calibration procedures. In this work, we establish the relationships between the camera model parameters of a SPC obtained by calibration and the metadata parameters provided by the camera manufacturer. These relationships are used to obtain an estimate of the camera model parameters for a given zoom and focus setting without having to acquire a calibration dataset. Experiments show that the parameters estimated by acquiring a calibration dataset and applying a calibration procedure are similar to the parameters obtained based on the metadata.

Keywords: Standard Plenoptic Camera · Calibration · Metadata parameters

1 Introduction

Plenoptic cameras are able to discriminate the contribution of each of the light rays that emanate from a given point in the scene. In a conventional camera, the contribution of the several rays is not distinguishable since they are collected on the same pixel. This discrimination on plenoptic cameras is possible due to the positioning of a microlens array between the main lens and the image sensor.

Plenoptic cameras sample the lightfield [8,9] which is a 4D slice of the plenoptic function [1]. There are several optical setups that are able to acquire the lightfield as camera arrays [15]. Here, we focus on compact and portable setups like the lenticular array based plenoptic cameras. More specifically, on the SPC [13] which has a higher directional resolution and produces images with lower

© Springer Nature Switzerland AG 2019
A. Morales et al. (Eds.): IbPRIA 2019, LNCS 11868, pp. 309–321, 2019.
https://doi.org/10.1007/978-3-030-31321-0_27

spatial resolution [7] when compared to the focused plenoptic camera (FPC) introduced by Lumsdaine and Georgiev [10, 14].

The camera models proposed for SPC [3, 4] are approximations of the real setup by considering the main lens as a thin lens and the microlenses as pinholes. There can be more complex models to describe the real setup. The SPC manufacturer provides metadata regarding the camera optical settings that help describing the camera. Namely, the metadata provided include the main lens focal length which is considered in [3, 4] to model the refraction of the rays by the main lens. On the other hand, the metadata also includes the distance at which a point is always in focus by the microlenses. Nonetheless, the assumption of pinhole like microlenses do not allow to incorporate directly this additional information on the camera models [3, 4].

The calibration procedures for SPCs [3, 4] do not consider the information provided by the camera manufacturer as metadata and therefore rely completely on the acquisition of a dataset with a calibration pattern for the specific zoom and focus settings to estimate the camera model parameters. Thus, in this work, we identify the relationships among the optical parameters provided as metadata as well as the relationships between these optical parameters and the entries of the camera model [4] for different zoom and focus settings of the camera. The relationships obtained are used to represent the camera model parameters based on the metadata parameters for a specific zoom and focus setting without having to acquire a novel calibration dataset.

In terms of structure, one presents in Sect. 2 a brief review of the camera models proposed for the SPC. In Sect. 3, the camera model [4] that describes the SPC by a 5 × 5 matrix that maps the rays in the image space to rays in the object space is summarized. In Sect. 4, one identifies the relationships among the parameters provided as metadata, and the relationships between the camera model entries and the metadata provided on the raw images. The results of estimating the camera model based on the metadata for a given zoom and focus setting are presented in Sect. 5. The major conclusions are presented in Sect. 6.

Notation: The notation followed throughout this work is the following: non-italic letters correspond to functions, italic letters correspond to scalars, lower case bold letters correspond to vectors, and upper case bold letters correspond to matrices.

2   Related Work

SPC allow to define several types of images by reorganizing the pixels captured by the camera on the 2D raw image (Fig. 1a) [13]. The raw image displays the images obtained by each microlens in the microlens array (Fig. 1b). There is another arrangement of pixels that is commonly used in SPC, the viewpoint or sub-aperture images. These images are obtained by selecting the same pixel position relatively to the microlens center for each microlens [13]. The microlens and viewpoint images exhibit different features due to the position of the microlens array on the focal plane of the main lens (Fig. 2). Thus, for these cameras, there

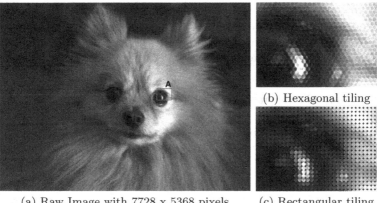

(a) Raw Image with 7728 x 5368 pixels   (b) Hexagonal tiling

(c) Rectangular tiling

Fig. 1. **(a)** Image captured on the sensor of a SPC. **(b)** Magnification of red box A in (a). This image depicts the hexagonal tiling of the microlenses images formed in the sensor. **(c)** Microlens images considered on the virtual plenoptic camera after the decoding process [4]. (Color figure online)

are mainly two calibration procedures, one based on viewpoint images [4] and other based on microlens images [3]. These consider camera models in which the main lens is modeled as a thin lens and the microlenses as pinholes.

The calibration based on viewpoint images [4] considers corner points as features and assumes a decoding process that transform the hexagonal tiling of the microlenses to a rectangular tiling (Fig. 1). This is done by interpolating the pixels of adjacent microlenses to get the missing ray information [5]. So in fact, this calibration procedure considers the calibration of a virtual SPC. There is an evolution of this work [16] that considers a better initialization for the camera model parameters. One of the disadvantages pointed out to this procedure is the fact of creating viewpoint images before a camera model is estimated.

On the other hand, the work of Bok *et al.* [3] allows to calibrate SPC directly from raw images using lines features. This procedure requires that line features appear on the microlens images which cannot be ensured when the calibration pattern is near the world focal plane of the main lens [3]. In this region, the microlens images consist of an image with very small deviations on the intensity values since these projections correspond to the same point in the scene [12] (Fig. 2).

The calibration procedures [3,4] assume that no information is known and therefore each of the parameters must be estimated by acquiring a dataset with a calibration pattern for a specific zoom and focus settings. A SPC provides metadata with information of the optical settings with the images acquired. Monteiro *et al.* [12] identified a relationship between the zoom and focus step provided in the metadata with the world focal plane of the main lens, but did not pursue this line of research. Here, we go a step further and identify the relationships of the metadata parameters among them and with the camera model parameters [4]. These relationships allow to obtain a representation of the camera model for an arbitrary zoom and focus settings based on the parameters provided by the manufacturer as metadata of the images acquired and without acquiring a calibration dataset.

3   Standard Plenoptic Camera Model

Let us consider a plenoptic camera that acquires a lightfield in the image space $L\left(\mathbf{\Phi}\right)$ with the plane Ω in focus, *i.e.* with the world focal plane of the main lens corresponding to the plane Ω (Fig. 2). The rays of the lightfield in the image space $\mathbf{\Phi} = [i, j, k, l]^{T}$ are mapped to the rays of the lightfield in the object space $\mathbf{\Psi} = [s, t, u, v]^{T}$ by a 5×5 matrix proposed by Dansereau *et al.* [4], the lightfield intrinsics matrix (LFIM) $\mathbf{H}$:

$$\tilde{\mathbf{\Psi}} = \mathbf{H}\tilde{\mathbf{\Phi}} \tag{1}$$

where $\tilde{(\cdot)}$ denotes the vector $(\cdot)$ in homogeneous coordinates. The rays in the image space are parameterized by pixels (i, j) and microlenses (k, l) indices while the rays in the object space are parameterized on a plane Π by a position (s, t) and a direction (u, v) in metric units [12]. Removing the redundancies of the LFIM with the translational components of the extrinsic parameters [2,4], one defines a LFIM with 8 free intrinsic parameters

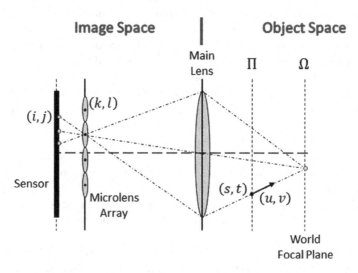

Fig. 2. Geometry of a SPC. The lightfield in the image space is parameterized using pixels and microlenses indexes while the lightfield in the object space is parameterized using a point and a direction. The lightfield in the object space is parameterized on plane Π regardless of the original plane Ω in focus.

$$\mathbf{H} = \begin{bmatrix} h_{si} & 0 & 0 & 0 & 0 \\ 0 & h_{tj} & 0 & 0 & 0 \\ h_{ui} & 0 & h_{uk} & 0 & h_u \\ 0 & h_{vj} & 0 & h_{vl} & h_v \\ 0 & 0 & 0 & 0 & 1 \end{bmatrix}. \tag{2}$$

This matrix does not provide a direct connection with the common intrinsic parameters defined within a pinhole projection matrix. The closer connection to the pinhole projection matrix is the one provided by Marto et $al.$ [11] regarding the representation of a camera array composed of identical co-planar cameras. In this setup, the LFIM can be represented as

$$\mathbf{H} = \begin{bmatrix} h_{si} & 0 & \\ 0 & h_{tj} & \mathbf{0}_{2\times 3} \\ \mathbf{0}_{3\times 2} & \mathbf{K}^{-1} \end{bmatrix} \quad \text{with} \quad \mathbf{K} = \begin{bmatrix} \frac{1}{h_{uk}} & 0 & -\frac{h_u}{h_{uk}} \\ 0 & \frac{1}{h_{vl}} & -\frac{h_v}{h_{vl}} \\ 0 & 0 & 1 \end{bmatrix} \tag{3}$$

where $\mathbf{0}_{n\times m}$ is the $n \times m$ null matrix, $[h_{si}, h_{tj}]^T$ corresponds to the baseline between consecutive cameras, and $\mathbf{K}$ corresponds to the intrinsics matrix that represents the cameras in the camera array defined using the LFIM (2) entries.

The LFIM introduced by Dansereau et $al.$ [4] describes a virtual plenoptic camera whose microlenses define a rectangular tiling (Fig. 1c) instead of the actual hexagonal tiling of a plenoptic camera (Fig. 1b). The rectangular tiling is a result of a decoding process [4] that corrects the misalignment between the image sensor and the microlens array, and removes the hexagonal sampling by interpolating the missing microlenses information from the pixels of the neighbouring microlenses [5].

4   Calibration on a Range of Zoom and Focus Levels

The metadata parameters (meta-parameters), provided by the camera manufacturer with the images acquired, are retrieved from the camera hardware. Here, we focus on the information that refers to the image sensor, main lens and microlens array. More specifically, meta-parameters that change with the zoom and focus settings of the camera, $i.e.$ the main lens world focal plane [12].

4.1   Camera Metadata Parameters

In [12], the influence of two meta-parameters in the definition of the main lens world focal plane was analyzed. Monteiro et $al.$ [12] identified that the world focal plane is mainly determined by a combination of the zoom and focus steps (Fig. 4b). Nonetheless, there are more parameters on the metadata of the images acquired that can determine the main lens world focal plane and that were not analyzed in [12]. For example, the main lens focal length that can be associated with changes on the zoom level or the infinity lambda that can be associated with the focus settings of the microlenses. Namely, the infinity lambda corresponds to the distance in front of the microlens array that is in focus at infinity. However,

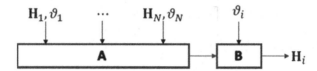

Fig. 3. Representation of a SPC based on meta-parameters provided in the images metadata. In step A, the affine functions $a(f)$, $b(f)$, $c(f)$, $d(f)$, $e(\lambda_\infty)$, and $g(\lambda_\infty)$ are estimated using several calibration datasets with different zoom and focus settings. These datasets are used to relate the entries of the LFIM $\mathbf{H}_{(\cdot)}$ (Sect. 3) and the meta-parameters $\vartheta_{(\cdot)}$ (Sect. 4). In step B, the LFIM $\mathbf{H}_i$ is estimated for an arbitrary zoom and focus settings using only the meta-parameters ϑ_i of a given image and without acquiring a calibration dataset for that specific zoom and focus settings (Sect. 5).

the microlenses optical settings are fixed. The optical settings are changed by modifying the main lens or the complex of lenses that compose the main lens. Thus, the infinity lambda describes the combined optical setup of the microlenses and main lens. In fact, representing the focal length, infinity lambda and target object depth (Fig. 4c), one finds a similar behavior to the one depicted in Fig. 4b. This shows that the world focal plane can also be defined by a combination of the focal length and the infinity lambda parameters.

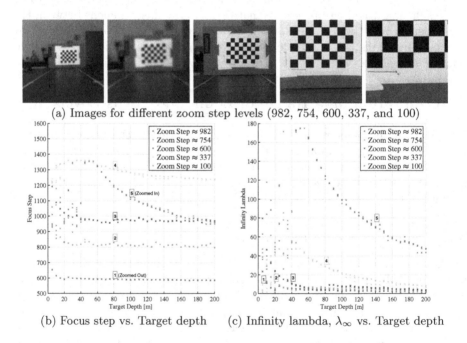

(a) Images for different zoom step levels (982, 754, 600, 337, and 100)

(b) Focus step vs. Target depth   (c) Infinity lambda, λ_∞ vs. Target depth

Fig. 4. Meta-parameters vs. Target depth. **(a)** represents the target object at depth 1.5 m for the different zoom steps. **(b)** represents the focus step with the depth of a target object for a selection of zoom steps. **(c)** represents the infinity lambda with the depth of a target object for a selection of zoom steps (or equivalently, focal lengths).

In order to identify and analyze the camera parameters depending on zoom and focus settings, we follow the same experimental approach defined in [12] and computed the Pearson correlation coefficient among the different meta-parameters [6]. In this experimental analysis, one identifies five parameters that vary with the main lens world focal plane: zoom step (zoom-stepper motor position), focus step (focus-stepper motor position), focal length, infinity lambda, and f-number. The first two parameters represent, up-to an affine transformation, optical parameters information. Namely, the zoom step is related with the focal length of the main lens (Fig. 5a) (correlation of 93.16%), and the focus step for a fixed zoom is related with the infinity lambda parameter (Fig. 5c) (correlation of 99.54%). On the other hand, the f-number is not used in the definition of the intrinsic parameters of a camera and it is normally described as the ratio f/D where f is the focal length and D is the diameter of the entrance pupil. This reduces the relevant metadata parameters to two, the focal length and the infinity lambda.

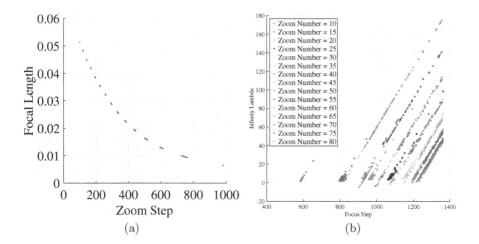

(a)   (b)

Fig. 5. Relationships among camera parameters provided on images metadata. The camera parameters were obtained experimentally by fixing the zoom number and autofocusing the camera to a target object placed at different depths. The zoom step **(a)** is related with the focal length of the main lens. **(b)** The focus step is related with the infinity lambda parameter. The zoom number corresponds to the number that appears on the interface of the camera.

4.2   Metadata Parameters Vs. LFIM

The LFIM depends on the optical settings of the camera. Let us now evaluate how the focal length and infinity lambda are related with the parameters of the LFIM described in Sect. 3. The derivation of Dansereau *et al.* [4] indicates how the LFIM parameters change with the focal length included in the images metadata. However, the assumption of microlenses as pinholes do not allow to introduce the concept of focus at infinity as a parameter of the LFIM. Thus, one wants to provide a relationship between the LFIM parameters and the camera parameters provided on the images metadata.

In order to evaluate these relationships, one needs multiple calibration datasets acquired under different zoom and focus settings. The datasets [12] were collected using a 1st generation Lytro camera and are summarized on Table 1. For establishing the relationships, we use 10 poses randomly selected from the acquired calibration pattern poses to estimate the LFIM [4] and repeated this procedure 15 times to get the mean and standard deviation values. Representing the entries of the LFIM and computing their Pearson correlation coefficients [6] against the focal length and infinity lambda, we found that the entries h_{si} and h_{tj}, which are related to the baseline, exhibit an affine relationship with the focal length (Fig. 6a–b) with a correlation coefficient of 99.97% and 99.98%, respectively. The entries h_{uk} and h_{vl}, which are related with the scale factors, exhibit a nonlinear relationship with the focal length (Fig. 6c–d) with a correlation coefficient of 84.94% and 84.75%, respectively. Furthermore, the remaining entries do not exhibit a correlation with any of the metadata parameters provided.

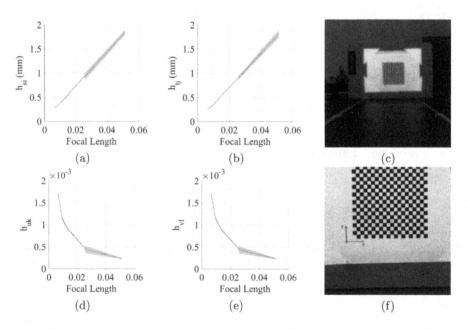

Fig. 6. Relationships of the LFIM entries with the focal length. The entries related with the baseline **(a)**–**(b)**, and with the scale factor **(d)**–**(e)** are represented against the focal length. The target object is depicted at 1 m with different focal lengths (0.0064 **(c)** and 0.0256 **(f)**).

Table 1. Information of the datasets [12] acquired under different zoom and focus settings. The meta-parameters are identified with the symbol *.

Dataset	Zoom step*	Focus step*	Focal length*	Infinity lambda*	Focus depth (m)	Calibration depth range (m)	Calibration poses
A	982	654	0.0064	23.5142	0.05	0.05–0.25	30
B	754	941	0.0094	47.5966	0.05	0.05–0.35	30
D	600	985	0.0130	8.7502	0.50	0.30–0.70	36
E	335	1361	0.0258	47.2068	0.50	0.30–0.80	36
F	337	1253	0.0256	12.8458	1.50	1.00–1.70	48
G	100	1019	0.0513	65.9678	1.50	1.00–1.80	51

If we consider the entries on the intrinsics matrix $\mathbf{K}$ (3), $1/h_{uk}$ and $1/h_{vl}$ exhibit an affine relationship with the focal length (Fig. 7a–b) with a correlation coefficient of 99.82% and 99.81%, respectively. On the other hand, the ratios h_{ui}/h_{uk} and h_{vj}/h_{vl} have an affine relationship with the infinity lambda (Fig. 7c–d) with a correlation coefficient of 99.55% and 99.83%, respectively. The principal point $[h_u/h_{uk}, h_v/h_{vl}]^T$ continue not having any relationship with the metadata parameters. The transformation to a pinhole like representation allows to simplify the relationships with the parameters provided by the manufacturer on the metadata of the images acquired.

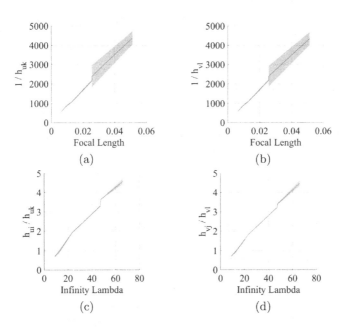

Fig. 7. Intrinsics matrix entries vs. focal length and infinity lambda. The entries related with the scale factor are represented against the focal length (a)–(b). The entries h_{ui}/h_{uk} and h_{vj}/h_{vl} are represented against the infinity lambda (c)–(d).

In summary, denoting f as the focal length (see sample values in Table 1 column 4), λ_∞ as the infinity lambda (sample values shown in Table 1 column 5), and $[c_u, c_v]^T$ as the principal point, one has

$$\mathbf{H} = \begin{bmatrix} a(f) & 0 & 0 & 0 & 0 \\ 0 & b(f) & 0 & 0 & 0 \\ \frac{e(\lambda_\infty)}{c(f)} & 0 & \frac{1}{c(f)} & 0 & \frac{c_u}{c(f)} \\ 0 & \frac{g(\lambda_\infty)}{d(f)} & 0 & \frac{1}{d(f)} & \frac{c_v}{d(f)} \\ 0 & 0 & 0 & 0 & 1 \end{bmatrix} \tag{4}$$

where $a(f)$, $b(f)$, $c(f)$, $d(f)$, $e(\lambda_\infty)$, and $g(\lambda_\infty)$ are the affine mappings identified earlier. In the next section, we detail the procedure followed to estimate the affine mappings and show numerical results for the datasets [12].

5   Experimental Results

In this section, we use the relationships established between the LFIM entries and the metadata parameters (Sect. 4) to obtain a representation for the parameters used to describe the camera for a specific zoom and focus settings.

The relationships $a(f)$, $b(f)$, $c(f)$, $d(f)$, $e(\lambda_\infty)$, and $g(\lambda_\infty)$, in Eq. (4), are estimated using the datasets in Table 1 except Dataset B. As in Sect. 4, one considered for each dataset 10 poses randomly selected from the acquired calibration pattern poses to estimate the camera model parameters [4] and repeated this procedure 15 times to get the mean values. The parameters of the affine mappings obtained using the mean values of the LFIM are summarized on Table 2.

Table 2. Line parameters estimated for the relationships between the LFIM entries and the focal length or the infinity lambda identified in (4).

Line parameters	$a(f)$	$b(f)$	$c(f)$	$d(f)$	$e(\lambda_\infty)$	$g(\lambda_\infty)$
Slope	35.1812	34.9393	85846.9190	84853.2935	0.0668	0.0655
y-Intercept	0.0281	0.0157	28.3403	48.7406	0.1793	0.1580

The Dataset B is not included in the previous analysis in order to be used to evaluate the accuracy of the camera representation (4) using the focal length and the infinity lambda meta-parameters. The LFIM entries are obtained by applying the affine mappings identified in Table 2. These entries are compared with the mean values obtained by repeating 15 times the calibration procedure [4] using 10 randomly selected poses of Dataset B and are summarized in Table 3. The principal point $[c_u, c_v]^T$ is assumed to be the center of the viewpoint image since no relationship was found with the metadata parameters. Table 3 shows that the entries obtained from the calibration are similar to the ones obtained

Table 3. LFIM entries estimated from focal length and infinity lambda using line parameters in Table 2 and from calibration procedure [4] for Dataset B.

Source	h_{si} (mm)	h_{tj} (mm)	$1/h_{uk}$	$1/h_{vl}$	h_{ui}/h_{uk}	h_{vj}/h_{vl}
From calibration	0.3702	0.3281	858.6118	859.7984	3.6389	3.4056
From metadata	0.3606	0.3459	839.5937	850.6042	3.3567	3.2778
Error (%)	2.6	5.4	2.2	1.1	7.8	3.8

from the metadata. Namely, the maximum deviation is 7.8% and occurs for the ratio h_{ui}/h_{uk}.

Additionally, one considered a set of 10 randomly selected images to evaluate the re-projection, ray re-projection [4], and reconstruction errors using the LFIM obtained from applying the calibration procedure [4] and from the metadata provided on the images acquired using the representation (4). The errors are summarized in Table 4. This table allows to have a more practical view of the difference between the two approaches considered. The errors presented are significant but is important to note that the extrinsic parameters are not tuned for the LFIM. The re-projection and ray re-projection errors are similar, being greater for the LFIM obtained from the metadata by 0.34 pixels and 0.14 mm, respectively. On the other hand, the reconstruction error for the metadata based estimation is significantly greater than the one obtained from calibration [4] but still lower than 65 mm. However, note that the LFIM representation using the focal length and the infinity lambda is based on a statistical analysis between the metadata parameters provided by the camera manufacturer and the parameters estimated from a calibration procedure that are affected by noise.

Table 4. Calibration errors associated with the estimation of the LFIM **H** from metadata and from the calibration procedure [4].

Source	Re-projection error (pixels)	Ray re-projection error (mm)	Reconstruction error (mm)
From calibration	5.7718	1.6172	10.0880
From metadata	6.1162	1.7617	61.3519

6 Conclusions

The different zoom and focus settings of the camera change the LFIM **H** used to describe the camera, so we proposed a representation based on the metadata parameters provided on the images acquired. We found that the main lens world focal plane can be determined by the focal length and the infinity lambda parameters. This allows to estimate the LFIM entries without requiring the acquisition of a calibration dataset for a specific zoom and focus settings.

Funding. This work was supported by the Portuguese Foundation for Science and Technology (FCT) projects [UID/EEA/50009/2019] and [PD/BD/105778/ 2014], the RBCog-Lab [PINFRA/22084/2016], and the E.U. Portugal 2020 / Project ELEVAR / I&D co-promotion 17924.

References

1. Adelson, E.H., Bergen, J.R.: The plenoptic function and the elements of early vision. In: Computational Models of Visual Processing, pp. 3–20. Vision and Modeling Group, Media Laboratory, Massachusetts Institute of Technology (1991)
2. Birklbauer, C., Bimber, O.: Panorama light-field imaging. Comput. Graph. Forum **33**(2), 43–52 (2014)
3. Bok, Y., Jeon, H.G., Kweon, I.S.: Geometric calibration of micro-lens-based light field cameras using line features. IEEE Transact. Pattern Anal. Mach. Intell. **39**(2), 287–300 (2017)
4. Dansereau, D.G., Pizarro, O., Williams, S.B.: Decoding, calibration and rectification for Lenselet-based plenoptic cameras. In: Proceedings of the IEEE Conference on Computer Vision and Pattern Recognition, pp. 1027–1034 (2013)
5. David, P., Le Pendu, M., Guillemot, C.: White Lenslet image guided demosaicing for plenoptic cameras. In: 19th International Workshop on Multimedia Signal Processing (MMSP), pp. 1–6. IEEE (2017)
6. Fisher, R.A.: Statistical methods for research workers. In: Kotz, S., Johnson, N.L. (eds.) Breakthroughs in Statistics. SSS, pp. 66–70. Springer, Heidelberg (1992). https://doi.org/10.1007/978-1-4612-4380-9_6
7. Georgiev, T., Zheng, K.C., Curless, B., Salesin, D., Nayar, S., Intwala, C.: Spatio-angular resolution tradeoffs in integral photography. In: Rendering Techniques pp. 263–272 (2006)
8. Gortler, S.J., Grzeszczuk, R., Szeliski, R., Cohen, M.F.: The Lumigraph. In: Proceedings of the International Conference on Computer Graphics and Interactive Techniques (SIGGRAPH), vol. 96, pp. 43–54. ACM (1996)
9. Levoy, M., Hanrahan, P.: Light field rendering. In: Proceedings of the International Conference on Computer Graphics and Interactive Techniques (SIGGRAPH), vol. 96, pp. 31–42. ACM (1996)
10. Lumsdaine, A., Georgiev, T.: The focused plenoptic camera. In: Proceedings of the International Conference on Computational Photography (ICCP), pp. 1–8. IEEE (2009)
11. Marto, S.G., Monteiro, N.B., Barreto, J.P., Gaspar, J.A.: Structure from plenoptic imaging. In: Proceedings of the Joint IEEE International Conference on Development and Learning and Epigenetic Robotics (ICDL-EpiRob), pp. 338–343. IEEE (2017)
12. Monteiro, N.B., Marto, S., Barreto, J.P., Gaspar, J.: Depth range accuracy for plenoptic cameras. Comput. Vis. Image Underst. **168**, 104–117 (2018)
13. Ng, R.: Digital light field photography. Ph.D. thesis, Stanford University (2006)

14. Perwass, C., Wietzke, L.: Single lens 3D-camera with extended depth-of-field. In: Proceedings of SPIE, Human Vision and Electronic Imaging XVII, vol. 8291, p. 829108. International Society for Optics and Photonics (2012)
15. Wilburn, B., et al.: High performance imaging using large camera arrays. In: Transactions on Graphics (TOG), vol. 24, pp. 765–776. ACM (2005)
16. Zhang, Q., Zhang, C., Ling, J., Wang, Q., Yu, J.: A generic multi-projection-center model and calibration method for light field cameras. IEEE Trans. pattern Anal. Mach. Intell. (2018). https://doi.org/10.1109/TPAMI.2018.2864617

Going Back to Basics on Volumetric Segmentation of the Lungs in CT: A Fully Image Processing Based Technique

Ana Catarina Oliveira[1], Inês Domingues[2(✉)], Hugo Duarte[2,3], João Santos[2,4], and Pedro H. Abreu[5]

[1] University of Coimbra, Coimbra, Portugal
anaf.oliveira95@gmail.com
[2] Medical Physics, Radiobiology and Radiation Protection Group,
Portuguese Institute of Oncology of Porto (IPO-Porto) Research Center,
Porto, Portugal
ines.domingues@isec.pt
[3] Nuclear Medicine Department, IPO-Porto, Porto, Portugal
hugo.duarte@ipoporto.min-saude.pt
[4] Medical Physics Department,
IPO-Porto Instituto de Ciências Biomédicas Abel Salazar,
Universidade do Porto, Porto, Portugal
joao.santos@ipoporto.min-saude.pt
[5] CISUC, Department of Informatics Engineering, University of Coimbra,
Coimbra, Portugal
pha@dei.uc.pt

Abstract. Radiotherapy planning is a crucial task in cancer patients' management. This task is, however, very time consuming and prone to a high intra and inter subject variance and human errors.

In this way, the present line of work aims at developing a tool to help the specialists in this task. The developed tool will consider the delimitation of anatomical regions of interest, since it is crucial to identify the organs at risk and minimize the exposure of these organs to the radiation.

This paper, in particular, presents a lung segmentation algorithm, based on image processing techniques, such as intensity projection and region growing, for Computed Tomography volumes. Our pipeline consists in first separating two halves of the volume to isolate each lung. Then, three techniques for seed placement are developed. Finally, a traditional region growing algorithm has been changed in order to automatically derive the value of the threshold parameter.

The results obtained for the three different techniques for seed placement were, respectively, 74%, 74% and 92% of DICE with the Iterative Region Growing algorithm.

This article is a result of the project NORTE-01-0145-FEDER-000027, supported by Norte Portugal Regional Operational Programme (NORTE 2020), under the PORTUGAL 2020 Partnership Agreement, through the European Regional Development Fund (ERDF).

A. Morales et al. (Eds.): IbPRIA 2019, LNCS 11868, pp. 322–334, 2019.
https://doi.org/10.1007/978-3-030-31321-0_28

Although the presented results have as use case the Hodgkin Lymphoma, we believe that the developed method is generalizable to any other pathology.

Keywords: Lung segmentation · Computed Tomography (CT) · 3D

1   Introduction

In 2019, 1,762,450 new cancer cases and 606,880 cancer deaths are projected to occur in the United States. For Hodgkin Lymphoma alone, 8,110 people are predicted to be diagnosed in 2019 and 1,000 people are likely to die [17].

Radiation therapy has a dominant role in cancer treatment and has always been a major part of the effort to cure cancer patients. The main goal of radiotherapy is to deliver a prescribed dose to the target volume, while sparing normal tissue [1]. Since radiotherapy is a personalised and localised treatment, the definition of tumour and target volumes is vital to its successful execution [3]. Contouring these regions is, however, a time consuming part of radiotherapy treatment planning [9] since in current clinical practice, this important task is typically performed visually on a slice-by-slice basis with very limited support of automated segmentation tools.

Computed Tomography (CT) is normally used as the basis for radiotherapy for two main reasons: (1), it can be used to improve the accuracy of dosimetry calculations, since it contains density information, allowing to calculate treatment beam, and (2) can be used to locate the patient with respect to the treatment machine, being more reliable in representing the shape and position compared with other image modalities [15]. Moreover, unlike other procedures, the patient can be scanned in the treatment position, which is an advantage [15].

We have thus developed an image processing pipeline to perform 3D segmentation of the lungs using CT information. The main contributions of the present work include:

– a new image processing method to identify volumes of interest for each half of the body containing the right or the left lung (Sect. 3.1);
– three new techniques to place a seed inside the lung (Sect. 3.2);
– a new, iterative, 3D region growing algorithm that automatically determines the threshold (Sect. 3.3);
– extensive evaluation of the results on a database with 132 lungs (Sect. 5).

The methods presented are simple, fast and do not need a training phase. Also, the parameter setting is made based on expert knowledge of the problem at hand. The three Seed Placement methods, combined with the Iterative Region Growing algorithm achieved, respectively, 74%, 74% and 92% of DICE and 72%, 72% and 90% of True Positive Rate.

This document is organised as follows. Section 2 presents the state of the art on lung segmentation in CT. Section 3 describes the here proposed method in all its components: separation of right and left body volumes in Sect. 3.1; three methods for seed placement in Sect. 3.2; and the new, iterative, region growing technique in Sect. 3.3. The experimental design is outlined in Sect. 4, while results are given in Sect. 5. The document finishes in Sect. 6 with some conclusions and directions for future work.

2   Related Work

A search was performed for works on lung segmentation in CT volumes published in the past 5 years. Most of the works [2, 7, 11, 20] use classifiers in their pipelines. This has the disadvantage of the need of a train phase and consequently, a (large) number of cases for the algorithms to learn with. Some other studies [6, 13, 21] make use of active contour techniques. These type of models typically need an arbitrary parameterization of the curves, thus, losing the opportunity to effectively use information present in the geometry of objects [4]. Other proposals include complex techniques such as Markov-Gibbs random field [18]. We believe that, with the increased use of digital imaging, and with its inherent higher quality, simpler approaches could be more adequate [5]. In the papers [8, 12], segmentation is performed in 2D and 3D connectivity is performed afterwards. This may lead to "jagged" and inconsistent final results.

Perhaps the most similar work to the one here proposed is the one presented in [14], where 3D Region Growing is also used. Their work, however, uses "of the shelf tools" belonging to ITK. Moreover, the experimental results are obtained on a database of only 30 full Chest CT exams.

As can be seen, several different algorithms have already been proposed. They are characterised by their complexity, high running time or segmentation in 2D, and need of large training database. This complexity is, in some cases, justified by the application on diseased lungs. No application on radiotherapy planning was found. Here, a new, tridimensional, simple method, based solely on image processing, is proposed. Not recurring to classification techniques, makes our proposal simpler, faster and there is no need of a large training set, requiring only some basic anatomic knowledge.

3   Lung Segmentation

The proposed algorithm is composed of three main blocks, as shown in Fig. 1. In "Laterality separation", a copy is made of the initial volume, with each new volume containing only one of the lungs. In "Seed definition", the initialisation of the following segmentation method is automatically determined. Three different methodologies are presented for this step. Lastly, in "Segmentation", the lung volumes are identified. A new, iterative, region-growing-based technique is proposed for this step.

Fig. 1. Lung segmentation pipeline.

3.1 Separation Between Right and Left Lungs

Gray values are first transformed to Hounsfield Units (HU), a measure of radiodensity. Rescale Slope and Intercept, needed for this transformation, were retrieved from the DICOM header. The idea is then to threshold the volume in a way that only the lungs are present. Literature, however, is slightly discordant on the HU values of the lung. For instance, in [8], the interval $[-700, -400]$ is given, in [14], the interval $[-1000, -500]$ is used, while in [21], the interval $[-1000, -400]$ is mentioned. For this part of the work, the interval $[-800, -500]$ showed to produce good results, given that the method is quite robust to this selection. In this way, a mask M_{HU} is created with zeros except in the voxels for which its values belongs to the interval HU, there being one. A sum projection of this mask is then made, creating a "cumulative transverse plane", as shown in the left part of Fig. 2. A sum projection of this plane is then done, creating a line profile, as shown in the right part of Fig. 2. A search for a local minimum gives the output of this part of the algorithm. Two volumes can now be created, by zeroing all the values to the left (or the right) of the found local minimum.

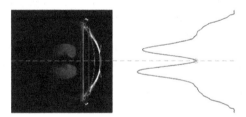

Fig. 2. Separation between right and left lungs. Left, cumulative transverse plane with separation line superimposed in dashed green; Right, cumulative profile, with local minimum as a green star and the separation line in dashed green. (Color figure online)

3.2 Placement of the Seed

Three different techniques are here proposed for the choice of a seed. These are described next.

Method 1. This technique uses anatomic and image acquisition knowledge and starts by placing the seed in the position $[\frac{1}{3}, \frac{2}{3}]$ for the right lung and $[\frac{2}{3}, \frac{2}{3}]$ for the left lung, of the central coronal plane. A vertical search is then performed until an intensity corresponding to the theoretical value of the HU of the lungs, that is, a value in the range $[-800, -500]$ is found. When the seed is not found,

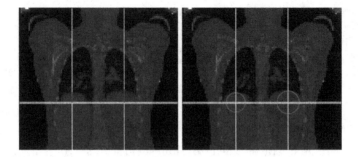

Fig. 3. Seed location by Methods 1 (left) and 2 (right). The first method consists on the search of an intensity value in the interval $[-800 - tolerance, -500 + tolerance]$ on the vertical axis, whereas the second method searches for the voxel in the interval that corresponds to the minimal distance to the initial seed.

it is added a tolerance to this interval, iteratively, until the search detects a value in the range $[-800 - tolerance, -500 + tolerance]$. An illustration of this method is given on the left side of Fig. 3.

Method 2. In this method, the seed initialisation is made as in the previous method. The closest point to each seed in the mask M_{HU} is then selected as the new seed. An illustration of this method is given on the right side of Fig. 3.

Method 3. The third method follows from the reasoning presented in Sect. 3.1. The z position of each seed (coronal plane) is given as the local maximum closest to the position chosen to separate the lungs (see right part of Fig. 2). Having this coordinate, the sagittal plane of M_{HU} can be retrieved for each lung, top-left plot of Fig. 4. Local maximum of the sum projection is chosen as the x coordinate of

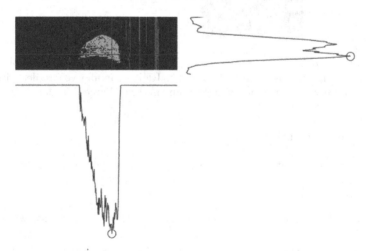

Fig. 4. Seed location by Method 3 (illustration for one of the lungs, only). See text for a more detailed explanation. (Color figure online)

the seed, as shown in the right plot of Fig. 4. For the y coordinates, the biggest connected component of the sagittal plane of M_{HU} is first selected (yellow region in top-left plot of Fig. 4), the sum profile computed, and the local maximum position is determined, bottom plot of Fig. 4.

3.3 Iterative Region Growing

Traditional region growing algorithm (Algorithm 1) starts with a given seed and adds a neighbour to the segmented region if the difference between the value of the neighbour pixel and the average value of the pixels already in the region does not exceed a threshold. This threshold needs to be manually set, case by case, as it often depends not only on the problem, but also on the image in question.

Algorithm 1. Traditional Region Growing Algorithm.

Inputs:
Seed vector, $s = [s_x, s_y, s_z]$
Volume to be segmented, CT
Tolerance threshold, Th
Output:
Volumetric mask, $Mask$, with the same size as CT

Initialise $Mask$ as a volume with the same size as CT, filled with *zeros*
Initialise $Checked$ as a volume with the same size as $Mask$, filled as *false*
Initialise $NeedsCheck$ as an *empty* stack

Set $Mask$ at s to *one*
Set $Checked$ at s to *true*
Add neighbour coordinates of s to $NeedsCheck$.

while $NeedsCheck$ is not empty **do**
    Pop a point p from $NeedsCheck$
    Set $Checked$ at p to *true*

    Calculate m_s, the average of CT grey values in the points where $Mask = 1$
    Retrieve m_p, the grey value of CT in p

    if $|m_s - m_p| > Th$ **then**
        Set $Mask$ at p to *one*
        Add neighbour coordinates of p to $NeedsCheck$
    end if

end while

We propose to automatically and iteratively update the value of the tolerance threshold, Th, as demonstrated in the Algorithm 2. In this version, we have now three parameters, the Tolerance Threshold Initialisation Th_0, Maximum Area

Threshold Th_H, and Minimum Area Threshold Th_L, which may, at first seem worse than the previous version with one parameter only. We note, however, that Algorithm 1 is extremely sensitive to the value of Th. In the iterative version, Th_H and Th_L can be set based on previous knowledge of the problem, in this case, biomedical knowledge on maximum and minimum lung volumes. Th is first initialised (Th_0), and is then iteratively adjusted accordingly to the volume resulting from the segmentation algorithm. If this volume is superior/inferior to Th_H/Th_L, then Th is be increased/decreased, respectively. In this way, the algorithm is very robust to the initial value of Th_0.

Algorithm 2. Iterative Region Growing Algorithm.

Inputs:
Seed vector, $s = [s_x, s_y, s_z]$
Volume to be segmented, CT
Tolerance threshold initialisation, Th_0
Maximum area threshold, Th_H
Minimum area threshold, Th_L
Output:
Volumetric mask, $Mask$, with the same size as CT

    Th initialisation, $Th = Th_0$
    while Algorithm 1 is iterating **do**
        if $Count(Mask == 1) > Th_H$ **then**
            Stop Algorithm 1
            Update Th, $Th = Th - 10$
            Restart Algorithm 1
        end if
        if $Count(Mask == 1) < Th_L$ **then**
            Update Th, $Th = Th + 10$
            Restart Algorithm 1
        end if
    end while

Before the application of the Region Growing algorithm, voxels have been resampled to an isomorphic resolution of [5, 5, 5] millimetres to remove variance in scanner resolution. Slice thickness and pixel spacing information present in the DICOM headers is used for this transformation. The values of the parameters were set as follows: $Th_0 = 225$ HU, $Th_L = 3000$ voxels (375 mL), and $Th_H = 40000$ voxels (5000 mL).

4 Experimental Methodology

Provided by Institute of Oncology of Porto (IPO), the private dataset of patients with Hodgkin Lymphoma used in this research work includes CT volumes used for radiotherapy planning, acquired after the frontline chemotherapy treatment and the corresponding ground truth contours delimited by experts. CT volumes were acquired with a pixel spacing of 1.0, 1.1, 1.2 or 1.3 mm and a difference of patient position between adjacent slices of 2.0, 2.5 or 5.0 mm, all in the DICOM format [16]. Information has been collected for a total of 69 patients (both adults and infants). However, the ground truth is only available for 132 lungs.

Four metrics were chosen, in the present work, to evaluate the results, Dice coefficient, Jacquard index, True Positive rate, and Volumetric Similarity. While the first three are overlap based, the last one is volume based [19]. These metrics were chosen due to their complementarity. Dice coefficient and Jacquard index are suitable when in the presence of outliers; True Positive rate for when recall is important; and Volumetric Similarity is appropriate both in scenarios with outliers and when the volume is important [19].

All of these metrics can be derived from the four basic cardinalities of the confusion matrix:

- TP: Voxels correctly considered to belong to the lung
- FP: Voxels incorrectly considered to belong to the lung
- TN: Voxels correctly considered not to belong to the lung
- FN: Voxels incorrectly considered not to belong to the lung

The Dice coefficient (DICE), also called the overlap index, is the most frequently used metric. It can be defined as:

$$DICE = \frac{2TP}{2TP + FP + FN} \tag{1}$$

The Jaccard index (JAC) is defined as the intersection divided by the union:

$$JAC = \frac{TP}{TP + FP + FN} \tag{2}$$

True Positive rate (TPr), also called Sensitivity or Recall, measures the portion of positive voxels in the ground truth that are also identified as positive by the segmentation being evaluated:

$$TPr = \frac{TP}{TP + FN} \tag{3}$$

This metric is sensible to segments size, and it penalises errors in small segments more than in large segments. Volumetric similarity (VS) is a measure that considers the volumes of the segments to indicate similarity:

$$VS = 1 - \frac{|FN - FP|}{2TP + +FP + FN} \tag{4}$$

5   Results and Discussion

Seed location performance was evaluated by checking if the automatic seed falls within the ground truth mask and by its distance to the centroid of the ground

Table 1. Seed placement performance. Best results in bold and signalled with "*" if statistically significant, according to the paired-sample t-test at the 5% level.

	Method 1	Method 2	Method 3
Percentage of valid seeds	78.03	78.03	**100.00***
Distance to centroid	21.47	18.61	**6.41***

truth masks[1]. It can be seen, from Table 1, that for method 3 all of the seeds fall inside the lung region. Moreover, they are significantly closer to the centroid of the ground truth mask, when compared with the seeds retrieved by the other techniques.

Performance of the segmentation is shown in Table 2 for each automatic seed finding method and also considering as seed the centroid of the ground truth mask. As a baseline segmentation technique, a HU threshold was considered, by retrieving the biggest connected component of M_{HU} (yellow region on Fig. 4).

Table 2. Segmentation performance. Best results in bold and signalled with "*" if statistically significant, according to the paired-sample t-test at the 5% level.

Segmentation	Seed	DICE	JAC	TPr	VS
HU threshold	Method 1	0.661	0.576	0.605	0.703
HU threshold	Method 2	0.656	0.571	0.600	0.691
HU threshold	Method 3	0.812	0.707	0.740	0.858
HU threshold	GT centroid	0.812	0.707	0.740	0.861
Region Growing	Method 1	0.736	0.703	0.716	0.779
Region Growing	Method 2	0.720	0.687	0.700	0.752
Region Growing	Method 3	0.894	0.853	0.871	0.926
Region Growing	GT centroid	0.872	0.833	0.850	0.916
Iterative Region Growing	Method 1	0.736	0.703	0.716	0.836
Iterative Region Growing	Method 2	0.741	0.707	0.721	0.841
Iterative Region Growing	Method 3	**0.923**	**0.882***	**0.900***	**0.956***
Iterative Region Growing	GT centroid	0.886	0.846	0.863	0.930

It is clear, from Table 2, that the proposed Iterative Region Growing outperforms the Standard Region Growing. Moreover, the seeds returned by method 3 originate the best segmentation results, as to be expected from the results in Table 1. In fact, results for Iterative Region Growing using the seeds returned by Method 3 are statistically significant, according to the paired-sample t-test

---

[1] We would like to note, however, that centroids do not necessarily fall inside the ground truth mask.

Table 3. Segmentation performance, considering valid seeds only. Best results are presented in bold.

Segmentation	Seed	DICE	JAC	TPr	VS
HU threshold	Method 1	0.639	0.559	0.587	0.671
HU threshold	Method 2	0.634	0.554	0.580	0.663
HU threshold	Method 3	0.812	0.707	0.740	0.858
HU threshold	GT centroid	0.786	0.685	0.717	0.834
Region Growing	Method 1	**0.925**	**0.883**	**0.900**	0.952
Region Growing	Method 2	0.895	0.854	0.870	0.922
Region Growing	Method 3	0.894	0.853	0.871	0.926
Region Growing	GT centroid	0.892	0.852	0.870	0.931
Iterative Region Growing	Method 1	**0.925**	**0.883**	**0.900**	0.952
Iterative Region Growing	Method 2	0.914	0.872	0.888	0.945
Iterative Region Growing	Method 3	0.923	0.882	**0.900**	**0.956**
Iterative Region Growing	GT centroid	0.900	0.859	0.877	0.939

at the 5% level, than all the other results, except the ones for Iterative Region Growing using as seed the centroid of the ground truth mask and only for DICE.

We were also interested in studying the performance for solely the cases where the seeds were placed inside the ground truth lungs. From Table 3, it can be observed that when it does provide a valid seed, method 1 generates seeds that lead to better segmentations according to DICE, JAC and TPr metrics. When looking at VS, method 3 is still the best. This leads us to the believe that the design of an algorithm that incorporates the ideas behind method 1 and method 3 is a possible future direction.When looking at the segmentation algorithms, differences between iterative and non-iterative versions are not significative according to the two-sample t-test at the 5% level, but they are significantly better according to the same test, when compared with HU threshold.

When the seed is placed inside the lung in a region with HU on the interval $[-800, -500]$, the segmentation is robust to the seed location. For example, in the case of one of the patients of our dataset, for method 1 the seed is located near the frontier of the lung, while with method 3 it is located in the centre, but both have very similar results on the segmentation metrics (see Fig. 5).

A fair comparison with the state of the art results is not possible due to differences in the used databases. We stress, for instance, that our database is composed of images from both adults and infants, increasing the segmentation difficulty. We note, however, that our method, besides having the advantages, when compared with the other state of the art techniques already described in Sect. 2, such as its simplicity, low running time, no need of a training phase, etc, achieves similar performances, even surpassing some of the recently proposed algorithms such as the ones presented in [2,21].

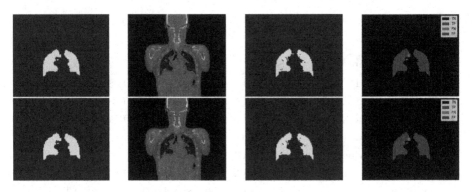

Fig. 5. Example of the results of lung segmentation for method 1 (top) and 3 (bottom). Ground truth on the far left; CT with superimposed ground truth contour (red) and seeds (blue) on the middle left; segmentation results on the middle right; and representation of the four cardinalities of the confusion matrix on the far right. (Color figure online)

6   Conclusions and Future Work

Segmentation is a problem present in several domains [4,5]. Here, a new, volumetric technique for lung segmentation in CT is presented and evaluated. This techniques are simple, fast, and does not need a training phase. Moreover, all the parameter setting is made based on expert knowledge of the problem at hand.

In this way, this completely automated method for lung segmentation may help to reduce the time dispensed by the clinicians when performing a manual analysis of the CT scan, assisting them in making better decisions when selecting the better treatment choice for the patient and/or evaluation of the effectiveness of the received treatment.

In the present work, three intensity-based seed finding methods were tested and an improvement of the typical Region Growing method was proposed. Seed finding methods achieved 84%, 84% and 96% of Volume Similarity for methods 1, 2, and 3 with Iterative Region Growing, respectively. If only the valid seeds are accounted for, the Iterative Region Growing method achieved results of 95%, 94% and 96%, respectively.

A possible improvement to the presented algorithm is to apply morphology to the results in order to close possible holes that might be present in the final segmentation. Another interesting approach would be to use this technique with a database containing lung abnormalities, such as pleural effusions, consolidations, and masses, since the current image segmentation approaches apply well only if the lungs exhibit minimal or no pathological conditions [10].

In the future, we intend to work on the segmentation of other organs, such as the esophagus. To develop iterative tools for better visualisation and manual adjust of the results is another line of interest.

References

1. Astaraki, M., et al.: Evaluation of localized region-based segmentation algorithms for CT-based delineation of organs at risk in radiotherapy. Phys. Imaging Radiat. Oncol. **5**, 52–57 (2018)
2. Birkbeck, N., et al.: Lung segmentation from CT with severe pathologies using anatomical constraints. In: Golland, P., Hata, N., Barillot, C., Hornegger, J., Howe, R. (eds.) MICCAI 2014. LNCS, vol. 8673, pp. 804–811. Springer, Cham (2014). https://doi.org/10.1007/978-3-319-10404-1_100
3. Burnet, N.G., Thomas, S.J., Burton, K.E., Jefferies, S.J.: Defining the tumour and target volumes for radiotherapy. Cancer Imaging **4**(2), 153–161 (2004)
4. Cardoso, J.S., Domingues, I., Oliveira, H.P.: Closed shortest path in the original coordinates with an application to breast cancer. Int. J. Pattern Recogn. Artif. Intell. **29**(01), 1555002 (2015)
5. Domingues, I., et al.: Pectoral muscle detection in mammograms based on the shortest path with endpoints learnt by SVMs. In: 32nd Annual International Conference of the IEEE Engineering in Medicine and Biology Society, pp. 3158–3161 (2010)
6. Filho, P.P.R., Cortez, P.C., da Silva Barros, A.C., Albuquerque, V.H.C., Tavares, J.M.R.S.: Novel and powerful 3D adaptive crisp active contour method applied in the segmentation of CT lung images. Med. Image Anal. **35**, 503–516 (2017)
7. Gill, G., Bauer, C., Beichel, R.R.: A method for avoiding overlap of left and right lungs in shape model guided segmentation of lungs in CT volumes. Med. Phys. **41**(10), 101908 (2014)
8. Lan, S., Liu, X., Wang, L., Cui, C.: A visually guided framework for lung segmentation and visualization in chest CT images. J. Med. Imaging Health Inform. **8**(3), 485–493 (2018)
9. Lustberg, T., et al.: Clinical evaluation of atlas and deep learning based automatic contouring for lung cancer. Radiother. Oncol. **126**(2), 312–317 (2018)
10. Mansoor, A., et al.: Segmentation and image analysis of abnormal lungs at CT: current approaches, challenges, and future trends. Radiogr. : A Rev. Publ. Radiol. Soc. North Am. **35**(4), 1056–1076 (2015)
11. Mansoor, A., et al.: A generic approach to pathological lung segmentation. IEEE Trans. Med. Imaging **33**(12), 2293–2310 (2014)
12. Nalepa, J., Czardybon, M., Walczak, M.: Real-time lung segmentation from whole-body CT scans using adaptive vision studio: a visual programming software suite. In: Kehtarnavaz, N., Carlsohn, M.F. (eds.) Real-Time Image and Video Processing, vol. 10670, p 11. SPIE (2018)
13. Nithila, E.E., Kumar, S.S.: Segmentation of lung from CT using various active contour models. Biomed. Signal Process. Control **47**, 57–62 (2019)
14. Nobrega, R.V.M., Rodrigues, M.B., Filho, P.P.R.: Segmentation and visualization of the lungs in three dimensions using 3D region growing and visualization toolkit in CT examinations of the chest. In: Bamidis, P.D., Konstantinidis, S.T., Rodrigues, P. (eds.) IEEE 30th International Symposium on Computer-Based Medical Systems (CBMS), vol. 2017, pp. 397–402 (2017)
15. Pereira, G.: Deep learning techniques for the evaluation of response to treatment in Hodgkin Lymphoma. Master in biomedical engineering, University of Coimbra (2018)
16. Pereira, G., Domingues, I., Martins, P., Abreu, P.H., Duarte, H., Santos, J.: Registration of CT with PET: a comparison of intensity-based approaches. In: International Workshop on Combinatorial Image Analysis (IWCIA) (2018)

17. Siegel, R.L., Miller, K.D., Jemal, A.: Cancer Statistics, 2019. CA: Cancer J. Clin. **69**(1), 7–34 (2019)
18. Soliman, A., et al.: Accurate lungs segmentation on CT chest images by adaptive appearance-guided shape modeling. IEEE Trans. Med. Imaging **36**(1), 263–276 (2017)
19. Taha, A.A., Hanbury, A.: Metrics for evaluating 3D medical image segmentation: analysis, selection, and tool. BMC Med. Imaging **15**(1), 29 (2015)
20. Xu, M., et al.: Segmentation of lung parenchyma in CT images using CNN trained with the clustering algorithm generated dataset. BioMed. Eng. OnLine **18**(1), 2 (2019)
21. Zhou, H., et al.: A robust approach for automated lung segmentation in thoracic CT. In: IEEE International Conference on Systems, Man, and Cybernetics (SMC), pp. 2267–2272 (2016)

Radiogenomics: Lung Cancer-Related Genes Mutation Status Prediction

Catarina Dias[1,2(✉)], Gil Pinheiro[1], António Cunha[1,3], and Hélder P. Oliveira[1,4]

[1] Instituto de Engenharia de Sistemas e Computadores, Tecnologia e Ciência,
Porto, Portugal
catfdias@gmail.com
[2] Faculdade de Engenharia, Universidade do Porto, Porto, Portugal
[3] Universidade de Trás-os-Montes e Alto Douro, Vila Real, Portugal
[4] Faculdade de Ciências, Universidade do Porto, Porto, Portugal

Abstract. Advances in genomics have driven to the recognition that tumours are populated by different minor subclones of malignant cells that control the way the tumour progresses. However, the spatial and temporal genomic heterogeneity of tumours has been a hurdle in clinical oncology. This is mainly because the standard methodology for genomic analysis is the biopsy, that besides being an invasive technique, it does not capture the entire tumour spatial state in a single exam. Radiographic medical imaging opens new opportunities for genomic analysis by providing full state visualisation of a tumour at a macroscopic level, in a non-invasive way. Having in mind that mutational testing of EGFR and KRAS is a routine in lung cancer treatment, it was studied whether clinical and imaging data are valuable for predicting EGFR and KRAS mutations in a cohort of NSCLC patients. A reliable predictive model was found for EGFR (AUC = 0.96) using both a Multi-layer Perceptron model and a Random Forest model but not for KRAS (AUC = 0.56). A feature importance analysis using Random Forest reported that the presence of emphysema and lung parenchymal features have the highest correlation with EGFR mutation status. This study opens new opportunities for radiogenomics on predicting molecular properties in a more readily available and non-invasive way.

Keywords: Radiogenomics · Mutation status · Predictive models

1 Introduction

Lung cancer is the most common cause of cancer death in the world, responsible for nearly 1.6 million deaths annually [10]. The main contributing factor for the high death rate of lung cancer is the late diagnosis [19]. Once diagnosed, lung cancer is often in an advanced stage, with 15% or less chance of a 5-year survival [15]. At that stage, tumours are already composed by multiple clonal subpopulations of cancer cells and, consequently, the treatment must be shaped based on the individual tumour heterogeneity. Precision medicine is the medical

© Springer Nature Switzerland AG 2019
A. Morales et al. (Eds.): IbPRIA 2019, LNCS 11868, pp. 335–345, 2019.
https://doi.org/10.1007/978-3-030-31321-0_29

field that tailors practices and/or therapies to individual patients by taking into account the individual variability of genes. The traditional method of analysing the tumour is by extracting tumour tissue in a biopsy, which is then characterised using genomic-based approaches. In spite of being a successful approach in clinical oncology, repeated biopsies tend to increase medical complications. Further, tumour characterisation usually demands several biopsies, since the results can vary depending on the part of the tumour that is analysed [23].

In Non-small cell lung cancer (NSCLC), which accounts 85% of all lung cancers [5], mutational testing of selected genes is a standard practice to determine how affected patients will respond to targeted therapy [9]. This includes determining the mutation status of epidermal growth factor receptor (EGFR), a cell receptor that activates growth and survival [24], and Kristen rat sarcoma viral oncogene homolog (KRAS), which activates the same pathway as EGFR when mutated [21]. Patients with mutant EGFR are sensitive to tyrosine kinase inhibitors (TKIs) *gefitinib* and *erlotinib*. Hence, patients with mutated EGFR lung cancer, who receive treatments with targeted TKIs are expected to have a longer progression-free survival in comparison to chemotherapy treatment. However, if *gefitinib* is administered in cases with non-mutated EGFR, the patient will undergo a shorter progression-free survival [18]. KRAS mutation status is also helpful for treatment planning. It has proven to be correlated with response to chemotherapy since patients with mutated KRAS which undergo chemotherapy have revealed inferior responses and shorter survival compared to patients with no KRAS mutation [13,26]. On that basis, identifying patients with mutated EGFR and KRAS is highly important in precision medicine.

As a less invasive technique compared to biopsy, radiographic medical imaging opens new opportunities for tumour characterisation. Images exhibit strong phenotypic differences between tumours, such as tumour size, presence of emphysema and/or fibrosis. Those differences normally fail to be recognised by the naked eye, thus they may have the potential to be valuable predictors of therapeutic benefit. Moreover, a great advantage of medical imaging is its ability to provide a full state visualisation of a tumour at a macroscopic level. Therefore, radiogenomics, the fusion of medical images and genomics, offers attractive opportunities for non-invasive treatment planning.

Given the relevance of the problem, in this paper, we propose predictive models for EGFR and KRAS mutation status, using a set of clinical and radiologist-observed qualitative imaging features, taking advantage of the learning capabilities of machine learning techniques.

The remainder of this paper is as follows. In Sect. 2, we present the related work which has been done so far. In Sect. 3, we present the proposed approach, while in Sect. 4 we detail and discuss the experimental results. We conclude the paper with Sect. 5, summarising its main contributions and findings.

2   Related Work

A thorough search of the relevant literature yielded only one related article which investigated whether EGFR and KRAS mutation status can be predicted using qualitative features obtained from imaging data. Gevaert et al. [11] used

89 qualitative image features of NSCLC patients tumours, annotated by a thoracic oncologist, to create models to predict EGFR and KRAS mutation status. A univariate correlation study was performed between mutation status and the qualitative imaging features and, afterwards, the most correlated features were used in a multivariate analysis using decision trees. Emphysema, airway abnormality, the percentage of ground glass component and the type of tumour margin reached the significance threshold of correlation with EGFR mutation status and they were used to build a decision tree model, which achieved an area under the ROC curve (AUC) of 0.89. With regard to KRAS mutation, no features reached the significance threshold of correlation and, consequently, the models built for KRAS were not considered useful (AUC = 0.55). Furthermore, some studies have investigated the association between EGFR mutation status and quantitative features, rather than qualitative features [7,16,17].

With a view on the advances that have been made, in this study, we propose different experimental methodologies that take advantage of powerful machine learning techniques to create predictive models using a new set of features. By using a different cohort of patients, as well as different features, one can further evaluate the relation between EGFR and KRAS mutation status and radiographic imaging data.

3   Methodology

This study aimed to investigate whether clinical and qualitative imaging features are advantageous mutation status predictors and build predictive models using two algorithms: Random Forest (RF) [3] and Multi-layer Perceptrons (MLP) [22] networks. The data was divided into a training set (80%) and a test set (20%), ensuring that each set maintains an equal proportion of instances of each class. It is important to mention that the set of subjects used for training and testing were kept constant for all the performed experiments. Hyper-parameters were chosen applying grid-search with 5-fold cross-validation to the training data and selecting the set of hyper-parameters of the model with the highest F-measure. The developed code and used data are available on *Github*.[1]

3.1   The Dataset

The study included a subset of 158 NSCLC patients tested for EGFR mutation status and 157 NSCLC patients tested for KRAS mutation status, characterised by qualitative and clinical features. The data was obtained from the open-access NSCLC-Radiogenomics dataset available at the cancer imaging archive (TCIA) database [2,6,12].

The qualitative features were obtained from an analysis of pre-treatment computed tomography (CT) images using a controlled vocabulary. The used terms are commonly used in radiology clinical practice and derive from descriptions in the radiology literature [1]. Definitions of some of the terms used in this

---

[1] https://github.com/catfdias/MutationStatus.git.

description can be found in [14]. The template of semantic terms was developed exclusively for nodules since it is the most prevalent expression of lung cancer. Therefore, other manifestations of lung cancer besides nodules (e.g. central obstructive tumours) are not included in this study.

From the 30 qualitative features available in the NSCLC-Radiogenomics dataset, some were discarded due to a large number of not applicable values (e.g. the fibrosis type field in a patient that has fibrosis absent), thus, a subset of 18 qualitative features was used in this study. The used set includes nodule and parenchymal features, which describe the nodules geometry, location, internal features and other related findings. Additionally, the patient's gender and smoking status were considered due to its significant association with mutation status prevalence, confirmed in recent studies [8,20,25]. From this point forward, gender and smoking status are designated as clinical features and the qualitative features extracted from the images as imaging features. Table 1 shows detailed information regarding the data distribution and the nomenclature used to classify the tumours.

The dataset comprises percentages of 26% and 25% mutated cases for EGFR and KRAS, respectively. Before feeding the data into the model, features were converted to binary vectors following a one-hot encoding strategy. Thereafter, the number of features increased from 20 to 73.

3.2 Random Forest

RF models were implemented for predicting EGFR and KRAS mutation status. As an algorithm based on ensemble learning, RF makes the predictions taking advantage of a group of models, instead of a single model. Random Forest samples both observations and features of training data in order to build independent decision trees which contribute by voting for the ensemble prediction. Bearing in mind that a decision tree is an unstable algorithm, by averaging the results of all decision trees, the variance component of the model will be minimised, which approximates the ensemble to an ideal model.

Due to the ability of RF to recognise the importance of the features for the problem in mind, it was conducted an analysis of the most valuable ones.

3.3 Multi-layer Perceptron

By virtue of the remarkable ability of MLP to extract patterns and detect trends, its performance on predicting the genes mutation status was tested.

The MLP assumes the distribution of classes is similar, which in this case would result in a model biased towards the negative class. However, in this study, the correct classification of both classes is equally important, since the classification of a patient with the wrong mutation status could lead to the administration of a less suitable treatment and, consequently, to shorter progression-free survival. To overcome class imbalance, it was conducted a Synthetic Minority Over-sampling Technique for Nominal and Continuous (SMOTE-NC) data approach, in which new instances are created based on the 5 nearest neighbours of the feature space that belong to the minority class. In comparison to traditional

Table 1. Distribution of the data used for KRAS and EGFR mutation status prediction experiments.

Feature	KRAS (%)	EGFR (%)
Clinical Features		
Gender		
Female	35.0	36.1
Male	65.0	63.9
Smoking Status		
Former	61.1	62.0
Non-smoker	24.2	22.8
Current	14.6	15.2
Imaging Features		
Axial Location		
Central	21.7	20.9
Peripheral	78.3	79.1
Nodule Attenuation		
Solid	56.1	57.6
Partially solid	40.8	39.2
Ground glass	3.2	3.2
Nodule Margins Primary Pattern		
Irregular	17.2	15.2
Lobulated	30.6	32.2
Spiculated	25.5	26.6
Poorly defined	16.6	15.8
Smooth	10.2	10.1
Nodule Margins Secondary Pattern		
Irregular	37.6	39.9
Lobulated	31.2	31.0
Spiculated	3.8	3.2
Poorly defined	16.6	15.2
Smooth	10.8	10.8
Nodule shape		
Complex	58.0	57.0
Round	29.3	29.7
Oval	12.1	12.7
Polygonal	0.6	0.6
Nodule calcification		
Peripheral	6.4	7.0
None	93.6	93.0
Nodule's findings		
Pleural retraction	10.2	10.1
Vascular convergence	20.4	20.3
Septal thickening	18.5	21.5
None	9.6	8.9
Attachment to pleura	7.6	7.6
Entering airway	21.7	20.3
Attachement to vessel	9.6	8.9
Bronchovascular bundle	2.5	2.5
Satellite Nodules in Primary Lesion Lobe (>4mm noncalcified)		
Absent	77.1	75.9
Solid	7.6	8.2
Partially-solid	5.1	5.1
Non-solid	10.2	10.8

Feature	KRAS (%)	EGFR (%)
Imaging Features		
Nodules in Non-Lesion Lobe Same Lung (>4mm noncalcified)		
Absent	82.8	81.6
Solid	5.7	6.3
Partially-solid	1.9	2.5
Non-solid	9.6	9.5
Nodules in Contralateral Lung (>4mm noncalcified)		
Absent	76.4)	75.3
Solid	5.1	5.7
Partially-solid	7.6	7.6
Non-solid	10.8	11.4
Centrilobular Nodules		
Absent	87.3	88.6
Present	12.7	11.4
Emphysema		
absent	47.8	48.7
present	52.2	51.3
Fibrosis		
absent	89.8	88.6
present	10.2	11.4
Nodule Internal Features		
Not applicable	50.3	51.3
Cavitated	8.9	8.2
Internal air bronchogram sign	28.0	27.8
Reticulation	7.0	7.0
Necrosis	3.2	3.2
Nodule cysts	2.5	2.5
Lung Parenchyma Features		
Bronchial wall thickening	33.8	32.9
Tree-in-bud sign	1.9	2.5
Airway ectasia	9.6	9.5
Mosaic oligemia	5.1	5.1
Bronchial prominence	0.6	1.3
Bronchiectasis	3.8	4.4
Normal	45.2	44.3
Primary Emphysema Pattern		
Paraseptal	12.7	12.0
Centrilobular	37.6	38.0
Panacinar	1.9	1.3
Not applicable	47.8	48.7
Secondary Emphysema Pattern		
Paraseptal	17.2	17.1
Centrilobular	8.3	7.6
Panacinar	0.6	0.6
Not applicable	73.9	74.7
Nodule Periphery		
Emphysema	26.1	26.6
Normal	70.1	69.6
Fibrosis	3.8	3.8

over-sampling, SMOTE-NC has the advantage of building a more general decision region of the minority class [4]. After applying SMOTE-NC, the training set contains the same number of mutated and wildtype samples. On the premise of keeping the fed data constant among the two algorithms, SMOTE-NC was applied to all the performed experiments.

4 Results and Discussion

The same experimental set-up was followed regarding EGFR and KRAS mutations. Interestingly, the average results were quite different, in a sense that it was possible to achieve models that reliably predict EGFR mutations but not KRAS mutation presence. From the experiments conducted, the predictive model for KRAS mutation with the best performance had an AUC of 0.56, having the RF as the classifier. Therefore, the following results are exclusively regarding EGFR.

4.1   EGFR Mutation Status Prediction

Three experiments were conducted in order to achieve the greatest set of features to predict EGFR mutation status: using clinical features, imaging features and both imaging and clinical features. Clinical features were only attempted with MLP, since the RF is not an ideal model for a set of 2 features, due to the lack of feature combinations. The range of values used in the grid search for RF and MLP for each hyper-parameter is presented in Tables 2 and 3, respectively.

Table 2. Set of hyper-parameter values used in the grid search for RF.

Hyper-parameter	Values
Maximum depth	3, 5, 10, 15, 20, 25, 30, n (n: number of features)
Maximum features	$\log_2 n$, $\sqrt{n}$, n (n: number of features)
Minimum samples split	2, 3, 5, 7, 9
Minimum samples leaf	1, 3, 5, 7
Number of estimators	30, 50, 70, 100, 300, 500, 700
Bootstrap	True, False

Table 3. Set of hyper-parameter values used in the grid search for MLP.

Hyper-parameter	Values
Alpha	0, 0.00001, 0.0001, 0.001, 0.01
Hidden layers	1, 2, 4, 6, 8, 10, 12
Learning rate	0.0001, 0.001, 0.01, 0.1
Momentum	0.0001, 0.001, 0.01, 0,1, 0.5, 0.9

The hyper-parameters of the models which achieved the highest F-measure in the 5-fold cross-validation for RF and MLP are included in Tables 4 and 5, respectively, for each experiment.

Table 4. Hyper-parameters of the RF model which achieved the highest F-measure in the 5-fold cross-validation.

Experiment	Maximum depth	Maximum features	Minimum samples split	Minimum samples leaf	Number of estimators	Bootstrap
Imaging features	5	$\log_2 n^*$	2	3	300	False
Imaging + Clinical features	15	$\log_2 n^*$	7	1	30	False

*n: number of features

Table 5. Hyper-parameters of the MLP model which achieved the highest F-measure in the 5-fold cross-validation

Experiment	Alpha	Hidden layers	Learning rate	Momentum	Activation	Optimiser
Clinical features	0.0	2	0.01	0.9	ReLU	SGD*
Imaging features	0.0	10	0.0001	0.9	ReLU	SGD*
Imaging + Clinical features	0.0	8	0.001	0.0001	ReLU	SGD*

*Stochastic Gradient Descent

Table 6 shows the results obtained on the test set by the RF and MLP models which provided the best results in the training set, in the three performed experiments.

Table 6. Results obtained in the different experiments.

Experiment	Recall	Precision	F-measure	AUC
Clinical features				
MLP	0.56	0.50	0.53	0.68
Imaging features				
RF	1.00	0.80	0.89	0.96
MLP	1.00	0.73	0.84	0.94
Imaging + Clinical features				
RF	1.00	0.80	0.89	0.96
MLP	1.00	0.80	0.89	0.96

On average, a better performance was achieved when both clinical and imaging features were used. The RF model achieved an equal performance using both clinical and imaging features or just imaging features; however, MLP reached a higher performance (AUC = 0.96) when clinical features were added to the feature set.

Focusing on the MLP experiments, clinical features achieved a satisfactory performance when predicting EGFR mutation status (AUC = 0.68), whereas, when using imaging data a reliable predictive model is obtained (AUC = 0.94). Further, when imaging and clinical data were combined, it was created a model which further increases the imaging data performance (AUC = 0.96). The RF model achieved the same performance in the two experiments potentially due to its intrinsic feature subsampling.

The ROC curves for the RF and MLP models using imaging and clinical features are presented in Fig. 1. The confusion matrix obtained by the two models using imaging and clinical features was identical, and it is presented in Fig. 2.

Feature Importance. Taking advantage of the RF ability to recognise features importance, it was conducted an analysis in order to find the most valuable

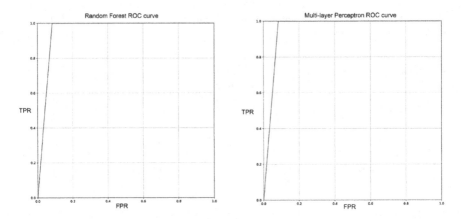

Fig. 1. ROC curve of the RF model (left) and MLP model (right).

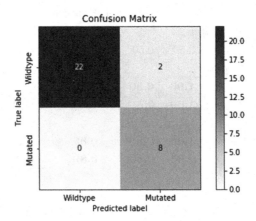

Fig. 2. Confusion matrix of the RF model and MLP models.

predictors amongst the used set of features. The RF model outputs a score for each feature which sums to one, and it describes the average decrease in impurity over trees.

Since the one-hot encoding approach was used to convert from categorical data to binary vectors, there is an importance score associated with each feature value and not a single score to the feature itself. Therefore, the feature score was considered to be the sum of the scores of each of its values. For instance, if having emphysema present had a score of 0.10 and emphysema absent a score of 0.08, the feature emphysema has a score of 0.18. Figure 3 shows the importance scores as a result of this analysis. Emphysema is plainly the feature with the highest score (0.18) followed by lung parenchymal features (0.16).

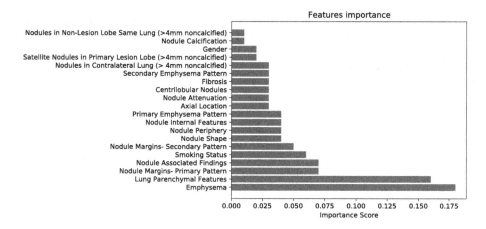

Fig. 3. Features importance scores.

5   Conclusions and Future Work

In this work, we report a radiogenomics model that is able to predict the mutation status of EGFR (AUC = 0.96) from CT scans in a less invasive procedure, compared to repeated biopsies during treatment. To the best of our knowledge, this work was the second attempt to create predictive models for EGFR and KRAS mutation status using qualitative radiographic image features. Even though our model outperforms the model created by Gevaert et al. [11] (AUC = 0.89), the present work did not use the same dataset as the first attempt and, consequently, results cannot be directly compared.

The results of this study suggest that an image signature exists that accurately predicts EGFR mutation status but not KRAS. Considering the fact that class proportions are similar in EGFR (26% mutated, 74 % wildtype) and KRAS (25 % mutated, 75% wildtype), the most reasonable hypothesis for this result is that KRAS mutations are not evident through radiographic qualitative features in the same extent as EGFR mutations, which appear to have particular patterns. Gevaert et al. [11], which was also able to create a predictive model for EGFR mutations but not for KRAS, also hypothesised that the results might result from different class proportions between EGFR and KRAS; however in the present study class proportions are similar, which enhances the hypothesis that KRAS mutations do not manifest through qualitative imaging features.

When clinical and imaging data were combined, the performance slightly increased, on average. These results have shown that, although the limited amount of data, images and clinical data combined are potential predictors of EGFR mutation status. However, it is important to further validate the results with a larger dataset, to clarify these features importance.

From the total set of features used in this work, emphysema and lung parenchymal features were the ones that presented a higher correlation with EGFR mutation status.

The model created in this study for the prediction of EGFR mutation status opens interesting opportunities for a better treatment planning and supports oncologists and radiologists with additional information at diagnosis. Key issues to be investigated in the future is whether quantitative features extracted directly from the image have the same predictive ability or even exceed qualitative features.

Acknowledgments. This work is financed by the ERDF – European Regional Development Fund through the Operational Programme for Competitiveness and Internationalisation - COMPETE 2020 Programme and by National Funds through the Portuguese funding agency, FCT - Fundação para a Ciência e a Tecnologia within project POCI-01-0145-FEDER-030263.

References

1. Bakr, S., et al.: A radiogenomic dataset of non-small cell lung cancer. Sci. Data **5**, 180202 (2018)
2. Bakr, S., et al.: Data for NSCLC Radiogenomics Collection (2017). https://doi.org/10.7937/K9/TCIA.2017.7hs46erv. https://wiki.cancerimagingarchive.net/x/W4G1AQ, type: dataset
3. Breiman, L.: Random forests. Mach. Learn. **45**(1), 5–32 (2001)
4. Chawla, N.V., Bowyer, K.W., Hall, L.O., Kegelmeyer, W.P.: Smote: synthetic minority over-sampling technique. J. Artif. Intell. Res. **16**, 321–357 (2002)
5. Chen, Z., Fillmore, C.M., Hammerman, P.S., Kim, C.F., Wong, K.K.: Non-small-cell lung cancers: a heterogeneous set of diseases. Nat. Rev. Cancer **14**(8), 535 (2014)
6. Clark, K., et al.: The cancer imaging archive (TCIA): maintaining and operating a public information repository. J. Digit. Imaging **26**(6), 1045–1057 (2013). https://doi.org/10.1007/s10278-013-9622-7
7. Digumarthy, S.R., Padole, A.M., Gullo, R.L., Sequist, L.V., Kalra, M.K.: Can CT radiomic analysis in NSCLC predict histology and EGFR mutation status? Medicine **98**(1) (2019)
8. Dogan, S., et al.: Molecular epidemiology of EGFR and KRAS mutations in 3,026 lung adenocarcinomas: higher susceptibility of women to smoking-related KRAS-mutant cancers. Clin. Cancer Res. **18**(22), 6169–6177 (2012)
9. Ellison, G., Zhu, G., Moulis, A., Dearden, S., Speake, G., McCormack, R.: EGFR mutation testing in lung cancer: a review of available methods and their use for analysis of tumour tissue and cytology samples. J. Clin. Pathol. **66**(2), 79–89 (2013)
10. Ferlay, J., et al.: Cancer incidence and mortality worldwide: sources, methods and major patterns in GLOBOCAN 2012. Int. J. Cancer **136**(5), E359–E386 (2015)
11. Gevaert, O., et al.: Predictive radiogenomics modeling of EGFR mutation status in lung cancer. Sci. Rep. **7**, 41674 (2017)
12. Gevaert, O., et al.: Non–small cell lung cancer: identifying prognostic imaging biomarkers by leveraging public gene expression microarray data—methods and preliminary results. Radiology **264**(2), 387–396 (2012). https://doi.org/10.1148/radiol.12111607, pMID: 22723499
13. Hames, M.L., Chen, H., Iams, W., Aston, J., Lovly, C.M., Horn, L.: Correlation between KRAS mutation status and response to chemotherapy in patients with advanced non-small cell lung cancer. Lung Cancer **92**, 29–34 (2016)

14. Hansell, D.M., Bankier, A.A., MacMahon, H., McLoud, T.C., Muller, N.L., Remy, J.: Fleischner society: glossary of terms for thoracic imaging. Radiology **246**(3), 697–722 (2008)

15. Janssen-Heijnen, M.L., Coebergh, J.W.W.: Trends in incidence and prognosis of the histological subtypes of lung cancer in North America, Australia, New Zealand and Europe. Lung Cancer **31**(2–3), 123–137 (2001)

16. Liu, Y., et al.: Radiomic features are associated with EGFR mutation status in lung adenocarcinomas. Clin. Lung Cancer **17**(5), 441–448 (2016)

17. Mei, D., Luo, Y., Wang, Y., Gong, J.: CT texture analysis of lung adenocarcinoma: can radiomic features be surrogate biomarkers for EGFR mutation statuses. Cancer Imaging **18**(1), 52 (2018)

18. Mok, T.S., et al.: Gefitinib or carboplatin-paclitaxel in pulmonary adenocarcinoma. N. Engl. J. Med. **361**(10), 947–957 (2009)

19. O'dowd, E.L., et al.: What characteristics of primary care and patients are associated with early death in patients with lung cancer in the UK? Thorax **70**(2), 161–168 (2015)

20. Papadopoulou, E., et al.: Determination of EGFR and KRAS mutational status in greek non-small-cell lung cancer patients. Oncol. Lett. **10**(4), 2176–2184 (2015)

21. Riely, G.J., Marks, J., Pao, W.: KRAS mutations in non–small cell lung cancer. Proc. Am. Thorac. Soc. **6**(2), 201–205 (2009)

22. Rosenblatt, F.: The perceptron: a probabilistic model for information storage and organization in the brain. Psychol. Rev. **65**(6) (1958)

23. Scrivener, M., de Jong, E.E., van Timmeren, J.E., Pieters, T., Ghaye, B., Geets, X.: Radiomics applied to lung cancer: a review. Transl. Cancer Res. **5**(4), 398–409 (2016)

24. Siegelin, M.D., Borczuk, A.C.: Epidermal growth factor receptor mutations in lung adenocarcinoma. Lab. Invest. **94**(2), 129 (2014)

25. Varghese, A.M., et al.: Lungs dont forget: comparison of the KRAS and EGFR mutation profile and survival of collegiate smokers and never smokers with advanced lung cancers. J. Thorac. Oncol. **8**(1), 123–125 (2013)

26. Zhou, H., et al.: Poor response to platinum-based chemotherapy is associated with KRAS mutation and concomitant low expression of BRAC1 and TYMS in NSCLC. J. Int. Med. Res. **44**(1), 89–98 (2016)

Learning to Perform Visual Tasks from Human Demonstrations

Afonso Nunes, Rui Figueiredo, and Plinio Moreno[✉]

Institute for Systems and Robotics, Instituto Superior Técnico,
Universidade de Lisboa, Lisbon, Portugal
afonsofrnunes@tecnico.ulisboa.pt
{ruifigueiredo,plinio}@isr.tecnico.ulisboa.pt

Abstract. The human visual system makes extensive use of saccadic eye movements to cope with decaying resolution towards the periphery of the visual field, and point the highest acuity region of the retina (i.e. the fovea) to regions of interest, while searching for objects in natural scenes. Experimental evidence has shown that, when searching for objects, humans exploit the advantage of *a priori* known relations between object classes (i.e. context) to prune the search space, which results in oculo-motor behaviours that are optimal both in terms of effectiveness (i.e. success rate) and efficiency (i.e. energy and time consumed) during search tasks execution. In this work we propose a biologically plausible system that learns from human demonstrations provided by eye tracking data to perform visual search tasks. The proposed framework leverages the recognition accuracy of state-of-the-art pre-trained CNNs with the sequential predictive power of RNNs, trained in an end-to-end manner to predict saliency maps (task-independent) and locate objects of interest (task-dependent).

Keywords: Active perception · Visual search · Eye tracking · Selective visual attention

1 Introduction

Humans move their eyes, via oculo-motor mechanisms, in order to sequentially improve their internal representation of the environment. By pointing the highest resolution region of the retina (i.e. the fovea) towards important sub-regions of the surrounding space, they avoid processing distracting stimuli in high detail.

Understanding and modeling attention mechanisms in biological visual systems, in particular gaze patterns, has an important impact in many computer vision applications including video surveillance and mobile robotics, where energy and computationally efficient solutions are a primordial requirement for many active perception problems [1,2].

The classical visual recognition models in the computer vision literature use salient features of the image, benefiting from information provided by hand-crafted filters (e.g. [3–5]). Recently, Deep Neural Networks (DNNs) have been

© Springer Nature Switzerland AG 2019
A. Morales et al. (Eds.): IbPRIA 2019, LNCS 11868, pp. 346–358, 2019.
https://doi.org/10.1007/978-3-030-31321-0_30

developed for recognizing thousands of objects and to autonomously generate visual characteristics that are optimized by training with large publicly available annotated datasets. These can implicitly learn, in an end-to-end data-driven manner, to tackle challenging visual tasks, including object detection [6,7], segmentation [8] and tracking [9], circumventing the feature engineering step, and having recently surpassed humans in classification tasks [10]. However, the main setback of deep learning techniques still resides on their computational and energetic requirements and the availability of expensive and parallel computing hardware (e.g. Graphical Processing Units (GPUs)) for real-time execution. Hence, it is still important to develop different ways of selectively processing visual information to further reduce the complexity of existing solutions, to computational levels processable by the human cognitive apparatus.

In this work we propose an eye fixation predictive model for free viewing and visual search tasks in natural scene analysis. The proposed model benefits from the recognition accuracy of Convolutional Neural Networks (CNNs), pretrained on large annotated datasets to extract visual features at hierarchical levels of increasing complexity and abstraction. These lower dimensional features are combined with the predictive power of Recurrent Neural Networks (RNNs), which are trained with eye tracking data to mimic human eye gaze behaviors. A set of experiments demonstrate that the proposed architecture can be used to predict gaze fixation sequences, both in free-viewing and visual search tasks.

The article is structured as follows: In Sect. 2, we overview the state-of-the-art on saliency and eye fixation predictive modeling. In Sect. 3, we describe in detail the proposed methodology. In Sect. 4, we experimentally validate our method and in Sect. 5 we wrap up with conclusions and future work.

2   Related Work

The concept of attention covers all factors that influence information selection mechanisms, whether they are driven by scene characteristics (bottom-up) or by task related expectations (top-down) [2]. In this section we briefly overview the state-of-the-art in saliency modeling which deals with the problem of extracting important image sub-regions and predicting where eye fixations should occur.

The most well known saliency-based visual attention model proposed in [11] is inspired by the seminal work of [12]. At first, spatial feature maps are built by extracting prominent local features from different low-level modalities (e.g. color, intensity, orientation) using center-surround operations at different scales. Then, each map is normalized and linearly combined in a single saliency map, to be further analyzed in detail, and in order of decreasing conspicuity.

The authors in [13] proposed a Bayesian framework for saliency learning using natural statistics (SUN). The most salient features are the ones with the highest point-wise self-information with features prior learned from a set of natural images (bottom-up modulation), i.e. features that are mostly distinct from the learned average, in a probabilistic sense, and the highest mutual information when searching for a given target object (top-down modulation).

One major limitation of the previous methodology resides in the fact that they compute the probability of regions being prominent, from visual characteristics, completely ignoring the sequential and temporal aspects of attention, in particular of eye fixations during visual search tasks.

Approaches based on information maximization principles [14,15] overcome these limitations by capturing spatio-temporal context in a probabilistic form and by actively integrating information across saccades, via iterative Bayesian inference. The authors in [16] extend the previous ideas by including behavioral aspects as optimization constraints, namely explicit energetic costs incurred when switching between target locations, with demonstrated advantages in visual search tasks with artificial foveal vision.

The visual scanpath prediction method proposed in [17] combines low-level features provided by hand-crafted filters with high-level object semantics, in a reinforcement learning framework, and uses a least-squares iteration algorithm to learn a policy that mimics the human gaze fixation behaviour.

The attention model proposed in [18] uses a recurrent neural network to control the fixation point of a low-dimensional, space-variant resolution retina-like image representation. The authors formulate the gaze control problem as a Partially Observable Markov Decision Process (POMDP) and employ reinforcement learning techniques to learn policies that maximize task-related rewards. The formulation is general enough with demonstrated applicability in a wide range of problems, including object classification in static images and object tracking in dynamic video sequences.

The work closest to ours is the one of [19] where the authors proposed a gaze prediction method that combines Convolution Neural Networks (CNNs) for feature extraction with Long Short Term Memory (LSTMs) networks for sequence modeling. Similar to the work of [17], the authors rely on human provided scanpath data, gathered with an eye tracking device for gaze scanpath prediction. However, instead of framing the problem within the reinforcement learning domain they learn in a supervised and end-to-end manner to infer eye fixation sequences directly from visual input, in free viewing. In this work, the method of [19] is extended to visual search with demonstrated application in a pedestrian search task.

3   Methodology

This section explains in detail our eye fixation prediction approach. Our method starts by using a CNN to reduce dimensionality and extract features from the visual input. We rely on the VGG16 model [20], pre-trained on the ImageNet dataset [21] for image classification, however we drop the 3 last fully-connected layers, since we are only interested in extracting feature maps. The feature maps consist of $16 \times 16 \times 512$ tensors that preserve the main characteristics of the original image. It corresponds to a lower-dimensional spatial quantization of the original image into a 16×16 grid, in which each of the 256 regions are represented by a 512-long feature vector. Thus, in the context of this work, a fixation sequence corresponds to an ordered sequence of CNN feature map elements, called feature vectors. In the example feature map of Fig. 1, the red vector corresponds to the feature vector of the left bottom corner region of the original image. These feature vectors are the input elements that our RNN model directly deals to learn to predict visual scanpaths. We chose a pre-trained CNN due to its state-of-the-art performance in extracting features that preserve images' main characteristics, while reducing input dimensionality, however other methods for feature extraction may be also viable.

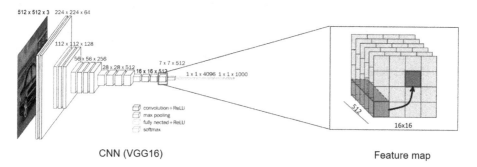

CNN (VGG16)   Feature map

Fig. 1. Feature extraction methodology using the VGG16 pre-trained CNN. (Color figure online)

Our RNN model consists of LSTM units, as depicted in Fig. 2, that were chosen due to their ability to retain past information through recurrent iterations. Our recurrent model is many-to-many, in the sense that it yields one output vector y_i for each input vector x_i. The x_i correspond to the CNN feature vectors while the y_i correspond to softmax 256-dimensional probability vector. In this sense, $y_i[k]$ corresponds to the predicted probability a person has to look at image region k ($k = 1..256$) at fixation i, given its present location feature vector x_i and considering some memory of the previous image fixations $j = 1..(i-1)$. An important advantage of our model is that it implies no compromise in terms of number of fixation steps. This model can be trained and tested with variable length scanpaths, since it works in a modular, self-recurring way.

The definition of the initial state vectors h_0 and c_0 is an important detail of our model. These variables correspond to the information our model has, prior to analyzing any image patch. In an attempt to make our model as fully trainable as possible, and to mimic some human biological inspiration, we use a low-resolution version of the complete feature map that provides context and environment of the image, and make its subsequent decisions considering this information. The way we do this is by introducing a simple fully connected layer between the low-resolution maps and the initial state vectors, whose weights are trained along with the LSTM weights, as a whole.

To train our RNN model we employ an algorithm that aims to find the best set of network parameters θ for a given image I, by minimizing the following log-likelihood formula

$$L(S|I,\theta) = \sum_{t=1}^{N} \log P(y_{t+1}|x_t, c_{t-1}, h_{t-1}) \tag{1}$$

where x represent feature vectors, y represent probability distribution maps and c and h represent the state vectors that connect sequential time steps. Thus, in the training phase, the network is given a sequence of x_t and y_t ($t = 1..N$) and the best parameters are estimated using standard backpropagation.

In summary, we present a recurrent model based on LSTM units, fully trainable both in making initial high-level estimations of the image and making informed decisions on where to attend next, in the case of visual search tasks.

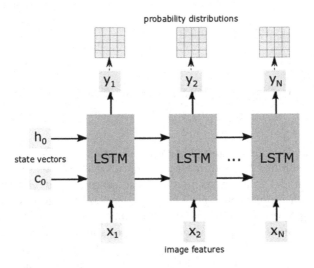

Fig. 2. Schematic drawing of our RNN model, with all relevant vectors labeled. x vectors are the feature vectors (with length 512) and y vectors are probability distributions (with length 256)

4   Experiments and Results

4.1   Evaluation Metrics

In order to evaluate the performance of our gaze prediction model, we have used two types of performance metrics, described below.

Similarity to Human Examples. The first one is concerned with the fixation sequences' similarity with human ground-truth sequences. It assesses quantitatively how close our model is to human sequences, and generates a score to express their relative similarity. More concretely, to assess the similarity of the sequences generated by our model to those that would be made by a human observer, we are interested in understanding if our network has learnt to look at the locations humans looked, with a similar order of fixations. To achieve this we start by using the *Meanshift* clustering algorithm. The bandwidth is chosen to be the one that maximizes the interaction rate I, according to $I = \frac{N_b - N_w}{C}$, where N_b represents the number of fixation transitions between clusters, N_w the number of fixation transitions within clusters, and C the total number of clusters.

We use this algorithm with all the testing data available for a given image, which results in a segmentation of the image based on human gaze data. Also, we can now represent the ground-truth sequences of image coordinates as sequences of clusters. Next we associate the generated sequence to the defined clusters, using Euclidean distance as criteria. Therefore, if we associate each cluster with a unique character, it is straightforward to compare the generated to (possibly several) ground-truth sequences using the *Needleman-Wunsch* [22] string matching algorithm. The scores retrieved from this method for the same image are then averaged across human subjects. We compare sequences with increasing number of fixation steps to assess our models performance for different sequence sizes.

Effectiveness in Goal. To assess task effectiveness we focus on the attainment of the desired goals. We designate by *accuracy* the percentage of images in the testing set in which our model was able to find the target. We consider the model to have found the target if in its scanpath it includes a location less than 2 units apart from the known location of the target. We also make an histogram with the number of fixations it took to find the target in the testing images and use them to extract some conclusions.

The second criterion is more task-oriented and evaluates the performance of the model to achieve the intended result in the least number of fixation steps. This one can only be applied in experiments with a specific task, whereas the previous one can be applied in more general cases. Since our work focuses on learning to perform visual tasks from human demonstration, these metrics analyze independently both the human similarity and effectiveness of our model.

4.2 Free-Viewing Task Performance

In the free-viewing task dataset of [23], the participants were asked to look at multiple images. The dataset comprises scanpaths of 15 subjects on a total of 1003 different images of varied contexts.

We use 90% of the dataset for training and the remaining 10% for testing. Furthermore, 10% of the training set is used as the validation set. We train our model using backpropagation with the *rmsprop* optimizer and with *categorical cross-entropy* loss [24]. The *learning rate* is fixed at 0.005 and we use image batches of size 64. We allow the loss to increase in the validation set for 5 consecutive epochs, otherwise the training process is stopped early, to prevent over-fitting the training set. Example images along with the scanpath generated by our model are presented in Fig. 3. We set the fixation sequence length to 6, to match the normal length of the ground-truth human fixation sequences.

Fig. 3. Example scanpaths that emerge with our method in free-viewing.

The results in Fig. 3 are illustrative of some human vision behavioral traits that were learnt by our model, namely the high number of visual fixations in regions where there are people (figure on the left), and the avoidance of flat, "less interesting" regions such as the sky (figure on the right).

Using the previously described metrics, we obtain similarity measures of the generated sequences to the ground-truth sequences. We plot these scores in Fig. 4 along with a measure of inter-human agreement. The line (*inter-subject*) corresponds to the average of the average scores a given sequence yields when compared to all other for the same image.

We observe that our network's similarity measures follow a similar tendency to the measures obtained when ground-truth scanpaths are compared among them. This shows that the training was effective, and that our model captured at least some human gaze behavioral traits while visually inspecting images. We notice that our eye scanpath prediction model has at its disposal considerably less information than humans do when performing similar tasks, since it relies on local image patches rather than on full resolution or human-like space-variant images, with resolution decaying away from the fixation point.

4.3   Task-Dependent Performance

After concluding that our model is suitable for learning human traits while visually inspecting images, we try to assess if it can be applied to specific search

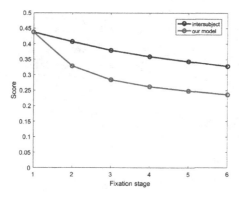

Fig. 4. Plot of the similarity measures obtained for the scanpaths generated by our model in the testing set images, along with inter-subject average similarity, for different fixation sequence lengths.

tasks. The eye tracking dataset used to assess our visual search approach was introduced in [25] and originally comprises 912 images. In our experiments we used only a subset containing 456 images of people, and eye tracking data recorded during pedestrian search tasks. The images with people are further divided according to the relative difficulty a human would have to find them, where 95 are labeled as EASY and 361 are labeled as HARD.

We train the same model as before, but with 3 different datasets: training set of EASY images, training set of HARD images, and dataset of mixed images (EASY+HARD). In these experiments we kept the relative size of training and testing subsets to 90% and 10% of the entire dataset, respectively, and using 10% of the training set as validation set.

Example images along with the scanpath generated by our model are presented in Fig. 5. As before, we generate fixation sequences of length 6 for each image in the testing set, since we have concluded from the data that the great majority of tasks can be concluded within this limit. It is noticeable in both examples that our model has learnt not to look at flat areas in the images, since people are rarely present in those.

In Fig. 5a we observe that our model successfully finds the pedestrians in 4 fixation steps. Furthermore, our model, after fixation 4, remains in the same patch. This is especially interesting since our model was not specifically programmed to do this - it learned from data to remain in the same image patch once it has believed to have found the pedestrians. In Fig. 5b the search task is also successfully completed. It is clear in this example that, rather than searching randomly through the image, our model makes "informed" decisions by avoiding improbable locations such as the sky or the treetops. Table 1 shows the accuracy obtained for the three trained models. The best accuracy corresponds to the model trained on the mixed dataset.

(a) EASY image   (b) HARD image

Fig. 5. Example scanpaths that emerge with our method during task-dependent visual search. The training set of this model is the mixed dataset.

Table 1. Person detection accuracy in several dataset partitions.

	Tested on	
Trained on	EASY	HARD
EASY	53.33%	34.92%
HARD	47.37%	38.07%
EASY+HARD	60.53%	38.89%

On the other hand we analyze the performance of our models in terms of number of fixation steps required to reach the target. We verify that, for each model, the average number of fixation steps required to reach the target is lower for the EASY images than for the HARD images. This is in accordance to what we observe to exist in the human ground-truth data. Once again the best performance is attained by the model trained on a mixed dataset, which highlights the importance of training our model with a diverse dataset. In Fig. 6 we present the corresponding histograms to the results on Table 1, with the actual distribution of fixation steps for each combination of train and test partition.

Results for the similarity measures are shown in Fig. 7. We observe that the inter-subject similarity is significantly higher when there exists a specific task. This is intuitively expected, since observers are aiming at a common goal, rather than just exploring the image, as before. Once again our model shows to behave in a human-like way, however the similarity difference in this case is more significant. We justify this difference with the shorter length of the training sequences, that is likely to be preventing our model to acquire its human traits.

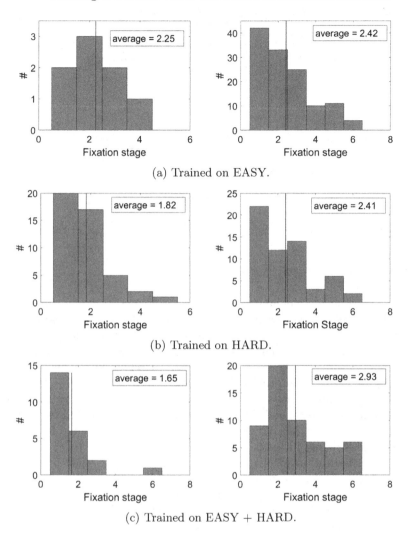

(a) Trained on EASY.

(b) Trained on HARD.

(c) Trained on EASY + HARD.

Fig. 6. Histograms of fixations for different training dataset partitions.

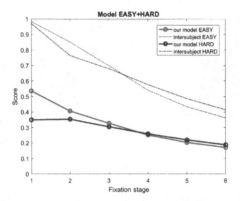

Fig. 7. Similarity results on a visual search task, trained on EASY+HARD.

5   Conclusions and Future Work

In this work we have extended a gaze sequence predictive model that leverages recent advantages on deep learning, with the availability of eye tracking datasets to visual search tasks. A set of experiments demonstrated that our model is successful at mimicking human gaze behaviors when visually inspecting images, both in free viewing scenarios as well as when searching for particular targets, even though contextual information is represented implicitly.

We expect to improve task performance in visual search by using space-variant and low-dimensional foveal image representations, to preserve important peripheral information, instead of localized image crops, extracted from uniform and fixed scale low-resolution feature grids.

The main limitation of our current implementation is that it only allows searching for a single known object class, which needs to be present in the training dataset. In the future, we intend to improve our model with the ability of searching for multiple known objects, by extending the input feature vector with the task, using for instance a "one hot encoding" of the class of interest.

Acknowledgments. This work has been partially supported by the Portuguese Foundation for Science and Technology (FCT) project [UID/EEA/50009/2019] and the grant program "New talents in Artificial Intelligence" from Fundação Gulbenkian. Finally, we gratefully acknowledge the support of NVIDIA Corporation with the donation of the Titan Xp GPU used for this research.

References

1. Begum, M., Karray, F.: Visual attention for robotic cognition: a survey. IEEE Trans. Auton. Ment. Dev. **3**(1), 92–105 (2011)
2. Borji, A., Itti, L.: State-of-the-art in visual attention modelling. IEEE Trans. Pattern Anal. Mach. Intell. **35**(1), 185–207 (2013)

3. Harris, C., Stephens, M.: A combined corner and edge detector. In: Alvey Vision Conference, vol. 15, no. 50, pp. 10–5244. Citeseer (1988)
4. Dalal, N., Triggs, B.: Histograms of oriented gradients for human detection. In: IEEE Computer Society Conference on Computer Vision and Pattern Recognition CVPR 2005, vol. 1. pp. 886–893. IEEE (2005)
5. Lowe, D.G.: Object recognition from local scale-invariant features. In: Proceedings of the 7th IEEE International Conference on Computer vision, vol. 2, pp. 1150–1157. IEEE (1999)
6. Redmon, J., Divvala, S., Girshick, R., Farhadi, A.: You only look once: unified, real-time object detection. In: Proceedings of the IEEE Conference on Computer Vision and Pattern Recognition, pp. 779–788 (2016)
7. Liu, W., et al.: SSD: single shot multibox detector. In: Leibe, B., Matas, J., Sebe, N., Welling, M. (eds.) ECCV 2016. LNCS, vol. 9905, pp. 21–37. Springer, Cham (2016). https://doi.org/10.1007/978-3-319-46448-0_2
8. He, K., Gkioxari, G., Dollár, P., Girshick, R.: Mask R-CNN. In: 2017 IEEE International Conference on Computer Vision (ICCV), pp. 2980–2988. IEEE (2017)
9. Held, D., Thrun, S., Savarese, S.: Learning to track at 100 FPS with deep regression networks. In: Leibe, B., Matas, J., Sebe, N., Welling, M. (eds.) ECCV 2016. LNCS, vol. 9905, pp. 749–765. Springer, Cham (2016). https://doi.org/10.1007/978-3-319-46448-0_45
10. He, K., Zhang, X., Ren, S., Sun, J.: Delving deep into rectifiers: surpassing human-level performance on imagenet classification. In: Proceedings of the IEEE International Conference on Computer Vision, pp. 1026–1034 (2015)
11. Itti, L., Koch, C., Niebur, E.: A model of saliency-based visual attention for rapid scene analysis. IEEE Trans. Pattern Anal. Mach. Intell. **11**, 1254–1259 (1998)
12. Koch, C., Ullman, S.: Shifts in selective visual attention: towards the underlying neural circuitry. In: Vaina, L.M. (ed.) Matters of Intelligence. SYLI, vol. 188. Springer, Dordrecht (1987). https://doi.org/10.1007/978-94-009-3833-5_5
13. Zhang, L., Tong, M.H., Marks, T.K., Shan, H., Cottrell, G.W.: Sun: a bayesian framework for saliency using natural statistics. J. Vis. **8**(7), 32–32 (2008)
14. Itti, L., Baldi, P.F.: Bayesian surprise attracts human attention. In: Advances in Neural Information Processing Systems, pp. 547–554 (2006)
15. Butko, N.J., Movellan, J.R.: Infomax control of eye movements. IEEE Trans. Auton. Ment. Dev. **2**(2), 91–107 (2010)
16. Ahmad, S., Yu, A.J.: Active sensing as bayes-optimal sequential decision making. arXiv preprint arXiv:1305.6650 (2013)
17. Jiang, M., Boix, X., Roig, G., Xu, J., Van Gool, L., Zhao, Q.: Learning to predict sequences of human visual fixations. IEEE Trans. Neural Netw. Learn. Syst. **27**(6), 1241–1252 (2016)
18. Mnih, V., Heess, N., Graves, A., et al.: Recurrent models of visual attention. In: Advances in Neural Information Processing Systems, pp. 2204–2212 (2014)
19. Ngo, T., Manjunath, B.: Saccade gaze prediction using a recurrent neural network. In: 2017 IEEE International Conference on Image Processing (ICIP), pp. 3435–3439. IEEE (2017)
20. Simonyan, K., Zisserman, A.: Very deep convolutional networks for large-scale image recognition. In: International Conference on Learning Representations (2015)
21. Russakovsky, O., Russakovsky, O., et al.: ImageNet large scale visual recognition challenge. Int. J. Comput. Vis. (IJCV) **115**(3), 211–252 (2015)

358   A. Nunes et al.

22. Needleman, S.B., Wunsch, C.D.: A general method applicable to the search for similarities in the amino acid sequence of two proteins. J. Mol. Biol. **48**(3), 443–453 (1970)
23. Judd, T., Ehinger, K., Durand, F., Torralba, A.: Learning to predict where humans look. In: IEEE 12th International Conference on Computer Vision, pp. 2106–2113. IEEE (2009)
24. Buja, A., Stuetzle, W., Shen, Y.: Loss functions for binary class probability estimation and classification: Structure and applications (2005)
25. Ehinger, K.A., Hidalgo-Sotelo, B., Torralba, A., Oliva, A.: Modelling search for people in 900 scenes: a combined source model of eye guidance. Vis. Cogn. **17**(6–7), 945–978 (2009)

Serious Game Controlled by a Human-Computer Interface for Upper Limb Motor Rehabilitation: A Feasibility Study

Sergio David Pulido, Álvaro José Bocanegra, Sandra Liliana Cancino, and Juan Manuel López[✉]

Escuela Colombiana de Ingeniería Julio Garavito,
AK 45 # 205-49, Bogotá D.C, Colombia
{sergio.pulido,alvaro.bocanegra}@mail.escuelaing.edu.co
{sandra.cancino,juan.lopez}@escuelaing.edu.co
https://sites.google.com/view/semilleropromise/investigadores

Abstract. Stroke affects the population worldwide, with a prevalence of 0.58% worldwide. One of the possible consequences is the negative impact in the motor function of the patient, limiting their quality of life. For these reason, Brain-Computer Interfaces are studied as a tool for improving rehabilitation processes. Nevertheless, to the best of our knowledge, there are no Brain-Computer Interface systems which use video-games for upper limb motor rehabilitation. This study aimed to design and assess a Human-Computer Interface that includes electroencephalography, forearm motion and postural analysis, with healthy subjects. This assessment was made by designing two scenarios in which the participant carried out exercises involving the mouth and the hand and forearm trajectory symmetry. Results show that the system is ready to be tested on patients, since the participants were comfortable using it. Also, the quantitative results, particularly, the metrics used in the video-game, are an important start for health professionals to characterize motor rehabilitation in stroke patients, enabling the path to the use of the designed system in motor rehabilitation therapies.

Keywords: Human-Computer Interface ·
Motor rehabilitation · Stroke · Serious games

1 Introduction

Every year, millions of people are affected by the consequences of having a stroke. According to the Observatorio Nacional de Salud, in 2014, the prevalence of stroke in Colombia was of 62582 cases [1]. A report made by the American Heart Association in 2018 [2] states that 795000 people experience a stroke every year

Supported by Escuela Colombiana de Ingeniería Julio Garavito. DII/C013 CI 2017.

in the United States. The report also shows that in 2015, worldwide prevalence of cerebrovascular disease was 42.4 million, affecting 0.58% of the world population. These concerning statistics demonstrate the need to find new ways to improve the quality of life of patients affected by stroke.

In the last few years, research in motor rehabilitation has focused on the use of new technologies. Johnson et al. [3] designed a Functional Magnetic Resonance Imaging (fMRI)-based Brain Computer Interface (BCI) in which stroke patients were given feedback in the means of repetitive Transcranial Magnetic Stimulation (rTMS), in order to enhance neural plasticity. The research exhibited benefits of the combination between BCIs and rTMS, as patients showed significant improvements over time. As well, Frolov et al. [4] designed a Motor Imagery-based BCI (MI-BCI) in which patients performed exoskeleton-driven movements of the hand; one group of patients controlled the movement with help of the MI-BCI and the other group of patients did not control the exoskeleton at all. While both groups showed important motor improvements, the results of the group that controlled the exoskeleton were significantly better.

A systematic review about BCIs for motor rehabilitation of stroke was performed in early 2018. The inclusion criteria were the following: papers written in English in the areas of engineering, computer science, medicine or neuroscience that were published from 2013 onwards. Overall, many types of BCI-based motor therapies have been subject of research. An option for improving brain plasticity in stroke patients is through the use of MI-BCI [5,6], as it has shown satisfactory results, and in some cases, not even requiring a training session [7]. Efficiency of different ways to give feedback has also been studied. In different projects, robotic aid was given to the patients whenever the systems detected movement intention [8], under the assumption of a superiority of somatosensory feedback over visual feedback [9]. While most researchers use subjective tests such as the Fugl-Meter Assessment (FMA) or the Modified Ashworth Scale (MAS) to detect motor improvements in patients, some others have opted for other methods, such as the analysis of functional connectivity of the brain [10] or Magnetoencephalography (MEG) for sensorimotor rhythms [11]. Whereas many projects regarding the use of BCIs for motor rehabilitation of stroke have been made, to the best of our knowledge, there is no research in this area related to the use of video-games as a means of upper limber motor rehabilitation. In addition, video-games have been used to improve some cognitive tasks showing promising results [16–19].

This work aimed to perform a feasibility study on healthy subjects of a system that provides entertainment to the patients as they carry out their motor rehabilitation therapies. The system is based on previous work of our research group [12], in which subjects achieved brain modulation while controlling a movement intention-based video-game, defining movement intention as both: motor imagery and movement. Through MI-BCI, the video-game is controlled by the subjects.

2 Methods

The proposed system is composed by a Brain-Computer Interface, a forearm motion tracking system and a postural tracking system. All of these modules allow the user to interact with two designed video-games.

2.1 Brain-Computer Interface

Through EEG signal analysis, an algorithm is capable of recognizing certain patterns. A g.Nautilus g.LADYBird with 32 active channels is used in this research to acquire the EEG signals. Electrodes were located near the motor cortex (Fz, FC1, FC2, C3, Cz, C4, CP1, CP2 and Pz), in order to acquire signals related to the motor imagery of the patient; this way, subjects receive online feedback regarding their movement intent. Signals were sampled at 250 sps, with a resolution of 16 bits. From each of the 9 channels, 13 features were extracted, obtaining a total of 117 features. The selected features were the following: Mean, variance, skewness, kurtosis, root mean square, relative power by frequency bands (Delta, Theta, Alpha, Beta, Gamma, Mu), Hjorth's mobility and Hjorth's complexity. These signals were split into windows of two seconds, in order to compute the selected features in each window. However, each window had an overlap of 75%; this way, signals were processed every 500 ms, making it an acceptable response time in order to give appropriate feedback to the users [13].

Finally, Support Vector Machine (SVM) models were used to classify the signals between motor imagery and standby state. For each session, two SVM models are generated: a quadratic SVM and a cubic SVM. These models were chosen in a way in which they are robust enough to classify these signals but simple enough for real-time processing. On the one hand, because of its lower complexity, the quadratic SVM does not achieve exceptional accuracy in individual trials but it allows for better generalization. On the other hand, the cubic SVM is able to reach higher accuracy values, but it is more prone to present over-fitting [14]. In order to decide which was the best model in each session, the cross-validation loss of the models was calculated. Particularly, the model with the lowest mean squared error during the training session was chosen.

2.2 Forearm Motion Tracking Module

In order to stimulate and quantify the motor progress during a rehabilitation process, we recorded the upper limb movement. Participants wear two Myo Armbands, one on each forearm to obtain movement data. Each device has a accelerometer and a gyroscope which are used to calculate, starting from a reference point, the positions of an forearms. The algorithm to estimate the position of the forearm is obtained from an open source code created by Thalmic Labs, developers of Myo Armband. By using this algorithm, the user can control an avatar in the video-game by performing arm movements, while the device records the signals of the user, for comparison between sessions and between both arms. Also, the device can use different vibration patterns as an haptic feedback for the video-game.

2.3   Postural Tracking Module

Patients with affected mobility in the upper limb tend to overcompensate their movements; for this reason, postural tracking can be useful during the rehabilitation sessions [15]. This is why, during the experiment, the subject is asked to use a special t-shirt with markers which a video acquisition and processing system use to track posture during the whole test. The position of the markers is analyzed in two different reference points: shoulders and spine. Whenever the system detects unnatural movements, the user is notified through different types of haptic feedback (provided by the Myo Armbands), depending on the axis of the overcompensation. In the first case, if the user performs shoulder overcompensation, the armbands have a strong vibration every 2 s. In the second case, if the user performs spine overcompensation, the armbands have a weak vibration every 0.5 s.

2.4   Experimental Protocol

Three stages were implemented during the tests:

- Preparation: During this stage, the acquisition devices are placed on the subject. The user is asked to put on the special t-shirt that allows the system to perform postural tracking. Additionally, the impedance of the electrodes is measured in order to assure quality of the signal.
- Training: For this stage, records of healthy subjects during both relaxation and movement imagery are acquired, as the module aimed to generate a decision model capable of classifying both states.
- Execution: The subjects play the designed video-game, which is composed of two scenarios or mini-games. For the first scenario, subjects are asked to move their forearms in a specific manner, while all the systems record their corresponding signals and help to offer feedback. For the second game, the subject controls an avatar through forearm movement. Since the system was developed for stroke patients and their movement may be limited, the system decides when can the users control it via forearm movement and when can they control it via BCI control. This way, the subjects control the avatar even if they can not move their compromised forearm.

2.5   Video-Game

The created video-game is named MindSense and it is composed by 2 mini-games, which were designed specially for this study in the game development platform Unity.

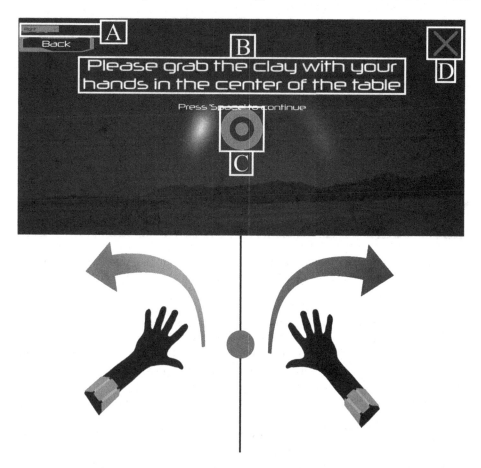

Fig. 1. Clay Game. The top of the figure shows user interface. The bottom shows the user movements. The lights in the top are equivalent to the arrows in the bottom. A. Selection of dominant hand. B. Description of the required action. C. Reference point from where movements are performed. D. Button to close the game.

The first mini-game, the Clay Game, consists on two lights controlled by the user, one light per arm. Subjects are asked to move their dominant arm as far as possible while pulling from a piece of clay placed in the center of a table. The mini-game uses an algorithm which detects whenever the forearm of the user stays still; once the algorithm detects that the arm stops, the subject is asked to imitate the movement with the other arm in the opposite direction. Depending on the symmetry of the paths, the user receives a score as feedback. A screenshot of the Clay Game is presented in Fig. 1, in which the center of the table is marked in a red circle, indicating that the user must perform the movements by taking this point as a reference.

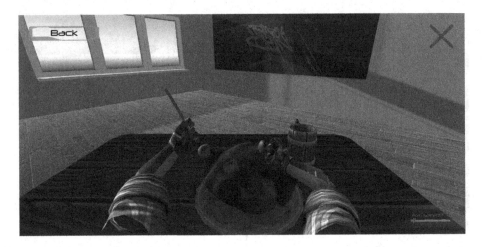

Fig. 2. Food Game. The user controls both arms through the Myo Armband or the BCI. The goal is to move the hand with the fork, from the food to the mouth.

The second mini-game, the Food Game, consists on controlling an avatar by moving both arms. The goal of the game is to feed the avatar, as it will imitate the forearm movements performed by the user. This game uses a movement recognition algorithm: initially, the Myo Armband provides signals related to forearm movement. Whenever the user stays still, the control of the avatar is transferred to the BCI; this way, the subject is able to control the avatar by using motor imagery. There is a plate of food in front of the avatar and the user must control it in order to feed it. However, if the BCI controls the system, the arm of the avatar follows a default path towards the mouth. A screenshot of the Food Game is presented in Fig. 2: the user must take a piece of food with the fork before eating. The scores of this game are a measurement of how fast can the user eat the food; both by moving the arm and by using motor imagery.

Both mini-games make use of the same algorithm in order to detect arm movement. First, the difference in position of the arm between two frames is recorded in a 2-second sliding window; if it surpasses certain threshold, it is registered as a frame in which the subject moved the arm, otherwise, it is registered as a frame in which the subject stayed still. If the subject moved during 20% of the 2-second window, it is considered that there is a general movement of the arm. Both the movement threshold and the percentage needed for the arm to be considered to be moving were defined empirically.

The experimental protocol is presented in Fig. 3.

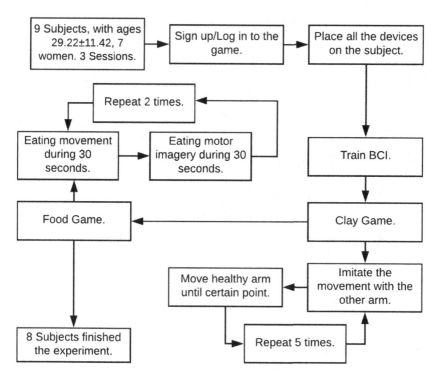

Fig. 3. Experimental protocol. First, subjects were selected. During each experiment, subjects were asked to sign up or log into the game (depending on the session number). After placing the devices and training the BCI, the subjects are asked to complete the mini-games multiple times.

3 Results

For each session, movement intention accuracy of the system was obtained. Results for each session and mean performance per subject are shown in Fig. 4. In the figure, the first three bars of each subject show accuracy of movement intention detection, while the last bar, shows the mean of these values. Despite the instability of the obtained accuracy, in most of the sessions an accuracy higher than 68% was reached, with atypical sessions with less than 45% of accuracy. Additionally, except for two subjects (51.01% for subject 3 and 56.06% for subject 6), all of the mean accuracy values are above 65%.

Results of the Clay Game are shown in Fig. 5. Each session, the subjects were asked to complete the exercise during five trials. The full path in each trial is divided in 10 equally distributed steps. A score is calculated for each step, with a total of 50 scores per session and 150 per subject. Each score was calculated depending on how symmetrical the motion signal of the dominant forearm was compared to the other forearm in a particular step of the path. In Fig. 5, scores of the subject 1 show a pattern in which the score decreased as the steps increased.

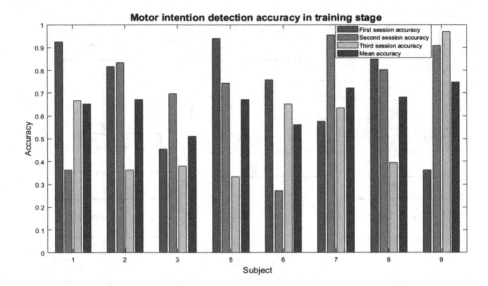

Fig. 4. Accuracy of the system during training stage.

This pattern is repeated in the sessions of most of the subjects, as can be seen in the mean scores. None of the subjects had a score below 60 in any step nor session.

For the Food Game, the results are quite different, since the game does not show scores directly to the user. Instead, it gives feedback to the users each time they complete the exercise. In this case, the scores of the game are represented as the interval between the moment in which the users grab food and the moment in which they feed the avatar. As mentioned before, the subjects were asked to do the exercise during 30 s. Then, during the next 30 s, motor imagery is used, and the process starts all over again. This way, intervals corresponding to motor imagery can be separated from intervals controlled by the movement of the forearm. The mean elapsed time for both motor imagery and forearm movement for all the subjects during the 3 sessions are shown on Table 1. It is clear that motor imagery intervals have higher values compared to intervals from forearm movement. Particularly, the elapsed time for motor imagery was 22.22 ± 11.37 s, whereas the elapsed time for forearm movement was 1.47 ± 0.62 s. As well, the number of repetitions for motor imagery was 5.75 ± 1.75, while the number of repetitions for forearm movement was 20.33 ± 6.75. In all the metrics, the coefficient of variation (standard deviation by mean) was less than 1, which signifies a low variation, indicating a relatively good stability. Furthermore, a tendency of changes in scores was not found between sessions, as scores decreased in some subjects while they increased in others.

Table 1. Mean of elapsed time for repetitions in-game during motor imagery (MI) and forearm movement (FM). S: Session, ET: Elapsed time in seconds, R: Repetitions, M ± STD: Mean ± Standard deviation

Subject	Motor imagery						Forearm movement					
	S1		S2		S3		S1		S2		S3	
	ET	R	ET	R	ET	R	ET	R	ET	R	ET	R
1	22.18	4	9.51	7	20.47	5	1.40	24	0.99	27	0.80	25
2	5.66	9	55.46	4	38.91	4	1.07	38	1.13	20	1.11	20
3	16.58	8	16.79	6	21.08	5	0.92	21	0.99	22	1.00	17
5	24.90	4	9.10	6	35.53	3	2.29	21	1.03	20	1.89	21
6	29.21	6	20.57	6	9.59	10	1.60	18	1.10	27	0.95	33
7	34.54	5	21.70	4	12.48	6	1.76	17	1.63	22	1.14	21
8	29.99	5	21.22	8	37.84	4	2.54	9	3.04	14	2.73	13
9	13.93	6	15.49	7	17.07	6	1.17	10	1.30	12	1.78	16
M ± STD ET	22.22 ± 11.37						1.47 ± 0.62					
M ± STD R	5.75 ± 1.75						20.33 ± 6.75					

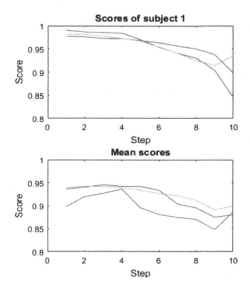

Fig. 5. Scores of subject 1 and average scores through subjects in Clay Game (blue: first session, red: second session, yellow: third session). (Color figure online)

4 Analysis

As noted before, the movement detection accuracy results were not as promising as hoped before for some sessions. One possible explanation is that subjects may have not been focused in the task at hand in some of the sessions. It is important to remark that, in systems where brain-wave modulation is key, the user must be compromised with the tasks for the movement intention detection to be adequate. If this is the case, results may improve in a stroke rehabilitation therapy, due to the compromise of the patients with the rehabilitation processes. Another possible explanation for the low results in some of the sessions may be that not enough machine learning models were considered. It is possible for different models to achieve a better performance.

Furthermore, an enhancement in the performance of the subjects was not expected. Instead, the aim of this study was to assess the feasibility of the system and find specific features that can be improved. However, there were important quantitative results.

For the Clay Game, it is clear that participants had struggle maintaining the symmetry between forearm movements as the distance between them increased. This tendency could have a relation with the eye-hand coordination of the participants, as it may be easier for them to imitate this movement when they are observing both hands. With respect to the scores between sessions, some subjects improved their scores, while others worsened them. The fact that none of the participants got scores below 0.6 is useful for a characterization process in order to monitor the rehabilitation course of stroke patients.

For the Food Game, intervals on the motor imagery sections were higher than those on the forearm movement sections. This is because it is easier to perform movement than imagining doing so (in a way in which the system detects it). For these same reasons, the amount of repetitions was lower on the motor imagery section. Furthermore, the found intervals are useful in order to start a characterization process for analyzing stroke patients performance on rehabilitation processes with this system. The same happens with repetitions, as its values tend to be at least three times higher on forearm movement sections for healthy subjects. Taking all results into account, the system was suitable for healthy subjects.

While no enhancement in performance of the subjects was expected, results of the subjects through the sessions point out that the system is usable, since all the subjects surpassed a score threshold on each mini-game. In the case of the Clay Game, the minimum score was 0.6, whereas for the Food Game, during the movement intention step, all the subjects completed at least 3 repetitions. Also, it is remarkable that all the subjects completed at least 3 repetitions during the first session. This means that all the subjects were capable of using the system since the first session.

5   Conclusions

For this project, a HCI system was designed and assessed. The system includes EEG, postural analysis, motion analysis and various types of feedback, on healthy subjects. The detection of movement intention is key for the development of the system, as, for stroke patients, it allows for improved brain plasticity, leading to motor rehabilitation. Concerning this, it is essential to find ways to improve this aspect of the system. An approach to this may be to enhance the experimental protocol in order to secure the total involvement of the subject in the tasks at hand. For instance, it is possible to suppress distractions by developing an adequate room in which external sound is nullified and by forbidding the entrance of electronic devices to the experimental area. Another approach to improving the detection of movement intention is to evaluate the use of more machine learning algorithms, taking into account that it is crucial to maintain a low processing time, in order for the system to work online.

Regarding the Clay Game, the participants seemed to be comfortable while using the system, since all the users were capable of reaching an acceptable result (more than 60 points). The minimum score reached for all the healthy subjects could be considered a beginning in order to characterize the used metric for future studies on stroke patients. Also, subjects tended to present a better performance during the first steps of the path. According to this, for stroke patients, the HCI should include a new score system in which a higher value is given to the first steps of the path. Another important result for future works are the intervals obtained in the Food Game, as the elapsed time and quantity of repetitions are relatively stable, making them possible metrics to characterize the rehabilitation processes for stroke patients. As a general outcome, the obtained results are essential to improve the designed system and apply it on motor rehabilitation processes of stroke patients.

References

1. Vargas, G., et al.: On behalf of Observatorio Nacional de Salud - Colombia: Carga de enfermedad por enfermedades crónicas, no transmisibles y discapacidad en Colombia. MinSalud (2015)
2. Benjamin, E.J., et al.: Heart Disease and Stroke Statistics - 2018 Update. Circulation, New York (2018)
3. Johnson, N.N., et al.: Combined rTMS and virtual reality brain-computer interface training for motor recovery after stroke. J. Neural Eng. **15**(1), 016009 (2018)
4. Frolov, A.A., et al.: Post-stroke rehabilitation training with a motor-imagery-based brain-computer interface (BCI)-controlled hand exoskeleton: a randomized controlled multicenter trial. Front. Neurosci. **11**, 400–411 (2017)
5. Li, M., et al.: Neurophysiological substrates of stroke patients with motor imagery-based brain-computer interface training. Int. J. Neurosci. **124**(6), 403–415 (2014)
6. Ibáñez, J., et al.: Low latency estimation of motor intentions to assist reaching movements along multiple sessions in chronic stroke patients: a feasibility study. Front. Neurosci. **11**, 126 (2017)
7. Mrachacz-Kersting, N., et al.: Efficient neuroplasticity induction in chronic stroke patients by an associative brain-computer interface. J. Neurophysiol. **115**(3), 1410–1421 (2016)
8. Marquez-Chin, C., Marquis, A., Popovic, M.R.: BCI-triggered functional electrical stimulation therapy for upper limb. Eur. J. Transl. Myol. **26**(3), 6222 (2016)
9. Ono, T., et al.: Brain-computer interface with somatosensory feedback improves functional recovery from severe hemiplegia due to chronic stroke. Front. Neuroeng. **7**, 19 (2014)

10. Young, B.M., et al.: Changes in functional connectivity correlate with behavioral gains in stroke patients after therapy using a brain-computer interface device. Front. Neuroeng. **7**, 25 (2014)

11. Foldes, S.T., Weber, D.J., Collinger, J.L.: MEG-based neurofeedback for hand rehabilitation. J. Neuroeng. Rehabil. **12**, 85–93 (2015)

12. Pulido, S.D., López, J.M.: Brain-computer interface based on detection of movement intention as a means of brain wave modulation enhancement. In: 13th International Symposium on Medical Information Processing and Analysis. Proceedings of SPIE, San Andrés, Colombia (2017)

13. Mrachacz-Kersting, N.: Efficient neuroplasticity induction in chronic stroke patients by an associative brain-computer interface. J. Neurophysiol. **115**(3), 1410–1421 (2016)

14. Bishop, C.M.: Pattern Recognition and Machine Learning. Oxford University Press, Oxford (2006)

15. Charness, A.: Stroke/Head Injury: A Guide to Functional Outcomes in Physical Therapy Management, 1st edn. Aspen Systems Corporation, Chicago (1986)

16. Kommalapati, R., Michmizos, K.P.: Virtual reality for pediatric neuro-rehabilitation: adaptive visual feedback of movement to engage the mirror neuron system. In: 2016 38th Annual International Conference of the IEEE Engineering in Medicine and Biology Society (EMBC), Orlando, FL, USA, pp. 5849–5852 (2016)

17. Bermúdez i Badia, S., Fluet, G.G., Llorens, R., Deutsch, J.E.: Virtual reality for sensorimotor rehabilitation post stroke: design principles and evidence. In: Reinkensmeyer, D.J., Dietz, V. (eds.) Neurorehabilitation Technology, pp. 573–603. Springer, Cham (2016). https://doi.org/10.1007/978-3-319-28603-7_28

18. Monteiro-Junior, R., Vaghetti, C.O., Nascimento, O.J., Laks, J., Deslandes, A.: Exergames: neuroplastic hypothesis about cognitive improvement and biological effects on physical function of institutionalized older persons. Neural Regen. Res. **11**(2), 201 (2016)

19. Nikolaidis, A., Voss, M.W., Lee, H., Vo, L.T.K., Kramer, A.F.: Parietal plasticity after training with a complex video game is associated with individual differences in improvements in an untrained working memory task. Front. Hum. Neurosci. **8**, 169 (2014)

Weapon Detection for Particular Scenarios Using Deep Learning

Noelia Vallez[(✉)], Alberto Velasco-Mata, Juan Jose Corroto, and Oscar Deniz

University of Castilla-La Mancha, ETSI Industriales, Avda Camilo Jose Cela sn,
13071 Ciudad Real, Spain
Noelia.Vallez@uclm.es

Abstract. The development of object detection systems is normally driven to achieve both high detection and low false positive rates in a certain public dataset. However, when put into a real scenario the result is generally an unacceptable rate of false alarms. In this context we propose to add an additional step that models and filters the typical false alarms of the new scenario while roughly maintaining the ability to detect the objects of interest. We propose to use the false alarms of the new scenario to train a deep autoencoder and to model them. The latter will act as a filter that checks whether the output of the detector is one of its typical false positives or not based on the reconstruction error measured with the Mean Squared Error (MSE) and the Peak Signal-to-Noise Ratio (PSNR). We test the system using an entirely synthetic novel dataset for training and testing the autoencoder generated with Unreal Engine 4. Results show a reduction in the number of FPs of up to 37.9% in combination with the PSNR error while maintaining the same detection capability.

Keywords: False positive ratio · Autoencoder ·
One-class classification · Object detection

1 Introduction

In security, detecting potentially dangerous situations as soon as possible is of vital importance. Constant human supervision of the images provided by Closed Circuit Television (CCTV) systems is non feasible. In the last decades, several efforts have been made to create automated video surveillance (AVS) systems that can locate potentially threatening objects or events in the video sequence [11]. With the extended use of modern deep learning techniques, these systems obtain promising results [10,19].

The development of these detection systems is normally driven to achieve both high detection and low false positive rates. Ideally, training data would contain representative instances from all possible application scenarios. In practice, obtaining such a huge amount of data is not feasible in terms of time and resources. This problem forces data scientists to be cautious about overfitting and poor generalization when training new models [20]. Some techniques such as dataset partitioning, L1 and L2 regularizations or early stopping are applied to

© Springer Nature Switzerland AG 2019
A. Morales et al. (Eds.): IbPRIA 2019, LNCS 11868, pp. 371–382, 2019.
https://doi.org/10.1007/978-3-030-31321-0_32

alleviate them [14]. However, misclassification of samples in new scenarios must be addressed. In this respect, it is conceivable that a weapon detector could be trained using a dataset containing instances from all possible weapons that provides accurate detections and a small number of false positives. Then, when put into a surveillance system in a real scenario the result is generally an unacceptable rate of false alarms [18]. This means that the system will almost certainly be switched off, specially in cases where the incidence of the event of interest is very low. In this context we propose to add an additional step that models and filters the typical false alarms that are particular of the new scenario while roughly maintaining the ability to detect the objects it was trained for.

We focus on a handgun detection problem [10]. When such detector runs in a new scenario over a period of time, all of those detections, most likely false positives, can be stored and leveraged. Here we propose to use those false positives to train a deep autoencoder to model the typical false alarms of the particular scenario (and this can be done down to the level of individual cameras). The one-class classification approach has been widely used in the literature to detect abnormal and extreme data [6]. Autoencoders have shown the ability to perform this task even where other techniques fail [5].

In addition, due to the large number of images required to train deep learning models, we propose to use an entirely synthetic dataset for training and testing the autoencoder. It consists of frames captured from a realistic 3D environment that resembles the new scenario in which the detector would be deployed.

The rest of the paper is organized as follows. Section 2 gives the details of the handgun detector used. Section 3 covers the generation of the synthetic dataset of the new detection scenario. Section 4 describes the procedure followed to filter the false positives of the new scenario. Finally, Sect. 5 shows the performance of the proposed method and Sect. 6 summarizes the main conclusions.

2   Handgun Detector

To address the reduction of the false positive rate of the handgun detector we need one as a starting point. Classical machine learning methods based on keypoint matching and feature extraction and classification have been extensively applied to RGB images taken by CCTV videocameras [10]. Most of these methods use the *sliding window* approach with which the detection problem is solved as a classification problem in every examined window. This approach not only works with traditional methods but also with the new deep learning classification architectures. Convolutional neural networks (CNNs) can be used in the same way as support vector machines (SVMs) or cascade classifiers without having to represent the image as a set of features before performing classification.

The problem with the *sliding window* approach is the variability in the object locations within the image and the large differences in their aspect ratios. Thus, the number of regions to be examined is huge. In [2] this problem is addressed by taking different regions of interest from the image using a selective search to extract a manageable number of regions called *region proposals*. R-CNN, Fast

R-CNN and Faster R-CNN are the main detection architectures based on such region proposals approach [13]. In addition, there are other detection networks like YOLO (You Only Look Once) [12] or SSD (Single Shot Detector) [8] that are able to predict the bounding boxes and the class probabilities for these boxes, examining the whole image only once.

Due to the unavailability of pretrained handgun detection models, we have trained one handgun detector with a dataset provided by the University of Seville, Spain. The dataset contains 871 images with a total of 177 annotated handguns. Images come from 2 different CCTV controlled scenarios. The CNN architecture selected was the Faster R-CNN (Fig. 1).

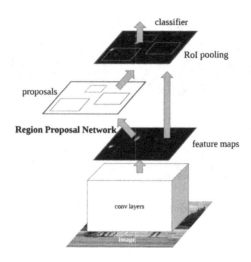

Fig. 1. Faster R-CNN architecture used for the handgun detector [13]

3   Synthetic Dataset

Collecting and annotating data to train deep learning networks is a tedious task. It is even more complex for detection and segmentation problems where someone selects not only the class but also a rectangle around every desired object in the image or pixel-level contour. The easiest solution would be to use an existing dataset for the problem addressed but this is not always possible.

To test the hypothesis in our work, we have used Unreal Engine 4 [17] to generate synthetic data from a school hall (Fig. 2). There are other popular alternatives such as Unity [16] or Lumberyard/CryEngine [9] that can be also used.

The synthetic scenario is similar to a high-school corridor. It is rendered with people walking across it generating a dataset of 3000 images where some people are carrying handguns, mobile phones or nothing in their hands. From these images, 2657 contains handguns, 343 do not contain the object of interest and the total number of handguns is 5437. Since we control the dataset generation, it is possible to automatically annotate where each handgun is. To store the annotations we have used XML files with the format defined in the Pascal VOC 2012 Challenge [1]. Although images containing the object of interest are not strictly necessary to train the autoencoder, they are needed to test the improvement of the detector+autoencoder system since reducing the false positive ratio might have a negative effect in the detection ratio.

Fig. 2. Synthetic scenario

4   Autoencoder

Autoencoder networks learn an approximation to an input data distribution. In other words, they learn how to compress input data into a short code and then reconstruct that code into something as close as possible to its original input. The structure of these networks consist of an encoder path that reduces the dimensionality ignoring signal noise and a decoder path that makes the reconstruction. Autoencoders are commonly used for anomaly detection or one-class classification [4].

In our case the autoencoder is trained to model one class: the typical false positives of the particular scenario (see Figs. 3 and 4).

Once the autoencoder is trained, it is applied to reconstruct images from a test dataset that contains also instances from the object of interest. If the reconstruction error is measured, it is lower for the class used to train the autoencoder and thus, a threshold can be established to separate the two classes (in this case, false positive or real positive). Therefore, the trained autoencoder will act as a filter that checks whether the output of the detector is one of its typical false positives (Fig. 5). In this work, two different error measures were applied, the

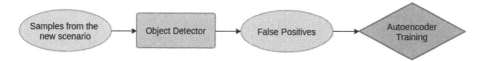

Fig. 3. Autoencoder training phase

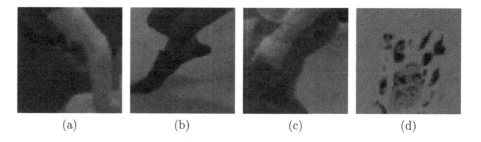

(a)   (b)   (c)   (d)

Fig. 4. Typical false positives of the handgun detector in this scenario (enlarged).

Mean Squared Error (MSE) [15] and the Peak Signal-to-Noise Ratio (PSNR) [7]. Figure 6 illustrates the reconstruction error of an autoencoder in a toy example using the MSE measure.

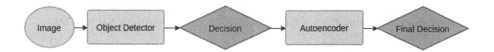

Fig. 5. Detector+Autoencoder system

The autoencoder structures used are shown in Fig. 7. In both cases, the input size is a 64 × 64 image with 3 channels. Their compressive paths consists of 1 or 6 convolutional and max-pooling layers. Similarly, the reconstruction paths have also 1 or 6 convolutional and up-sampling layers.

The dataset used is composed of 2100 images from the synthetic scenario. From them, 1200 images where used to generate the false positives to train the autoencoder and the remaining 900 to validate it.

5   Results

The Faster R-CNN model trained obtained an mAP of 79.33%. Its training process took around 2 days to complete 62 epochs in 2 nVIDIA Quadro M4000 cards using Keras with TensorFlow backend and CUDA 8.0 installed in a PC running Ubuntu 14.04 LTS.

We applied both the handgun detector and the detector+autoencoder app-roach to a test dataset composed of 900 images from the synthetic scenario,

376 N. Vallez et al.

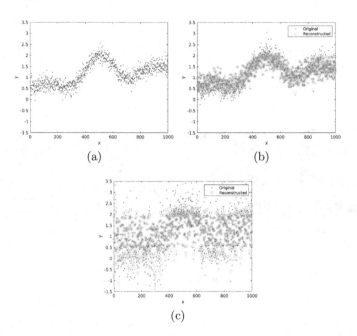

(a)   (b)

(c)

Fig. 6. Reconstruction error comparison. The more the data differs from the training set the higher the error is. (a) Autoencoder training data from the function $y = 1 + 0.05x + sin(x)/x + 0.2 * r$ where r is a normally distributed random number. (b) Reconstructed data from $y = 1 + 0.05x + sin(x)/x + 0.3 * r$ with $MSE = 0.533$. (c) Reconstructed data from $y = 1 + 0.05x + sin(x)/x + 0.8 * r$ with $MSE = 1.099$.

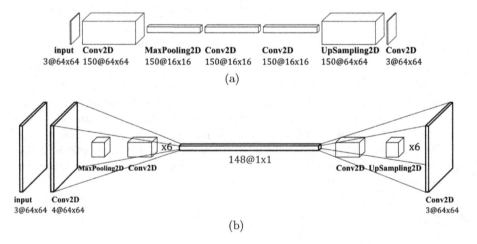

Fig. 7. Autoencoder architectures used.

containing 808 instances of the object. The histograms of the reconstruction error of the autoencoders by error measure are depicted in Figs. 8 and 9. The best separation of the FPs from the TPs was obtained with the larger autoencoder structure when using MSE and the smaller when using PSNR. Although the TPs are overlapped with the FPs in both cases, the first part of the histograms, that do not contain TPs, contain the 26.5% of all the FPs when the MSE is used and the 37.9% when the PSNR is used. This means that those FPs can be potentially filtered with the autoencoder by selecting a threshold based on the reconstruction error. Furthermore, if the image reconstructions of the FPs and the TPs are compared, it is possible to notice how the FPs are reconstructed better than the TPs (see Fig. 10).

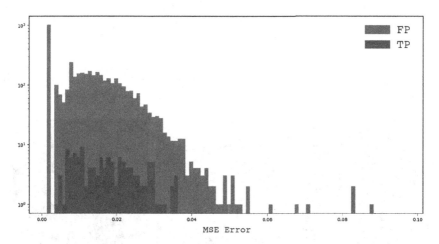

Fig. 8. Histogram of the reconstruction errors using MSE (y-axis uses logarithmic scale). FP - false positives of the detector, TP - true positives of the detector.

In addition, the corresponding precision-recall curves were also obtained. Considering the possible outputs of the detector, where TP and TN represent the number of true positives and true negatives and FP and FN stand for the number of false positives and the number of false negatives, the precision (p) and the recall (r) values can be calculated as shown in Eq. 1 [3].

$$p = \frac{TP}{TP + FP} \qquad r = \frac{TP}{TP + FN} \tag{1}$$

The experimental results show a reduction in the number of false positives while maintaining the detection capabilities (Figs. 11 and 12). The figures depict the precision-recall curves of both the detector and the detector+autoencoder approaches. Those curves were obtained varying the detector confidence threshold and fixing a certain threshold for the autoencoder. When compared, they show a maximum increase in the precision of 0.015 at the same recall values when the autoencoder and the MSE are used (maximum distance between precisions without displacing the recall values to the left) and a maximum increase

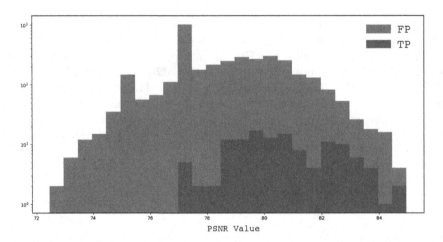

Fig. 9. Histogram of the reconstruction errors using PSNR (y-axis uses logarithmic scale). FP - false positives of the detector, TP - true positives of the detector.

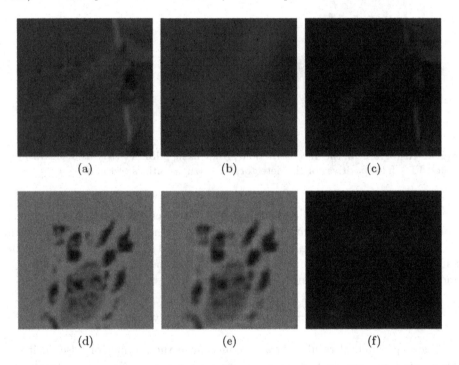

Fig. 10. Reconstruction made by the deepest autoencoder of a TP and a FP of the detector. (a) TP, (d) FP, (b) and (e) are the reconstructions, and (c) and (f) are the difference between the original image and the reconstruction.

in the precision of 0.020 at the same recall values when the autoencoder and the PSNR are used (threshold = 77.24).

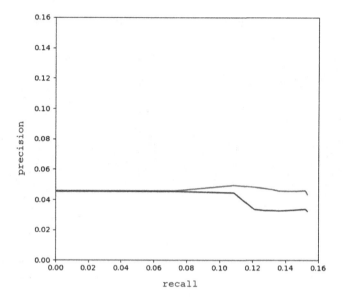

Fig. 11. Precision-recall curves using MSE. Detector in blue, Detector+Autoencoder in green. Best viewed in color. MSE threshold = 0.0044.

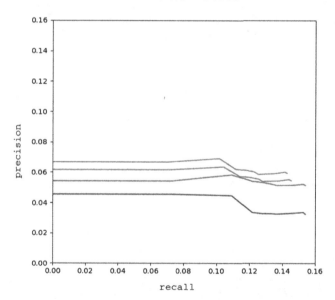

Fig. 12. Precision-recall curves using PSNR. Detector in blue, Detector+Autoencoder in green. Best viewed in color. Thresholds used, from top to bottom, 78.5, 78 and 77.24.

In addition, it is also worth noting that, although the results with the shortest autoencoder architecture and the PSNR error measure are superior to those obtained by the deepest architecture and the MSE error measure, the reconstructed image is noisier (see Fig. 13).

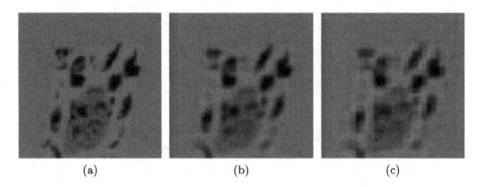

(a)   (b)   (c)

Fig. 13. Reconstruction with the two autoencoder architectures. (a) Original image of a FP of the detector. (b) Reconstruction with the deepest autoencoder architecture. (c) Reconstruction with the shortest autoencoder architecture.

With respect to the computational times, the deep autoencoders training processes took only 45 min to complete 500 epochs in an nVIDIA GTX 1060 MaxQ card using Keras with TensorFlow backend and CUDA 9.0 installed in a Windows 10 PC.

6   Conclusions

This work focuses on reducing the number of false alarms when a surveillance application is run in a new particular scenario. A synthetic scenario has been generated with the game engine Unreal Engine 4, resembling a surveillance camera inside a high-school hall where people is walking by. Images coming from this scenario were used as input of a pretrained handgun detector. Its false positive detections where then used to train a deep autoencoder to model them and act like a filter, removing those false positives.

The autoencoder has demonstrated to be able to reduce the number of FPs in up to 37.9% in combination with the PSNR error while maintaining the same capability of detecting handguns. Notice that the number of FPs and the detection ability of the detector is a trade-off that should be always considered. The detector+autoencoder approach proposed helps keep a good balance between them since it is possible to not affect both at the same time for the first range of the autoencoder thresholds.

Overall, our approach can be used with generic detectors (i.e. a generic handgun detector) and with different particular scenarios (down to the level of individual cameras, which depending on the point of view, lighting, etc, will produce different false positives). Thus, in practice we would only need the generic detector and one trained autoencoder per camera feed.

As future work, it would be useful to consider other detection architectures to check if they can influence the autoencoder. In addition, other error metrics can be used to get the autoencoder threshold.

Acknowledgments. We thank Professor Dr. J.A. Alvarez for the surveillance images provided for training the handgun detector. This work was partially funded by projects TIN2017-82113-C2-2-R by the Spanish Ministry of Economy and Business and SBPLY/17/180501/000543 by the Autonomous Government of Castilla-La Mancha and the ERDF.

References

1. Everingham, M., Van Gool, L., Williams, C.K.I., Winn, J., Zisserman, A.: The PASCAL Visual Object Classes Challenge 2012 (VOC 2012) Results. http://www.pascal-network.org/challenges/VOC/voc2012/workshop/index.html
2. Girshick, R.B., Donahue, J., Darrell, T., Malik, J.: Rich feature hierarchies for accurate object detection and semantic segmentation. CoRR abs/1311.2524 (2013). http://arxiv.org/abs/1311.2524
3. Goutte, C., Gaussier, E.: A probabilistic interpretation of precision, recall and F-Score, with implication for evaluation. In: Losada, D.E., Fernández-Luna, J.M. (eds.) ECIR 2005. LNCS, vol. 3408, pp. 345–359. Springer, Heidelberg (2005). https://doi.org/10.1007/978-3-540-31865-1_25
4. Gutoski, M., Ribeiro, M., Aquino, N.M.R., Lazzaretti, A.E., Lopes, H.S.: A clustering-based deep autoencoder for one-class image classification. In: 2017 IEEE Latin American Conference on Computational Intelligence (LA-CCI), pp. 1–6 (2017)
5. Hofer-Schmitz, K., Nguyen, P.H., Berwanger, K.: One-class Autoencoder approach to classify Raman spectra outliers. In: European Symposium on Artificial Neural Networks, ESANN 2018, pp. 189–194 (2018)
6. Khan, S.S., Madden, M.G.: One-class classification: taxonomy of study and review of techniques. Knowl. Eng. Rev. **29**(3), 345–374 (2014). https://doi.org/10.1017/S026988891300043X
7. Kotevski, Z., Mitrevski, P.: Experimental comparison of PSNR and SSIM metrics for video quality estimation. In: Davcev, D., Gómez, J.M. (eds.) ICT Innovations 2009, pp. 357–366. Springer, Heidelberg (2010). https://doi.org/10.1007/978-3-642-10781-8_37
8. Liu, W., et al.: SSD: single shot multibox detector. CoRR abs/1512.02325 (2015). http://arxiv.org/abs/1512.02325
9. Lumberyard. https://aws.amazon.com/es/lumberyard. Accessed 09 Apr 2019
10. Olmos, R., Tabik, S., Herrera, F.: Automatic handgun detection alarm in videos using deep learning. CoRR abs/1702.05147 (2017). http://arxiv.org/abs/1702.05147
11. Raghunandan, A., Mohana, M., Pakala, R., Aradhya, H.V.R.: Object detection algorithms for video surveillance applications. In: IEEE - 7th International Conference on Communication and Signal Processing, April 2018. https://doi.org/10.1109/ICCSP.2018.8524461
12. Redmon, J., Divvala, S.K., Girshick, R.B., Farhadi, A.: You only look once: unified, real-time object detection. CoRR abs/1506.02640 (2015). http://arxiv.org/abs/1506.02640

13. Ren, S., He, K., Girshick, R.B., Sun, J.: Faster R-CNN: towards real-time object detection with region proposal networks. CoRR abs/1506.01497 (2015). http://arxiv.org/abs/1506.01497
14. Srivastava, N., Hinton, G., Krizhevsky, A., Sutskever, I., Salakhutdinov, R.: Dropout: a simple way to prevent neural networks from overfitting. J. Mach. Learn. Res. **15**(1), 1929–1958 (2014)
15. Tan, C.C., Eswaran, C.: Performance comparison of three types of autoencoder neural networks. In: 2008 Second Asia International Conference on Modelling Simulation (AMS), pp. 213–218, May 2008. https://doi.org/10.1109/AMS.2008.105
16. Unity. https://unity.com. Accessed 09 Apr 2019
17. Unreal Engine 4. https://www.unrealengine.com. Accessed 09 Apr 2019
18. Vállez, N., Bueno, G., Déniz, O.: False positive reduction in detector implantation. In: Peek, N., Marín Morales, R., Peleg, M. (eds.) AIME 2013. LNCS, vol. 7885, pp. 181–185. Springer, Heidelberg (2013). https://doi.org/10.1007/978-3-642-38326-7_28
19. Xu, D., Yan, Y., Ricci, E., Sebe, N.: Detecting anomalous events in videos by learning deep representations of appearance and motion. Comput. Vis. Image Underst. **156**, 117–127 (2017). https://doi.org/10.1016/j.cviu.2016.10.010. http://www.sciencedirect.com/science/article/pii/S1077314216301618. Image and Video Understanding in Big Data
20. Zhang, C., Bengio, S., Hardt, M., Recht, B., Vinyals, O.: Understanding deep learning requires rethinking generalization. arXiv pre-print (2016). http://arxiv.org/abs/1611.03530

Hierarchical Deep Learning Approach for Plant Disease Detection

Joana Costa[1,2(✉)], Catarina Silva[1,2], and Bernardete Ribeiro[2]

[1] School of Technology and Management, Polytechnic Institute of Leiria,
Leiria, Portugal
[2] CISUC - Center for Informatics and Systems, University of Coimbra,
Coimbra, Portugal
{joanamc,catarina,bribeiro}@dei.uc.pt

Abstract. In this paper we propose a hierarchical deep learning approach for plant disease detection. The detection of diseases in plants using deep image approaches is attracting researchers as a way of taking advantage of cutting-edge learning techniques in scenarios where major benefits can be achieved for mankind. In this work, we focus on diseases of three major different agricultural crops: apple, peach and tomato. Using a real-world dataset composed of nearly 24,000 images, including healthy examples, we propose a hierarchical deep learning approach for plant disease detection and compare it with the standard deep learning approaches. Results permit the conclusion that hierarchical approaches can overcome standard approaches in terms of detection performance.

1 Introduction

One of the greatest technological advances that humans have ever made has been the domestication of plants during the agricultural revolution 8 to 12,000 years ago at multiple sites around the world [5]. These and the following breakthroughs allowed for the evolution of human population to the current remarkable 7.7 billion people[1].

Such evolution has put colossal pressure on agriculture technology, that is nowadays under immense challenges due to infectious diseases and pests spread by the globalization and compounded by climate change [1].

Hence, major benefits can arise from more accurate and faster detection of plant diseases, which is being increasingly enabled by sensors that obtain real time information from crops plantation sites.

Such approaches, usually coined as smart farming since they constitute an evolution of traditional techniques, are also important to tackle the challenges of agricultural production in terms of productivity, environmental impact, food security and sustainability [7].

---

[1] http://www.worldometers.info/world-population/.

© Springer Nature Switzerland AG 2019
A. Morales et al. (Eds.): IbPRIA 2019, LNCS 11868, pp. 383–393, 2019.
https://doi.org/10.1007/978-3-030-31321-0_33

To tackle these challenges, more information on agricultural ecosystems is needed, which is being obtained through new technologies achieved by monitoring, measuring and continuously analyzing various physical aspects and phenomena. Such scenarios have become possible through the development of Internet of Things (IoT), allowing that remote and often autonomous sensors provide information, whether sensory as humidity and temperature, or imaging as snapshots taken by fixed cameras or drone-based imaging.

This setup permits the gathering of huge amounts of data over large geographical areas, thus originating problems that can be framed into big data scenarios, which have not yet been widely applied in agriculture [6].

A recent technique that can deal with such challenges is Deep Learning (DL) [10] that by using "deeper" neural networks better represents data by means of various convolutions, allowing larger learning capabilities and higher accuracy [7].

In this paper we propose a two-level deep learning hierarchical approach for plant disease detection. The first level starts to determine the crop in the image and the second level focuses on learning the specific diseases of that specific crop. Results show the overall advantages of the approach and allow for the application in real scenarios in which the crop is a given.

The rest of the paper is organized as follows. In the next section we will present the current state of the art approaches to plant disease detection with deep learning, including a brief introduction to existing deep learning implementation strategies. In Sect. 3 we will introduce the hierarchical approach for plant disease detection starting from a standard baseline approach. Section 4 will detail on the experimental setup, namely on the dataset used, the deep learning model setup and the performance metrics. Section 5 will present and discuss the results obtained, and in Sect. 6 final conclusions and future lines of research are pointed out.

2   Plant Disease Detection with Deep Learning

2.1   Deep Learning Strategies

Deep learning is extremely good for visual feature extraction from images, audio signals, or text, which makes it very attractive to be used today. Given that specific datasets of plant disease images can be easily processed by deep neural networks without the need to extract manually hand-crafted features, we posit that higher performance than traditional methods can be obtained from these models. In particular, Convolutional Neural Networks (CNN) (and its variants) are becoming the current state of art in many image processing problems.

Deep neural network architectures fall into the four main types (for a recent survey see [12]):

1. Stacked Auto-Encoder (SAE)
2. Convolutional Neural Networks (CNN)
3. Resctricted-Boltzmann Machines (RBM)

4. Deep Belief Networks (DBN)

Despite the diversity of models, improved learning algorithms and new application players are exponentially increasing.

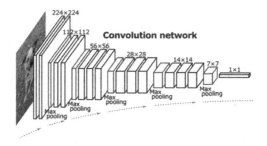

Fig. 1. Convolutional neural network [16]

CNN are depicted in Fig. 1. Being biologically inspired variants of Multilayer perceptrons. From Hubel and Wiesel's early work on the cat's visual cortex [3,4], we know that the visual cortex contains a complex disposition of cells, called receptive fields, that are sensitive to small regions of the visual field. These simple cells act as filters that respond to edge-like patterns and form complex hierarchies that are invariant to the position of the pattern. Being the visual cortex the most powerful visual processing system known, its behavior has inspired many models, for instance, the CNN.

2.2 Image-Based Approaches

Convolutional Neural Networks (CNN) are multilayer neural networks designed to recognize visual patterns directly from image pixels. The work by [9,11] has been pioneer for the current CNN that are researched today. As a consequence, CNN have been catapulted to the center of object recognition research. The rekindled interest in CNN is largely attributed to [8] CNN model, that showed significantly higher image classification accuracy on the 2012 ImageNet Large Scale Visual Recognition Challenge (ILSVRC). Their success resulted from a model inspired by LeCun's previous work and a few twists that enabled training with 1.2 million labeled images (*e.g.* GPU programming, max(x,0) rectifying non-linearities and dropout regularization). Likewise, the emergence of the giant processing power from parallel programming models developed by large HPC (High Performance Computing) teams working in the industry also contributed by a large amount to their success.

2.3 Disease Detection

Although it is still a developing research field, different successful approaches have been pursued as we will show in this section.

In [2] a deep-learning-based approach is presented to detect diseases and pests in tomato plants using images captured in-place. Three main families of DL detectors were considered and combined with additional information to effectively recognize nine different types of diseases and pests, with the ability to deal with complex scenarios from a plant's surrounding area.

In [14] DL is applied to Cassava disease detection. Cassava is the third largest source of carbohydrates for human food in the world, hence any improvement in the disease control can prove precious specially in less developed countries. In this work, the authors have applied transfer learning to train a deep convolutional neural network to identify three diseases and two types of pest damage, achieving accuracy of up to 98%.

In [13] the authors take advantage of increasing global smartphone penetration and recent advances in computer vision made possible by deep learning to propose smartphone-assisted disease diagnosis. Using a deep convolutional neural network to identify 14 crop species and 26 diseases, the trained model achieves an accuracy of 99.35% on a held-out test set, demonstrating the feasibility of this approach.

In [7] it is possible to find a survey on deep learning in agriculture, giving a broad view of the wide range of applications and methods that are being pursued.

3   Hierarchical Approach for Plant Disease Detection

To introduce the proposed approach, we will firstly define the problem of plant disease detection using images. It can be stated as follows.

Consider a set I of N images of C agricultural crops. For each crop c there are different target classes, i.e it is a multiclass problem. Those classes include healthy crops and specific diseased crops.

The goal can then be stated as, given a set of images I,

$$I = \{i_1, i_2, \ldots, i_N\}, \tag{1}$$

identify, for each image i_n the target class T_d^c, i.e. make the correspondence of each image i_n to a specific disease d of crop c or just the correspondence to a healthy state.

3.1   Standard Baseline Approach

The standard baseline approach is depicted in Fig. 2 and consists of building a model that uses all images and detects for each input simultaneously the crop and the disease. This approach is the typical deep learning approach in which all preprocessing efforts are supported by the neural network and has usually good results.

However, in plant disease detection, specially in approaches based in the crop/disease pair, difficulties can arise for instance due to the same disease in different crops, or different diseases producing similar damaged patterns but in different crops and hence a more structured approach may become appropriate.

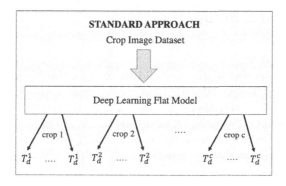

Fig. 2. Standard baseline deep learning flat approach.

3.2   Hierarchical Approach

Figure 3 shows the proposed two-step hierarchical deep learning approach.

The rationale is to divide the problem of crop disease detection in two sub-problems. The first (STEP 1) consists in identifying the crop and the second (STEP 2) consists in determining for a specific crop which, if any, disease in the dataset is identified/recognized. Such division may prove reasonable for various reasons and applications.

It also allows for the reuse of the models in STEP 2 in scenarios where we have plantations of only one crop and the goal is to detect diseases in that specific crop.

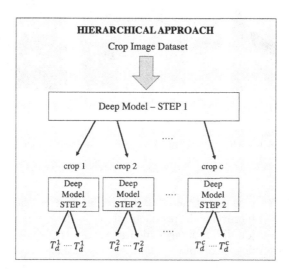

Fig. 3. Hierarchical deep learning approach.

4   Experimental Setup

4.1   Dataset

PlantVillage is an online platform dedicated to crop health and crop diseases (http://www.plantvillage.org) with more than 2 million site visits reported [5]. It started as an online crowdsourced forum where users could put questions and receive answers from other users. Additionally, a library of open access information of images of over 150 crops and over 1,800 diseases was also constructed and curated by plant pathology experts.

(a)   (b)   (c)

(d)   (e)   (f)

Fig. 4. Example images of some classes in the dataset: (a) Apple healthy; (b) Peach healthy; (c) Tomato healthy; (d) Apple black rot; (e) Peach bacterial spot; and, (f) Tomato leaf mold.

In this work, we have constructed a dataset using the PlantVillage images to test the proposed hierarchical approach for plant disease detection. We have focused on three major different agricultural crops: apple, peach and tomato. Figure 4 shows example images of the classes in the dataset, namely three examples of healthy apple, peach, and tomato crops, and three examples of diseases from the same crops.

Table 1 presents the classes associated with these three crops, consisting in 16 classes and nearly 24,000 image examples. We have used 70% of each class

Table 1. Dataset classes and training/testing sets

Class	Number of samples	Train samples	Test samples
Apple___Apple_scab	630	441	189
Apple___Black_rot	621	435	186
Apple___Cedar_apple_rust	275	193	82
Apple___healthy	1645	1152	493
Peach___Bacterial_spot	2297	1608	689
Peach___healthy	360	252	108
Tomato___Bacterial_spot	2127	1489	638
Tomato___Early_blight	1000	700	300
Tomato___healthy	1591	1114	477
Tomato___Late_blight	1909	1336	573
Tomato___Leaf_Mold	952	666	286
Tomato___Septoria_leaf_spot	1771	1240	531
Tomato___Spider_mites Two-spotted_spider_mite	1676	1173	503
Tomato___Target_Spot	1404	983	421
Tomato___Tomato_mosaic_virus	373	261	112
Tomato___Tomato_Yellow_Leaf_Curl_Virus	5357	3750	1607
Total	23988	16793	7195

for training purposes and 30% for testing. Notice that both classes per crop and examples per class are heterogeneous.

4.2   Deep Learning Models

The evaluation of our approach was performed with a fine-tuned InceptionV3 CNN proposed in [17]. In an Inception CNN, the input is split into a few lower-dimensional embeddings, transformed by a set of specialized filters, and merged by concatenation. The solution space of this architecture is a strict subspace of the solution space of a single large layer operating on a high-dimensional embedding, and this it is expected the representational power of large and dense layers, but at a considerably lower computational complexity [18].

We have used the implementation of the InceptionV3 distributed in Keras[2] with the TensorFlow[3] backend. The InceptionV3 was initialized with weights pre-trained on the ImageNet dataset, and fine-tuned to transfer learning.

We have empirically defined our experimental configurations, and thus all CNN run for 30 epochs and 50 steps each. They consistently converged to a better accuracy performance.

The main goal was to compare the performance of the standard and hierarchical approaches using identical models as base classifiers, not targeting improvements in the classification performance of the models.

---

[2] http://keras.io.
[3] http://tensorflow.org.

4.3   Evaluation Metrics

In this setup a multiclass single label problem is defined, i.e., each image belongs to just one of the specified classes. One of the most common evaluation strategies is to calculate the accuracy of a classifier, i.e., the percentage of image instances that are correctly assigned to their class.

We can also simplify our problem into a binary decision problem, as each image can be classified as being in a given class, or not. In order to evaluate the binary decision task, a contingency matrix can be defined to represent the possible outcomes of the classification, as shown in Table 2.

Table 2. Contingency table for binary classification

	Class positive	Class negative
Assigned Positive	a	b
	(True Positives)	(False Positives)
Assigned Negative	c	d
	(False Negatives)	(True Negatives)

Traditional measures can be defined based on this contingency table, such as error rate ($\frac{b+c}{a+b+c+d}$) and the above-mentioned accuracy ($\frac{a+d}{a+b+c+d}$). However, for unbalanced problems, i.e., problems where the number of positive examples is rather different among classes, or in the case of a binary problem, the number of positive examples is rather different from the negative examples more specific measures should be defined to capture the performance of each model. Typical examples include recall ($R = \frac{a}{a+c}$) and precision ($P = \frac{a}{a+b}$). Additionally, combined measures that give a more holistic view of performance in just one value, like the van Rijsbergen F_1 measure [15], which combines recall and precision:

$$F_1 = \frac{2 \times P \times R}{P + R}. \tag{2}$$

Two conventional methods are widely used in multiclass scenarios, namely macro-averaging and micro-averaging. Macro-averaged performance scores are obtained by computing the scores for each class and then averaging these scores to obtain the global means. Differently, micro-averaged performance scores are computed by summing all the previously introduced contingency matrix values (a, b, c and d), and then use the sum of these values to compute a single micro-averaged performance score that represents the global score.

5   Experimental Results and Analysis

Table 3 summarises the performance results obtained by classifying the testing set, with the macro-averaged performance measures, namely recall, precision, F1,

Table 3. Standard and hierarchical results for each crop/disease pair

Class	Standard approach				Hierarchical approach			
	Recall	Precision	F1	Accuracy	Recall	Precision	F1	Accuracy
Apple_Apple_scab	91.53%	77.93%	84.18%	99.10%	97.88%	72.55%	83.33%	98.97%
Apple_Black_rot	91.40%	92.92%	92.64%	99.62%	98.39%	91.96%	95.06%	99.74%
Apple_Cedar_apple_rust	71.95%	95.16%	81.94%	99.64%	95.12%	95.12%	95.12%	99.89%
Apple_healthy	93.31%	99.78%	96.44%	99.53%	99.80%	77.48%	87.23%	98.00%
Peach_Bacterial_spot	99.98%	96.88%	97.92%	99.60%	95.50%	95.50%	95.50%	99.14%
Peach_healthy	99.07%	96.40%	97.72%	99.93%	80.56%	95.60%	87.44%	99.65%
Tomato_Bacterial_spot	65.05%	93.47%	76.71%	95.50%	72.41%	98.93%	83.62%	97.48%
Tomato_Early_blight	84.67%	51.31%	63.90%	96.01%	78.67%	65.92%	71.73%	97.41%
Tomato_healthy	57.44%	92.57%	70.89%	96.87%	98.32%	89.85%	93.89%	99.15%
Tomato_Late_blight	72.60%	97.20%	83.12%	97.65%	95.99%	90.76%	93.30%	98.90%
Tomato_Leaf_Mold	80.77%	84.62%	82.65%	98.65%	87.06%	96.51%	91.54%	99.36%
Tomato_Septoria_leaf_spot	96.42%	76.42%	85.26%	97.54%	88.51%	93.81%	91.09%	98.72%
Tomato_Spider_mites 2-spotted_spider_mite	41.15%	84.15%	55.27%	95.34%	94.63%	90.84%	92.70%	98.96%
Tomato_Target_Spot	72.21%	80.21%	76.00%	97.33%	74.11%	86.91%	80.00%	99.83%
Tomato_Tomato_mosaic_virus	57.14%	25.10%	34.88%	96.68%	91.07%	86.44%	88.70%	99.64%
Tomato_Tomato_Yellow_Leaf_Curl_Virus	98.88%	80.74%	88.90%	94.48%	93.96%	98.63%	96.24%	98.36%
Average	79.54%	82.87%	79.28%	97.78%	90.12%	89.18%	89.16%	98.83%

and accuracy. Each row refers to one of the 16 classes considered. Results are presented both for the standard flat approach and for the hierarchical approach.

A global analysis of macro-average shows the potential benefits of the proposed method. Specifically in classes where the flat model underperforms in terms of recall and precision, the hierarchical approach shows significant improvement. Notice that the accuracy often used as the only metric, in fact offers a rather biased notion of the performance obtained.

Table 4 shows the macro-averaged performance measures for the three crops in STEP 1: Apple, Peach, and Tomato. This remarkably easier problem exhibits an almost perfect result. This achievement allows for the improvement that STEP 2 presents over the flat approach.

Table 4. Hierarchical approach STEP 1 results

Class	Hierarchical STEP 1			
	Recall	Precision	F1	Accuracy
Apple	100.00%	81.13%	89.58%	96.93%
Peach	94.98%	97.05%	96.01%	99.12%
Tomato	96.26%	100.00%	98.09%	97.16%
Average	97.08%	92.73%	94.56%	97.74%

6   Conclusions and Future Work

In this paper we have proposed a two-level deep learning hierarchical approach
for plant disease detection. The first level starts to determine the crop in the
image and the second level focuses on learning the specific disease of that specific
crop. Results show the overall advantages of the approach and are expected to
be used in real scenarios applications in which the crop is given.

Results show the effectiveness of the proposed hierarchical approach in terms
of overall performance. When targeting specific crop diseases such approach can
bring significant advantages, when considering other measures than accuracy
that only provides a partial representation of the resulting performance.

Future work is foreseen on further exploring the hierarchical nature of the
proposal and additional research on different deep architectures.

References

1. Bourne, J.E.: The End of Plenty: The Race to Feed a Crowded World. W. W. Norton & Company, New York (2015)
2. Fuentes, A., Yoon, S., Kim, S.C., Park, D.S.: A robust deep-learning-based detector for real-time tomato plant diseases and pests recognition. Sensors **17**(9), 2022 (2017)
3. Hubel, D.H., Wiesel, T.N.: Receptive fields, binocular interaction and functional architecture in the cat's visual cortex. J. Physiol. **160**, 106–154 (1962)
4. Hubel, M., Wiesel, T.N.: Brain and Visual Perception. Oxford Univeristy Press, Oxford (2005)
5. Hughes, D.P., Salathé, M.: An open access repository of images on plant health to enable the development of mobile disease diagnostics through machine learning and crowdsourcing. CoRR (2015)
6. Kamilaris, A., Kartakoullis, A., Prenafeta-Boldú, F.: A review on the practice of big data analysis in agriculture. Comput. Electron. Agric. **143**(1), 23–37 (2017)
7. Kamilaris, A., Prenafeta-Boldú, F.X.: Deep learning in agriculture: a survey. Comput. Electron. Agric. **147**, 70–90 (2018)
8. Krizhevsky, A., Sutskever, I., Hinton, G.E.: Imagenet classification with deep convolutional neural networks. In: Advances in Neural Information Processing Systems, pp. 1097–1105 (2012)
9. LeCun, Y., Boser, B., Denker, J.S., Henderson, D., Hubbard, R.E.H.W., Jackel, L.D.: Backpropagation applied to handwritten zip code recognition. Neural Comput. **1**(4), 541–555 (1989)
10. LeCun, Y., Bengio, Y., Hinton, G.: Deep learning. Nature **521**(7553), 436–444 (2015)
11. LeCun, Y., Bottou, L., Bengio, Y., Haffner, P.: Gradient-based learning applied to document recognition. Proc. IEEE **86**(11), 2278–2324 (1998)
12. Liu, W., Wang, Z., Liu, X., Zeng, N., Liu, Y., Alsaadi, F.E.: A survey of deep neural network architectures and their applications. Neurocomputing **234**, 11–26 (2017)
13. Mohanty, S.P., Hughes, D.P., Salathe, M.: Using deep learning for image-based plant disease detection. Frontiers Plant Sci. **7**, 1419 (2016)

14. Ramcharan, A., Baranowski, K., McCloskey, P., Ahmed, B., Legg, J., Hughes, D.P.: Deep learning for image-based cassava disease detection. Frontiers Plant Sci. **8**, 1852 (2017)
15. van Rijsbergen, C.: Information Retrieval. Butterworths, London (1979)
16. Simonyan, K., Zisserman, A.: Very deep convolutional networks for large-scale image recognition. CoRR abs/1409.1556 (2014)
17. Szegedy, C., Vanhoucke, V., Ioffe, S., Shlens, J., Wojna, Z.: Rethinking the inception architecture for computer vision. In: Proceedings of the IEEE Conference on Computer Vision and Pattern Recognition, pp. 2818–2826 (2016)
18. Xie, S., Girshick, R., Dollár, P., Tu, Z., He, K.: Aggregated residual transformations for deep neural networks. In: 2017 IEEE Conference on Computer Vision and Pattern Recognition (CVPR), pp. 5987–5995. IEEE (2017)

An Artificial Vision Based Method for Vehicle Detection and Classification in Urban Traffic

Camilo Camacho[1]([✉]) [ID], César Pedraza[2] [ID], and Carolina Higuera[1] [ID]

[1] Universidad Santo Tomás, Bogotá, Colombia
edgarcamacho@usantotomas.edu.co
[2] Universidad Nacional de Colombia, Bogotá, Colombia

Abstract. This paper proposes a system to analyze urban traffic through the using of artificial vision, in order to get reliable information about the traffic flow in cities with severe traffic jam, as in Bogotá, Colombia. It was proposed a method efficient enough to be implemented in an embedded system, in order to process the images captured by a local camera and send the synthesized information to the cloud. This approach would allow spending fewer data transference, because it would not be necessary to send the video of each camera in the city via streaming, instead, each camera would send only the relevant traffic information. The system is able to calculate traffic flow, classified in motorbikes, buses, microbuses, minivans, sedans, SUVs and trucks.

The detection was implemented using a cascade classifier that evaluates HAAR features, providing a detection rate of 74.9% and a false positive rate of 1.4%. A Kalman filter was used to track and count the detected vehicles. Finally, a Convolutional Neural Network performing as classifier, with accuracies around 88%. The complete system presented errors around 2.5% in contrast with the manual counting in traffic of Bogotá, Colombia.

Keywords: Intelligent transportation systems ·
Convolutional Neural Networks · Artificial vision · Road car counting

1 Introduction

An intelligent transportation system (ITS) is the operational combination of a set of technologies that, when combined and managed in the right way, improve the performance and capabilities of the entire system. An ITS improves the transport system of a city or an intercity highway, making it safer and more efficient [16].

In intelligent transportation systems, as in traditional systems, the decisions taken to improve the mobility on public roads are based on traffic variables

Supported by Universidad Santo Tomás.

such as traffic flow and average speed, which at the same time are the objective of optimization. Some of these decisions can be the change of green times of traffic lights or the allowing of the right-of-way of lanes in specific directions, or merely the measured variables can be used to establish the performance of transit policies, analyzing their values before and after their implementation.

In some countries the vehicle counting is still performed manually to calculate urban flows. This process is performed by staff located at intersections on the main roads of the city, which draw strokes in formats of vehicular capacity, where each stroke represents a car that travels by that location [2]. This method presents an unknown level of error, increasing the uncertainty of the results. In order to improve the reliability of the data obtained, redundant measures are used; then, several people perform the same work at the same location and then their results are averaged. This method does not establish a known level of error; on the contrary, it increases the costs of the measurement.

This work presents a method for detection, tracking, counting and classification of urban traffic through digital image processing and convolutional neural networks, which aims to improve the disadvantages presented by the manual process. In Sect. 2 it is presented a state of the art that introduces the three main stages that make up the proposed method: detection, tracking and classification. Section 3 explains the techniques used in each stage of the algorithm. Finally, Sect. 4 presents the results obtained from the tests carried out in different locations in the city of Bogotá. Conclusions are presented in Sect. 5.

2   Related Work

Typically, the process of detecting and counting vehicles with artificial vision is divided into three main sections: detection, which establishes the location of the vehicles present in each frame of a video; tracking, which calculates the route followed by each of the vehicles in the scene, and classification, which identifies different types of vehicles. There are a large number of methods used to perform these tasks, here are some of them as well as several previous works:

2.1   Detection

A widely used method for object detection is background subtraction, which is based on the background modelling in a sequence of images. The segments corresponding to moving objects are calculated through the difference between each frame and the previously calculated model, which is later updated with the same frame. In [3], a vehicular detection is done using this method, carrying out the background modeling from a Gaussian Mixed Model (GMM). This technique models each of the background pixels as a mixture of Gaussian distributions. Therefore, pixels that do not adjust to such distributions are considered as moving objects [15].

In [19] the potential location of the vehicles in the image is made by previously placing the shadows of the objects in the scene, in order to corroborate the detection from symmetry and edge detection. A similar approach that adds texture analysis to the image is presented in [9]. Another approach for detection of objects is the eigen component search of the element to be detected; for instance, in [11] these components are searched from basic forms, such as circles to find the tires of a bicycle. In the case of vehicles, in [13] was presented a method based on the detection of the back of the car, specifically from the rear lights. The detection method used in this work is based on a cascade classifier previously trained through the algorithm presented by Viola and Jones in [17]. Some work done with this method is presented in [14,18,20].

2.2 Tracking

Tracking corresponds to trace the path of a moving object as it changes its location in a scene. The tracking process is mainly divided into three types: based on region, in which the deviation of the segment corresponding to the moving object is calculated; based on active contour, which calculates and updates a box enclosing the detected object, and based on characteristics, which traces and follows specific characteristics of the object [12].

In addition, a method must be established to estimate future positions of the object, in order to perform the tracking even when the detection step does not deliver the position in a frame. The Kalman filter is an estimator that uses measures observed in the past to make future predictions of the variable. [8] presents satisfactory results in the tracking of multiple moving objects. The particle filter can be similarly used to this application, which allows to simplify the traditional methods used by Kalman filter [1].

2.3 Classification

The classification of vehicles is made from machine learning algorithms, which are trained with large datasets of different types of cars, followed by the extraction of different characteristics, such as dimensions (length, width, area), border orientation histogram, HAAR features, color, among others. These characteristics are applied to a classifier like a neural network, a support vector machine, a boost classifier, among others. Ojha y Sakhare present in [10] a summary of works done with different types of classifiers and characteristics.

Sometimes, the choosing of the correct type of feature presents a challenge, takes a long time, and does not always give good results. The Convolutional Neural Networks (CNNs) handle that problem using convolutional layers, extracting features through spatial filters, whose weights are learned in the same way as the other networks parameters. [4,5] present some works where CNNs are used for the classification of vehicles.

3 Proposed Method

The proposed procedure for the detection, tracking, counting and classification of vehicles is summarized in Fig. 1. This procedure is performed independently for cars and motorcycles.

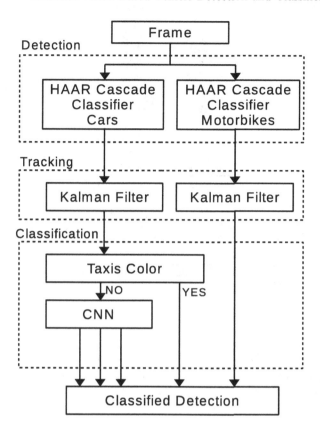

Fig. 1. Proposed method.

Viola-Jones Algorithm. The first stage of detection is based on the Viola-Jones algorithm [17], that is based on a cascade classifier that uses *HAAR* features, as those shown in the Fig. 2, where the feature value is the difference between the sum of the pixels under the black and white regions.

The set of characteristics are preselected through an *AdaBoost* learning algorithm. The process is carried out by sweeping a detection window along the image, and processing it through the stages of the classifier as shown in the Fig. 3, establishing whether the object is in a specific position from the result of the classifier. This algorithm offers high detection rates with very short processing times.

Kalman Filter. Each of the detected positions are applied to a Kalman filter, in order to track the vehicle and estimate the unknown locations in the frames where the detector does not deliver them. The counting of a vehicle is done when its position leaves a region of pre-established interest.

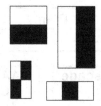

Fig. 2. Example of *Haar* features

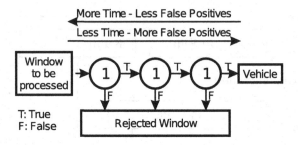

Fig. 3. Cascade classifiers.

Classification. Classification is only performed for cars. A color classifier is applied to determine if a detection corresponds to a taxi (taxis are yellow in Colombia), if not, the detection (snapshot of the detected car) is applied to the convolutional neural network, that classifies it between bus, microbus, minivan, sedan, SUV and truck.

For the purpose of this work, it was used a variation of the AlexNet Convolutional Neural Network [6], pretrained with the ImageNet dataset [7], which contains millions of images from 1000 categories. The net architecture is shown in the Fig. 4.

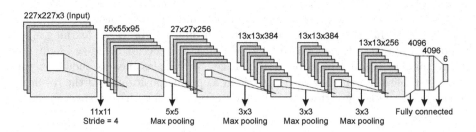

Fig. 4. Variation of the AlexNet Convolutional Neural Network [6].

4   Experiments and Results

The following subsections present the trainings and results obtained from the stages of the algorithm, as well as the total performance.

4.1   Cascade Classifier Training

The training of the cascade classifier was performed with a bank of images of 12500 positive samples (vehicles) and 14000 negative samples (houses, buildings, people, animals, empty streets, etc.). Figure 5 presents some examples of positive samples.

Fig. 5. Examples of vehicle dataset.

In order to find the appropriate training parameters, sweeps were performed on the following parameters. The chosen parameters gave a detection rate of 74.9% and a false positive rate of 1.4%.

– **Number of stages**: from 15 to 30. Chosen: **20**
– **Types of features**: HAAR, HOG and LBP. Chosen: **HAAR**
– **Detection window** size: 12 × 24, 18 × 18, 18 × 24, 18 × 36, 24 × 24. Chosen: **18 × 24**
– **Type of boosting**: DAB, RAB, LB, GAB. Chosen: **GAB**.

4.2   Classification

The training of the color classifier was done with a bank of images of 295 positive (taxis) and 713 negative (non-taxis), obtaining an accuracy of 99.3% with the test dataset.

According to the Fig. 4, the input of the CNN classifier image must be 224 × 224, and the original output size is 100. However, the last fully connected layer was changed to have 6 outputs, and a *fine-tuning* was performed with the BIT dataset [4] (558 buses, 883 microbuses, 476 minivans, 5922 sedans, 1392 SUVs and 822 trucks). It was used 80% of the dataset for training and validation, and the remaining 20% for testing, getting the learning curve presented in the Fig. 6, and the confusion matrix of the Table 1. There, can be observed that around 300 iterations the training converged with loss around 0.1 and training precision around 97%. The testing accuracy was 88% on average for all the classes.

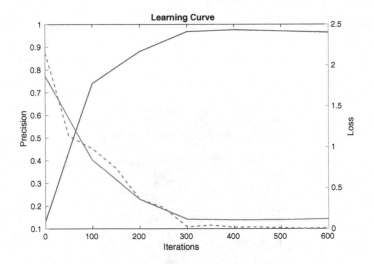

Fig. 6. Learning curve of the CNN fine-tuning.

Table 1. Confusion matrix got from the CNN testing.

	Bus	Microbus	Minivan	Sedan	SUV	Truck	Precision
Bus	157	0	10	0	0	0	94%
Microbus	4	240	4	3	14	0	90%
Minivan	1	8	118	3	1	12	82%
Sedan	0	17	0	1684	76	0	94%
SUV	0	21	7	47	343	0	82%
Truck	1	1	30	0	0	215	87%
						Mean	88%

4.3 Performance of the Complete Method

In order to evaluate the performance of the complete method, traffic videos were captured at different locations in the city of Bogotá. Then, the total number of vehicles was counted manually, how many of them were taxis, and the number of motorcycles that crossed the captured road area. Subsequently, the videos were processed by the proposed method. Table 2 shows the results obtained for each of the videos with 640 × 480 video resolution, regarding the manual count. Some time and site conditions were changed in order to show the working of the method in different situations. These features are:

- Video 1- Main 4 lane road. Hour: 15:20. Weather: cloudy. Average speed: 27 km/h.
- Video 2- Main 4 lane road. Hour: 10:30. Weather cloudy. Average speed: 12 km/h.
- Video 3- Main 4 lane road. Hour: 14:12. Weather: sunny. Average speed: 17 km/h.
- Video 4- Main 2 lane road. Hour: 11:00. Weather: cloudy. Average speed: 22 km/h.
- Video 5- Main 2 lane road. Hour: 15:00. Weather: cloudy. Average speed: 24 km/h.

Table 2. Performance of the complete method with the 640 × 480 video.

Video	Total cars			Only taxis			Motorbikes		
	Manual count	Ours	Error	Manual count	Ours	Error	Manual count	Ours	Error
1	342	344	0,6%	87	81	−6,9%	149	139	−6,7%
2	168	173	3,0%	48	42	−12,5%	80	77	−3,8%
3	240	247	2,9%	59	48	−18,6%	102	95	−6,9%
4	268	281	4,9%	52	49	−5,8%	178	177	−0,6%
5	214	216	0,9%	61	51	−16,4%	59	54	−8,5%
AVG			**2,5%**			**−12,0%**			**−5,3%**

The system was implemented in a Cubieboard 4 and a surveillance camera with 640 × 480 pixels of resolution. It was obtained detection and tracking processing times around 11.11 ms, appropriate for a real time application. In addition, it was got classification times around 76 ms, which allowed implementing it with a FIFO approach in an independent thread.

5 Conclusions

A method of measuring urban traffic through digital image processing was proposed, implemented and verified, which is divided into three main stages. The detection was based on a cascade classifier and HAAR characteristics, obtaining a detection rate of 74,9% and a false positive rate of 1,4%. The tracking and counting stage, performed from the implementation of a kalman filter, allows to increase the previous detection rate, obtaining an average error of 2,0% in the total vehicle count and 5,5% in motorcycles count. The Convolutional Neural Network presented an average precision of 88,0% in the tests performed. Additionally, the execution times allow the implementation of the system in a commercial embedded platform.

In general, it can be concluded that the proposed method presents appropriate results without the need of high image resolution, allowing its execution on platforms that do not have a high computational capacity, generating the possibility of implementing low cost traffic measurement systems, in which the analysis can be performed locally.

References

1. Arulampalam, M.S., Maskell, S., Gordon, N., Clapp, T.: A tutorial on particle filters for online nonlinear/non- tracking. IEEE Trans. Signal Process. **50**(2), 174–188 (2002). https://doi.org/10.1109/78.978374
2. Bañón, L., García, B., Francisco, J.: Manual de carreteras. elementos y proyecto. Ortiz e Hijos, Contratista de Obras, S.A. (2000). http://rua.ua.es/dspace/handle/10045/1788
3. Chen, Z., Ellis, T., Velastin, S.A.: Vehicle detection, tracking and classification in urban traffic. In: 2012 15th International IEEE Conference on Intelligent Transportation Systems, pp. 951–956 (2012). https://doi.org/10.1109/ITSC.2012.6338852
4. Dong, Z., Wu, Y., Pei, M., Jia, Y.: Vehicle type classification using a semisupervised convolutional neural network. IEEE Trans. Intell. Transp. Syst. **16**(4), 2247–2256 (2015). https://doi.org/10.1109/TITS.2015.2402438
5. Kim, P.K., Lim, K.T.: Vehicle type classification using bagging and convolutional neural network on multi view surveillance image. In: 2017 IEEE Conference on Computer Vision and Pattern Recognition Workshops (CVPRW), pp. 914–919 July 2017. https://doi.org/10.1109/CVPRW.2017.126
6. Krizhevsky, A., Sutskever, I., Hinton, G.E.: Imagenet classification with deep convolutional neural networks. In: Pereira, F., Burges, C.J.C., Bottou, L., Weinberger, K.Q. (eds.) Advances in Neural Information Processing Systems 25, pp. 1097–1105. Curran Associates, Inc. (2012). http://papers.nips.cc/paper/4824-imagenet-classification-with-deep-convolutional-neural-networks.pdf
7. Lab, S.V.: Imagenet. http://www.image-net.org/
8. Li, X., Wang, K., Wang, W., Li, Y.: A multiple object tracking method using Kalman filter. In: The 2010 IEEE International Conference on Information and Automation, pp. 1862–1866 (2010). https://doi.org/10.1109/ICINFA.2010.5512258
9. Li-sheng, J., Bai-yuan, G., Rong-ben, W., Lie, G., Yi-bing, Z., Lin-hui, L.: Preceding Vehicle Detection Based on Multi-characteristics Fusion. In: 2006 IEEE International Conference on Vehicular Electronics and Safety, pp. 356–360 (2006). https://doi.org/10.1109/ICVES.2006.371615
10. Mendoza-Schrock, O., Bourbakis, N., Rizki, M., Velten, V.: Vehicle classification for civilian and non-civilian applications: survey. In: NAECON 2014-IEEE National Aerospace and Electronics Conference, pp. 163–168 (2014). https://doi.org/10.1109/NAECON.2014.7045796

11. Mikolajczyk, K., Zisserman, A., Schmid, C.: Shape recognition with edge-based features. In: Harvey, R., Bangham, A. (eds.) British Machine Vision Conference (2003), vol. 2, pp. 779–788. The British Machine Vision Association, Norwich (2003). https://hal.inria.fr/inria-00548226

12. Ojha, S., Sakhare, S.: Image processing techniques for object tracking in video surveillance- survey. In: 2015 International Conference on Pervasive Computing (ICPC), pp. 1–6 (2015). https://doi.org/10.1109/PERVASIVE.2015.7087180

13. dos Santos, D.J.a.A.a.: Automatic vehicle recognition system: an approach using car rear views and backlights shape. Ph.D. thesis, s.n., Lisboa (2008)

14. Shujuan, S., Zhize, X., Xingang, W., Guan, H., Wenqi, W., De, X.: Real-time vehicle detection using mixed features and gentle AdaBoost classifier. In: The 27th Chinese Control and Decision Conference (2015 CCDC), pp. 1888–1894, May 2015. https://doi.org/10.1109/CCDC.2015.7162227

15. Stauffer, C., Grimson, W.E.L.: Adaptive background mixture models for real-time tracking. In: Proceedings. Proceedings of 1999 IEEE Computer Society Conference on Computer Vision and Pattern Recognition (Cat. No PR00149), vol. 2, p 252 (1999). https://doi.org/10.1109/CVPR.1999.784637

16. U.S. Department of Transportation: History of Intelligent Transportation Systems. http://www.its.dot.gov/history/

17. Viola, P., Jones, M.: Rapid object detection using a boosted cascade of simple features. In: Proceedings of the 2001 IEEE Computer Society Conference on Computer Vision and Pattern Recognition. CVPR 2001, vol. 1, pp. I-511–I-518 (2001). https://doi.org/10.1109/CVPR.2001.990517

18. Wen, X., Yuan, H., Yang, C., Song, C., Duan, B., Zhao, H.: Improved Haar wavelet feature extraction approaches for vehicle detection. In: 2007 IEEE Intelligent Transportation Systems Conference, pp. 1050–1053 (2007). https://doi.org/10.1109/ITSC.2007.4357743

19. Wen, X., Zhao, H., Wang, N., Yuan, H.: A rear-vehicle detection system for static images based on monocular vision. In: 2006 9th International Conference on Control, Automation, Robotics and Vision. 2006, pp. 1–4 (2006). https://doi.org/10.1109/ICARCV.2006.345157

20. Xiang, X., Bao, W., Tang, H., Li, J., Wei, Y.: Vehicle detection and tracking for gas station surveillance based on and optical flow. In: 2016 12th World Congress on Intelligent Control and Automation (WCICA), pp. 818–821 (2016). https://doi.org/10.1109/WCICA.2016.7578324

Breaking Text-Based CAPTCHA with Sparse Convolutional Neural Networks

Diogo Daniel Ferreira[1], Luís Leira[1], Petya Mihaylova[2],
and Petia Georgieva[1]($\boxtimes$)

[1] Department of Electronics, Telecommunications and Informatics,
University of Aveiro, Aveiro, Portugal
petia@ua.pt
[2] Technical University of Sofia, Sofia, Bulgaria
petya.petkova@tu-sofia.bg

Abstract. CAPTCHA is an automated test designed to check if the user is human. Though other approaches are explored (such as object recognition), the text-based CAPTCHA is still the main test used by many web service providers, to separate human users from bots. In this paper, a sparse Convolutional Neural Network (CNN) to break text-based CAPTCHA is proposed. Unlike previous CNN solutions, which mainly use fine-tuning and transfer learning from pre-trained models, the proposed framework does not require a pre-trained model. The sparsity constraint deactivates connections between neurons in the CNN fully connected layers that leads to improved accuracy compared to the baseline approach. Visualization of the spatial distribution of neuron activity shed light on the internal learning and the effect of the sparsity constraint.

Keywords: Text-based CAPTCHA · Convolutional Neural Networks · Sparsity constraint · Neuron activity visualization

1 Introduction

CAPTCHA (Completely Automated Public Turing Test to tell Computers and Humans Apart) is an automated test designed to check if the user is human. The test is made using a challenge-response approach, where the challenge is easy for a human to solve, but hard for a machine. If the response is correct, the machine assumes that the user is human. CAPTCHAs are used by most service providers, such as email or online shopping, to prevent bots from abusing their online services. For example, to prevent a botnet to create hundreds of new email accounts per second, a CAPTCHA can be used to assure that the users creating the email accounts are humans.

This Research work was funded by National Funds through the FCT - Foundation for Science and Technology, in the context of the project UID/CEC/00127/2013.

A. Morales et al. (Eds.): IbPRIA 2019, LNCS 11868, pp. 404–415, 2019.
https://doi.org/10.1007/978-3-030-31321-0_35

The construction of CAPTCHAs is not an easy task because it is difficult to create challenges hard for machines but easily solvable by humans. Over the years, the most used CAPTCHAs are based on visual-perception tasks. Distorted characters are presented that must be typed correctly by the user. Background and foreground noise is usually added, making it almost impossible for a computer to automatically recognize the characters. However, for humans, the characters are relatively easy to recognize, due to our brain's capacity for recognizing patterns. There are three characteristics for a modern text-based CAPTCHA to be resilient:

- The large variation in the shape of letters. While there is an infinite variety of versions for the same character that the human brain can recognize, the same is not true for a computer. If all the versions of a character are different, it is hard for a computer to recognize any version not previously seen.
- Due to the large variation in the shape of characters, it can also be hard to perform segmentation for each character, mainly when the characters have no space in between.
- In specific CAPTCHAs, the context may be the key to answer correctly to the task. When the word is taken into context, it is easier for a human to answer what are the characters in the challenge, even if some of them are dubious.

The conjugation of these three characteristics makes a CAPTCHA hard to solve by a machine. However, over the last few years a number of techniques to break the text-based CAPTCHA have been proposed, [1]. Most of the solutions are based on deep (neural network) learning models trained on millions of images using clusters of computers or alternatively using fine-tuning and transfer learning from pre-trained models. The models are usually designed with huge number of parameters to account for the complexity of large scale data that they learn from. However, when it comes to production deployment on embedded or mobile devices, the network size, speed, and power consumption become an issue.

In this paper, we propose a strategy for limiting the neuron activity and show that this improves and speed up the learning compared to the baseline approach. The strategy is illustrated on Kaggle text-based CAPTCHA data set.

The rest of the paper is organized as follows. Section 2 reviews related works. Section 3 explains the proposed sparse CNN framework. Simulation results and discussions are presented in Sects. 4 and 5 summarizes the work.

2   Related Work

In 2014, the authors of [2] for the first time stated that text distortion-based CAPTCHAs schemes should be considered insecure due to technological advances. They presented a general framework for solving text-based CAPTCHAs, with a multi-step algorithm based on reinforcement learning with joint phases of segmentation and text recognition. Since then, alternative CAPTCHA schemes based on object recognition have been proposed, [3],

[4], making it harder for machines to solve them, but remaining easily solvable by humans. However, there are still many implementations of insecure text-based CAPTCHAs on the web. In [5], the authors propose two approaches for text-based CAPTCHA recognition, based on pattern matching and hierarchical algorithms. Both approaches attempt to find the shape of the objects by defining key points in their structure using the Canny Edge Detector and then comparing it to the structure of each character in a local database. The first approach tries to find words in images starting with visual cues, and incorporates lexical information later (the CAPTCHAs texts are words from the dictionary). The second approach searches for entire words at once using a dictionary with all 411 words that the considered CAPTCHAs contain. The second algorithm achieved better results, with an accuracy of 92% on the EZ-Gimpy dataset. This study showed that, algorithms that deal with the whole CAPTCHA at once tend to output higher accuracy than algorithms that deal with each character separately.

In [6], the authors break the Microsoft CAPTCHA, used for systems such as MSN or Hotmail, with image segmentation and pattern matching. The segmentation and recognition combined achieved 60% accuracy. Other approaches, such as [7] or [8], also use image segmentation and pattern matching to break specific CAPTCHAs datasets.

In [9], the authors propose a two-step approach to recognize text-based CAPTCHAs. The data include CAPTCHAs from the most visited websites, like MSN, Yahoo, Google/Gmail or TicketMaster. First, segmentation is applied to separate the characters and then a Convolutional Neural Network (CNN) to recognize them. It is shown that most of the errors derive from a bad segmentation. The highest accuracy is close to 90%.

In [10], a CNN is used for CAPTCHA recognition. The network is composed by three convolution layers, three pooling layers and two fully-connected layers. The network recognizes the sequence without pre-segmentation, and with a fixed size of six characters. The problem of CNN requiring a very large training set is solved with an Active Learning mechanism. To prevent from feeding the neural network with millions of CAPTCHAs, each CAPTCHA is recognized with a certain measured uncertainty. Only the most uncertain CAPTCHAs on the test set are used for retraining the model. The algorithm reaches an accuracy of almost 90%.

In [11], a novel approach is taken to break text-based CAPTCHAs. It introduces the Recursive Cortical Network (RCN), a hierarchical probabilistic generative model with an outstanding capacity of generalization based on the human brain, designed to be trained with few examples. This model achieved an accuracy of 94.3% on character recognition on the reCAPTCHA algorithm, created and currently used by Google.

Neural networks are often over-parameterized with significant redundancy among the weights and the CNNs do not make an exception. To address this problem we propose in this paper sparsity constraint approach originated in deep autoencoders training. The idea is to limit the neuron activity and enforce learning of non-redundant information.

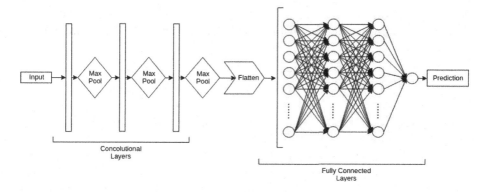

Fig. 1. CNN framework

3   The Proposed Framework

CNN combined with large-scale labeled data has become a standard recipe for achieving state-of-the-art performance on computer vision tasks in recent years. The general architecture of the CNN is given in Fig. 1). Typically, a CNN alternatively stacks convolutional (C) and sub-sampling (e.g. max-pooling) (M) layers. In a C layer, small feature extractors (kernels) sweep over the topology and transform the input into feature maps. In a M layer, activations within a neighborhood are abstracted to acquire invariance to local translations. After several C and M layers, feature maps are flattened into a feature vector, and followed by fully-connected (FC) layers. In this paper Rectified Linear Units (ReLU) are applied in the convolutional layers [12]. ReLU is formally defined as $f(x) = max(0, x)$. It has become very popular in the past couple of years since it improves significantly the convergence speed and avoids the vanishing gradient problem. The inputs of the last FC layer are passed through a softmax function, to compute the class probabilities. Given an input $x^{(i)}$ with a label $y^{(i)}$, the softmax function estimates the probability that this example belongs to each of the class labels $j = 1, 2, \ldots c$

$$p(y^{(i)} = j | x^{(i)}; \theta) = \frac{e^{\theta_j^T x^{(i)}}}{\sum_{j=1}^c e^{\theta_j^T x^{(i)}}} \tag{1}$$

The network outputs c dimensional vector of the estimated probabilities, where θ is a matrix of parameters connecting the softmax layer with the previous (hidden) layer.

$$\hat{y}_{cnn}(x^{(i)}) = \frac{1}{\sum_{j=1}^c e^{\theta_j^T x^{(i)}}} \begin{bmatrix} e^{\theta_1^T x^{(i)}} \\ e^{\theta_2^T x^{(i)}} \\ \ldots \\ e^{\theta_c^T x^{(i)}} \end{bmatrix} \tag{2}$$

The denominator in Eq. (2) normalizes the distribution to sum to one. Given a batch of m training examples, the baseline softmax cost function to be minimized is

$$J(\theta) = -\frac{1}{m}[\sum_{i=1}^{m}\sum_{j=1}^{c} 1\{y^{(i)} = j\} \log \frac{e^{\theta_j^T x^{(i)}}}{\sum_{j=1}^{c} e^{\theta_j^T x^{(i)}}}]. \tag{3}$$

We propose a strategy for limiting the neuron activity in the FC layers which is imposed by sparsity constraints on the hidden units.

Let a_k denotes the activation of hidden unit k and $a_k(x)$ denotes the activation of hidden unit k when the network is given a specific input x. Further, let $\hat{\rho}_k = \frac{1}{m}\sum_{i=1}^{m}[a_k(x^{(i)})]$ be the average activation of hidden unit k (averaged over the training set). We would like to (approximately) enforce the constraint $\hat{\rho}_k = \rho$, where ρ is a sparsity parameter, typically a small value. In other words we would like the average activation of each hidden unit j to be close to ρ. This is enforced by an extra penalty term in the cost function that penalizes $\hat{\rho}_k$ if deviating significantly from ρ. Many choices of the penalty term will give reasonable results, here we choose the Kullback-Leibler (KL) divergence which is a standard function for measuring how different two distributions are [13]. KL-divergence measure between a Bernoulli random variable with mean ρ and a Bernoulli random variable with mean $\hat{\rho}_k$ is given as

$$KL(\rho||\hat{\rho}_k) = \sum_{k=1}^{s} \rho \log \frac{\rho}{\hat{\rho}_k} + (1-\rho) \log \frac{(1-\rho)}{(1-\hat{\rho}_k)} \tag{4}$$

Here s is the number of the units in the hidden layer, and the index k is summing over the hidden units of the network. The choice of ρ expresses the desired level of sparsity, here we set it to a common value of 0.1.

In the sparse cost function J_{sparse}, β controls the importance of the sparsity penalty term $KL(\rho||\hat{\rho}_k)$

$$J_{sparse} = J + \beta KL(\rho||\hat{\rho}_k)] \tag{5}$$

The intuition behind this optimization framework is to specialize the neurons in learning specific patterns and as a consequence enforce compression to happen.

4   Experiments and Results

4.1   Dataset

The Kaggle CAPTCHA dataset has been used to evaluate the proposed framework[1]. A few examples are given in Fig. 2(a). Each image consists of five random characters from a set of 19 characters: 2, 3, 4, 5, 6, 7, 8, b, c, d, e, f, g, m, n, p, w, x, y. The characters have the same font but rotated in different angles. Since

---

[1] https://www.kaggle.com/fournierp/captcha-version-2-images.

(a) Examples of Kaggle text-based CAPTCHA images.

(b) CAPTCHA sample denoising. On the top, from left to right: the original image, the image after Otsu thresholding, the image after one dilation. On the bottom, from left to right: the image after the erosion, the image after the second dilation and finally the contouring of the characters.

(c) Samples of single character images.

Fig. 2. CAPTCHA images and prepossessing

the images have been heavily corrupted by noise (black lines over the characters), a few denoising steps are taken to remove or alleviate the noise as shown in Fig. 2(b). First, the Otsu method [14] is applied to perform clustering-based image thresholding and transform the CAPTCHA into a binary image. Next, morphological transformations are applied to the image. A dilation and an erosion are applied sequentially with 3×3 kernel, followed by a second dilation with 3×1 kernel, in an attempt to eliminate the horizontal lines that create the noise in the image. Single character images were then extracted by segmentation as shown in Fig. 2(c).

Since the original dataset has been small (1070 images) we applied rotation and shifting operations to augment it. Combinations of rotations $(-10°, 0°, 10°)$, vertical $(-3\,\text{px}, 0\,\text{px}, 3\,\text{px})$ and horizontal $(-3\,\text{px}, 0\,\text{px}, 3\,\text{px})$ shifts were applied to generate 27 variations of each original single character training image. After the augmentation, data grew to 5350 images per single character and 87048 examples in total. For the training 69638 items were randomly selected and the test set was limited to 17409 items.

4.2 Performance Evaluation

CNNs with varying depth have been trained, with architectures shown in Table 1. Stochastic gradient descent optimization was applied with learning rate of 0.0001 and dropout step of 0.5 to prevent overfitting. Figures 3, 4 and 5 illustrate the performance of the proposed method (with sparsity constraint) and the baseline approach (without sparsity constraint) in terms of training and testing accuracy. The testing accuracy is evaluated in each training epoch with new test data. Some observations can be made from the figures. The sparsity constraint makes the learning models less sensitive to the network dimension compared to the baseline approach. Note the similar behavior of the right side plots in Figs. 3, 4 and 5. In contrast, the left side plots show lack of learning (CNN1), overfitting (CNN2) and finally achieved a good performance while increasing the CNN complexity. The maximum testing accuracy in the baseline is 90.2 % (for CNN3). The maximum accuracy of the proposed framework is 95.7% (for CNN2). The performance of the lower complexity model (CNN1) is significantly more affected by the sparsity constraints.

The capability of CNN to model highly nonlinear functions comes with high computational and memory demands both during the model training and inference. The sparsity constraints impose connection between the neurons in the FC layers, which enforces the neurons to learn non redundant information and therefore reduces the amount of information processed, [15].

4.3 Neuron Activity Visualization

One way to understand what the CNN is learning and to asses the effect of the sparsity constraint on the internal learning process, is to visualize the representations captured by the hidden units, [16]. These representations are not always easy to understand [17], therefore we illustrate here only those that we found interpretable.

The matrix of weights between the flatten layer (256 units) and the first hidden FC layer (512 units) is denoted as $\theta^{(1)}$ (dimension 256×512). We visualize the weights collected in $\theta^{(1)}$ as representation images. Each column of $\theta^{(1)}$ is reshaped into a square 16×16 pixels image and visualized on one cell of the visualization panel shown in Fig. 6. Figure 6 illustrates the representation images

Table 1. CNN architectures

Layer type	Neurons	Kernel size
CNN1		
2 Conv layers & pooling layers	256	3×3 & 2×2
2 Fully-connected (FC) layers	512	
Softmax layer	19	
CNN2		
4 Conv layers & pooling layers	256	3×3 & 2×2
2 Fully-connected (FC) layers	512	
Softmax layer	19	
CNN3		
6 Conv layers & pooling layers	256	3×3 & 2×2
4 Fully-connected (FC) layers	512	
Softmax layer	19	

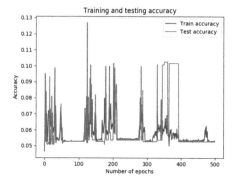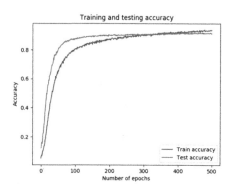

Fig. 3. CNN1: left (baseline), right (proposed framework)

of all 512 hidden units for a given visible input (the image of the character 4) as a visualization panel of 32 by 16 cells. On the left side of Fig. 6 are the results from the CNN trained with the baseline cost function J (Eq. 3) and on the right side the weights obtained after training with the sparse cost function J_{sparse} (Eq. 5). The proposed framework resulted in more "blank" cells which means that those neurons are not activated by the input image. Since the weights are fit in such a way not to violate the constraints on the neuron activity, the 'blank' neurons are not specialized in the specific patterns of the character '4' image. The sparsity constrain acts as an inhibitor/promoter for particular stimulus and therefore favor a non-uniform distribution of the neuron activity over training examples. This hypothesis (in agreement with [18]), may explain the uniform spatial distribution of the neuron activity with the baseline training.

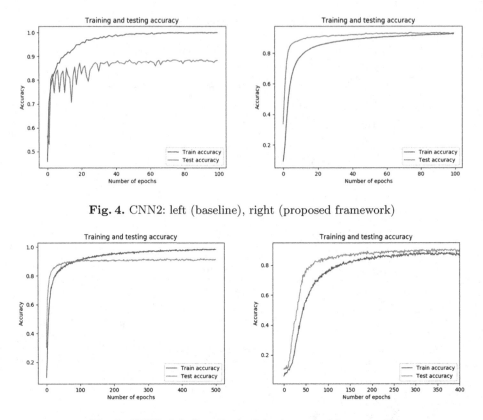

Fig. 4. CNN2: left (baseline), right (proposed framework)

Fig. 5. CNN3: left (baseline), right (proposed framework)

Curiously, results that support the inhibitor/promoter effect of the sparsity constraint were observed also with the convolutional filters. We applied the same strategy for visualizing as before and the visualization panels (8 by 8 cells) of the 64 filters in the 3rd and the 4th convolutional layers are illustrated in Figs. 7 and 8. The differences are more distinct applying gray heat map. The rough granulated cells indicate specific features learned by the filters, while the smooth gray cells indicate inactivate filters. Though the convolutional filters were not explicitly constrained in Eq. (5), the constrained neuron activity in the FC layers backpropagate this effect and force the convolutional filters to be more selective to a particular stimulus. There are much more inactive filters to the specific patterns of the image of the character '4' in the right plots of Figs. 7 and 8 compared to the baseline trained filters in the left plots of the same figures.

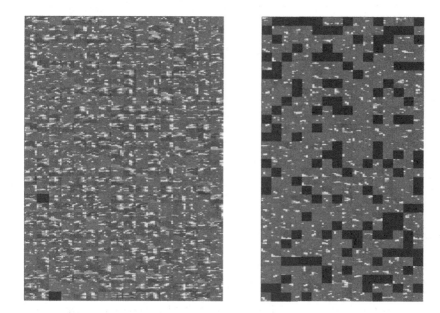

Fig. 6. CNN2. Features learning in the 1st hidden FC layer given image of character '4': left (baseline); right (proposed method)

Fig. 7. CNN2. Filter patterns in 3rd conv layer: left (baseline); right (proposed framework)

Fig. 8. CNN2. Filter patterns in 4th conv layer: left (baseline); right (proposed framework)

5 Conclusions

In this paper, CNN learning framework with sparse cost function has been proposed to demonstrate a possible attack that can be made to a text-based CAPTCHA. The proposed model has a low complexity which makes its training transparent and fully controlled. Although the quantity and quality of the image dataset is not very high, the framework shows good results. The concept of sparsity, widely applied in deep autoencoders, is a good alternative to account for parameter redundancy and compressed sensing. The outcomes of experiments suggest that adding sparsity constraint can improve the network accuracy and convergence speed. Future extension of this work would be upgrading this framework to non-text based CAPTCHAs also.

References

1. Chen, J., Luo, X., Guo, Y., Zhang, Y., Gong, D.: A survey on breaking technique of text-based CAPTCHA. Secur. Commun. Netw. **2017**, 1–15 (2017)
2. Bursztein, E., Aigrain, J., Moscicki, A., Mitchell, J.C.: The end is nigh: generic solving of text-based CAPTCHAs. In: 8th USENIX Workshop on Offensive Technologies, San Diego, CA (2014)
3. Sivakorn, S., Polakis, I., Keromytis, A.D.: I am robot: (deep) learning to break semantic image CAPTCHAs. In: IEEE European Symposium on Security and Privacy (2016)
4. Zhao, B., et al.: Towards evaluating the security of real-world deployed image CAPTCHAs. In: 11th ACM Workshop on Artificial Intelligence and Security, New York, NY, USA, pp. 85–96 (2018)
5. Mori, G., Malik, J.: Recognizing objects in adversarial clutter: breaking a visual CAPTCHA. In: IEEE Computer Society Conference on Computer Vision and Pattern Recognition (2003)
6. Yan, J., Salah El Ahmad, A.: A low-cost attack on a Microsoft CAPTCHA. In: 5th ACM Conference on Computer and Communications Security, CCS 2008, New York, NY, USA (2008)
7. Yan, J., Salah El Ahmad, A.: Breaking visual CAPTCHAs with Naive pattern recognition algorithms. In: Twenty-Third Annual Computer Security Applications Conference (2007)
8. Li, S., Amier Haider Shah, S., Asad Usman Khan, M., Khayam, S.A., Sadeghi, A.-R., Schmitz, R.: Breaking e-banking CAPTCHAs. In: 26th Annual Computer Security Applications Conference, pp. 171–180 (2010)
9. Chellapilla, K., Simard, P.Y.: Using machine learning to break visual human interaction proofs (HIPs). In: Advances in Neural Information Processing Systems (NIPS), pp. 265–272. MIT Press (2005)
10. Stark, F., Hazırbas, C., Triebel, R., Cremers, D.: CAPTCHA recognition with active deep learning. In: German Conference on Pattern Recognition (2015)
11. George, D., et al.: A generative vision model that trains with high data efficiency and breaks text-based CAPTCHAs. Science **358**(6368), eaag2612 (2017)
12. Xu, B., Wang, N., Chen, T., Li, M.: Empirical evaluation of rectified activation in convolution network. In: ICML Deep Learning Workshop (2015)

13. Kullback, S., Leibler, R.A.: On information and sufficiency. Ann. Math. Stat. **22**, 79–86 (1951)
14. Otsu, N.: A threshold selection method from gray-level histograms. IEEE Trans. Sys. Man. Cyber. **9**, 62–66 (1979)
15. Bozhkov, L., Georgieva, P.: Overview of deep learning architectures for EEG-based brain imaging. In: IEEE World Congress on Computational Intelligence - IJCNN (2018)
16. Yosinski, J., Clune, J., Nguyen, A.M., Fuchs, T.J., Lipson, H.: Understanding neural networks through deep visualization. In: 31st International Conference on Machine Learning (2015)
17. Zeiler, M.D., Fergus, R.: Visualizing and understanding convolutional networks. In: Fleet, D., Pajdla, T., Schiele, B., Tuytelaars, T. (eds.) ECCV 2014. LNCS, vol. 8689, pp. 818–833. Springer, Cham (2014). https://doi.org/10.1007/978-3-319-10590-1_53
18. Selvaraju, R.R., Cogswell, M., Das, A., Vedantam, R., Parikh, D., Batra, D.: Grad-CAM: visual explanations from deep networks via gradient-based localization. In: 2017 IEEE International Conference on Computer Vision (ICCV) (2017)

Image Processing Method for Epidermal Cells Detection and Measurement in *Arabidopsis Thaliana* Leaves

Manuel G. Forero[1]([⊠]) [iD], Sammy A. Perdomo[2],
Mauricio A. Quimbaya[3], and Guillermo F. Perez[3]

[1] Universidad de Ibagué, Ibagué, Colombia
manuel.forero@unibague.edu.co
[2] Servicio Nacional de Aprendizaje (SENA)-ASTIN, Cali, Colombia
sam.perdomo16@gmail.com
[3] Department of Natural Science and Mathematics, Pontificia Universidad
Javeriana, Cali, Colombia
maquimbaya@javerianacali.edu.co,
gf_perez101@hotmail.com

Abstract. *Arabidopsis thaliana* is the most important model specie employed for genetic analysis in plants. As it has been extensively proven, the first pair of extended leaves and its cellular and morphological changes during *Arabidopsis* development, is and accurate model to understand the molecular and physiological events that control cell cycle progression in plants. Nevertheless, cell analysis on leaves depends significantly on images acquired from a microscopy coupled to a drawing tube, where cells are traced by hand for posterior digitalization and analysis. This process is tedious, inaccurate and highly temporally inefficient. A new image processing method for cell detection in leaves of *Arabidopsis thaliana* is presented. Using complementary image processing techniques, we introduce a good way to obtain the original cell contour shapes, surpassing the limitations given by factors like noise, stomata, blurred edges, and non-uniform illumination. Results show the new methodology minimizes considerably the time of cell detection compared with the microscopy coupled tube method, and produces matching percentages over 80%.

Keywords: *Arabidopsis thaliana* · Cell drawings ·
Epidermal cells image detection · Image analysis method

1 Introduction

In the model plant *Arabidopsis thaliana*, the analysis of epidermal cells belonging to the first pair of extended leaves is a fairly widespread technique in the scientific community that uses this model plant to understand different biological phenomena, since it provides truthful information, easily quantifiable and comparable in relation to the phenotypic response of the plant to physiological stimuli, environmental alterations and to genetic modifications directly induced in the plant genome. Thus, *Arabidopsis* is

© Springer Nature Switzerland AG 2019
A. Morales et al. (Eds.): IbPRIA 2019, LNCS 11868, pp. 416–428, 2019.
https://doi.org/10.1007/978-3-030-31321-0_36

an accurate model to understand the molecular and physiological events that control cell cycle progression in plants [1].

Given that a good percentage of the academic community that does research in plants uses *Arabidopsis thaliana* as a biological model of experimentation, and in addition, given that the analysis of epidermal cells in this model plant is a generalized technique for obtaining biological information that can support a particular biological phenomenon, actually the *Arabidopsis Thaliana* epidermal cells quantification and characterization is a highly laborious and temporarily inefficient process because this methods are based in free hand drawings performed using an optical microscopy coupled to a tube drawing where cells are traced by hand for posterior digitalization and analysis, which is clear from recent academic publications still describing the drawing-based methodology [2–7].

The generation of a specific application that can replace the drawings of epidermal cells made upon leaves structure, turns into a great technical facilitator for the rapid collection and analysis of images that can sustain or distort a specific hypothesis. Additionally, an automated tool for epidermal cells analysis will deliver trustable results about cell number and area in a reduced period of time and with more reproducible results.

In our knowledge, currently, there is not software or plug-in, available to the scientific community, that performs the cellular detection of *Arabidopsis* epidermal cells based on pictures. Some applications described in [8–12], shows detection strategies for different kind of cells; each one depends of image acquisition and image properties. However, no one of these methods consider images with similar properties of *Arabidopsis* epidermal cells images. The present research presents a new method to detect *Arabidopsis thaliana* epidermal cells improving precision, consumed time and reproducibility of the results compared with the manual technique.

2 Methodology

2.1 Samples Processing and Image Acquisition

For the acquisition of abaxial cell surface of epidermal tissue from *Arabidopsis thaliana*, plants with 21 days of post-germination growth (DAG) were used. Plants were processed by immersing the first real pair of leaves overnight in acetone and transferred to a 1:9 acetic acid: 100% ethanol solution afterwards. Subsequently, cleared leaves were stored in lactic acid for microscopy. Epidermal cells were photographed under an optical microscope Carl Zeiss AXIO-Lab-A1 with a 10X magnification coupled to a Carl Zeiss-AxioCam ERc 5s camera. Forty images with a resolution of 2560 × 1920 pixels were acquired (see Fig. 1) and employed for testing. Ten of them were randomly chosen to validate the generated method for epidermal cells detection. The ground-truth of two images was traced by hand by a specialist.

2.2   Algorithm Development

For the automatic detection of epidermal cells based on leaves pictures, here we propose a multistep technique that includes two filtering processes to eliminate the undesired structures of the leaf and the acquisition noise, a contrast enhancement procedure, an edge detection step and the implementation of some editing tools. As a final result an application written in Java, compatible with the freely available ImageJ software, was developed.

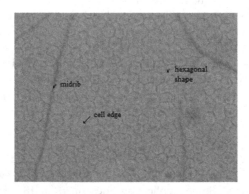

Fig. 1. Sample of microscopy *Arabidopsis Thaliana* image.

Figure 2 depicts the different steps that were implemented for the development of the method for automatic epidermal cells detection.

In the first step the acquired image is pre-processed to reduce noise, which is composed of undesired physical structures of the leaf epidermis and the inherent acquisition noise. In order to eliminate these undesired structures a morphological filtering is done. Given that these structures have a stronger signal than the wall cells, we look for eliminating the wall cells instead, so the undesired noise remains in the resulting images. Then, wall cells are obtained by just subtracting the filtered image from the original one. Given that cell walls are seen as thin dark lines (see Fig. 1), they are eliminated by applying a morphological opening [13], where the side of the structural element is wider than the thickness of the walls. Figure 3 shows the image obtained after filtering with a structural element of radius 2.

Once the undesired structures have been eliminated, it is observed than the cell intensity is quite low, therefore the image is treated in such a way that the contours of the cell walls can be accentuated, since, being a very weak signal, it must be conditioned for the detection process to ensure high detection performance. Therefore, a contrast adjustment is applied using the CLAHE (Contrast Limited Adaptive Histogram Equalization) technique [14], which equalizes the image by regions increasing the intensity difference between the background and the walls. Figure 4a illustrates the result of adjusting the contrast (CLAHE) on Fig. 3.

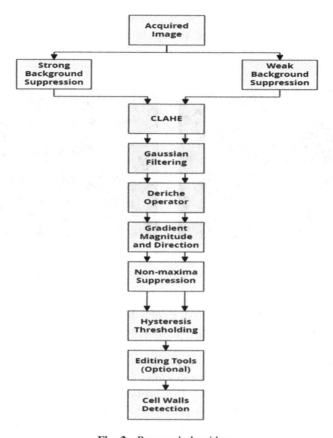

Fig. 2. Proposed algorithm.

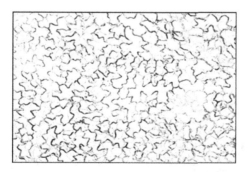

Fig. 3. *Arabidopsis* image after subtracting the background (Contrast was adjusted for visualization).

With the morphological filtering noise due to undesired structures is greatly attenuated, but the acquisition noise still remains. To eliminate it a Gaussian filter is used, given that it provides noise reduction without introducing ringing artefacts. The standard deviation of the filter was 2; enough to eliminate the acquisition noise without notably blurring the cell walls. Figure 4b depicts the results after Gaussian filtering is applied.

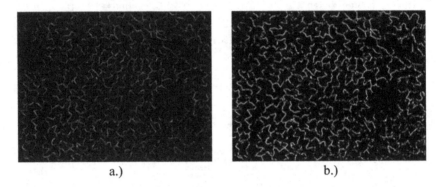

a.)   b.)

Fig. 4. (a) Enhanced contrast by using the CLAHE method, (b) Gaussian filter (r = 2).

Although a non-linear edge-preserving smoothing filter was tested [15], it did not improve the quality of the following segmentation. At this point the pre-processing stage ends.

Edge operators are appropriate to detect cell walls, given that they appear as thin lines in a clear dark background. Under this idea, the Deriche operator is used; which is still considered in the state of art [16]. This operator is based on the Canny principle, but has a better behavior in order to get image segmentation [17]. In this way, the derivatives for the x and y coordinates are obtained and shown in Fig. 5a, and b respectively.

a.)   b.)

Fig. 5. (a) Derivate x, (b) Derivate y.

The Deriche operator provides the gradient of the image but does not deliver edge pixels as such. Thus, it is necessary to employ contour selection techniques to find the cell walls. Then the selection of contours proposed by Canny is employed [18]. It consists of two stages: non maxima suppression and hysteresis thresholding.

The suppression of pixels that do not correspond to a local maximum makes possible to eliminate all those pixels that are not part of the true border. To do this, the magnitude and direction of the gradient are calculated and results are depicted on Fig. 6a and b.

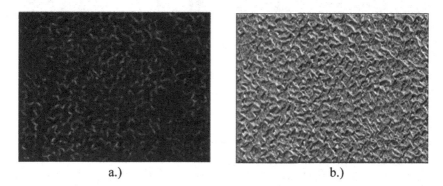

a.)   b.)

Fig. 6. (a) Gradient magnitude, (b) Gradient direction.

For each pixel, the magnitude of the gradient to both sides of each pixel to a distance d in the direction of the gradient of the pixel is evaluated, and compared against the magnitude of the pixel (see Fig. 6b). Those pixels identified as local maxima, that is, when the gradient magnitude is greater than both of its neighbors, are retained, eliminating those others.

As displayed in Fig. 7a, the edges corresponding to the cell walls are better defined; however not all pixels correspond to the true edge. Therefore, a hysteresis threshold is applied, in which two thresholds are defined, one high, and one low. Thus, when the image is evaluated, the pixels that exceed the high threshold are classified immediately as true edge pixels, and those connected to them, which magnitude gradient exceeds the low threshold and are located in the perpendicular direction to the direction of the gradient at that point, are also classified as true, the remaining ones are eliminated. To ensure good classification, the chosen high and low thresholds were 20 and 250 respectively. These values were obtained experimentally by exploring in the histogram of the images the average gray level ranges of the edge pixels. Figure 7b illustrates the result of the hysteresis thresholding with low value of 1 and high value of 30.

As shown in Fig. 7b, most of the cellular contours were detected; however some edges are still missing. Some of the missing borders are caused by the presence of the undesired physical structures at the leaf that are overlapped to the cell walls. Therefore, in the first step where these structures were removed, the overlapped cell walls were also removed. In order to recover as many missing edges as possible, the following strategy is proposed: considering the original image that contains the whole information about the edges, it is possible to start from it again in order to recover the missing information connected to the borders already found. Therefore, it is necessary to use

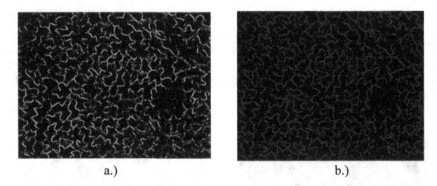

Fig. 7. (a) Non-maxima suppression, (b) Hysteresis thresholding.

different pre-processing parameters to get weak noise reduction, i.e., more noise is present, but also more edges are retained.

Therefore, the original image is smoothed again but using a weaker filter, i.e., a bigger structural element, allowing having more information about the cell walls, but also more noise. Figure 8a illustrates the result of subtracting the background from the original figure with a radius value of 10. Then, the same steps previously proposed after filtering are applied until the non-maxima suppression.

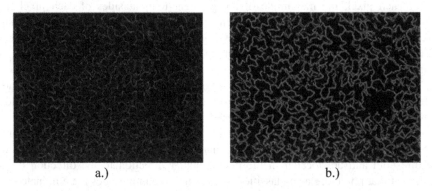

Fig. 8. (a) Weak background subtraction, (b) Final hysteresis thresholding result

As it can be seen in Fig. 8a, comparing with Fig. 4a, many edges, previously missing, appears now in the image, but also some objects that are not part of the cell contours. The edges detected in Fig. 7b are now used to extract the missing ones from Fig. 8a, while rejecting the undesired leaf structures. This is done, based on the functional principle of the hysteresis thresholding, because the magnitude and direction of the true edge pixels, found in Fig. 7b are now used to select from Fig. 8a those pixels connected to the true edge pixels, perpendicular to the gradient direction and similar magnitude, while rejecting false cell edges. Thus, the result obtained recovers edge pixels lost during the first process. Figure 8b illustrates the final hysteresis thresholding of the process with a low value of 15 and high value of 250.

Although the proposed method is quite efficient, some contours can still be missing because some edges are diffused due to poor focus or presence of stomata. Thus, in order to give to the user the possibility of manually editing the results of the detection of epidermal cells in *Arabidopsis thaliana* leaves, the developed software *Arabidopsis Leaf Cell Detection* (ALCD), includes some editing tools for the modification of results after detection, as well as a cell characterization function that allows to obtain the number of cells and their area and label each cell with a different color according to its label or area.

In order to edit the images the following steps are suggested: first the walls are thinned. Before edge thinning an iterative dilation process is repeatedly applied until the double contour of the cell walls is connected. Once the dilatation is done, a thinning or skeletonization technique is applied to the image, in which each edge contour get one pixel width. This tool facilitates the process of cells labeling.

The next step consists in debugging the detected borders with the objective of completing those contours of cells that were impossible to recover by the proposed method, or where the image presents a focus or illumination problem and removing those lines or small branches connected to edges that resulted from the thinning process. Once the image is prepared, the cell analysis tool from the user interface is applied, so each cell is represented with a unique label value that identifies it, and assigned a color that differentiates it from the rest in order to facilitate visualization. Subsequently a table is obtained including the corresponding tag number for each cell and its corresponding area in pixels. The number of tags relates the approximate number of cells in the image. Figure 9 illustrates the result obtained by operating the cell analysis tool.

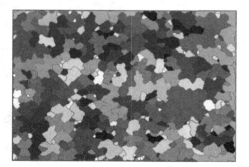

Fig. 9. Labeled image. (Color figure online)

Table 1 specifies the output format of the interface data, where the column on the left represents the label of the cell and the one on the right its area in pixels.

Table 1. Label and cell area specification

Label	Area (pixels)
0	93788
1	202371
2	6014
3	3945
4	4830

3   Method Validation

To validate the method ten *Arabidopsis thaliana* leaves images were chosen with different physical characteristics, belonging to the group of samples acquired. As is shown in Fig. 10, each of the images taken for evaluation has different physical characteristics allowing us to demonstrate the performance of the proposed method. The detection of the cell contours provided satisfactory results even in presence of non-uniform illumination or blurry regions resulting from the acquisition. The edges that cannot be recovered by the development of the method correspond to those in which undesired structures appear quite accentuated, where there is not presence of undesired objects, are completed and differentiable, showing that better acquisition conditions of the samples, means better edge segmentation. On average the necessary time to detect epidermal cells under our approach is 7 ± 1.5 min.

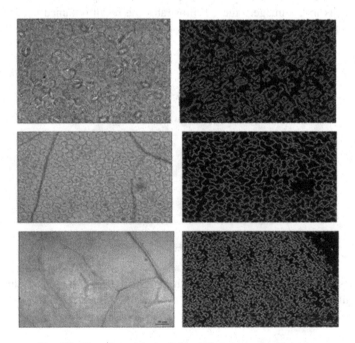

Fig. 10. Experimental result for different acquired images.

Two measures were taken into account to validate the method: percentage of time optimization used in detection and percentage of edge matching of cell detected with corresponding ground truth images. The ground truth images were generated by tracing manually the pixels that corresponded to *Arabidopsis Thaliana* cell walls from two acquired images. With the percentage of optimization, it is possible to describe a quantitative notion of the total time improvement that is used to get a satisfactory detection of the epidermal cells, between the proposed method in relation to the time demanded for the same purpose but through the classic freehand drawing method.

According to the records studied from specialized personnel in this area, the average time taken to draw a significant number (100–200 cells) of epidermal cells, varies from one hour to two hours. The calculation of the percentage of time optimization follows the next expression:

$$t_{op}(\%) = \frac{t_m - t_{prom}}{tprom} \times 10 \tag{1}$$

Where t_m corresponds to the total employed time in the proposed method, including both the processing time and editing time, and t_{prom} is the average time spent in the freehand detection methods.

Complementarily, the matching percentage is determined by the number of pixels that correspond between the resulting images of the developed method and the ground truth. The ground truth image refers to the result of manually tracing the cellular contours of the same processed image, in which it is ensured that the drawn edges are correct. The matching percentage is calculated as follows:

$$m(\%) = p/c \times 100 \tag{2}$$

Where p is the number of pixels in the processed image that correspond spatially to those in the ground truth, and c refers to the total number of edge pixels of the ground truth image.

For validation purposes, two samples from the previously analyzed images were evaluated measuring the time demanded during the detection of the cellular contours, and their correspondence of edges with each ground truth. Table 2 summarizes the percentages obtained for each case. The validated samples are illustrated in Fig. 11; the image on the right corresponds to the result of the proposed method, and the one on the left to its corresponding ground truth.

Table 2. Percent detection time optimization and matching of true edges with ground truth

Validation	$t_{op}(\%)$	$m(\%)$
1	91.66	86.11
2	90	80.4

As it is shown in Table 2, the percentages of optimization in time and matching are quite high, so it is verified that the proposed method does offer improvements in the

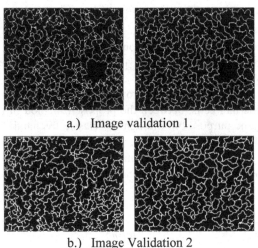

a.) Image validation 1.

b.) Image Validation 2

Fig. 11. Validation images.

time consumption for detection of cellular contours in microscopic images of *Arabidopsis thaliana*, compared to the tedious and inefficient freehand tracing techniques commonly used for these purposes.

4 Conclusions

For cell cycle analysis in *Arabidopsis thaliana*, characterization of epidermal cells in number and size is one of the most important inputs to take into account in order to properly describe a specific phenomenon. For this reason, this depiction implies that all cells must be detected from the acquired microscopy image. The protocol of acquisition and preparation of the proposed method ensures that each sample is taken under the same conditions, thus reducing the probability of variation from one type of sample to another.

The main problem to carry out this kind of analysis in plants is based on the tedious process of being able to detect all the cells, because it is necessary to use a drawing tube device coupled to the microscope, to trace by freehand each one of the cellular contours. This takes a long time, between one or two hours, so the whole process is inefficient in time and resources. The techniques of digital image processing help to get a good solution to this problem. The proposed method of detecting epidermal cell borders in *Arabidopsis Thaliana* leaves uses complementary techniques such as background subtraction, noise filtering, contrast enhancement, edge detection, non-maximal suppression and hysteresis thresholding, in a strategic way to detect the contours of the cells in just a few minutes, making this process faster than the classical form with freehand tracing, improving the efficiency of cellular analyzes.

Experimental results show that the proposed method detects most of the cell borders, even in the presence of undesired factors such as noise, blurring, non-uniform illumination, presence of undesired morphological structures as trichomes. It is possible to modify any result and adjust it to the needs of the researcher, thanks to the graphical user interface developed, which provides all the necessary tools for this purpose, making it possible to complement the analysis processes with approximate information of the number of cells and their area.

These results will facilitate the cell cycle associated research, where the methods of tracing cells contours by tubes coupled to the microscope can be replaced, thus facilitating the process of accurately quantifying phenotypic characteristics such as leaf area, number of cells per leaf, leaf cells area and number of stomata, basic but sufficient measures to analyze the phenotypic effect of a particular gene on the control of the cell cycle in this model plant.

References

1. Koornneef, M., Meinke, D.: The development of Arabidopsis as a model plant. Plant J.: Cell Mol. Biol. **61**, 909–921 (2010)
2. Gonzalez, N., Pauwels, L., Baekelandt, A., Milde, D., et al.: A repressor protein complex regulates leaf growth in Arabidopsis. Plant Cell **27**, 2273–2287 (2015)
3. Juraniec, M., Heyman, J., Schubert, V., Salis, P., De Veylder, L., Verbruggen, N.: Arabidopsis Copper Modified Resistance1/Patronus1 is essential for growth adaptation to stress and required for mitotic onset control. New Phytol. **209**, 177–191 (2016)
4. Van Dingenen, J., et al.: Chloroplasts are central players in sugar-induced leaf growth. Plant Physiol. **171**, 590–605 (2016)
5. Van Leene, J., Blomme, J., Kulkarni, S.R., Cannoot, B., et al.: Functional characterization of the Arabidopsis transcription factor bZIP29 reveals its role in leaf and root development. J. Exp. Bot. **67**, 5825–5840 (2016)
6. Saini, K., et al.: Alteration in auxin homeostasis and signaling by overexpression of PINOID kinase causes leaf growth defects in *Arabidopsis thaliana*. Front. Plant Sci. **8**, 1009 (2017)
7. Zhao, L., Li, Y., Xie, Q., Wu, Y.: Loss of CDKC; 2 increases both cell division and drought tolerance in *Arabidopsis thaliana*. Plant J.: Cell Mol. Biol. **91**, 816–828 (2017)
8. Fatma, U., Kutay, I., Kasim, T., Bulent, Y.: Automated quantification of immunomagnetic beads and leukemia cells from optical microscope images. Biomed. Sig. Process. Control **49**, 473–482 (2019)
9. Sajjad, S., Mohsen, M., Dana-Cristina, T.: A new method of SC image processing for confluence estimation. Micron **101**, 206–212 (2017)
10. Han, H., Wu, G., Li, Y., Zi, Z.: eDetect: a fast error detection and correction tool for live cell imaging data analysis. iScience **13**, 1–18 (2019)
11. Kevin, S., Filippo, P., Tamas, B., Krisztian, K., Tivadar, D., et al.: Phenotypic image analysis software tools for exploring and understanding big image data from cell-based assays. Cell Syst. **6**(6), 636–653 (2018)
12. Osowskio, S., Les, T., Markiewicz, T., Jesiotr, M.: Automatic reconstruction of overlapped cells in breast cancer FISH images. Expert Syst. Appl. **137**, 335–342 (2019)
13. Li, N., et al.: STERILE APETALA modulates the stability of a repressor protein complex to control organ size in *Arabidopsis thaliana*. PLoS Genet. **14**, e1007218 (2018)
14. Sternberg, S.R.: Biomedical image processing. Computer **16**, 12 (1983)
15. Zuiderveld, K.: Contrast Limited Adaptive Histogram Equalization. Academic Press Professional Inc., San Diego (1994)

16. Wang, B., Fan, S.: An improved Canny edge detection algorithm. In: IWCSE 2009 Proceedings of the 2009 Second International Workshop on Computer Science and Engineering, vol. 01, p. 4 (2009)
17. Deriche, R.: Using Canny's criteria to derive a recursively implemented optimal edge detector. Int. J. Comput. Vis. **1**, 20 (1987)
18. Canny, J.: A computational approach to edge detection. IEEE Trans. Pattern Anal. Mach. Intell. **8**, 679–698 (1986)

User Modeling on Mobile Device Based on Facial Clustering and Object Detection in Photos and Videos

Ivan Grechikhin[1,2(✉)] and Andrey V. Savchenko[1,2]

[1] National Research University Higher School of Economics,
Laboratory of Algorithms and Technologies for Network Analysis,
Nizhny Novgorod, Russia
{igrechikhin,avsavchenko}@hse.ru

[2] Samsung-PDMI Joint AI Center, St. Petersburg Department of Steklov Institute
of Mathematics, Fontanka Str., St. Petersburg, Russia

Abstract. The article describes an approach for extraction of user preferences based on the analysis of a gallery of photos and videos on mobile device. It is proposed to firstly use fast SSD-based methods in order to detect objects of interests in offline mode directly on mobile device. Next we perform facial analysis of all visual data: extract feature vectors from detected facial regions, cluster them and select public photos and videos which do not contain faces from the large clusters of an owner of mobile device and his or her friends and relatives. At the second stage, these public images are processed on the remote server using very accurate but rather slow object detectors. Experimental study of several contemporary detectors is presented with the specially designed subset of MS COCO, ImageNet and Open Images datasets.

Keywords: User modelling · Object detection · Mobile systems · Facial clustering · Convolutional neural network

1 Introduction

Nowadays the world lives through time, when social networks and mobile devices create a vast stream of multimedia data including photos and videos [1]. Such visual data contains specific information about a user, which might be used to improve quality of user modeling and, consequently, accuracy of recommender systems. Recently, deep learning techniques including convolutional neural networks (CNNs) are applied in many image processing tasks [2]. In particular, CNN-based object detection for discovering of particular categories of interests, e.g., interior objects, food, transport, sports equipment, animals, etc., can be used to extract information from the user's photos and videos about his or her preferences [3]. Applications of computer vision in recommender systems becomes all the more attractive. For example, visual recommender systems are used in PInterest services in order to provide relevant photos based on their

© Springer Nature Switzerland AG 2019
A. Morales et al. (Eds.): IbPRIA 2019, LNCS 11868, pp. 429–440, 2019.
https://doi.org/10.1007/978-3-030-31321-0_37

content [4] with Web-scale object detection and indexing. Visual search and recommendation of clothing, shoes and jewelry with the VisNet architecture [5] and its modification [6] using extraction of visual features and a parallel shallow net.

However, as the photos and videos on mobile devices are created by a user, they potentially contain sensitive personal information, and there is a need for protections of users' privacy. As a result, these multimedia data should not be transferred to remote servers for analysis with the state-of-the-art complex methods [2]. That is why there is a significant demand for developing efficient architectures of CNNs [3,7], which can be implemented directly on a mobile device. There exist a number of architectures that computationally effective and at the same time have good accuracy [8]: SSD (Single shot detector) [9], SSDLite [10], YOLO [11], with different variations of MobileNet [10,12] in a backbone CNN. Unfortunately, if it is necessary to detect small objects (road signs, food, fashion accessories, etc.), the accuracy of such computationally efficient detectors is usually much lower when compared to Faster R-CNN [13] with very deep backbone CNN, such as ResNet [14] or InceptionResNet [15].

In this paper we exploit the fact that not all images can be considered as personal data. For example, panoramic photos, images of food, interiors and showplaces may be sent to remote server for object detection. These types of photos usually contain important information about user's preferences. Hence, we propose here to automatically select private and public visual data. All public photos and videos can be processed remotely with more accurate but slow models. At the same time, private data must be processed offline on mobile device with less precise but lightweight detectors. Here we assume that private image data contains faces of a user (owner of a mobile device) and his or her relatives and friends. Such private photos are chosen using known face recognition [16,17] and clustering techniques [18,19].

The rest of the paper is organized as follows. In Sect. 2 we introduce the proposed pipeline for user preference prediction based on object detection and facial clustering. Experimental results for subsets of MS COCO, ImageNet and Open Images datasets are presented in Sect. 3. Concluding comments and future works are discussed in Sect. 4.

2   Materials and Methods

The quality of recommender systems significantly depends on the solution of the user modeling problem [20]. In this task it is required to develop the user's profile, i.e., predict his or her interests in $C > 1$ categories of preferences, e.g., food, animals, sports equipment, etc. In this paper we assume that a collection of photos and videos is given, and the profile can be defined as a histogram, i.e., frequencies of C categories, which are observed in the gallery of mobile device.

If each category corresponds to some specific objects, this problem can be solved with existing object detection techniques. In order to train the detection models, a balanced dataset with 146 categories of objects with known bounding boxes was created. The balance is chosen at a level of 5000 images per category.

If a category has less pictures, all of them are used, otherwise the images are randomly selected. The list of categories was split into the following high-level interests: activity, appliances, children, indoors, fashion, food, musical instruments, pets/animals, planting, product/services, outdoors, sport and transport. Each high-level group contains 2–15 different types of particular objects from MS COCO [22], ImageNet [21] and Open Images Dataset (OID) [3]. Sample images are given in Fig. 1.

Fig. 1. Example of images from different categories used for training: (a)–(e) animals (bear, cock, polar bear, dog, giraffe), (f)–(j) transport (plane, bus, train, truck, car), (k)–(o) fresh food (salad, strawberry, orange, plum, banana), (p)–(t) mixture from various high-level groups (drink, wine, tennis, dumbbell, tower).

Computational complexity and memory costs of the most accurate object detectors [13], e.g., Faster R-CNN [13] with ResNet or InceptionResNet backbones, do not allow them to be implemented even in the most contemporary mobile devices. At the same time, the usage of remote server to process *all* multimedia data of an user is not permitted because of the privacy constraints. That is why development of fast object detection models is so important [8,9,13]. Unfortunately, fast SSD-based methods are known to be much less accurate when compared to Faster R-CNN, if the small objects should be detected [13]. However, namely such small objects usually defines the user preferences on his or her photos and videos. Hence, in this paper we propose to implement the multi-stage procedure of user modeling (Fig. 2).

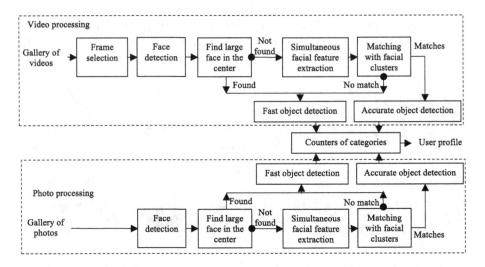

Fig. 2. Proposed pipeline for visual user modeling based on object detection and facial clustering.

Here public data is extracted from the whole collection of images and videos using facial analytic techniques implemented in offline mode on a mobile device. Namely, private photos and videos are assumed to contain faces of an owner of mobile device and his or her friends, relatives, etc. At first, $N \geq 0$ faces are detected on all photos using, e.g., lightweight SSD-based object detectors.

As the facial images $X_n, n \in \{1, ..., N\}$ do not contain labels of concrete subjects on the photos, the problem of extracting people from the gallery should be solved by clustering methods. Namely, every face on image should be assigned to one of the labels $1...K$, where K is a number of people on images in the user's gallery. As K is usually unknown [23], hierarchical agglomerative or density based spatial clustering methods should be used [24]. In order to apply these methods, numerical feature vector should be extracted from each detected facial image. As the faces are observed in *unconstrained* conditions, modern transfer learning and domain adaptation techniques can be used for this purpose [2]. According to these methods, the large external dataset of celebrities is used to train a deep CNN [25,26]. The outputs of one of the last layers of this CNN form D-dimensional ($D \gg 1$) off-the-shelf features [27] $\mathbf{x}_n = [x_{n;1}, ..., x_{n;D}]$ of the photos X_n from the gallery [17]. These feature vectors are L_2-normed in order to provide additional robustness to variability of observation conditions.

The same procedure is implemented for a set of frames from each *video* from a gallery (Fig. 2), and features of the obtained cluster centers are appended to the set of all facial feature vectors. After that, the final clustering is conducted to detect clusters with sufficient number of faces in the gallery. We assume that these faces belong to owner of the device and to his relatives and friends, and all information on corresponding visual data is personal. A user may also manually choose additional private photos.

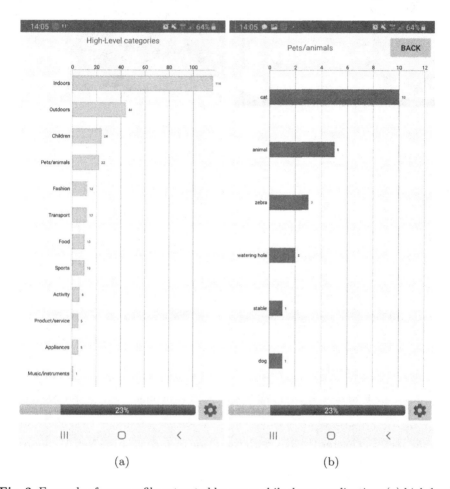

Fig. 3. Example of user profile extracted by our mobile demo application: (a) high-level groups, (b) detailed object categories

We try to predict if each photo can be loaded to external server or it contains private information to see if the processing should be implemented on a mobile device in the offline mode. The photo is considered to be private if it contains faces either in the middle (portrait photo) or a face from rather large facial clusters (family members or close friends). After that the "private" photos are fed to the input of simple TensorFlow object detector, e.g., SSDLite with MobileNet backbone. The "public" photos are processed by rather complex and accurate

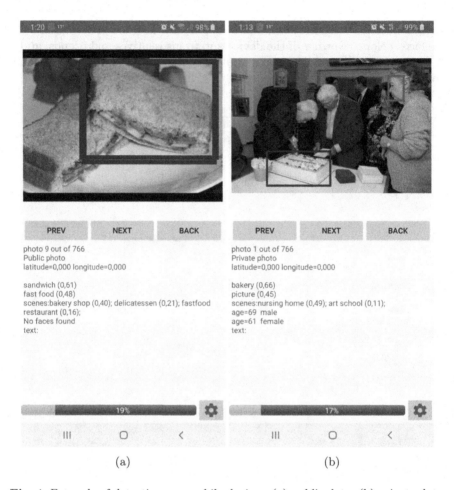

Fig. 4. Example of detections on mobile devices: (a) public data, (b) private data.

detector, e.g., the Faster R-CNN with Inception or InceptionResNet. We map the detected objects on the predefined list of categories and obtain the most frequent categories filling the user profile. The videos are processed similarly: each of 3–5-th frames in each video is selected and the same procedure is repeated, though the decision about private/public status is made for a whole video by considering the video public only if all frames are marked as "public". Then the list of obtained user preferences from public photos and videos is sent back to the mobile device, where it is combined with the results of offline detector in order to obtain the final user profile, i.e., the total preferences histogram.

The proposed pipeline (Fig. 2) is implemented in a demo mobile application for Android. It can work in offline mode but supports object detection with Faster R-CNN with InceptionResNet v2 backbone on a remote Flask server. This application sequentially processes all photos from the gallery in the background thread. However, the intermediate results are available, so it is not necessary to wait for a long time. The resulted profile (histogram of detected objects for each high-level group) is displayed in the main window (Fig. 3a). It is possible to tap any bar in this histogram, and a new form with detailed categories is shown (Fig. 3b). If the user taps a concrete category, a special form appears, which contains a list of all photos from the gallery, in which the corresponding object was found. Sample processing of public and private photos are given in a Fig. 4.

3   Experimental Results

In the first experiment we examined facial clustering results [18] using the GFW (Grouping Faces in the Wild) dataset [28]. It contains facial images from albums of 60 real users from a Chinese social network portal. The size of an album varies from 120 to 3600 faces, with average number of subjects $C = 46$. Three CNNs were used for feature extraction: VGGFace [25], ResNet-50 fine-tuned on VGGFace2 [26] and MobileNet [18,29], trained by us on the same VGGFace2 dataset [26]. The VGGFace, VGGFace2 and MobileNet extract $D = 4096$, $D = 2048$ and $D = 1024$ non-negative features in the output of "fc7", "pool5_7x7_s1" and "reshape_1/Mean" layers from 224×224 RGB images, respectively. The estimates of relation of the number of clusters to the number of different people in the dataset K/C, ARI (Adjusted Rand Index), AMI (Adjusted Mutual Information), homogeneity, completeness and BCubed F-measure for different cluster linkages are shown in Table 1. The best results are marked by bold.

Table 1. Facial clustering results, GFW dataset.

Method	Features	K/C	ARI	AMI	Homogeneity	Completeness	F-measure
Single	VGGFace	4.10	0.440	0.419	0.912	0.647	0.616
	VGGFace2	3.21	0.580	0.544	0.942	0.709	0.707
	MobileNet	4.19	0.492	0.441	0.961	0.655	0.636
Average	VGGFace	1.42	0.565	0.632	0.860	0.751	0.713
	VGGFace2	1.59	0.603	**0.663**	0.934	0.761	0.746
	MobileNet	1.59	**0.609**	0.658	0.917	**0.762**	**0.751**
Complete	VGGFace	**0.95**	0.376	0.553	0.811	0.690	0.595
	VGGFace2	1.44	0.392	0.570	0.916	0.696	0.641
	MobileNet	1.28	0.381	0.564	0.886	0.693	0.626
Weighted	VGGFace	1.20	0.464	0.597	0.839	0.726	0.662
	VGGFace2	1.05	0.536	0.656	0.867	**0.762**	0.710
	MobileNet	1.57	0.487	0.612	0.915	0.727	0.697
Median	VGGFace	5.30	0.309	0.307	0.929	0.587	0.516
	VGGFace2	4.20	0.412	0.422	0.929	0.639	0.742
	MobileNet	6.86	0.220	0.222	**0.994**	0.552	0.411

Here the specially-trained MobileNet [18] is in most cases more accurate than the widely-used VGGFace. As expected, the quality of this model is slightly lower when compared to much deeper ResNet-50 trained on the same VGGFace2 dataset. However, this MobileNet with average-linkage clustering is characterized by the highest BCubed F-measure, which is slightly higher than the best BCubed F-measure (0.745) reported by the authors [28]. The most important advantage of MobileNet model is an excellent speed (5–10-times than VGGFace and VGGFace2) appropriate for offline mobile processing. Moreover, the dimensionality of the feature vector is 2–4-times lower leading to the faster computation of distances during clustering.

In the next experiments the gathered dataset (Fig. 1) was used to train several object detection architectures using the Tensorflow Object Detection API, namely, SSDLite (with image size 300×300 and 512×512) with MobileNet backbone, Faster R-CNN with Inception v2, ResNet-50/101 and InceptionResNet v2 with Atrous convolutions. The size of the model files and average inference time on a laptop (MacBook Pro 2015) and mobile phone (Samsung S9+) are presented in Table 2. Our SSDLite models are rather fast (200 ms and 500 ms per photo on Samsung S9+ for SSDLite-300 and SSDLite-512, respectively). However, Faster R-CNN models are inappropriate for offline detection on mobile devices (1.2 s per image for simple Inception v2 model).

Table 2. Performance analysis of object detection models.

CNN	Model size, MB	Average inference time, s.	
		Laptop	Mobile phone
SSDLite-300 (MobileNet v2)	31.83	0.16	0.30
SSDLite-512 (MobileNet v2)	31.83	0.21	0.52
Faster R-CNN (Inception v2)	64.91	0.40	1.25
Faster R-CNN (InceptionResNet v2)	204.34	1.01	2.39

The detection results for several testing images are shown in Fig. 5. As expected, a simple SSD-based model here misses many small objects relevant for user modeling. For example, this method does not detect anything on Fig. 5a.

The quantitative results of object detection models were obtained using 5000 testing images in each of 146 categories. In particular, we computed recall (an average rate of detected objects from one class) and precision (average rate among categories of correctly detected object for one category to all detected objects of this category). Additionally, some of the categories are considered as "family" categories to each other. For example, sometimes it is not mistake if a category "animal" is detected instead of concrete objects, e.g., "cat" or "dog", detection of "building" instead of "skyscraper" is also suitable, etc. Because of this, precision and recall were calculated "as-is" (exact matching) and with taking into account these family categories. The results are presented in Table 3.

Fig. 5. Sample detection results: (a), (c) SSDLite (MobileNet v2); (b), (d) Faster R-CNN (InceptionResNet).

Here there exist three models with the highest quality: Faster R-CNN with both versions of InceptionResNet (original and quantized by standard procedure from TensorFlow) and ResNet-101. Overall, InceptionResNet returns more detections than the model with Inception or ResNet backbones.

In addition, the most reliable categories were selected for each model so their average recall was more than 0.75 and their precision was calculated (Table 4). The asterisks here mean that some of the important categories were not included in selected categories e.g., faces and buildings. Here the main quality criterion is the number of selected categories, which characterizes how stable is the model for a variety of categories. ResNet-101 on average has higher recall and precision in every taken measurement, however, the number of selected categories is smaller.

Table 3. Estimates of object detection precision/recall.

CNN	Exact match		Family match	
	Recall	Precision	Recall	Precision
Faster R-CNN (InceptionResNet v2)	0.425	0.477	0.448	0.534
Faster R-CNN (InceptionResNet v2 quantized)	0.425	0.471	0.448	0.528
Faster R-CNN (Inception v2)	0.393	0.537	0.414	0.593
Faster R-CNN (ResNet-50)	0.332	0.583	0.35	0.636
Faster R-CNN (ResNet-101)	0.465	0.562	0.485	0.618
SSDLite (MobileNet v2)	0.149	0.465	0.166	0.525
SSDLite (MobileNet v2 quantized)	0.149	0.463	0.166	0.524

Table 4. Results of object detection with selection of reliable classes with recall ≥ 0.75.

CNN	Number of selected categories	Precision
Faster R-CNN (InceptionResNet v2)	78	0.662
Faster R-CNN (InceptionResNet v2 quantized)	79	0.663
Faster R-CNN (Inception v2)	44	0.762
Faster R-CNN (ResNet-50)*	30	0.838
Faster R-CNN (ResNet-101)	67	0.760
SSDLite (MobileNet v2)*	3	0.773
SSDLite (MobileNet v2 quantized)*	3	0.768

This is due to lower precision/recall for categories that are considered to be important. The lowest precision is for such categories as house, animal, car and, especially, face.

4   Conclusion

In this paper we proposed the novel algorithm for user modeling (Fig. 2) based on detection of special objects in photos and videos in a gallery of mobile device. It is known that lightweight SSD-based models cannot be used for small object detection (Fig. 5), so the processing on a remote server with contemporary GPU is needed. However, transfer of *all* photos to the server is usually undesirable due to the presence of personal information in many photos. Hence, we implemented the heuristical rule-based classification of private and public photos/videos based on the presence of faces of relatives/friends on private images. We implemented an efficient facial analysis in offline mode using specially trained MobileNet [18], which was practically as accurate as the state-of-the-art facial feature extractors (Table 1).

In order to train various object detectors and compare their performance (Tables 2, 3 and 4), we gathered a special dataset (Fig. 1). As expected, lightweight models, e.g, SSDLite, are faster and take less memory than Faster R-CNN with different backbones, however their accuracy is also much lower (Table 3). The proposed approach was implemented in a special mobile application. In future, it is important to detect private photos more accurately by, e.g., using text recognition techniques for processing of scanned documents (tickets, passports, etc.). Moreover, other backbones for SSD should be also examined as current MobileNet v2 leads to rather low quality (Fig. 5).

Acknowledgements. This research is based on the work supported by Samsung Research, Samsung Electronics and was prepared within the framework of the Academic Fund Program at the National Research University Higher School of Economics (HSE University) in 2019 (grant No. 19-04-004) and by the Russian Academic Excellence Project "5-100".

References

1. Harrison, G.: Next Generation Databases: NoSQL, NewSQL, and Big Data. Springer, Heidelberg (2015). https://doi.org/10.1007/978-1-4842-1329-2
2. Goodfellow, I.: Deep Learning (Adaptive Computation and Machine Learning series). MIT Press, Cambridge (2016)
3. Kuznetsova, A., et al.: The open images dataset v4: unified image classification, object detection, and visual relationship detection at scale. arXiv preprint arXiv:1811.00982 (2018)
4. Zhai, A., et al.: Visual discovery at Pinterest. In: Proceedings of the 26th International Conference on World Wide Web Companion, pp. 515–524 (2017)
5. Shankar, D., Narumanchi, S., Ananya, H.A., Kompalli, P., Chaudhury, K.: Deep learning based large scale visual recommendation and search for e-commerce. arXiv preprint arXiv:1703.02344 (2017)
6. Andreeva, E., Ignatov, D.I., Grachev, A., Savchenko, A.V.: Extraction of visual features for recommendation of products via deep learning. In: van der Aalst, W.M.P. (ed.) AIST 2018. LNCS, vol. 11179, pp. 201–210. Springer, Cham (2018). https://doi.org/10.1007/978-3-030-11027-7_20
7. Huang, J., et al.: Speed accuracy trade-offs for modern convolutional object detectors. arXiv preprint arXiv:1611.10012 (2016)
8. Qin, Z., Zhang, Z., Chen, X., Wang, C., Peng, Y.: FD-MobileNet: improved MobileNet with a fast downsampling strategy. In: Proceedings of the 25th International Conference on Image Processing (ICIP), pp. 1363–1367. IEEE (2018)
9. Liu, W., et al.: SSD: single shot multibox detector. In: Leibe, B., Matas, J., Sebe, N., Welling, M. (eds.) ECCV 2016. LNCS, vol. 9905, pp. 21–37. Springer, Cham (2016). https://doi.org/10.1007/978-3-319-46448-0_2
10. Sandler, M., Howard, A., Zhu, M., Zhmoginov, A., Chen, L.C.: Inverted residuals and linear bottlenecks: mobile networks for classification, detection and segmentation. arXiv preprint arXiv:1801.04381 (2018)
11. Redmon, J., Farhadi, A.: YoloV3: an incremental improvement. arXiv preprint arXiv:1804.02767 (2018)
12. Howard, A.G., et al.: MobileNets: efficient convolutional neural networks for mobile vision applications. arXiv preprint arXiv:1704.04861 (2017)

13. Ren, S., He, K., Girshick, R., Sun, J.: Faster R-CNN: towards real-time object detection with region proposal networks. In: Advances in Neural Information Processing Systems (NIPS), pp. 91–99 (2015)
14. He, K., Zhang, X., Ren, S., Sun, J.: Identity mappings in deep residual networks. In: Leibe, B., Matas, J., Sebe, N., Welling, M. (eds.) ECCV 2016. LNCS, vol. 9908, pp. 630–645. Springer, Cham (2016). https://doi.org/10.1007/978-3-319-46493-0_38
15. Szegedy, C., Ioffe, S., Vanhoucke, V., Alemi, A.A.: Inception-v4, Inception-ResNet and the impact of residual connections on learning. In: Proceedings of Thirty-First AAAI Conference on Artificial Intelligence, vol. 4, p. 12 (2017)
16. Savchenko, A.V., Belova, N.S.: Unconstrained face identification using maximum likelihood of distances between deep off-the-shelf features. Expert Syst. Appl. **108**, 170–182 (2018)
17. Savchenko, A.V.: Efficient statistical face recognition using trigonometric series and CNN features. In: Proceedings of the 24th International Conference on Pattern Recognition (ICPR), pp. 3262–3267. IEEE (2018)
18. Savchenko, A.V.: Efficient facial representations for age, gender and identity recognition in organizing photo albums using multi-output ConvNet. PeerJ Comput. Sci. **5**, e197 (2019). https://doi.org/10.7717/peerj-cs.197
19. Pan, S.J.: A survey on transfer learning. IEEE Trans. Knowl. Data Eng. **22**(10), 1345–1359 (2010)
20. Lakiotaki, K., Matsatsinis, N.F., Tsoukias, A.: Multicriteria user modeling in recommender systems. IEEE Intell. Syst. **26**(2), 64–76 (2011)
21. Deng, J., Dong, W., Socher, R., Li, L.J., Li, K., Fei-Fei, L.: ImageNet: a large-scale hierarchical image database. In: Proceedings of the International Conference on Computer Vision and Pattern Recognition (CVPR), pp. 248–255. IEEE (2009)
22. Lin, T.-Y., et al.: Microsoft COCO: common objects in context. arXiv preprint arXiv:1405.0312 (2014)
23. Sokolova, A.D., Kharchevnikova, A.S., Savchenko, A.V.: Organizing multimedia data in video surveillance systems based on face verification with convolutional neural networks. In: van der Aalst, W.M.P., et al. (eds.) AIST 2017. LNCS, vol. 10716, pp. 223–230. Springer, Cham (2018). https://doi.org/10.1007/978-3-319-73013-4_20
24. Han, J., Pei, J., Kamber, M.: Data Mining: Concepts and Techniques. Elsevier, Burlington (2011)
25. Parkhi, O.M., Vedaldi, A., Zisserman, A.: Deep face recognition. In: Proceedings of the British Conference on Machine Vision (BMVC), vol. 1, p. 6 (2015)
26. Cao, Q., Shen, L., Xie, W., Parkhi, O.M., Zisserman. A.: VGGFace2: a dataset for recognizing faces across pose and age. In: Proceedings of the International Conference on Automatic Face and Gesture Recognition (FG 2018), pp. 67–74 (2018)
27. Sharif, R.: CNN features off-the-shelf: an astounding baseline for recognition. In: Proceedings of the Conference on Computer Vision and Pattern Recognition Workshops, pp. 806–813. IEEE (2014)
28. Yue, H., Kaidi, C., Cheng, L., Chen, C.L.: Merge or not? Learning to group faces via imitation learning. arXiv preprint arXiv:1707.03986 (2017)
29. Kharchevnikova, A.S., Savchenko, A.V.: Neural networks in video-based age and gender recognition on mobile platforms. Opt. Mem. Neural Netw. **27**(4), 246–259 (2018)

Gun and Knife Detection Based on Faster R-CNN for Video Surveillance

M. Milagro Fernandez-Carrobles$^{(\boxtimes)}$, Oscar Deniz, and Fernando Maroto

ETSI Industriales, VISILAB, University of Castilla-La Mancha, Ciudad Real, Spain
MMilagro.fernandez@uclm.es
http://visilab.etsii.uclm.es

Abstract. Public safety in public areas is nowadays one of the main concerns for governments and companies around the world. Video surveillance systems can take advantage from the emerging techniques of deep learning to improve their performance and accuracy detecting possible threats. This paper presents a system for gun and knife detection based on the Faster R-CNN methodology. Two approaches have been compared taking as CNN base a GoogleNet and a SqueezeNet architecture respectively. The best result for gun detection was obtained using a SqueezeNet architecture achieving a 85.44% AP_{50}. For knife detection, the GoogleNet approach achieved a 46.68% AP_{50}. Both results improve upon previous literature results evidencing the effectiveness of our detectors.

Keywords: Object detection · Guns · Knives · Video surveillance

1 Introduction

The use of weapons in public places has become a major problem in our society. These situations are more frequent in countries where weapons are legally purchased or their use is not controlled [10]. Crowded places are specially vulnerable. Unfortunately, mass shootings have become one of the most dramatic problems we face nowadays [20].

Video surveillance systems, typically based on classic closed circuit television (CCTV) are especially useful for intruder detection and remote alarm verification [6]. However, these systems need to be continuously supervised by a human operator. In this respect, it is estimated that the concentration of a security guard watching a camera panel decreases catastrophically after 20 min.

Security can be increased applying artificial vision algorithms on images obtained from video surveillance systems. Another advantage of these algorithms is the possibility of monitoring larger spaces using fewer devices thus requiring less dependence on the human factor.

Machine learning techniques have been widely used in the field of video surveillance. The prevalent paradigm of deep learning has but increased the potential of machine learning in automatic video surveillance. The objective of

© Springer Nature Switzerland AG 2019
A. Morales et al. (Eds.): IbPRIA 2019, LNCS 11868, pp. 441–452, 2019.
https://doi.org/10.1007/978-3-030-31321-0_38

this work is the development of two novel weapon detectors, for guns and knives, applying deep learning techniques and assess their performance.

This paper is organised as follows. Section 2 describes related work for video surveillance. The datasets used are detailed in Sect. 3 where gun and knife datasets are described in detail. Section 4 presents the object detector approaches and Convolutional Neural Networks (CNNs) used for this work. Results for gun and knife detection are shown in Sect. 5. Finally, Sect. 6 is devoted to the Conclusions.

2   Previous Work

The applications of the deep learning paradigm for weapon detection are still rather limited. The seminal work of Olmost et al. [14] presented an automatic handgun detection system for video surveillance. This system was based on a Faster R-CNN with a VGG16 architecture trained using their own gun database. Results provided zero false positives, 100% recall and a precision (IoU = 0.5) value of 84.21%.

In Valldor et al. [17] a firearm detector for application to social media was presented. The detector employed a Faster R-CNN and an Inception_v2 network for feature extraction. A public database of images containing several firearms was manually labelled and used for training. Benchmarking was performed on the COCO dataset obtaining a ROC curve that showed usable results.

Verma et al. [18] used the Internet Movie Firearm Database (IMFDB) to generate a handheld gun detector. For that purpose, a Faster R-CNN based on a VGG16 architecture was applied only for feature extraction. Classification was performed using three different classifiers: a Support Vector Machine (SVM), a K-Nearest Neighbor (KNN) and a Ensemble Tree classifier. The best result achieved was 93.1% accuracy, using a Boosted Tree classifier. We have to note that the IMFDB dataset contains mostly profile images of pistols and revolvers at high resolution with homogeneous background, which is not a realistic situation.

The work of Akcay et al. [5] presented a detection and classification system for X-ray baggage security imagery. The work explored the applicability of multiple detection approaches based on sliding window CNN, Faster R-CNN, Region-based Fully Convolutional Networks and YOLO. Their system was composed by images divided into six classes: camera, laptop, gun, gun component, knife and ceramic knife. The best results for firearm detection were achieved with a YOLO architecture obtaining a 97.4% AP_{50}. For knife cases, the best results were obtained using a Faster R-CNN based on a ResNet-101 architecture with a 73.2% AP_{50}.

Finally, in Kanehisa et al. [11] the YOLO algorithm was applied to create a firearm detection system. The firearm dataset used for this study was extracted from the IMFDB website. Detection results obtained a 95.73% of sensitivity, 97.30% of specificity, 96.26% of accuracy and 70% of mAP_{50}.

Regarding knife detection, the most relevant results have been obtained in the context of the COCO (Common Objects in Context) Challenges. COCO is a large-scale object detection dataset focused on detecting objects in context [13]. Each year COCO launches a challenge based on any of the following artificial vision tasks: detection, segmentation, keypoints or scene recognition. The last object detection challenge using bounding boxes was released in 2017 where the best result for knife detection was obtained by the Intel Lab team. Employing a Faster R-CNN and a HyperNet architecture this team achieved 36.6% AP_{50}. In Yuenyong et al. [19] knife detection was explored using a dataset of 8,527 infrared (IR) images. A GoogleNet architecture was applied to classify IR images as person or person carrying hidden knife. The classification accuracy reported was 97.91%.

In summary, the Faster R-CNN seems to be the prevalent deep architecture for gun and knife detection. This work also focuses on that architecture. As a novelty, this paper uses other CNNs not previously applied for this purposed and focused on the generation of lightweight models specially tailored for embedded, constrained and distributed systems operating in real environments with noisy and sometimes missing data.

3   Materials

In this section we describe the training and test datasets used for both object detection tasks. The data augmentation techniques used are also described.

3.1   Training Datasets

The gun dataset has been extracted from [14]. The dataset is composed by 3,000 images of guns from different views and scenarios. In order to increase the accuracy of the detector, a data augmentation technique was applied to the dataset. The aim is to perform transformations that simulate realistic views of the object to be detected, see Fig. 1:

– Increasing brightness (10%) in order to simulate different illuminations
– Image scaling to simulate different distances to the object
– Mirroring and rotations (5°) to create different canonical views of the object

With these transformations, the dataset was increased to a total of 15,000 images.

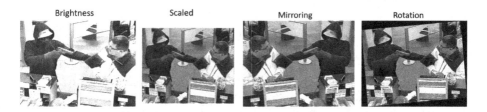

Fig. 1. Data augmentation transformations for the gun dataset

On the other hand, the COCO 2017 dataset [1] has been used to train the knife detector. COCO is a large dataset for object detection and segmentation tasks. The full dataset has a total of 330,000 images with 1.5 million objects divided into 80 classes, one of them knives. In the dataset there are a total of 4,326 images of knives, with a total of 7,770 knives labelled. This dataset has been extended by applying mirroring and scaling transformations for data augmentation so that a total of 12,978 knife images were obtained.

3.2   Benchmarking Datasets

The gun test set was generated leveraging several existing gun datasets with a total of 1,303 images:

- The Olmos et al. test set [14] which is composed by a total of 608 images and 304 weapon images.
- The small_gun category of the Gupta dataset [4] with a total of 80 images of guns.
- The handgun class of the Open Images Dataset V4 [3] with 89 images of guns [12].
- Finally, 526 random images from the COCO dataset [1] without weapon instances.

Regarding knives, the test set was generated using 169 images from the knife and kitchen knife classes of the Open Images Dataset V4 [3, 12] and 526 random images from the COCO dataset [1] without knives. Thus, the knife test dataset had a total of 695 images.

4   Methodology

As mentioned above, the main objective of this work is the development of an object detector that efficiently locates guns and knives in real-time video. For that purpose, an approach based on deep learning techniques and more specifically through the Faster R-CNN methodology will be adopted. This object detection approach uses internally a CNN and a Regional Proposal Network (RPN) for the classification and location processes respectively. In order to better understand this methodology, a brief description of its evolution and performance is described below.

4.1   Evolution of R-CNN Object Detectors

The Regions-CNN method was developed in 2014. The processing of a R-CNN can be divided into three steps [8]. Firstly, an algorithm called selective search generates approximately 2,000 region proposals (or regions of interest). These region proposals are independent divisions of the image where an object could be located. Secondly, a CNN extracts features individually from each region proposal. Finally, the object is classified using a Support Vector Machine (SVM) methodology. Region proposals are considered as positive when their Intersection over Union (IoU) measure against the ground truth exceeds an arbitrary value. Later, the object bounding box localization is calculated by overlapping the selected region proposals.

One of the main disadvantages of the R-CNN was its slow execution time. Fast R-CNN was proposed in 2015 as an improvement of R-CNN [7]. Fast R-CNN is twenty five times faster than its predecessor mainly due to two modifications. Feature extraction is performed using a CNN on the whole input image. Region proposals are selected as in the R-CNN approach by an external selective search method and included in the last layers of the network as projections on the feature map. The SVM classifier in this approach is replaced by a softmax layer. Although the Fast R-CNN was a breakthrough compared to the R-CNN, it still relied on algorithms such as the selective search that formed bottlenecks and slowed down the execution time of the detector.

In 2016 the Faster R-CNN method introduced a new region proposal extraction method called Regional Proposal Network (RPN) [15]. The idea of a RPN is to take advantage of the convolutional layers to obtain region proposals directly. Consequently, a sliding window is applied on the CNN feature map in order to extract region proposals of different sizes. The RPN is not responsible for classifying localized objects, this task is subsequently carried out by a Fast R-CNN. Therefore, a Faster R-CNN is a Fast R-CNN plus a RPN.

4.2   Faster-CNN Base Architectures

The CNN selected as Faster R-CNN base architecture should depend on its final purpose. Many CNNs employ a very deep architecture with the aim of obtaining a higher accuracy at a high computational cost. On the other hand, other architectures can be used that sacrifice precision in order to obtain models that can be integrated into embedded systems. In this work, GoogleNet and SqueezeNet CNN architectures have been tested and compared with the purpose of exposing their advantages and disadvantages for our task of weapon detection in video.

GoogleNet. The GoogleNet network [16] is a CNN developed in 2014. This network demonstrated high accuracy for object detection in the ImageNet contest Large-Scale Visual Recognition Challenge 2014, being the winning architecture with a 6.66% error rate. The architecture of this CNN is mainly composed by Inception layers which are based on covering large image areas while keeping a high resolution for small areas with high feature density. For that, the network applies convolutions in parallel with different filter sizes. The GoogleNet architecture is composed by a total of 22 layers, see Table 1.

Training using GoogleNet for the Faster R-CNN was carried out applying a stochastic gradient descent optimization algorithm with a momentum of 0.9 to accelerate gradient vectors, a L2 regularization method and an initial learning rate of $1e-3$. The optimization was run for 30 epochs.

Table 1. GoogleNet architectural dimensions [16]

Layer name/type	Output size	Filter size/stride	Depth	#1 × 1	#3 × 3 reduce	#3 × 3	#5 × 5 reduce	#5 × 5	Pool proj	Params	Ops
Convolution	112 × 112 × 64	7 × 7/2	1							2.7K	34M
Max pool	56 × 56 × 64	3 × 3/2	0								
Convolution	456 × 56 × 192	3 × 3/1	2		64	192				112K	360M
Max pool	28 × 28 × 192	3 × 3/2	0								
Inception	28 × 28 × 256		2	64	96	128	16	32	32	159K	128M
Inception	28 × 28 × 480		2	128	128	192	32	96	64	380K	304M
Max pool	14 × 14 × 480	3 × 3/2	0								
Inception	14 × 14 × 512		2	192	96	208	16	48	64	364K	73M
Inception	14 × 14 × 512		2	160	112	224	24	64	64	437K	88M
Inception	14 × 14 × 512		2	128	128	256	24	64	64	463K	100M
Inception	14 × 14 × 528		2	112	144	288	32	64	64	580K	119M
Inception	14 × 14 × 832		2	256	160	320	32	128	128	840K	170M
Max pool	7 × 7 × 832	3 × 3/2	0								
Inception	7 × 7 × 832		2	256	160	320	32	128	128	1072K	54M
Inception	7 × 7 × 1024		2	384	192	384	48	128	128	1388K	71M
Avg pool	1 × 1 × 1024	7 × 7/2	0								
Dropout (40%)	1 × 1 × 1024		0								
Linear	1 × 1 × 1000		1							1000K	1M
Softmax	1 × 1 × 1000		0								

SqueezeNet. The SqueezeNet network [9] is a CNN developed in 2016. The main goal of this network was the deployment of a small CNN architecture with fewer parameters instead of improving the accuracy. SqueezeNet achieved the same accuracy than AlexNet on the ImageNet dataset obtaining a model size with 50x fewer parameters. Therefore, it is a valuable alternative for embedded systems, field-programmable gate arrays (FPGAs) and other constrained systems. The SqueezeNet architecture follows three strategies to reduce the number of parameters while maintaining the accuracy level:

- Most convolutions replace 3 × 3 filters for 1 × 1 filters. As a consequence, the number of parameters is reduced 9× by convolution
- Decrease the number of channels using squeeze layers
- Downsample late in the network so that convolution layers have large activation maps. Downsampling is performed reducing the size of the input data or selecting those layers in which downsampling is going to be carried out. Most of the layers have a stride of 1 and layers with stride larger than 1 are accumulated at the end of the network. That produces large activation maps improving accuracy levels

SqueezeNet applies fire modules to achieve the previous strategies. A fire module is composed by a squeeze convolution layer (1 × 1 filters) and an expand layer (mixture of 1 × 1 and 3 × 3 convolution filters). Three parameters are included in a fire module: $s1 \times 1$ (from squeeze layer), $e1 \times 1$, and $e3 \times 3$ (from expanded layer). All are related to the number of filters used in these layers.

Table 2. SqueezeNet v1.0 architectural dimensions [9]

Layer name/type	Output size	Filter size/stride	Depth	s1 × 1 (#1 × 1 squeeze)	e1 × 1 (#1 × 1 expand)	e3 × 3 (#3 × 3 expand)	1 × 1 sparsity	e1 × 1 sparsity	e3 × 3 sparsity	Parameter after pruning
Convolution	111 × 111 × 96	7 × 7/2(×96)	1				100% (7 × 7)			14,208
Max pool	55 × 55 × 96	3 × 3/2	0							
Fire	55 × 55 × 128		2	16	64	64	100%	100%	33%	5,746
Fire	55 × 55 × 128		2	16	64	64	100%	100%	33%	6,258
Fire	55 × 55 × 256		2	32	128	128	100%	100%	33%	20,646
Max pool	27 × 27 × 256	3 × 3/2	0							
Fire	27 × 27 × 256		2	32	128	128	100%	100%	33%	24,742
Fire	27 × 27 × 384		2	48	192	192	100%	50%	33%	44,700
Fire	27 × 27 × 384		2	48	192	192	50%	100%	33%	46,236
Fire	27 × 27 × 512		2	64	256	256	100%	50%	33%	77,581
Max pool	13 × 13 × 512	3 × 3/2	0							
Fire	13 × 13 × 512		2	64	256	256	50%	100%	30%	77,581
Convolution	13 × 13 × 1000	1 × 1/1(×1000)	1				20% (3 × 3)			103,400
Avg pool	1 × 1 × 1000	13 × 13/1	0							
Softmax	1 × 1 × 1000		0							

Fire module sets that s1 × 1 must be less than the sum of e1 × 1 and e3 × 3, so the squeeze layer helps to limit the number of input channels to the 3 × 3 filters. The SqueezeNet architecture is composed by a total of 13 layers, see details in Table 2.

In this approach, training using a SqueezeNet for the Faster R-CNN was carried out applying a stochastic gradient descent optimization algorithm with a momentum of 0.9 to accelerate gradient vectors, a L2 regularization method and an initial learning rate of $1e-4$. As with the GoogleNet approach, 30 epochs were used to train the classifier.

5   Results

In a detection task there are two possible results, positive and negative. Some positive cases can be classified as negative and vice versa. These cases are called false positives (type I error) and false negatives (type II error), respectively. Thus, the following four cases are considered: True Positive (TP), True Negative (TN), False Positive (FP) and False Negative (FN). However, in an object detection task object localization must be considered too. The accuracy achieved in an object detector is commonly evaluated using the mean average precision (mAP). This measurement is defined as the average of the maximum precisions at different recall values. Therefore, the three main concepts considered within this measurement are: precision, recall and Intersection over union (IoU).

– Precision measures the likelihood of a positive case being classified as such. This value is estimated using the amount of real positive cases which were classified as positive. Then, it is the percentage of correct positive predictions.

Precision is calculated as in Eq. 1.

$$Precision = \frac{TP}{TP + FP} \qquad (1)$$

- Recall (or sensitivity) measures the likelihood of classifying the object as positive. In other words, it measures how good is the network finding positives cases, see Eq. 2.

$$Recall = \frac{TN}{TN + FP} \qquad (2)$$

- The Intersection over Union (IoU) quantifies the overlapping between 2 regions. This measures how valuable the prediction is with respect to the ground truth (the real object boundary). A prediction is usually considered to be correct when the IoU is equal or greater than 0.5.
- The Precision-Recall Curve summarizes the trade-off between precision and recall values using different probability thresholds. The area under this curve is known as average precision (AP), a value between 0 and 1 which evaluates the quality of the model. When there is more than one object to be detected, the average precision is calculated for each object resulting in the mean average precision (mAP).

For the problem of gun detection, Faster R-CNN trained using GoogleNet obtained a 55.45% of AP_{50} (AP at IoU $= 0.50$). Faster R-CNN using a SqueezeNet obtained 85.44% of AP_{50}, a significant difference over GoogleNet. The precision-recall curve acquired for SqueezeNet is shown in Fig. 2. This detector achieved good results, obtaining similar or even improving upon previous results described in the literature. A comparison between our results and another similar work are shown in Table 3.

Table 3. Results comparison for weapon detection based on a Faster R-CNN

Gun Detector performance measurement			
	Our approaches		Olmos et al. [14]
Faster R-CNN methodology	GoogleNet	SqueezeNet	VGG16
AP_{50}	55.45%	**85.44%**	84.21%
Knife Detector performance measurement			
	Our approaches		COCO Challenge17 (Intel Lab) [2]
Faster R-CNN methodology	GoogleNet	SqueezeNet	HyperNet
AP_{50}	**46.68%**	1.1%	36.6%

Regarding knife detection, Faster R-CNN based on GoogleNet achieved an AP_{50} of 46.68% and a 1.1% using SqueezeNet, being these results much lower than expected. Nevertheless, results obtained using GoogleNet architecture improve previous knife detection results reported in the literature, see Table 3. Figure 3 shows the precision-recall curve for the GoogleNet approach.

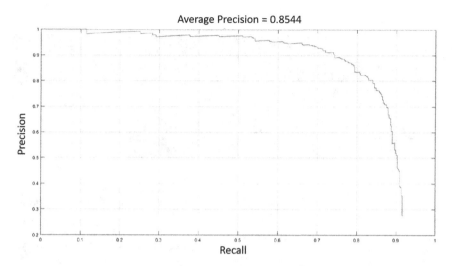

Fig. 2. Precision-recall curve for gun detection using Faster R-CNN and SqueezeNet architecture v1.1

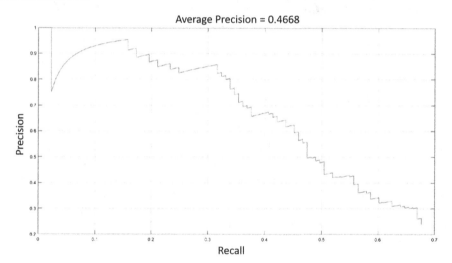

Fig. 3. Precision-recall curve for knife detection using Faster R-CNN and GoogleNet architecture

Some visual results obtained for weapon detection are shown in Fig. 4. These results demonstrate the capability of our weapon detectors to locate guns and knives on the test dataset.

Gun detections

Knife detections

Fig. 4. Scores obtained for gun (Faster R-CNN/SqueezeNet) and knife (Faster R-CNN/GoogleNet) detectors

6   Conclusions

Public and crowded areas are still the target of many violent acts. Video surveillance can be helped by automatic image analysis using artificial vision. This paper describes the implementation of several weapon detectors for video surveillance based on Faster R-CNN methodologies. Several previous studies have applied the Faster R-CNN methodology but, as far as the authors know, none of them have been actually focused on the development of lightweight models that could be later used in constrained and real-time devices. GoogleNet or SqueezeNet are architectures for that purpose. For training, gun and knife images from the work of Olmos et al. and COCO dataset have been used. Several transformations such as rotations, scaling or brightness were applied in order to augment the datasets. Detectors were developed using the GoogleNet and SqueezNet architectures as CNN base on a Faster R-CNN. The best result for gun detection was obtained using a SqueezeNet architecture achieving a 85.44% AP_{50}. For knife detection, GoogleNet approach accomplished a 46.68% AP_{50}. Both detector results improve upon previous literature from similar studies evidencing the effectiveness of our detectors.

Acknowledgments. This work was partially funded by projects TIN2017-82113-C2-2-R by the Spanish Ministry of Economy and Business and SBPLY/17/180501/000543 by the Autonomous Government of Castilla-La Mancha and the ERDF.

References

1. COCO dataset 2017. http://cocodataset.org. Accessed 04 May 2019
2. COCO detection leaderboard. http://cocodataset.org/#detection-leaderboard. Accessed 04 May 2019
3. Open images dataset v4. https://storage.googleapis.com/openimages/web/index.html. Accessed 04 May 2019
4. Weapon detection by neural network. https://github.com/Shubham02gupta/Weapon-Detection-by-Neural-network/tree/master/train. Accessed 04 May 2019
5. Akcay, S., Kundegorski, M.E., Willcocks, C.G., Breckon, T.P.: Using deep convolutional neural network architectures for object classification and detection within X-ray baggage security imagery. IEEE Trans. Inf. Forensics Secur. **13**(9), 2203–2215 (2018). https://doi.org/10.1109/TIFS.2018.2812196
6. Dastidar, J.G., Biswas, R.: Tracking human intrusion through a CCTV. In: 2015 International Conference on Computational Intelligence and Communication Networks (CICN), pp. 461–465, December 2015. https://doi.org/10.1109/CICN.2015.95
7. Girshick, R.: Fast R-CNN. In: 2015 IEEE International Conference on Computer Vision (ICCV), pp. 1440–1448, December 2015. https://doi.org/10.1109/ICCV.2015.169
8. Girshick, R., Donahue, J., Darrell, T., Malik, J.: Rich feature hierarchies for accurate object detection and semantic segmentation. In: Proceedings of the 2014 IEEE Conference on Computer Vision and Pattern Recognition, CVPR 2014, pp. 580–587. IEEE Computer Society, Washington, DC (2014). https://doi.org/10.1109/CVPR.2014.81
9. Iandola, F.N., Moskewicz, M.W., Ashraf, K., Han, S., Dally, W.J., Keutzer, K.: SqueezeNet: Alexnet-level accuracy with 50x fewer parameters and <1mb model size. CoRR abs/1602.07360 (2016). http://arxiv.org/abs/1602.07360
10. Jehan, F., et al.: The burden of firearm violence in the United States: stricter laws result in safer states. J. Inj. Violence Res. **10**(1), 11–16 (2018). https://doi.org/10.5249/jivr.v10i1.951
11. Kanehisa, R., Neto, A.: Firearm detection using convolutional neural networks. In: Proceedings of the 11th International Conference on Agents and Artificial Intelligence, ICAART, vol. 2, pp. 707–714, January 2019. https://doi.org/10.5220/0007397707070714
12. Kuznetsova, A., et al.: The open images dataset V4: unified image classification, object detection, and visual relationship detection at scale. CoRR abs/1811.00982 (2018). http://arxiv.org/abs/1811.00982
13. Lin, T., et al.: Microsoft COCO: common objects in context. CoRR abs/1405.0312 (2015). http://arxiv.org/abs/1405.0312
14. Olmos, R., Tabik, S., Herrera, F.: Automatic handgun detection alarm in videos using deep learning. Neurocomputing **275**, 66–72 (2018). https://doi.org/10.1016/j.neucom.2017.05.012. http://www.sciencedirect.com/science/article/pii/S0925231217308196
15. Ren, S., He, K., Girshick, R., Sun, J.: Faster R-CNN: towards real-time object detection with region proposal networks. In: 28th International Conference on Neural Information Processing Systems, NIPS 2015, vol. 1, pp. 91–99. MIT Press, Cambridge (2015). http://dl.acm.org/citation.cfm?id=2969239.2969250
16. Szegedy, C., et al.: Going deeper with convolutions. In: 2015 IEEE Conference on Computer Vision and Pattern Recognition (CVPR), pp. 1–9, June 2015. https://doi.org/10.1109/CVPR.2015.7298594

17. Valldor, E., Stenborg, K.G., Gustavsson, D.: Firearm detection in social media images. In: Swedish Symposium on Deep Learning 2018, September 2018
18. Verma, G.K., Dhillon, A.: A handheld gun detection using Faster R-CNN deep learning. In: Proceedings of the 7th International Conference on Computer and Communication Technology, ICCCT-2017, pp. 84–88. ACM, New York (2017). https://doi.org/10.1145/3154979.3154988
19. Yuenyong, S., Hnoohom, N., Wongpatikaseree, K.: Automatic detection of knives in infrared images. In: 2018 International ECTI Northern Section Conference on Electrical, Electronics, Computer and Telecommunications Engineering (ECTI-NCON), pp. 65–68, February 2018. https://doi.org/10.1109/ECTI-NCON.2018.8378283
20. Zhang, Y., Wang, Y., Foley, J., Suk, J., Conathan, D.: Tweeting mass shootings: the dynamics of issue attention on social media. In: Proceedings of the 8th International Conference on Social Media and Society, #SMSociety 2017, pp. 59:1–59:5. ACM, New York (2017). https://doi.org/10.1145/3097286.3097345

A Method for the Evaluation and Classification of the Orange Peel Effect on Painted Injection Moulded Part Surfaces

Atae Jafari-Tabrizi$^{(\boxtimes)}$ (iD), Hannah Luise Lichtenegger, and Dieter P. Gruber

Polymer Competence Center Leoben GmbH, Roseggerstraße 12, 8700 Leoben, Austria
{atae.jafari,hannah.lichtenegger,dieter.gruber}@pccl.at

Abstract. Orange peel effect, the wavy appearance on the surface, is one of the frequently encountered defects on the painted injection moulded parts. In this work, a method for evaluation and classification of the orange peel effect on a black-painted high-gloss surface using image processing and frequency analysis is presented. A monochrome camera is used to acquire images from the surface of the part, while an LED-bar is illuminating it. Because the part is complex shaped, in order to inspect the whole surface, a robotic manipulator is used to handle the part in front of the camera. After taking images, the region of interest (the region in the image illuminated by the light source) is selected manually. Based on the result of the image processing and frequency analysis assessing the waviness of the ROI, a score to evaluate and classify the intensity of the orange peel effect is calculated. Finally, the outcomes of this method are compared with the subjective evaluation of the experts, in order to examine the reliability of the evaluation method.

Keywords: Image processing · Orange peel effect · Surface inspection

1 Introduction

In today's highly competitive industry, automotive and other large industrial manufacturers intend to bring their products to perfection; both functionally and cosmetically. Painted (or unpainted) injection moulded parts have an extensively high application in the goods that are being used in everyday life. Hence it is necessary to inspect them for cosmetic defects before putting them into service [11]. Nowadays machine vision is on the way to become a standard in quality inspection. Its use brings many advantages, while the following are the most important: High reliability and repeatability, saving production time, and minimizing errors caused by humans.

Since end users of injection moulded parts are humans, "defects" defined on these parts are all in correspondence with the human vision. Thus, it is important for the machine vision to be able to mimic the human perception of the surface in the sense of defect detection. This point was addressed in previous

© Springer Nature Switzerland AG 2019
A. Morales et al. (Eds.): IbPRIA 2019, LNCS 11868, pp. 453–464, 2019.
https://doi.org/10.1007/978-3-030-31321-0_39

work such as for instance in [3], where a machine vision system is used to develop a model parameter to simulate the human perception of the sink marks (local disturbances in curvature of the surface [6]) that happen frequently in injection moulded parts. A morphological method is proposed in [2] to detect surface properties such as scratches. In [4] a model based on skewed Gauss function is developed in order to quantitatively evaluate the visually perceptible sink marks on the surface of the injection moulded parts. Phase Measuring Deflectometry was used in [6, 7] to detect the sink marks.

Although collaboration of machine learning with machine vision showed outstanding performance in detection and evaluation of the cosmetic defects, and can inspect parts with virtually zero error, in some cases it still lacks the power of experience and subjective evaluation that a skilled operator has. The frequently encountered defects on the surface of a painted injection moulded part which can be detected without difficulty are paint dots, scratches, and paint flows. The orange peel effect occurs frequently too, and can cover a relatively large area on the surface. It can have several degrees of intensities, which makes it difficult for machine learning algorithms to distinguish a defective orange peel surface from a non-defective one.

The orange peel effect can be described as a series of peaks and valleys on the surface, which their combination creates a look similar to that of the skin of an orange. An example of a surface with orange peel effect can be seen in Fig. 1. Unless it is created intentionally in the given case, it is one of the most important defects in the paint and polishing industries. It is a result of improper application of spray paint on the surface, and causes distortions in the light that reflects to its surroundings, as mentioned by Konieczny [5]. According to Konieczny [5], and Sone and Watanabe [12] on dark-painted high-gloss surfaces this effect becomes more significant.

In literature there is work dealing with the prediction and evaluation of the orange peel effect. For the prediction of the visual appearance of a painted steel surface, Scheers et al. [10] have proposed a method to correlate the waviness of the painted steel surfaces to the surface roughness parameters. In order to measure and characterize the orange peel defects in the nanometre range on polished metallic surfaces, in [9] the interferometric microscopy was used. In [8], Miranda-Medina et al. have proposed a correlation between surface roughness parameters and degree of orange peel effect (which is determined by visual inspection for each surface) on highly polished steel surfaces. The aim is to find the surface roughness parameters which can be used to distinguish between different orange peel grades. In this paper, phase shifting interferometry has been used to measure the surface topography. Sone and Watanabe [12] have used a spectral camera and a suitable illumination to project light patterns on the surface. Then, the frequency analysis of the projected pattern and human visual inspection results have been used to find a correlation between measurements and subjective evaluation. A commercial device [1] is able to characterize the surface by magnitude of the distortions. It simulates the human eye's resolution at different distances. The signal received from subject surface is being processed by a mathematical filter and magnitude of wavelengths at various ranges are used to determine the degree of the orange peel.

The aim of this work is to evaluate and classify the waviness due to the orange peel effect on a painted surface. The selected surface is from a part that is being used in automotive industry. It is black-painted and high-gloss, which makes it an interesting testing case for surface inspection.

In the rest of the paper first the experimental setup and the evaluation algorithm are explained. Then the feasibility of developing such an automatic orange peel evaluation framework is addressed by looking for a correlation between the result of the machine vision measurements and the subjective evaluation by the experts. This is done by classifying 30 surfaces with different orange peel degrees by the developed algorithm, and comparing the results with the classifications done by visual testing experts.

In addition to the fact that this evaluation method eliminates human errors which are due to the qualitative nature of the visual inspection, it has also the following advantages compared to the other machine vision results.

- Compared to the commercial product mentioned above [1], the advantage of this method is that it is non-contact, eliminating the risk of damaging the surface.
- The evaluation process is computationally inexpensive, and therefore it is fast.
- This approach directly utilizes the dynamics of the anomaly itself (waviness), and since there are no learning processes involved, it does not require any prior training phase, or large amounts of labelled data.
- The algorithm is flexible. By changing the parameters of the algorithm it is possible to apply it to other surfaces or other inspection environments; For instance for surfaces with shorter or longer orange peel wavelengths, or different illuminating methods.

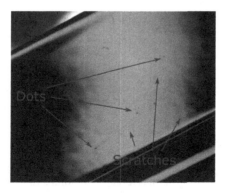

Fig. 1. A surface with orange peel effect. Other defects (e.g., scratches, dots) are also visible.

2   Experimental Setup

In experiments samples of 3D-shaped black-painted high-gloss parts were inspected for the orange peel effect. The setup consists of a camera, a white-coloured LED-bar, and a robotic manipulator to handle the part in front of the camera. The camera has a resolution of 1.3 MP but its image acquisition rate can reach up to 168 frames per second. The lens has a fixed focal length of 25 mm. The testing sample manipulator used is a 6-axis industrial robotic arm which can be pre-programmed to follow a demanded path. Its maximum speed during automatic operation can reach up to 2 m/s. A sketch of the setup can be seen in Fig. 2.

During the experiments the aperture is set to maximum and the focus is adjusted to the surface. Setting the aperture to the maximum has the following benefits.

- Maximum aperture makes it possible to set the exposure time of the camera low, which makes it possible to increase the frame rate of the camera. This in turn helps to reduce the inspection time of the whole surface. This is a demanded quality for an in-production inspection process.
- Another effect of maximum aperture is to have a smaller depth of field, which results in a smaller area on the surface in focus. Having a small sharp region on the surface may count as a challenge, if the minimization of the inspection time is intended: A smaller sharp region (which ROI is a part of) means it will take a longer time to inspect the whole surface of the part. This brings the necessity of handling the part with a high-speed robot.

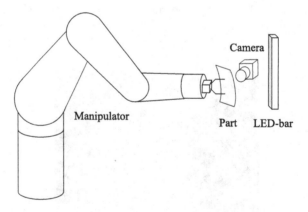

Fig. 2. A sketch of the experimental setup showing the configuration of the camera and the LED-bar, and the robotic manipulator handling the part

The experimental setup is constructed the way that, in addition to the orange peel effect other defect types can also be detected with little additional effort.

In this configuration the orange peel effect is visible at the edge of the reflection of the LED-bar from the surface. As seen in Fig. 1, it is possible to detect other anomalies such as scratches and dots at the centre of the ROI, while the orange peel effect can be seen at the edge.

3 Evaluation Algorithm

A decisive parameter in classifying the orange peel effect comes from the nature of the anomaly itself: The wavelengths of the wavy structure on the surface. By illuminating the subject surface from the correct angle these waves are clearly visible.

In the image containing the region of interest, each pixel column contains the waviness information caused by the orange peel effect. Therefore, the idea is to analyse each pixel column separately and calculate an "orange peel intensity score". Finally, the scores are combined to give a final result for the whole ROI. A diagram showing the steps taken during the analysis of a single pixel column is shown in Fig. 3.

1. After acquisition of the image, a ROI is selected.
2. A pixel value matrix is created from ROI, containing values between 0 and 255.
3. A frequency analysis is performed for every column of the matrix (the analysis includes the following steps: Filtering out the low frequencies, obtaining the frequency spectrum by Fast Fourier Transformation, and picking out the dominant frequency).
4. Calculation of the intensity score of the orange peel (the higher the value, the more defective the anomaly is).
5. Steps 3 and 4 are performed for every pixel column, and an average of the intensity scores of all columns is taken for the evaluation of the orange peel effect.

The ROI is selected manually from the captured greyscale image and the pixel values (between 0 and 255) of each pixel column is extracted by using an implemented Python script.

In order to find the frequencies corresponding to the orange peel, first high-pass filtering must be applied. Cutting off the very low frequencies eliminates the interference caused by bright to dark transitions in the image, which do not correspond to the orange peel effect. There might also be some other shape-errors on the surface which are not related to the orange peel effect. These shape-errors can occur due to a failure in the painting process, or a fault from the injection moulding process.

Applying the high-pass filter masks out all the features which have a wavelength larger than 1.5 mm from the pixel array. This length is displayed in Fig. 4. It must be noted that the algorithm is designed to evaluate the orange peel effect on this specific surface. Depending on the needs of the manufacturer or specifications of the production process the wavelengths comprising the orange peel

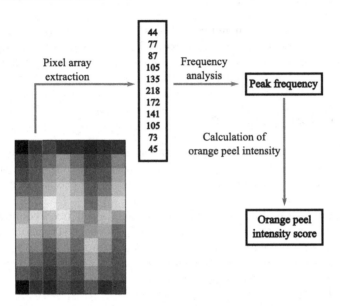

Fig. 3. Calculation procedure of orange peel intensity score of a single pixel column: after extraction of the pixel array, it goes through a frequency analysis to obtain the peak frequency, then based on the peak frequency and its amplitude the orange peel intensity score is calculated.

effect can differ from part to part. Therefore another surface with a different painting and production process will need another cut-off frequency. This cut-off frequency is to be determined depending on the needs of the manufacturer and the application areas of the subject surface. As an example, a surface with an application in a range not too close to the human's field of view will have different cut-off frequency requirements than a surface which is used very close to human's field of view.

After cutting off the lower frequencies, the dominant frequency corresponding to the waviness of the surface is found. For this purpose a Fast Fourier Transform (FFT) function is used. The dominant frequency, and its amplitude are used as two parameters to calculate the final intensity score of the orange peel anomaly. It is known that the higher the frequency, the more "defective" the orange peel looks. In addition, the amplitude of the dominant frequency represents a "weight" that amplifies the effect of this frequency on the perceived surface structure. Therefore, at each column if there is a dominant frequency found, it is multiplied by its amplitude, as shown in Eq. 1,

$$I_i = A \times f \tag{1}$$

where I_i is the orange peel intensity score of the ith column, f is the dominant frequency of the pixel column, and A is the amplitude of the dominant frequency.

In a final step, these values are added for the whole image, and divided by the number of the columns to get a normalized value. This makes the score of

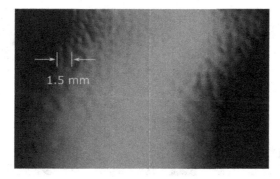

Fig. 4. The distance on the part over which any orange peel anomaly is ignored

the evaluation to be independent of the size of the ROI. Equation 2 shows the formula used to calculate the orange peel intensity score of the whole ROI,

$$I = \frac{\sum_{i=1}^{n} A_i \times f_i}{n} \qquad (2)$$

where f_i is the dominant frequency of the ith column, A_i is the amplitude of the dominant frequency, and n is the number of columns.

4   Results

In order to assess the relevance between the calculated orange peel intensity score and the human perception, the following approach was followed: 30 images from the same surface area of different parts having different degrees of orange peel were selected. Each image corresponds to an area of approximately 14 mm × 23 mm on the surface of the part. First, an orange peel intensity score was calculated using the proposed algorithm for each of these images. Then 5 experts were asked to evaluate the same images and classify them to 3 predefined levels of orange peel intensity: "weak or no orange peel", "medium orange peel", and "strong orange peel". In Fig. 5 examples for weak, medium, and strong orange peel can be seen.

In order to compare the calculated scores with the evaluations of the experts, the average value of the experts' scores were taken. The results of the two evaluation approaches are shown in Fig. 6. The images are named from 1 to 30. The scores of the algorithm are red, while the average of the scores given by the experts are blue. The scale on the right hand side of the image corresponds to the human perception results, and the scale on the left hand side corresponds to the score calculated by the algorithm.

After comparing the results, it can be said that in the cases of "weak" and "strong orange peel" there is a strong consistency between the algorithm and the evaluation of the experts. For the five images having the lowest amount of orange peel according to the algorithm (images 1 to 5 in Fig. 6), there is a 100%

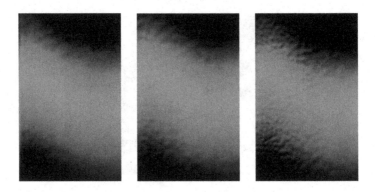

Fig. 5. The examples for different degrees of orange peel on the black-painted high-gloss injection moulded part. The left image corresponds to a "weak orange peel", the middle image is an example for "medium orange peel", and the right image has a "strong orange peel".

correspondence between the two evaluation methods; all of the experts have labelled these images as "weak or no orange peel". The surface, which according to the algorithm has the strongest orange peel (image 30 in Fig. 6), has been evaluated to have "strong orange peel" by all experts.

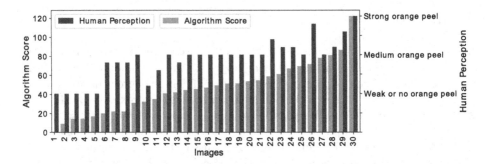

Fig. 6. Results of the two approaches to evaluate the orange peel appearance in 30 images (Color figure online)

For "medium orange peel" images, the result of the algorithm shows deviations from the evaluations of the volunteers. Out of these images, image 10 and image 26 are noticeable outliers and will be further analysed. According to the algorithm, image 9 and image 10 have similar orange peel degrees. However according to the experts' estimation image 9 has a stronger orange peel. These two images can be seen in Fig. 7.

Algorithm has calculated image 26 and 25 to have similar orange peel degrees, while according to the experts the orange peel effect in image 26 is stronger. The two exemplary images can be seen in Fig. 8.

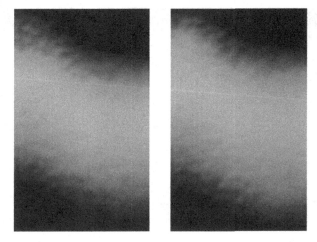

Fig. 7. Image 9 (left) and image 10 (right). According to the algorithm the two images have a similar orange peel, while according to the volunteers image 9 has a stronger orange peel.

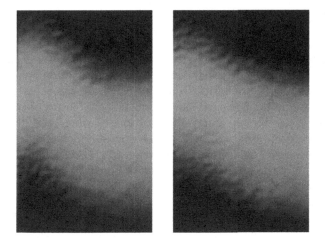

Fig. 8. Image 25 (left) and image 26 (right). According to the algorithm the two has a similar orange peel, while according to the volunteers image 26 has a stronger orange peel.

Another important difference between the results of the two evaluation approach can be noticed in images 6, 7, and 8. The scores of these images calculated by the algorithm are close to that of image 5. This image is evaluated by the experts to have "weak or no orange peel". On the other hand, the outcome of the experts' evaluation of images 6, 7, and 8 have put them in "medium orange peel" range. Looking at the results in Fig. 6 it can be seen that image 13 has also been evaluated by the experts to have exactly the same orange peel level as

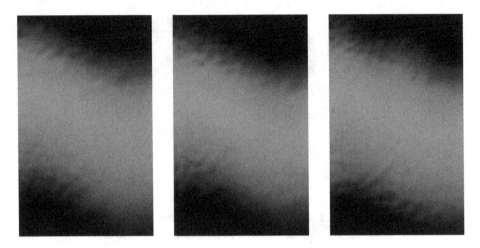

Fig. 9. From left: Image 5, 6, and 13. According to the algorithm, image 6 has similar degree of orange peel as image 5, while according to the experts its orange peel degree is substantially higher and is similar to that of image 13.

images 6, 7, and 8. However, according to the algorithm it has a higher orange peel score. Images 5, 6, and 13 are shown in Fig. 9. From images 6, 7, and 8, in order to compare them with images 5 and 13, only image 6 is picked, since all three of them have received similar scores by both evaluation approaches.

The deviations of the quantitative evaluation from the experts' evaluations is an expected result. This is particularly true if the intensity of the orange peel on the surface is in the "medium" range. In cases where the distinction of a defective surface from a not defective one becomes difficult, it is an advantage to utilize the presented quantitative evaluation method to avoid reliance on the subjective nature of the evaluation by humans. The evaluation by a human can be especially prone to errors when the assessor is doing the same thing over for a long period of time.

In order to further attest the robustness and flexibility of the algorithm, it is planned to perform more experiments with different surface characteristics, and compare the results with expert opinions.

5   Conclusion

In this paper a new method for quantitative evaluation and classification of the orange peel effect on painted injection moulded parts was presented. Images were captured using a monochrome camera from the surface of a black-painted high-gloss injection moulded part while it was being handled by an industrial robot, then using image processing techniques it was looked for waviness on the surface. Next, the results of the comparison between the orange peel intensity scores of the 30 images with different degrees of orange peel effects calculated by the algorithm and the evaluations of the experts were presented. Although the results of the two approaches for "weak" and "strong orange peel" appearances were similar, there were differences when orange peel effect was in "medium" range. The reason for this is assumed to be due to the nature of the human perception, which is based on subjective evaluation, therefore can differ from the quantitative evaluation.

In order to further develop this work, it is planned to use quantitative correlation methods in order to be able to compare the results of the algorithm even more precisely with the expert evaluations. In addition, a method will be implemented to automatically select the regions of interest from the image before evaluation. This process will make it possible to implement the algorithm automatically, without any need to manually select the ROI. For this purpose, machine learning methods can be used. This will make it possible to apply this evaluation method as an in-line inspection method. The experimental setup can also be used to detect and evaluate defects other than orange peel. For this purpose a larger framework is under development to utilize machine learning methods to inspect the surface for other anomalies. As a next step, the method proposed here can be integrated to this framework so that a complete in-line inspection system can be constructed.

Acknowledgements. The research work of this paper was performed at the Polymer Competence Center Leoben GmbH (PCCL, Austria) within the framework of the COMET-program of the Federal Ministry for Transport Innovation and Technology and the Federal Ministry of Digital and Economic Affairs with contributions by Flex. The PCCL is funded by the Austrian Government and the State Governments of Styria, Lower Austria and Upper Austria.

References

1. Orange peel/DOI meters. https://www.byk.com/en/instruments/products/appearance-measurement/orange-peel-doi-meter.html
2. Gruber, D.P., Wallner, G., Buder-Stroißnigg, M.: Method for analysing the surface properties of a material (2006). pCT-Patent, WO2006135948
3. Gruber, D.P., Berger, G., Pacher, G., Friesenbichler, W.: Novel approach to the measurement of the visual perceptibility of sink marks on injection molding parts. Polym. Test. **30**(6), 651–656 (2011). https://doi.org/10.1016/j.polymertesting.2011.04.013. http://www.sciencedirect.com/science/article/pii/S0142941811000717
4. Gruber, D.P., Macher, J., Haba, D., Berger, G.R., Pacher, G., Friesenbichler, W.: Measurement of the visual perceptibility of sink marks on injection molding parts by a new fast processing model. Polym. Test. **33**, 7–12 (2014). https://doi.org/10.1016/j.polymertesting.2013.10.014. http://www.sciencedirect.com/science/article/pii/S0142941813002158
5. Konieczny, J., Meyer, G.: Computer rendering and visual detection of orange peel. J. Coat. Technol. Res. **9**(3), 297–307 (2012). https://doi.org/10.1007/s11998-011-9378-2

6. Macher, J., Gruber, D.P., Altenbuchner, T., Pacher, G.A., Berger, G.R., Friesen-bichler, W.: Detection of visually perceptible sink marks on high gloss injection molded parts by phase measuring deflectometry. Polym. Test. **34**, 42–48 (2014). https://doi.org/10.1016/j.polymertesting.2013.12.008. http://www.sciencedirect.com/science/article/pii/S014294181300250X

7. Macher, J., Gruber, D.P., Altenbuchner, T., Pacher, G.A., Berger, G.R., Friesen-bichler, W.: A novel sink mark model for high gloss injection molded parts - correlation of deflectometric and topographic measurements. Polym. Test. **39**, 12–19 (2014). https://doi.org/10.1016/j.polymertesting.2014.07.001. http://www.sciencedirect.com/science/article/pii/S0142941814001469

8. Miranda-Medina, M.L., et al.: Characterisation of orange peel on highly polished steel surfaces. Surf. Eng. **31**(7), 519–525 (2015). https://doi.org/10.1179/1743294414Y.0000000407

9. Miranda-Medina, M.L., Somkuti, P., Steiger, B.: Detection and classification of orange peel on polished steel surfaces by interferometric microscopy. J. Phys.: Conf. Ser. **450**, 012009 (2013). https://doi.org/10.1088/1742-6596/450/1/012009

10. Scheers, J., Vermeulen, M., Maré, C.D., Meseure, K.: Assessment of steel surface roughness and waviness in relation with paint appearance. Int. J. Mach. Tools Manuf. **38**(5), 647–656 (1998). https://doi.org/10.1016/S0890-6955(97)00113-2. http://www.sciencedirect.com/science/article/pii/S0890695597001132. International Conference on Metrology and Properties of Engineering Surfaces

11. Smith, M., Smith, G., Hill, T.: Gradient space analysis of surface defects using a photometric stereo derived bump map. Image Vis. Comput. **17**(3), 321–332 (1999). https://doi.org/10.1016/S0262-8856(98)00136-X. http://www.sciencedirect.com/science/article/pii/S026288569800136X

12. Sone, T., Watanabe, S.: Measurement and evaluation method of orange peel. Electron. Imaging **2017**(8), 62–65 (2017). https://doi.org/10.2352/ISSN.2470-1173.2017.8.MAAP-283. https://www.ingentaconnect.com/content/ist/ei/2017/00002017/00000008/art00010

A New Automatic Cancer Colony Forming Units Counting Method

Nicolás Roldán[1,2]([⊠]) [iD], Lizeth Rodriguez[1,2] [iD],
Andrea Hernandez[3] [iD], Karen Cepeda[3] [iD],
Alejandro Ondo-Méndez[3] [iD], Sandra Liliana Cancino Suárez[1] [iD],
Manuel G. Forero[4] [iD], and Juan M. Lopéz[1] [iD]

[1] Department of Biomedical Engineering,
Escuela Colombiana de Ingeniería Julio Garavito, Bogotá, Colombia
{nicolas.roldan,lizeth.rodriguez-r}@mail.escuelaing.
edu.co, {sandra.cancino,juan.lopezl}@escuelaing.edu.co
[2] Department of Biomedical Engineering,
School of Medicine and Health Sciences, Universidad del Rosario,
Bogotá, Colombia
[3] Clinical Research Group, School of Medicine and Health Sciences,
Universidad del Rosario, Bogotá, Colombia
{andrea.hernandez,karen.cepeda}@urosario.edu.co,
alejandro.ondo@mail.urosario.edu.co
[4] Facultad de Ingeniería, Universidad de Ibagué, Ibagué, Colombia
manuel.forero@unibague.edu.co

Abstract. Clonogenic assays are an essential tool to evaluate the survival of cancer cells that have been exposed to a certain dose of radiation. Its result can be used in the generation of strategies for the optimization of radiotherapy treatments. The analysis of this type of data requires that the specialist performs the manual counting of colony forming units (CFU), i.e., find every cell that retains the ability to produce a large progeny. This task is time consuming, prone to errors and the results are not reproducible due to specialist subjective assessment. Digital image processing tools can deal with the flaws described above. This article presents a new technique for automatic CFU counting. The proposed technique extracts the regions of interest (ROIs), where a local segmentation algorithm finds and labels the CFUs in order to quantify them. Results show good sensitivity and specificity performance compared to state-of-the-art software used for CFU detection and counting.

Keywords: CFU · Colony counting · Cell counting · Cancer ·
Cell proliferation · Image analysis

1 Introduction

Clonogenic cell survival assay is a widely used technique employed to analyze the effects of physical, chemical or environmental conditions on cell survival and proliferation [1]. This method has been used in a large number of studies with diverse kind of

A. Morales et al. (Eds.): IbPRIA 2019, LNCS 11868, pp. 465–472, 2019.
https://doi.org/10.1007/978-3-030-31321-0_40

cells and has contributed with information about the effect of ionizing radiation on mammalian cells. The assay measures the cell number that retain the ability to produce a large number of progeny from a single cell, i.e. the cell that achieve at least five or six generations of successive replications to form a colony, composed of, at least, fifty cells, which is called Colony Forming Unit (CFU) [2].

Up to day, the analysis of clonogenic assay data is made primary by an ordinal approach, i.e., comparing two or more culture discs and establishing which has the higher number of colonies [3] or by manual counting of CFUs [4]. Some important technical improvements have been developed to facilitate hematopoietic cells counting, as the STEM Vision[tm] system, which combines special culture medium, dedicated six well culture plates, and an automated acquisition and image analysis system in order to obtain a fully automated colony counting [5]. Nevertheless, this type of system is expensive, highly consuming in laboratory reagents and materials, and hence, its use is very limited.

Digital image processing is widely used in the automation of processes, allowing to obtain fast, reproducible and unbiased results, and improving the ecosystem of the sample analysis service in cancerology and others research fields. In this way, image processing techniques can be employed to increase the efficiency on the analysis of clonogenic assays, allowing to reduce the time spent in the analysis of the assays, and convert the technique into a high-throughput data analysis tool, increasing in the number of trials and samples processed per experiment. Therefore, in this paper a new methodology for CFU automatic counting is proposed.

2   Materials

For this project, glioblastoma (U87) and breast cancer (MCF-7) cell lines were used. Cells were cultured in a DMEM medium (Biowest®) supplemented with 10% Bovine Fetal Serum (BFS, Hyclone®), and seeded in Petri culture dishes of 100 mm diameter until reaching a confluence of 70% to 80%. For the clonogenic assay we followed the Plating After Treatment protocol [2]. Briefly, the cells were harvested by incubation with trypsin/EDTA (Gibco®) and counted. 50.000 cell were transferred to 5 mL of medium in 15 mL Falcon tubes. Tubes were irradiated with 6 MV x-ray photons, using a clinical-use linear accelerator, in doses of 0, 1, 2, 4, 6, 8 and 10 Gy. To ensure the electronic balance and a better dose homogeneity in the whole volume, tubes were irradiated in a paraffin physical simulator. After irradiation, 200–2000 cells were seeded per well, in 6-well culture plates, depending on the radiation dose used. Seeding was made by triplicate for every dose. Cells were incubated at 37 °C and 5% CO_2, to allow colony formation. After 14 days of incubation, the formed colonies were fixed and stained with a mixture of 4% formaldehyde and 1% violet crystal.

Each culture plate was scanned using a HP Scanjet Pro with a resolution of 944 × 636 pixels. To develop the identification process, 216 wells were extracted from

36 RGB images. Wells were manually analyzed and these results were employed as ground truth. 66 images belonged to the U87 and 150 to the MCF-7 cell line.

3 Development

A CFU occupy a very small fraction of the total image area (See Fig. 2a). Considering the counted CFUs, the mean area ratio of a colony to the full image size represents only a $6 * 10^{-5}$ factor. In addition, the background can be very complex due to debris. The proposed method consist of four stages, illustrated in Fig. 1: Preprocessing and noise removal, contrast enhancement and background correction, segmentation and, labeling and counting.

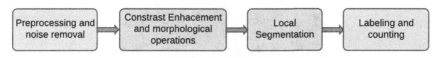

Fig. 1. General flowchart of the algorithm.

3.1 Pre-processing

As can be seen in Fig. 2a, image background is not uniform, having a greater intensity in the centre and presenting vignetting at the corners. Therefore, images were pre-processed to correct uneven lighting (Fig. 2b). For this purpose, images were converted to grayscale coding scheme by computing a weighted sum of the linear color components. A gamma correction using $\gamma = 0.4$ (empirically found), was applied, to increase the separation between CFUs and background. Since some CFUs were not well defined, a morphological opening, using a diamond-shaped structuring element, was employed to obtain the result shown in Fig. 2b. The structural element shape was selected based on the form of colonies and its size (7 pixels) was determined by the smallest colony in the image according to user criteria.

Black top-hat, which is the difference between the original grayscale image and a morphological closing, was applied to images, using a structuring element smaller or equal than the smallest CFU (Fig. 2c). The same diamond-shaped structuring element was used to remove non-significant objects and attenuate background noise. As a result, current image has a bimodal distribution. To enlarge separation between the ROI and the background, another contrast enhancement with $\gamma = 0.8$ was implemented.

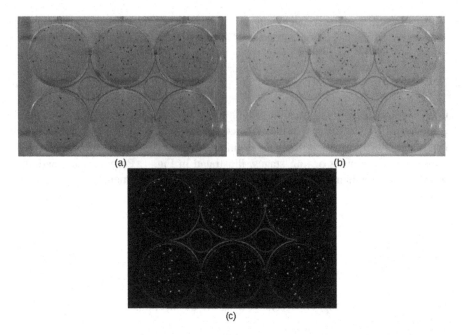

Fig. 2. (a) Original grayscale image. (b) Gamma correction and morphological opening. (c) Top-hat filtering and gamma correction.

3.2   Wells Detection

From the above stage, the images contain a better differentiated ROI from background. However, each image contains one plate of six wells. By taking advantage of the circular shaped form of each well, circular Hough Transform was used for its segmentation (see Fig. 3a). Well radius range is calculated using the relation between wells area and image dimensions. Subsequently, information about centre coordinates and radius r from each well is used with an Euclidean distance criteria, for keeping pixels located inside the well as shown in Fig. 3b.

One of the main problems in colony counting is the misclassification of noise out of the ROI as colonies, due to intensity similarity of some areas of the recipient, where the wells are placed. This condition can increase the number of false positives and therefore may affect specificity. Hence, isolating the wells from the background eliminates possible false positives detected outside them. At the same time, a one by one well colony counting is required. This is achieved by labeling and counting only the elements that are inside the ROI.

3.3   Colonies Segmentation, Labeling and Counting

After applying the processing techniques described above, gray level image showed a well-defined bimodal intensity distribution. Therefore, Otsu's method was used to find optimal threshold for each well separately, in order to achieve its colonies segmentation.

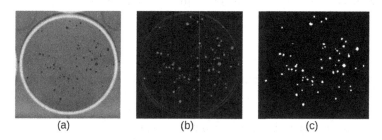

(a)   (b)   (c)

Fig. 3. (a) Petri dishes detection by Hough Transform. (b) Euclidean Distance removal. (c) Local thresholding using Otsu's Method, elimination based on region properties criteria and processing using watershed algorithm.

It was statistically found that a colony, composed of 50 cells, has an average size greater than 7 pixels. Therefore, this value was employed as threshold to reject the regions with a smaller area. In addition, some dish edges were identified as colonies. Given that colonies have a circular shape, objects which eccentricity was greater than 0.9, value found empirically, were eliminated.

In some assays, CFUs overlapped, leading to an underestimated counting. A marker-controlled Watershed algorithm is applied to separate each one of these overlapped CFUs. This algorithm uses as a marker the maxima found by Distance Transform. A general flowchart of preprocessing and noise removal step is shown in Fig. 4.

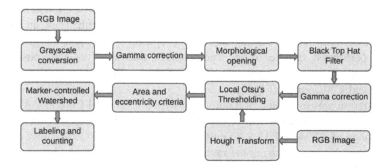

Fig. 4. Detailed flowchart of the algorithm.

4 Results and Discussion

It is worth mentioning that global or local thresholding techniques based on intensity histogram are not the best solution for CFUs segmentation from background, given the illumination gradient and the presence of artifacts (i.e., shadow) that can be observed on the inferior part of the dishes. One solution to this particular problem is to improve the acquisition system, which guarantees an even light distribution over the well's area using a chamber with LEDs arranged under [6–8] and/or around them [9], and a digital camera support on the top of it.

4.1 Validation

Some general cell counting techniques have been developed, but several parameters must still be adjusted manually. OpenCFU [10] is an open-source software programmed in C++ which uses OpenCV framework functions. OpenCFU has a graphical interface where counting results are shown, but also it allows the user to choose the region of interest, avoiding the detection of false positives outside and at the edges of the discs. It is one of the fastest programs processing an image of 1.6 × 1.6 kpx in approximately 0.69 s. However, in order to get the results, the user has to choose various parameters like the type of threshold (bilateral, inverted or regular) or if a color filter may be used. Image J plugin, CellCounter, allows the user to select a well and adjust a binary threshold for colonies segmentation and counting. In this kind of tools, the user has to follow a process of try and error, and/or modify various parameters to obtain an acceptable result, which can be really difficult if there is no digital image processing background.

Validation of the system is done by comparing the results obtained by the proposed system and a manual labelling and counting. At the same time, two software programs are used for CFU counting over the same image sets. In OpenCFU, processing was made with a regular threshold and no filter configuration, adjusting the radius in such a way that the best result is obtained for each image. In CellCounter, a circular ROI was selected to be segmented by a threshold adjusted to give the best result. Although other software was employed to carry out this task, they did not give the best results [11].

Sensitivity and specificity are selected as the main validation metrics. Manually labelling consist on placing red marks in image objects that were CFUs, and blue marks in the ones that were not. The resulting image was the ground truth. Then, an automatic comparison between results coming from the proposed system and the ground truth image was performed, in order to find true positives, true negatives, false positives and false negatives. Same process was repeated with the result images obtained by using OpenCFU and CellCounter (Fig. 5).

The program was evaluated in a computer with a CPU Intel Core i5 2.4 GHz, OS Windows 8.1 Pro and 4 GB RAM. It was able to process all the dataset with an average time per image of 12.94 ± 1.13 s. Even though OpenCFU and CellCounter are faster, taking less than 1 s in processing the image, and CellCounter has the better sensitivity and OpenCFU the better specificity of all three software respectively, only our system keeps good average sensitivity (>0.78) and specificity (>0.78) for the whole image sets analyzed, i.e., our system is better rejecting those objects that do not have at least 50

Table 1. Performance of colony counting methods.

Folder	Method	Sensitivity	Specificity
05.04.2016	CellCounter	0.88 ± 0.12	0.22 ± 0.16
	OpenCFU	0.29 ± 0.11	0.99 ± 0.02
	Method proposed	0.86 ± 0.11	0.78 ± 0.14
25.09.2015	CellCounter	0.61 ± 0.32	0.51 ± 0.39
	OpenCFU	0.47 ± 0.29	0.93 ± 0.09
	Method proposed	0.83 ± 0.20	0.79 ± 0.27
26.04.2015	CellCounter	0.83 ± 0.22	0.11 ± 0.21
	OpenCFU	0.47 ± 0.27	0.94 ± 0.08
	Method proposed	0.78 ± 0.35	0.73 ± 0.21

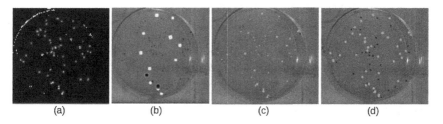

(a)   (b)   (c)   (d)

Fig. 5. Comparison of results from each algorithm. (a) CellCounter, (b) OpenCFU (c) Propused system (d) Manual labelling and counting.

cells, in other words, that fulfill the requirements of area and shape simultaneously, preserving the ability to identify those objects that are colonies (true positives) and at the same time, rejecting those that are not (Table 1).

5 Conclusions and Future Work

An automatic system able to segment colonies from CFU assay images with multiple wells was successfully developed, keeping a good sensitivity and specificity rather than a good sensitivity with poor sensibility, as in the case of CellCounter, or the opposite, as in the case of OpenCFU, used to validate the results of this work. Future works should aim to validate the algorithm in bigger and different datasets in order to improve it, replacing or adding steps in the processing to make a more robust algorithm.

References

1. Unkel, S., Belka, C., Lauber, K.: On the analysis of clonogenic survival data: statistical alternatives to the linear-quadratic model. Radiat. Oncol. **11**(1), 11 (2016)
2. Franken, N., Rodermond, H., Stap, J., Haveman, J.: Clonogenic assay of cells in vitro. Nat. Protoc. **1**(5), 2315-9 (2006)

3. Hingorani, M., et al.: Therapeutic effect of sodium iodide symporter gene therapy combined with external beam radiotherapy and targeted drugs that inhibit DNA repair. Mol. Ther.: J. Am. Soc. Gene Ther. **18**(9), 1599–1605 (2010)
4. Godínez, R.R., Gómez, A.E.Z., Ocampo, E.A., del Moral Hernández, O.: "Modulación de la apoptosis por la oncoproteína E6 de la variante E-G350 del VPH-16 en células HaCaT en respuesta a quimioterapia y radioterapia. Tlamati **7**(3), 5–10 (2016)
5. Brown, M.E., Rondon, E., Rajesh, D., Mack, A., Lewis, R., Feng, X., et al.: Derivation of induced pluripotent stem cells from human peripheral blood T lymphocytes. PLoS ONE **5** (6), e11373 (2010)
6. Chunhachart, O., Suksawat B.: Construction and validation of economic vision system for bacterial colony count. In: International Computer Science and Engineering Conference (ICSEC). IEEE, Thailand (2016). https://doi.org/10.1109/ICSEC.2016.7859888
7. Chiang, P.J., Tseng, M.J., He, Z.S., Li, C.H.: Automated counting of bacterial colonies by image analysis. J. Microbiol. Methods **108**, 74–82 (2015). https://doi.org/10.1016/j.mimet. 2014.11.009
8. Siragusa, M., Dall'Olio, S., Fredericia, P.M., Jensen, M., Groesser, T.: Cell colony counter called CoCoNut. PLoS ONE **13**(11), e0205823 (2018). https://doi.org/10.1371/journal.pone. 0205823
9. Carpenter, A., et al.: CellProfiler: image analysis software for identifying and quantifying cell phenotypes. Genome Biol. (2006). https://doi.org/10.1186/gb-2006-7-10-r100
10. Geissmann, Q.: OpenCFU, a new free and open-source software to count cell colonies and other circular objects. PLoS ONE **8**(2), e54072 (2013). https://doi.org/10.1371/journal.pone. 0054072
11. Matić, T., Vidović, I., Silađi, E., Tkalec, F.: Semi-automatic prototype system for bacterial colony counting. In: 2016 International Conference on Smart Systems and Technologies (SST), pp. 205–210. IEEE, Croatia (2016). https://doi.org/10.1109/SST.2016.7765660

Deep Vesselness Measure
from Scale-Space Analysis of Hessian
Matrix Eigenvalues

Ricardo J. Araújo[1,2]([✉]) [ID], Jaime S. Cardoso[1,3] [ID], and Hélder P. Oliveira[1,2] [ID]

[1] INESC TEC, Porto, Portugal
ricardo.j.araujo@inesctec.pt
[2] Faculdade de Ciências da Universidade do Porto, Porto, Portugal
[3] Faculdade de Engenharia da Universidade do Porto, Porto, Portugal

Abstract. The enhancement of tubular structures such as vessels in medical images has been addressed in the past, aiming for easier extraction and or visualization of such structures by professionals. Some literature methodologies propose vesselness measures whose design is motivated by local properties of vascular networks and how these influence the eigenvalues of the Hessian matrix. However, past work fails to combine properly the scale-space and neighborhood information, thus leading to the proposal of suboptimal vesselness measures. In this paper, we show that a shallow convolutional neural network is able to learn more optimal embedding spaces from the eigenvalue analysis at different scales, thus leading to a stronger vessel enhancement. Additionally, we also show that such a system maintains one of the biggest advantages of Hessian-based vesselness measures, which is the robustness to data with varying statistics.

Keywords: Blood vessel enhancement · Computer vision · Deep learning

1 Introduction

Blood vessels have high relevance in many clinical practices and diseases diagnoses. A few examples are the diagnosis of diabetic retinopathy, eligibility for liver transplant, and detection of aneurysms. Imaging data is quickly generated, especially in screening programs, and practitioners often become overwhelmed by the volume of data to analyze. Thus, some computer vision researchers naturally started targeting the automation of blood vessel analysis, and the first advances were already published by the end of the past century.

The first methodologies to emerge encoded prior knowledge of blood vascular networks in different ways, either in the design of probing filters [1], model fitting [2], or tracing algorithms that iteratively estimate a new point in the center axis of the vessel [3]. When datasets containing retinal fundus images started to

© Springer Nature Switzerland AG 2019
A. Morales et al. (Eds.): IbPRIA 2019, LNCS 11868, pp. 473–484, 2019.
https://doi.org/10.1007/978-3-030-31321-0_41

become publicly available [4,5], an important branch of algorithms emerged, the ones using supervision in order to find more complex mappings from the original data to a vessel probability map [6,7]. Recently, deep learning methodologies have also been applied to this scenario, raising the performance of automated blood vessel analysis to a new level [8,9]. Nonetheless, good generalization of such methodologies to data coming from different distributions is still hard to achieve, a scenario where unsupervised methods, which are based on strong but intuitive priors, frequently work better.

Vessel enhancement based on Eigen decomposition of the Hessian matrix is one of the most widely used enhancement processes, due to natural formulation for both 2D and 3D scenarios, good generalization to different data distributions, and also high noise suppression capabilities. Several hand-designed metrics combining the eigenvalue information were proposed in the past. Frangi's vesselness measure [10] is one of the most used. Recently, Jerman et al. [11] proposed a vesselness measure that tries to correct some deficiencies of past ones, such as low responses at aneurysms due to suppression of rounded structures, and poor enhancement of bifurcations due to local deviation from a piecewise linear structure, which is the profile being targeted in other hand-designed metrics. Even though these metrics rely on strong prior knowledge of the problem, they usually end missing much information due to how they combine multiscale information, thus being suboptimal.

In this work, we aim at finding a more complex and optimal vesselness measure mapping the eigenvalue information at different scales into the final vessel enhanced image, by using a Deep Neural Network (DNN). By using supervision, our goal is to obtain a deep vesselness measure that combines the advantages of both deep learning methodologies (finding deep complex functions) and using prior knowledge (increased robustness to data coming from different distributions). Recent research considered the implementation of Frangi's algorithm as a neural network, by carefully initialization of its weights [12]. The authors then used supervision to update weights responsible for the computation of the Hessian, and coefficients controlling the relevance of the different eigenvalue ratios used in Frangi's vesselness. Note, however, that the first option strongly relaxes the use of prior knowledge, as the network is able to learn features completely different from the Hessian, thus regularization may be lost. Additionally, the authors do not consider exploring other functions mapping the Eigen maps to the final vesselness, being restricted to the use of the maximum operator across the responses obtained at different scales, which is suboptimal.

The paper is structured as follows: this Section introduced the relevance of vessel enhancement and our contribution to this problem; Sect. 2 briefly describes traditional frameworks relying on the multiscale eigen-analysis of the Hessian matrix, and how we propose to obtain a deep vesselness measure through supervision; in Sect. 3 we explain the conducted experiments, and present and discuss results; finally, Sect. 4 concludes the paper and points some directions for future work.

2   Methodology

Vessel enhancement methodologies based on the Eigen decomposition of the Hessian explore the local intensity curvature of images. Such analysis is performed at different scales in order to find structures of different sizes. We start by describing in detail the pipeline these approaches follow. Then, we point out its main limitations and propose a novel design by switching its final part by a DNN.

2.1   Eigen Decomposition of the Hessian for Vessel Enhancement

Given an D-dimensional image I, the type of structure present at a given location $x = (x_1, x_2, \ldots, x_D)$ may be inferred through the analysis of the Hessian matrix at x, a $D \times D$ matrix encoding the second order derivatives of the intensities:

$$H_{ij}(x, \sigma) = \sigma^\gamma I(x) * \frac{\partial^2 G(x, \sigma)}{\partial x_i \partial x_j}, \quad i, j = 1, \ldots, D \tag{1}$$

where G is a D-variate Gaussian, σ denotes its standard deviation, dictating the scale at which the image is being analyzed, γ is a constant that normalizes responses obtained at different scales, allowing a fair comparison [13], and $*$ represents the convolution operation.

The Eigen analysis of $H(x, \sigma)$ produces D eigenvectors representing the principal directions that decompose the second order structure of the image at x. Each of them has an eigenvalue associated, a scalar whose magnitude and signal allow to characterize the intensity curvature along the corresponding eigenvector. From now on, let us consider that the Eigen decomposition of the Hessian at a location x:

$$L(x, \sigma) = \text{eig}(H(x, \sigma)) \tag{2}$$

produces a set of eigenvalues $\lambda_1, \lambda_2, \ldots, \lambda_D$, such that, $|\lambda_1| \leq |\lambda_2| \leq \ldots \leq |\lambda_D|$. These provide a concise description of the local geometry at x, allowing the design of functions that respond to particular geometries.

In this context, a vesselness measure is any function f of the eigenvalues that is suited for the enhancement of blood vessels:

$$V(x, \sigma) = f(L(x, \sigma)) \tag{3}$$

A common assumption on vessel geometry is it being piecewise linear, that is, locally, it resembles a cylinder. As in this work we deal only with 2D images, we restrict the following discussion to this scenario. Nonetheless, extension to 3D is straightforward and addressed in the aforementioned works. The most commonly used vesselness measure is Frangi's [10], which for the 2D case is given by:

$$V_F = \begin{cases} 0 & \text{if } \lambda_2 \leq 0, \\ \exp\left(-\frac{R_B^2}{2\beta^2}\right) \cdot \left(1 - \exp\left(-\frac{S^2}{2c^2}\right)\right) & \text{otherwise} \end{cases} \tag{4}$$

where $R_B = |\lambda_1|/|\lambda_2|$ is a ratio measuring local similarity to a blob through eccentricity of the second order ellipsis, $S = \sqrt{\lambda_1^2 + \lambda_2^2}$ is the amount of local structure, and β and c, control the relevance of those quantities, respectively. This measure assumes that vessels are darker than the background, but inverting the conditions of Eq. (4) is enough to detect brighter vessels instead.

In the case of Jerman's vesselness [11], assumptions are slightly relaxed in order to better model aneurysms and bifurcations:

$$\mathcal{V}_J = \begin{cases} 0 & \text{if} \quad \lambda_2 \leq 0 \vee \lambda_p \leq 0, \\ 1 & \text{if} \quad \lambda_2 \geq \lambda_p/2 > 0, \\ \lambda_2^2 \cdot (\lambda_p - \lambda_2) \cdot \left[\frac{3}{\lambda_2 + \lambda_p}\right]^3 & \text{otherwise} \end{cases} \tag{5}$$

where λ_p is a regularized eigenvalue for ensuring that robustness to noise is achieved in regions with uniform intensity.

Regardless of the used vesselness measure, the final enhanced image, V, is obtained by combining the responses obtained at different scales σ, through a pixelwise maximum operation:

$$V(\boldsymbol{x}) = \max_{\sigma_1, \dots, \sigma_n} \mathcal{V}(\boldsymbol{x}, \sigma) \tag{6}$$

The traditional pipeline here described is represented in Fig. 1.

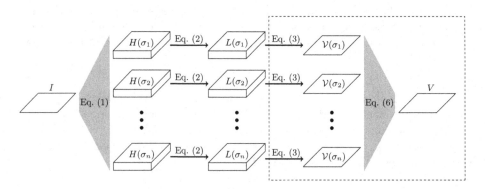

Fig. 1. Multiscale pipeline of traditional vessel enhancement methodologies, where n denotes the number of scales. Pixelwise Hessian matrix and corresponding eigenvalues are represented as feature vectors for convenience. In this paper, we propose a design using a DNN replacing the functions delimited by dashed lines.

2.2   Deep Vesselness Measure

In this work, we replace hand designed vesselness measures (see region delimited by dashed lines in Fig. 1) by a DNN. Our motivation is twofold. First, mapping eigenvalue information into a vessel probability (Eq. (3)) through hand-designed functions, despite being based on prior intuition, is likely suboptimal. Second, combining the responses at different scales by a pixel-wise maximum operation (Eq. (6)) discards much of the information encoded at all scales and fails to capture high-level local information that may be useful in challenging regions. Thus, we replace those functions by a neural network having as input a concatenation of the eigenvalue description, and as output a vessel probability map, as represented in Fig. 2. We use label supervision in order to update its weights, aiming to obtain a more optimal deep vesselness measure, which is still regularized as we only provide the scale-space eigenvalue description.

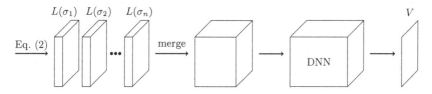

Fig. 2. Proposed pipeline for vessel enhancement. A DNN learns a more complex vesselness measure from the eigenvalue information.

Neural Network Considerations. We model our DNN as a Fully Convolutional Network (FCN) [14], such that images seen at train and test phases may have different sizes. This also allows us to train our model in small patches of blood vessel images, and still later obtain predictions for entire images at a single pass. This may be relevant due to memory issues and to avoid training with unnecessary data, such as the black regions in retinal fundus images. In this work, we consider patch-based training, which is not expected to affect the performance of a FCN.

Batch normalization [15] is especially useful in very deep networks, which is not the case we seek here. Additionally, care must be taken when the statistics of the data are not the same in the train and test sets. Such difference may be a result of performing patch-based training, where, for example, entire images of the retina have different statistics than small patches that were just taken from the retinal fundus area. It may also naturally arise from training and testing in different datasets. This last scenario is very relevant as one of the main advantages of traditional Hessian-based methods is their good generalization to other distributions of data. Thus, we did not consider batch normalization.

Recent findings [8] seem to support that reducing space resolution via max pooling or strided convolution does not improve the performance of networks trying to capture small details, as is the case of blood vessels. Preliminary experiments that we conducted support this, such that slightly increasing the kernel dimension and keeping spatial resolution equal across the entire network proved to be more effective. Having features already encoding neighborhood information as the input of our neural network may also contribute to learn interesting deep vesselness measures by only looking at a relatively small neighborhood. Even then, we found that using dilated convolution [16] in the intermediate layers improved the performance of the system.

An ideal vessel enhancement algorithm would output probability of 1 for every pixel belonging to a vessel and probability of 0 otherwise. However, note that, for an adequate scale σ, and when analyzing pixels over the cross section of a vessel, it is expected that the Eigen decomposition of the center pixel is the one matching better the Eigen description of an ideal vessel ($|\lambda_1| << |\lambda_2|$, in the 2D case). This is the reason why vesselness measures such as Frangi's enhance more the central regions of vessels. Even though a DNN is able to find complex relations between eigenvalues and thus learn effectively when hard labels are provided, in preliminary experiments, we found that using soft labels (obtained by blurring the hard labels with a standard normal distribution) was helpful. Nevertheless, as will be shown in Sect. 3, our design is still capable of enhancing the peripheral regions of vessels extremely well.

The Rectified Linear Unit function was used as an activation function throughout the network and the last non-linearity was a Sigmoid function. Regarding the loss function, we considered the binary cross entropy, which is also adequate when soft labels are given. The Adam optimizer [17] was used to update the weights. Other state-of-the-art considerations such as Dropout were also tested. The final FCN design is represented in Fig. 3. More information on the tuning procedure is given in Sect. 3.

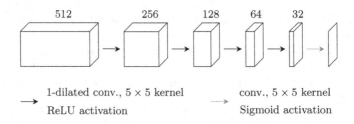

Fig. 3. Fully convolutional design used in the experiments after tuning the network architecture. The first set of features is obtained by doing convolution over the eigenvalue pile of features with 5×5 kernels and no dilation.

3   Experiments and Results

In this Section, we start by briefly describing the datasets we use in our experiments. Afterwards, we present the setting we followed to tune the network architecture and its hyperparameters, finally reaching the design presented in Fig. 3. Finally, we detail the different experiments taken into account to show the properties of the proposed methodology and present their results.

3.1   Datasets

To the best of our knowledge, blood vessel 2D imaging datasets containing the ground truth vessel masks only exist for the retinal case. This means that supervision may only be done recurring to this type of vascular network. To conduct the experiments, we resorted to 4 complete retinal datasets, DRIVE [5], STARE [4], CHASEDB1 [18], and IOSTAR [19].

DRIVE is a result of a diabetic retinopathy screening program conducted in The Netherlands comprising 40 images of size 584×565 split in the same proportion into train and test sets, with 7 of them showing signs of early diabetic retinopathy. A Canon CR5 non-mydriatic 3CCD camera with a $45°$ field of view was used. STARE is a dataset containing 20 images of size 605×700, half of which showing pathology. They were obtained with a TopCon TRV-50 fundus camera. CHASEDB1 was compiled after the Child Heart and Health Study in England. It comprises 28 images of size 960×999, where central reflex is particularly abundant. IOSTAR includes 30 images of size 1024×1024 and a field of view of $45°$ acquired with the EasyScan camera from i-Optics B.V., The Netherlands.

3.2   Implementation Details

Having in mind the considerations discussed in Sect. 2.2, we tuned the architecture and hyper-parameters using the DRIVE dataset. We randomly set aside three images from DRIVE's training set for validation purposes and used the remaining ones for training different model configurations. In this work we did not conduct any preprocessing step, we just selected the green channel information and normalized it to the range $[0, 1]$. We considered $\sigma \in [1, 11]$ with steps of 2. At each training epoch, we give the models 300 batches of 8 patches of size 64×64. A total of 100 epochs were conducted. These values were empirically found to be appropriate in preliminary experiments but their variation do not yield significant performance alterations. Patches were randomly extracted from the field of view region of images. We considered data augmentation via random vertical or horizontal flipping, and rotations in the range $[-\pi/2, \pi/2]$. The parameters of the Adam optimizer were initialized as described in [17]. The best performance in the validation set was obtained when using the design represented in Fig. 3.

With the exception of DRIVE, the datasets do not have a prior split into train and test sets. Thus, we consider the first 10 images of each for training purposes and the remaining are set aside for testing. According to the experiment we conduct, different sets are used for training and testing but details will be provided as necessary. The training procedure is conducted as described before for network and hyperparameter tuning, but this time all the available training data is used. Frangi's and Jerman's enhancement responses were obtained using their Matlab implementations.

3.3   Results and Discussion

We start by considering the scenario where we train our model using a given dataset and test on the images set aside from the same dataset. With this experiment, we aim to show that for a specific distribution of data, it is possible to use deep learning to obtain more complex and optimal vesselness measures for such data. The ROC curves of our method, are shown, along with the ones from the baselines, in Fig. 4.

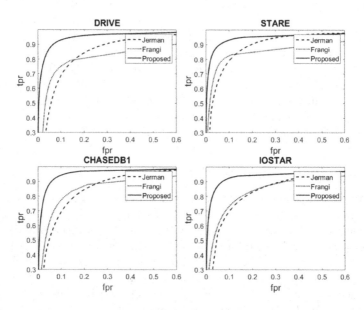

Fig. 4. ROC curve of the proposed methodology when trained and tested in the same dataset. The ROC curves of the baseline methods are presented for comparison.

This shows that, when we target specific data distributions, we are able to obtain more optimal vesselness functions than the traditional ones. This was expected, but note that traditional methods do not target specific dataset distributions, but instead a representation that generalizes well. Obviously, the more interesting scenario is to analyze what occurs when we use the proposed methodology to enhance blood vessels in images coming from datasets that were not accessed during training. Thus, we now consider the scenario where we set the test dataset aside and train using the remaining ones. The ROC curves of the system in such conditions are again compared against the baselines in Fig. 5. It is possible to conclude that our system is indeed capable of generalizing well to data coming from other distributions than the ones available during training. For very high false positive ratios, Jerman's vesselness occasionally achieves higher true positive ratio, however note that such region is not ideal for enhancement functions as it already comprises a large portion of noise. This is clearly seen in Fig. 7, where our method proves to be much more robust to noise than Jerman's one.

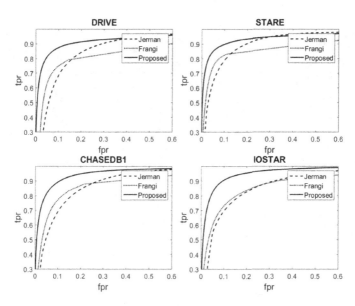

Fig. 5. ROC curve of the proposed methodology when trained and tested in different datasets. The ROC curves of the baseline methods are presented for comparison.

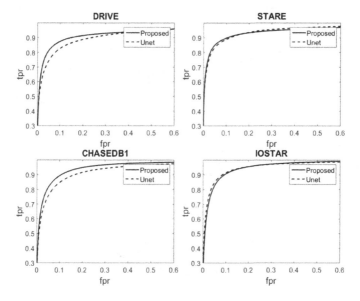

Fig. 6. ROC curves of the proposed methodology and a regular Unet, when trained and tested in different datasets.

Finally, we compare the generalization capability of the proposed method against the well-established Unet [20] for biomedical image segmentation. Such network has much more capacity and has increased flexibility as it is not

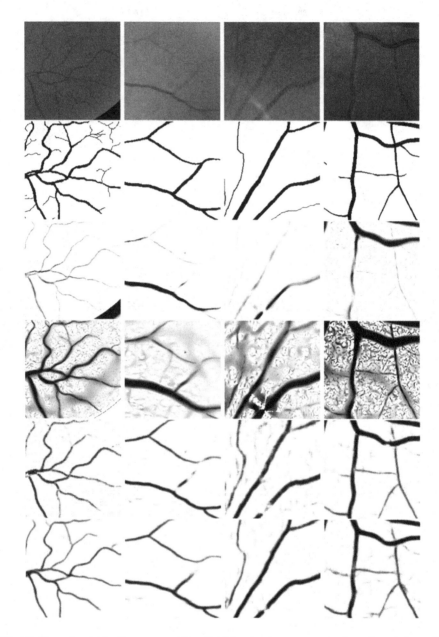

Fig. 7. From top to bottom, example retinal fundus images, ground truth vessel masks, and corresponding enhancements by, respectively, Frangi's, Jerman's, Unet, and proposed vesselness. From left to right, image from DRIVE, STARE, CHASEDB1, and IOSTAR. Concerning the Unet and proposed method, training was conducted in all datasets, except the testing set. Frangi's vesselness was rescaled for visualization purposes, since the signal at narrow vessels is usually small.

restricted to use a given set of features, as we do in the proposed methodology. Instead, it creates representations from the image itself. Figure 6 shows the ROC curves of both methods. Both methodologies have similar performance in STARE and IOSTAR but the proposed approach generalized better for DRIVE and CHASEDB1. This shows that regularizing deep neural networks, as we do in this work, may be a proper way of achieving designs that generalize better, even using less parameters. We provide visual results of our method and the baselines in Fig. 7.

4   Conclusion and Future Work

In this work we extended traditional Hessian-based methodologies for the enhancement of blood vessels in medical images. By replacing hand-design functions mapping eigenvalue descriptions to the final output with a DNN, we were able to learn more optimal functions than traditional algorithms. At the same time, when comparing with DNNs which are fed the images instead of an eigenvalue description, our methodology generalized better to data coming from distributions other than the ones used at training. This shows that our methodology continues to embed significant prior knowledge, thus helping to achieve good abstraction of what a blood vessel is.

Regarding future work, we would like to extend the framework to 3D data, possibly exploring synthetic datasets during the training procedure, as it becomes trivial to obtain the ground truth.

Acknowledgments. This work was financed by National Funds through the Portuguese funding agency, FCT - Fundação para a Ciência e a Tecnologia within PhD grant number SFRH/BD/126224/2016 and within project UID/EEA/50014/2019.

References

1. Chaudhuri, S., Chatterjee, S., Katz, N., Goldbaum, M.: Detection of blood vessels in retinal images using two-dimensional matched filters. IEEE Trans. Med. Imaging **8**(3), 263–269 (1989)
2. Niessen, W.J., van Bemmel, C.M., Frangi, A.E., Siers, M.J.A., Wink, O.: Model-based segmentation of cardiac and vascular images. In: Proceedings IEEE International Symposium on Biomedical Imaging, pp. 22–25. IEEE, Washington, DC (2002)
3. Sun, Y.: Automated identification of vessel contours in coronary arteriograms by an adaptive tracking algorithm. IEEE Trans. Med. Imaging **8**(1), 78–88 (1989)
4. Hoover, A., Kouznetsova, V., Goldbaum, M.: Locating blood vessels in retinal images by piece-wise threshold probing of a matched filter response. IEEE Trans. Med. Imaging **19**(3), 203–210 (2000)
5. Staal, J., Abràmoff, M.D., Niemeijer, M.: Ridge-based vessel segmentation in color images of the retina. IEEE Trans. Med. Imaging **23**(4), 501–509 (2004)
6. Lupascu, C.A., Tegolo, D., Trucco, E.: FABC: retinal vessel segmentation using AdaBoost. IEEE Trans. Inf Technol. Biomed. **14**(5), 1267–1274 (2010)

7. Fraz, M.M., et al.: An ensemble classification-based approach applied to retinal blood vessel segmentation. IEEE Trans. Biomed. Eng. **59**(9), 2538–2548 (2012)
8. Liskowski, P., Krawiec, K.: Segmenting retinal blood vessels with deep neural networks. IEEE Trans. Med. Imaging **35**(11), 2369–2380 (2016)
9. Hu, K., et al.: Retinal vessel segmentation of color fundus images using multiscale convolutional neural network with an improved cross-entropy loss function. Neurocomputing **309**(2), 179–191 (2018)
10. Frangi, A.F., Niessen, W.J., Vincken, K.L., Viergever, M.A.: Multiscale vessel enhancement filtering. In: Wells, W.M., Colchester, A., Delp, S. (eds.) MICCAI 1998. LNCS, vol. 1496, pp. 130–137. Springer, Heidelberg (1998). https://doi.org/10.1007/BFb0056195
11. Jerman, T., Pernuš, F., Likar, B., Špiclin, Ž.: Enhancement of vascular structures in 3D and 2D angiographic images. IEEE Trans. Med. Imaging **35**(9), 2107–2118 (2016)
12. Fu, W., Breininger, K., Würfl, T., Ravikumar, N., Schaffert, R., Maier, A.: Frangi-Net: a neural network approach to vessel segmentation. arXiv preprint arXiv:1711.03345 (2017)
13. Lindeberg, T.: Edge detection and ridge detection with automatic scale selection. Int. J. Comput. Vision **30**(2), 117–156 (1998)
14. Long, J., Shelhamer, E., Darrell, T.: Fully convolutional networks for semantic segmentation. In: Proceedings of the IEEE Conference on Computer Vision, pp. 3431–3440. IEEE, Chile (2015)
15. Ioffe, S., Szegedy, C.: Batch normalization: accelerating deep network training by reducing internal covariate shift. arXiv preprint arXiv:1502.03167 (2015)
16. Yu, F., Koltun, V.: Multi-scale context aggregation by dilated convolutions. arXiv preprint arXiv:1511.07122 (2015)
17. Kingma, D.P., Ba, J.: Adam: a method for stochastic optimization. arXiv preprint arXiv:1412.6980 (2014)
18. Owen, C.G., et al.: Measuring retinal vessel tortuosity in 10-year-old children: validation of the computer-assisted image analysis of the retina (CAIAR) program. Investig. Ophthalmol. Vis. Sci. **50**(5), 2004–2010 (2009)
19. Abbasi-Sureshjani, S., Smit-Ockeloen, I., Zhang, J., Ter Haar Romeny, B.: Biologically-inspired supervised vasculature segmentation in SLO retinal fundus images. In: Kamel, M., Campilho, A. (eds.) ICIAR 2015. LNCS, vol. 9164, pp. 325–334. Springer, Cham (2015). https://doi.org/10.1007/978-3-319-20801-5_35
20. Ronneberger, O., Fischer, P., Brox, T.: U-Net: convolutional networks for biomedical image segmentation. In: Navab, N., Hornegger, J., Wells, W.M., Frangi, A.F. (eds.) MICCAI 2015. LNCS, vol. 9351, pp. 234–241. Springer, Cham (2015). https://doi.org/10.1007/978-3-319-24574-4_28

Segmentation in Corridor Environments: Combining Floor and Ceiling Detection

Sergio Lafuente-Arroyo$^{(\boxtimes)}$, Saturnino Maldonado-Bascón,
Hilario Gómez-Moreno, and Cristina Alén-Cordero

GRAM, Department of Signal Theory and Communications, University of Alcalá,
Alcalá de Henares, Spain
sergio.lafuente@uah.es
http://agamenon.tsc.uah.es/Investigacion/gram

Abstract. Automatic segmentation from indoor images has several applications for mobile platforms. We address the problem of corridor segmentation and propose an approach by combining floor and ceiling detection. However, different difficulties may limit the accuracy of the system. To overcome these difficulties, a strategy is used in this paper to evaluate the degree of consistency of ceiling and floor guidelines. The method is based on computing the disparity between the hypothesized vanishing points by intersecting the boundaries par-wise. The approach is evaluated in a novel dataset. Our experimental validation confirms that the integration of floor and ceiling detection with the consistency model performs effectively and robustly. Because of the simplicity of the method, the image processing is quite fast and robust.

Keywords: Semantic segmentation · Corridor structure ·
Vanishing point

1 Introduction

In indoor environments mobile platforms have to navigate along corridors to reach a room in order to perform a specific task. For this reason, the extraction of visual information can provide rich knowledge. The importance of scene understanding is a core computer vision problem for robot navigation in corridors. By inferring labels each pixel is associated with the class of its enclosing region (floor, ceiling or wall). A fundamental part of the indoor segmentation process is the floor detection step. However, several difficulties appear in visual floor detection associated to common specular reflections. The reflections on the floor may come from the ceiling lights, outdoor lighting from windows and doors, that even make it difficult sometimes for a human observer to distinguish the floor area. We tackle these problems as a challenge to propose a model based on the complementary detection of the floor and the ceiling in order to ensure

© Springer Nature Switzerland AG 2019
A. Morales et al. (Eds.): IbPRIA 2019, LNCS 11868, pp. 485–496, 2019.
https://doi.org/10.1007/978-3-030-31321-0_42

a valid model for the corridor structure. How can we efficiently integrate both detections into the model? This is the question we want to answer with this work.

In man-made environments, such as corridors, sets of parallel lines intersect at points at infinity. Their projections in an image are called vanishing points (VPs). In this paper we use a priori knowledge about the 3D scene in the sense that the corridor guidelines (wall-floor and wall-ceiling boundaries) intersect in the image at the vanishing point, which is located somewhere along the horizon line. Our objective is to estimate the common image intersection for the four corridor guidelines. Due to the noisy detection the imaged boundaries will generally not intersect in an unique point and the VP can be computed by intersecting the boundaries par-wise. The disparity of these points is an indicator of the degree of validation of the detected boundaries.

The paper is structured as follows. In Sect. 2, a review of related works is introduced. Section 3 addresses our proposed method of detecting floor and ceiling. Section 4 discusses an algorithm to verify the consistency of candidate corridor guidelines. Section 5 demonstrates the experimental results with real images and Sect. 6 draws the final conclusions about the research performed in this paper.

2   Related Work

Existing literature contains several works on indoor floor segmentation based on computer vision, which can be easily classified depending on whether they use a purely appearance or geometric/homographic standpoint or those which combine both. In those approaches based on appearance, multiple visual clues from the environment are used for detection. In [1], a combination of color and gradient histograms to distinguish free navigable space is used. Due to over reliance on color based descriptors, the approach fails in homogeneous environments. A different approach in [2] uses a combination of vertical edges, thresholding and segmentation to approximate the wall-floor boundaries and then classify horizontal edges that lie on that boundary. This approach gives good results, robustly dealing with specular reflection on floor which is common in indoor environments. Nevertheless, it fails either when vertical edges are missed in the lower half of the image or when side walls are close to the robot.

Different approaches based on geometry exploit the ground plane constraint and focus on just finding the ground plane [3–6]. In [3], the motion between two images is modeled by a homography constraint as a criterion for ground plane detection. Optical flow is also used in [4] for ground plane detection. Both researches [3,4] used a monocular camera, while dense point correspondences relied on stereo homographies in [5]. More recently, in [6] a combination of sparse optical flow and planar homography for ground plane detection is used. Aforementioned methods are computationally intense.

While purely appearance based approaches fail under homogeneity of appearance, geometric methods are robust enough at detecting features that define the floor. However, geometry based approaches need extra hints to segment the boundary that include the floor features. The approach of [7], which applies one of the methods that encompasses both geometry and appearance, is able of developing geometrical reasoning by searching the best fitting model, which is transformed into a full 3D model. Other interesting strategy [8] creates valid box layout hypotheses by using detected line segments and virtual rays from orthogonal vanishing points. With the same approach of considering geometric backgrounds to improve scene interpretation, a method for supporting relations of indoor scenes from an RGBD (Red-Green-Blue-Depth) image is proposed in [9].

Recently, more and more computer vision tasks such as image classification have been solved by Convolutional Neural Network (CNN). As an example of application to indoor environments, we can find the work of Hazirbas [10]. This technique broadly surpasses other conventional approaches in terms of accuracy but it usually needs more proccessing time and memory. Therefore, further effort is required to explore new architectures in order to make semantic segmentation more efficient.

3   Floor and Ceiling Detection

This section describes a completely automated process of floor and ceiling detection in the image. The stages include detection of line segments, clustering and detection of boundaries.

3.1   Detection of Line Segments

Edges convey essential information for distinguishing separations. The popular Canny detector is used for this purpose in this work. Probabilistic Hough transform is then applied to the resulting edge image.

Unlike some previous work [2], which establishes an unique set of values for parameters, the proposed line extraction consists of two detectors. If no lines are extracted with the first detector, a new detection is applied with more flexible conditions. The second detector focuses on no remarkable lines. We tune the Canny threshold, t, in order to maintain a compromise between accuracy and robustness to noises and outliers. As the value of this parameter decreases, no remarkable lines can be detected but instead outliers may appear. The existence of outliers provoke a major dispersion of crossing points between pairs of lines. In Fig. 1 we plot the dispersion of crossing points and the percentage of images without detection and as functions of the Canny threshold. In terms of compromise, this parameter has been set to $t_1 = 30$ for the first detector and $t_2 = 15$ for the second one. By applying the above procedure, we obtain a set of line segments $\mathcal{L} = \{l_1, l_2, ..., l_N\}$ defined by their two endpoints.

Man-made environments include a lot of regularities due to their intrinsic structure. Detected lines in corridor images can be grouped into three categories regarding to the angle range: vertical lines, transversal lines and horizontal lines.

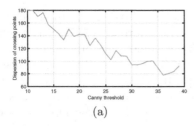 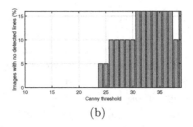

(a)   (b)

Fig. 1. The impact of Canny threshold on the accuracy of line detectors. We show (a) the average dispersion of crossing points between pairs of lines and (b) the percentage of images without detection as a function of the Canny threshold.

(a)   (b)

Fig. 2. Line extraction and classification: vertical (green), floor (blue) and ceiling (red). (Color figure online)

The last ones are discarded for our purpose. Vertical lines correspond, in general, to walls boundaries such as doors and windows whereas transversal lines identify, in general, wall-floor and wall-ceiling boundaries. The angle ranges have been established by considering the appearance of imaged wall-floor and wall-ceiling boundaries with the camera fixed at different heights from the ground. At this point the transversal lines are classified in four sets: C_l and C_r includes, respectively, the line segments which correspond to potential left and right ceiling boundaries whereas the sets F_l and F_r include, respectively, the potential left and right floor boundaries. Figure 2 shows two examples of line detection, where the color of each line represents its category. It is worth noting that, in general, detected transversal lines include spurious edges. The sets F_l and F_r may contain line segments as effect of reflections, shadows and tile joints. Beside this, the sets C_l and C_r may include segments from the structure of the ceiling, upper doorframes and ceiling lights.

3.2 Clustering

Once detected the meaningful line segments, an agglomerative clustering scheme is used for each one of the four sets. Clustering is based on two features: slope and

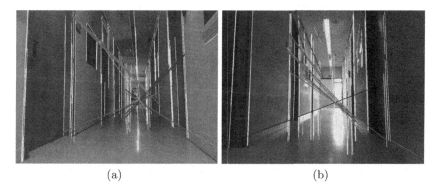

(a)   (b)

Fig. 3. Illustration of the clustering process where each cluster is depicted with a different color. (Color figure online)

bias of each segment. In order to compute the bias it is considered the intersecting points of the detected segments with the borders of the image. Specifically, intersections of floor segments with the bottom image border and, on the other hand, intersections of ceiling segments with the upper image border. Both features slope and bias are integrated in a function to compute pairwise distances between two clusters C_m and C_n as:

$$d_{C_m,C_n} = \frac{1}{N_{C_m}} \frac{1}{N_{C_n}} \sum_{i \in C_m} \sum_{j \in C_n} \sqrt{\left(\frac{\Delta\theta_{ij}}{\pi}\right)^2 + \left(\frac{\Delta x_{ij}}{W}\right)^2} \qquad (1)$$

where $\Delta\theta_{ij}$ is the angle difference between each pair of line segments l_i and l_j with range of values $[-\pi, \pi]$, Δx_{ij} is the X-coordinate step between intersection points of l_i and l_j with the image borders and W is the width of the image. The parameters $\frac{1}{N_{C_m}}$ and $\frac{1}{N_{C_n}}$ are the number of points of C_m and C_n.

The two clusters with the smallest distance are merged in each iteration and the operation is repeated until the distance d_{C_m,C_n} between the two closest clusters is larger than a certain threshold, which has been adjusted experimentally. Thus, the final clusters $\mathcal{C} = \{C_1, C_2, ..., C_N\}$ are obtained and each cluster C_j is characterized by a prototype line $\mathbf{t}_j$. In order to compute each cluster prototype, we give more relevance to longer lines. The slope and the bias of each $\mathbf{t}_j$ are computed as the weighted average of the segments of the cluster C_j. Figure 3 shows the resulting clusters for two examples, where lines of each cluster are depicted with the same color. The prototype lines of the different clusters become the candidate floor and ceiling boundaries for the following stage (see Fig. 4).

3.3 Detection of Boundaries

Given a set of candidate boundaries $\{\mathbf{t}_j\}_{j=1}^N$ generated by the clustering step, the goal is to find a function which estimates the strength of each candidate. In order to deal with this problem, it is proposed a weighted sum of scores based on two individual visual aims for each prototype line $\mathbf{t}_i$ of F_l and F_r:

$$\phi(\mathbf{t}_i) = w_1\phi_1(\mathbf{t}_i) + w_2\phi_2(\mathbf{t}_i)) \qquad (2)$$

where w_1 and w_2 are the weights and ϕ_1 and ϕ_2 are the individual scores which we describe in the following items:

- Intersections with vertical lines (ϕ_1). It is inspired by the fact that low endpoints of vertical lines delimit theoretically the floor boundaries in the image. Thus, the algorithm favours transversal lines with a higher number of intersections with vertical lines in the proximity of their low endpoints. As intersections of transversal lines with vertical lines may extend beyond the floor boundaries, due to reflections or shadows, a tolerance margin for intersections needs to be defined.
- Length of the prototype line (ϕ_2). Intuitively, very short line segments are frequently noisy, hence, their contribution should be constrained by comparison with long segments. This score favours longer line segments as candidates to floor boundaries.

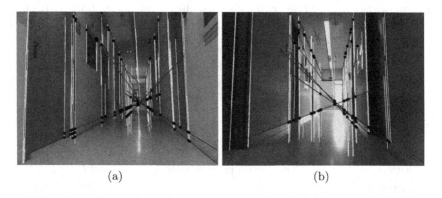

(a)   (b)

Fig. 4. Prototype lines and their intersection points with vertical lines.

In a similar way, the weighted sum of scores for the sets C_l and C_r introduces an extra visual aim (ϕ_3) corresponding to the maximum angle. The objective of ϕ_3 is to give more relevance to those line segments with higher slope. The slope of ceiling boundaries in the image is higher than the corresponding to those spurious line segments from upper doors or windows sides in the walls. Figure 4 shows the prototype lines of the different clusters and their intersection points with vertical lines. The output of this stage returns an array of weights associated to each one of the four sets (C_l, C_r, F_l and F_r). Then, it can be defined the boundary $\mathbf{b}$ as the candidate line whose weighted sum $\phi(\mathbf{t}_j)$ is the highest among all candidates being also higher than certain threshold T:

$$\mathbf{b} = \mathbf{t}_j^* = \underset{\phi(\mathbf{t}_j)>T}{\operatorname{argmax}}\left(\phi(\mathbf{t}_j)\right). \qquad (3)$$

(a)   (b)

Fig. 5. Examples of detection of wall-floor and wall-ceiling boundaries.

This process is applied independently for each one of the four sets C_l, C_r, F_l and F_r. Then the left and right floor boundaries, denoted, respectively, as $\mathbf{b}_1$ and $\mathbf{b}_2$, and the left and right ceiling boundaries, denoted, respectively, as $\mathbf{b}_3$ and $\mathbf{b}_4$, are obtained. Figure 5 illustrates the four detected boundaries.

4   Consistency of Boundaries

Ideally, assuming perfect imaging condition and line segment extraction, parallel lines should intersect at a dominant VP as is shown in Fig. 6(a). However, in the real world, there are pixel noise, image distortion, discretization errors, and line segment extraction errors, which make the problem much more challenging. In addition, an incorrect detection of floor and ceiling boundaries could cause disparity over all possible VP locations (see Fig. 6(b)). From the four boundaries $\mathbf{b}_1, \mathbf{b}_2, \mathbf{b}_3, \mathbf{b}_4$, we can compute the VP and use it for validating the floor and ceiling boundaries. Each pair of boundaries $\mathbf{b}_i$ and $\mathbf{b}_j$ defines a hypothesis VP as $\mathbf{v}_{ij} = \mathbf{b}_i \times \mathbf{b}_j$. Then, six VP estimations can be determined, denoting the set as $\mathcal{V} = \{\mathbf{v}_1, \mathbf{v}_2, ..., \mathbf{v}_6\}$.

It is proposed a new strategy which models the impact of an incorrect boundary detection and is able to correct it. That is, the method detects whether one of the four boundaries does not fit well to the candidate VP. The strategy is based on building four partition sets $\{S_1, S_2, S_3, S_4\}$, where each one of them does not take into account one of the four boundaries. Thus, we define $S_1 = \{\mathbf{b}_2, \mathbf{b}_3, \mathbf{b}_4\}$, $S_2 = \{\mathbf{b}_1, \mathbf{b}_3, \mathbf{b}_4\}$, $S_3 = \{\mathbf{b}_1, \mathbf{b}_2, \mathbf{b}_4\}$ and $S_4 = \{\mathbf{b}_1, \mathbf{b}_2, \mathbf{b}_3\}$. For each one of the four sets the algorithm determines three VP hypotheses for each pair of boundaries (see Fig. 6(c)–(f)). Let $\{d_1, d_2, d_3, d_4\}$ be the sum of distances among the three VP hypotheses associated respectively with the partitions $\{S_1, S_2, S_3, S_4\}$. In case of having a boundary $\mathbf{b}_j$ not parallel to the other ones, its impact is included in all the partitions except in the partition S_j. Due to the method relies on distances to identify the incorrect boundary, if all distances $\{d_i\}_{i=1}^4$ with $i \neq j$ are greater than a fixed fraction of d_j, the corresponding boundary

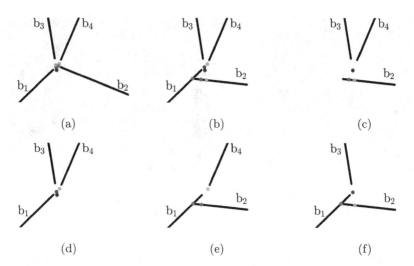

Fig. 6. Hypothesis VP based on detected boundaries. (a) Example with parallel boundaries. (b) Example with a not parallel boundary respect to the remaining ones. (c)
Partition S_1. (d) Partition S_2. (d) Partition S_3. (e) Partition S_4.

b_j would be replaced by the prototype line with the second highest weighted
sum ϕ. We apply recursively the procedure until reaching a dominant VP. The
method ensures that the four boundaries that define the corridor structure are
parallel.

5   Results

In order to check the performance of the proposed approach, we have generated a
test dataset from different locations within the Politechnic School of the University of Alcala. The dataset consists of 106 frames distributed in three sequences
through the corridors of the building. For the camera, a 1024×768 resolution is
selected.

Figure 7 shows the output results of some images from our dataset. Each
corridor image is segmented into four possible regions corresponding to floor,
ceiling and walls (left and right). Each one of these regions is enclosed by two
boundaries and represented by a polygon of three vertexes: the points in which
both boundaries intersects with the borders of the image and the vanishing
point. Different colors are used to identify these regions. The black point is the
detected vanishing point in which converge the four boundaries. We test the
algorithm under different conditions. Thus, Fig. 7(e)–(f) shows the robustness of
the algorithm to changes of perspective. The performance of the system has been
tested with partial occlusions in presence of persons (Fig. 6(g)–(h)). It is noticeable that even when the ceiling-wall boundary is occluded partially, the weight
of the occluded boundary is even greater than the corresponding to remaining

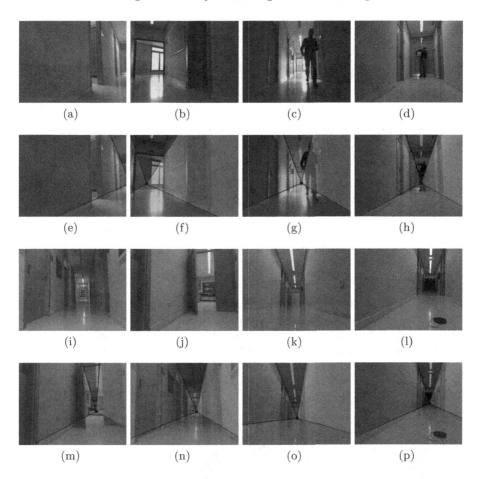

Fig. 7. Output segmentation results (Best viewed in color): floor (red), ceiling (blue), left wall (green) and right wall (orange). First and third rows are the original images. Second and fourth rows are the output images. (a), (b), (e), (f) Different perspectives. (c), (d), (g), (h) Occlusions of boundaries due to the presence of persons. (i), (j), (m), (n) Different distances to the end of the corridor. (k), (l), (o), (p) Images captured with different heights of the camera from the floor.

candidates. Figure 7(m)–(n) includes two images with different distances to the end of the corridor. In addition, we have tested the robustness of the system to images captured with different camera heights and Fig. 7(m)–(n) shows two examples. On the other hand, some isolated images in different environments, such as hospitals, have been tested with the same approach. Figure 8 illustrates some results.

When the robot is close to side walls (Fig. 9(a)–(d)), the ceiling region is reduced or does not exist in the image. It may cause errors as in (Fig. 9(e)–(f)), where several guidelines are incorrectly detected. However, two criteria have been

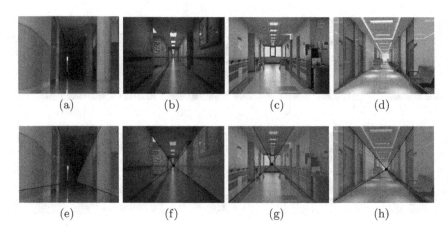

Fig. 8. Output segmentation results in different corridor environments. Top row: original frames. Second row: segmentation results.

established in order to validate the ceiling detection: (1) ceiling area in the image must be more reduced than floor area and (2) ceiling boundaries must extend to both sides of the vertical line that passes through the vanishing point. If the ceiling detection does not follow one of both conditions, ceiling detection is not valid. In these cases, the system only depicts the floor region, as we can see in Fig. 9(g)–(h).

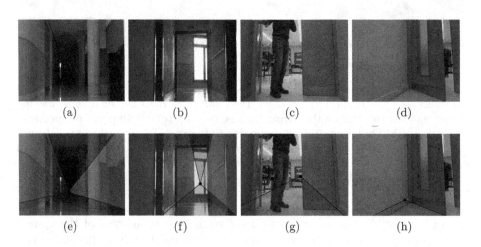

Fig. 9. Examples of views close to side walls. Top row: original frames. Second row: segmentation results.

In order to obtain quantitative results, we have labelled the ground truth of floor and ceiling in the images of the dataset. For this purpose we have developed an application Python that allows to define manually the ceiling and floor boundaries. In the test phase a mask is generated for each image by comparison of both manual and automatic segmentation, as we can see in Fig. 10. Blue pixels represent those ones that were classified correctly, red correspond to misclassified pixels and green are not detected pixels with respect to the ground truth. On average, 93.3% of ceiling and floor pixels labelled manually were classified correctly in our dataset and only 4.15% of the pixels of the image were incorrectly segmented.

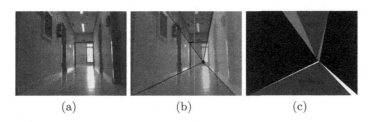

(a)   (b)   (c)

Fig. 10. Example of segmentation mask. (a) Original image. (b) Segmentation. (c) Output mask: correct segmentation (blue), incorrect segmentation (red), not detected (green). (Color figure online)

All processing steps have been implemented in Python, making use of the Numpy package and the OpenCV library. In order to decrease the processing time, we have reduced the image size to 512×384 while keeping the same results reported above. On an Intel Core i5-7500 CPU, the implementation takes only 30.78 ms on average per image.

6   Conclusions

In this paper we have presented a corridor segmentation algorithm based on a combined floor and ceiling detection. An important advantage over traditional floor-based detectors is the fact that ceiling detection can help to correct floor detection in some cases. The disparity of vanishing points hypotheses is used as indicator of the degree of validation of the boundaries that define the corridor structure. The quantitative experiments in a novel and challenging dataset are conducted to validate the effectiveness of the proposed method.

Acknowledgements. This work has been supported by project PREPEATE, with reference TEC2016-80326-R, of the Spanish Ministry of Economy, Industry and Competitiveness, and by the project with reference CCGP2017-EXP/054 of the Government of the Community of Madrid.

References

1. Lorigo, L., Brooks, R., Grimsou, W.: Visually-guided obstacle avoidance in unstructured environments. In: Intelligent Robots and Systems IROS 1997, vol. 1, pp. 373–379 (1997)
2. Yinxiao, L., Birchfield, S.T.: Image-based segmentation of indoor corridor floors for a mobile robot. In: 2010 IEEE/RSJ International Conference on Intelligent Robots and Systems, Taipei, Taiwan (2010)
3. Zhou, J., Li, B.: Robust ground plane detection with normalized homography in monocular sequences from a robot platform. In: IEEE International Conference on Image Processing, pp. 3017–3020 (2006)
4. Cui, X.N., Kim, Y.G.: Floor segmentation by computing plane normals from image motion fields for visual navigation. Int. J. Control Autom. Syst. **7**(5), 788–798 (2009)
5. Fazl-Ersi, E., Tsotsos, J.: Region classification for robust floor detection in indoor environments. In: IEEE Image Analysis and Recognition, pp. 717–726 (2009)
6. Kumar, S., Dewan, A., Madhava Krishna, K.: A Bayes filter based adaptive floor segmentation with homography and appearance cues. In: 8th Indian Conference on Computer Vision Graphics and Image Processing (ICVGIP), p. 54. ACM (2012)
7. Lee, D., Hebert, M., Kanade, T.: Geometric reasoning for single image structure recovery. In: IEEE Conference on Computer Vision and Pattern Recognition (CVPR), pp. 2136–2143 (2009)
8. Baligh-Jahromi, A., Sohn G.: Geometric context and orientation map combination for indoor corridor modeling using a single image. In: International Archives of the Photogrammetry, Remote Sensing and Spatial Information Sciences. XXIII ISPRS Congress, Prague, vol. XLI-B4, pp. 295–302 (2016)
9. Silberman, N., Hoiem, D., Kohli, P., Fergus, R.: Indoor segmentation and support inference from RGBD images. In: Fitzgibbon, A., Lazebnik, S., Perona, P., Sato, Y., Schmid, C. (eds.) ECCV 2012. LNCS, vol. 7576, pp. 746–760. Springer, Heidelberg (2012). https://doi.org/10.1007/978-3-642-33715-4_54
10. Hazirbas, C., Ma, L., Domokos, C., Cremers, D.: FuseNet: incorporating depth into semantic segmentation via fusion-based CNN architecture. In: Lai, S.-H., Lepetit, V., Nishino, K., Sato, Y. (eds.) ACCV 2016. LNCS, vol. 10111, pp. 213–228. Springer, Cham (2017). https://doi.org/10.1007/978-3-319-54181-5_14

Development of a Fire Detection Based on the Analysis of Video Data by Means of Convolutional Neural Networks

Jan Lehr[1]($\boxtimes$)(iD), Christian Gerson[1], Mohamad Ajami[1], and Jörg Krüger[2]

[1] Fraunhofer IPK, Pascalstr. 8-9, 10587 Berlin, Germany
jan.lehr@ipk.fraunhofer.de
[2] TU Berlin, Straße des 17. Juni 135, 10623 Berlin, Germany

Abstract. Convolutional Neural Networks (CNNs) have proven their worth in the field of image-based object recognition and localization. In the context of this work, a fire detector based on CNNs has been developed that detects fire by analyzing video sequences. The major additions of this work will primarily be realized through the use of temporal information contained in the video sequences depicting fire. In contrast to state of the art fire detectors, a large image database with 160,000 images with an even distribution of positive and negative samples has been created. To be able to compare image-based and video-based approaches as objectively as possible, different image-based CNNs will be trained under the same conditions as the video-based networks within the scope of this work. It will be shown that video-based networks offer an advantage over conventional image-based networks and therefore benefit from the temporal information of fire. We have achieved a prediction accuracy of 96.82%.

Keywords: Deep learning · Convolutional Neural Networks · Fire detection · Video-based networks

1 Introduction

Every year, fires cause extensive damage to property and persons. In Germany alone, 343 people lost their lives due to a fire in 2015 [1]. In 2017, forest fires with an area of more than one million hectares were registered in Europe. According to insurance company AON plc, the major forest fires in Portugal in 2017 caused economic loss of almost 1.2 billion USD. 111 people lost their lifes [2]. In order to reduce this extent of damage and tragedy, an outdoor fire detection system would be necessary. Classical fire and smoke detectors have proven their worth in detection tasks and the triggering of various alert mechanisms. At the same time, they have a significant disadvantage because they were developed for indoors rather than outdoors use. To solve this problem, a fire detector based on optical sensors is to be developed. It offers the possibility to cover a large area cost-efficiently. Furthermore, there already exist cameras in the public and private

A. Morales et al. (Eds.): IbPRIA 2019, LNCS 11868, pp. 497–507, 2019.
https://doi.org/10.1007/978-3-030-31321-0_43

sectors to monitor indoor and outdoor areas. These systems can be realized either as a system on chip (SoC) solution or as a centralized computing solution.

2   Related Works

The state of the art will be described in two subsections. The first subsection refers to general image processing methods that are considered and evaluated. In the second subsection, related work in the field of deep learning will be considered in connection with the detection of fire.

2.1   Color Rules Methods

Earlier work focused on detecting fire based on its specific color or color spectrum as well as its movement. Chen et al. [3] detect fire using its chromatic features in the RGB color space along with its saturation in the HSI color space according to its temperature. They define three different rules that must apply to fire and define threshold values that must be empirically selected on a case-by-case basis depending on the application. Other works such as those by Rinsurongkawong et al. [13] broaden this concept to include the optical flow. They also define one more rule for the intensity and look at the growth of fire across frames. However, these concepts are unsuitable for an outdoor fire detector due to the fact that the lighting conditions may change rapidly depending on the time of the day and on the weather conditions. In addition, the above thresholds need to be readjusted for each environment and time of the day. Another classic approach uses only color features of fire for indoor fire detection for a real-time application [11].

2.2   Deep Learning Methods

Recent works in the field of fire and smoke detection are based on Convolutional Neural Networks (CNNs). Tao et al. [14] use a classical AlexNet architecture with minor changes to detect smoke in images. The thus modified AlexNet is trained end-to-end on a comparatively small dataset. They also trained and tested the network on a second, slightly larger dataset with uneven distribution (2,201 positive, 8,511 negative). Zhang et al. [15] not only detect fire, but also determine its position in the image itself. They train a binary classifier at the patch level as the entire image is divided into smaller patches, with a patch size of 32×32 pixels. In addition, they train a classical Support Vector Machine (SVM) classifier to compare both approaches. Since the patch level classifier only has access to local image information, they train an AlexNet as a second approach. The position of the fire is then determined using the patch level classifier. Since the dataset is very small (178 images), they use a pre-trained AlexNet that was trained on the ImageNet dataset. The patch level classifier achieves an accuracy of 97.3%. The CNN classifier also achieves a higher accuracy (97.3%) than the SVM classifier (95.6%). As the dataset was created out of video frames, a large diversity of training and test data is not given which is why it lacks the needed generalization. This assumption is confirmed in the authors' statement that their trained AlexNet achieves an accuracy of 100% in both testing and training. Muhammad et al. [16] present a cost-effective fire detection CNN architecture based on GoogLeNet for surveillance videos. There is no mention of the ratio of fire to non-fire images. Lee et al. [17] design a fire detector especially for fire detection in forests. The accuracy of 99% depends on a relatively small dataset consisting of 289 images (Table 1).

Table 1. Related works on fire detection using deep learning

Author	Network	Dataset			Acc.	FP
		Fire	No fire	Note		
Tao et al. [14]	AlexNet	2,201	8,511	Images	99.4%	0.44%
Zhang et al. [15]	SVM	1,307	11,153	Patches	95.6%	2.1%
Zhang et al. [15]	AlexNet PT	"	"	"	97.3%	1.2%
Muhammad et al. [16]	GoogLeNet PT	68.690		Videos	94.43%	0.05%
Lee et al. [17]	GoogLeNet	10,985	12,068	Videos	99%	-

3 Fire Dataset

A look at known datasets gives an indication of the minimum size of a fire dataset. ImageNet [5] with 1.2 million images divided into 1,000 categories is one of the most popular and common. There are about 1,000 images per category. The CIFAR-10 and the CIFAR-100 [9] datasets consist of 60,000 images divided into 10 and 100 categories, respectively, with an image size of 32×32 pixels. There are 6,000 images per category in the CIFAR-10 dataset. In the CIFAR-100, there are only 600 images per category. MNIST [10], on the other hand, consists of 60,000 small images of size 28×28 pixels depicting handwritten numbers with 6,000 images per number. The EMNIST dataset [4] is an extension of the MNIST dataset. It contains 240,000 images which have an even distribution over all categories. The main objective of the present work should, therefore, be a size of at least 80,000 images per category in order to measure up to the state of the art. Moreover, the dataset has to contain scenes with hard negatives such as sunrises and sunsets, lights at night and other objects similar to fire in order to decrease the percentage of false positives. For the analysis of video sequences, a total of 59 videos with and without fire are added to the dataset. This video dataset is splitted into 3,550 video sequences with twelve frames each (corresponding to a time of two seconds). For the creation of the database, the photo service Flickr offers its own API. The keywords that were used to collect the images for the dataset are listed in Figs. 1 and 2.

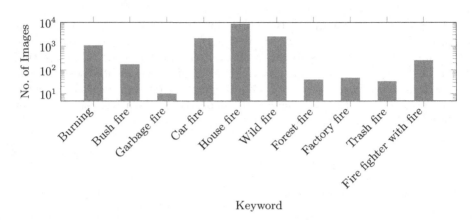

Fig. 1. Used keywords to search for images and image distribution for positive examples. The main keyword is "fire" with 41.527 images. Total number of images: 84,964.

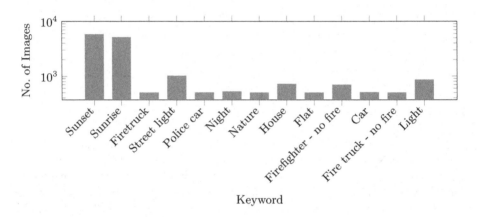

Fig. 2. Used keywords to search for images and image distribution for positive examples. In addition there are 61.691 arbitrary images. Total number of images: 85,685.

4   Methods Used

In this section, two neural network concepts are presented that have already been used successfully for the video classification.

Stacked CNN Network. In [8] three different architectures for video analysis are presented. For this purpose, the features of the individual images of a sequence are combined in different ways (see Fig. 3). The Slow-Fusion model is the only model that performs better than the usual and well-known Single Frame model.

Furthermore, in [12] the presented models are extended to use different pooling possibilities to extract the temporal information from the image sequence

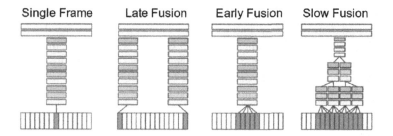

Fig. 3. Different concepts for the fusion of time information. [8]

(see Fig. 4). In comparison, the Conv Pooling method performs best. It starts where the image processing in the CNNs ends. The feature vectors of the single frames of a video sequence are combined with a max pooling. However, this does not happen locally in each frame but across the image sequence. The resulting feature vector is used to make a classification decision via subsequent fully connected layers.

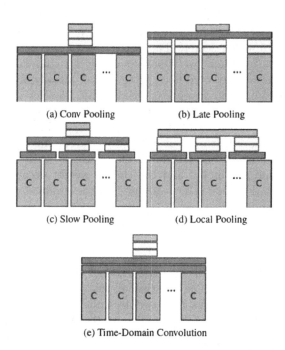

Fig. 4. Different concepts for the local and temporal pooling. [12]

LSTM Network. The Long Short-Term Memory (LSTM), developed by Hochreiter and Schmidhuber [6], is another concept for the analysis of temporal information in image sequences. This is done through the internal ability of a LSTM cell to store information. Classical neural networks use perceptrons as their basic element. If these are replaced by a LSTM cell, the neural network can also process temporal information. A LSTM cell essentially consists of three gates:

- the input gate: it determines to what extent new information is presented to the cell
- the forget gate: it determines to what extent a value remains or is forgotten in a cell
- the output gate: it determines to what extent the internal value of the cell is passed on to the next LSTM cell.

4.1 Baseline Network

The baseline networks will provide a basis for comparison for the subsequent network improvements and will serve as a feature extractor for the video classifiers. As a first step, different architectures were trained from scratch by means of the database created in Sect. 3. The architectures AlexNet, GoogLeNet, SqueezeNet, Cifar10 and VGG16 were selected. The latter two were chosen because of their small size which makes them particularly suitable for embedded systems. The first three networks were selected because they have already achieved very good classification results in various applications. Finally, AlexNet was chosen as feature extractor for the video classifiers. The network was trained using 256×256 pixel images differing from the 224×224 pixel images which are normally used in order to extract more features out of the images. For the training, a learning rate of 0.02 was chosen and then reduced exponentially with a gamma of 0.99. The chosen momentum was 0.9 and the chosen weight decay was 0.002. The Stochastic gradient descent (SGD) was used as the gradient method. Figure 5 shows the learning curve.

4.2 Tower Networks

The basic architectures mentioned in the previous section form the basis of the tower networks (see Fig. 6). The pre-trained image-processing (convolutional) layers of the baseline networks are used to build a tower network of twelve architecturally identical towers. The videos in the dataset are used to extract frames from them every 0.2 s. The tower network is always presented with twelve frames at the same time. The extracted features of the twelve frames are then combined in a temporal pooling layer. The resulting feature vector is classified by three fully connected layers. Here, a complete new training of the last two fully connected layers takes place. The first fully connected layer is also initialized with the weights of the baseline networks and continues learning with a very small learning rate. This is necessary so that this layer can change from the features

Training of AlexNet using Fire Dataset

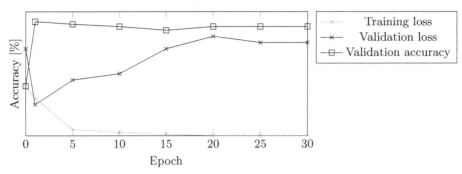

Fig. 5. Learning curve of AlexNet for single image classification. For illustration reasons, not all values are displayed.

of the individual images to features of an entire image sequence. In contrast, the pre-trained convolutional towers are no longer trained as they serve as feature extractors.

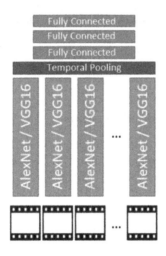

Fig. 6. Structure of the tower network.

4.3 LSTM Networks

The baseline networks are also used in the LSTM networks to utilize the benefit of the pre-trained knowledge for fire detection (see Fig. 7). In contrast to the tower networks, only one baseline network (rather than twelve parallel networks) is used. The frames of a video sequence are presented to the network one by one.

The image characteristics of the convolutional layers are post-processed with a fully connected layer. In contrast to the tower networks, the weights of this layer is not trained further. The LSTM layer is then used. Here, the temporal information, i.e. the information of the frame sequence, is processed.

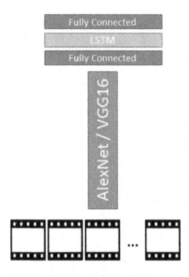

Fig. 7. Structure of the LSTM network.

4.4   Classifier for Embedded Systems

A Raspberry Pi 3 was used as an embedded system. Here, the AlexNet required up to seven seconds to classify one image. This value is too high for system critical to safety. The SqueezeNet [7] needs only 0.7 s. With this architecture, neither a tower network nor an LSTM network is possible. The principle of majority voting is therefore applied. Four sequential images are classified. If a fire is detected on more than two images, the system raises an alarm.

5   Results

In this section, the fire detection methods presented above are evaluated.

5.1   Baseline Networks

During training (see Fig. 5) an overfitting was detected. For this reason, the networks were evaluated for each episode. The best classification results are compared in Table 4 together with the results of the video-based networks. To validate the trained classifiers with regard to their ability to correctly recognize fire, the images are analyzed in a classification heatmap. In this heatmap, the areas that have primarily contributed to the activation of the network are marked in red (cf. Fig. 8).

Fig. 8. The heatmap for the image classification. (Color figure online)

Furthermore, the baseline networks were evaluated regarding their capability to classify video sequences. The concept of majority voting (Sect. 4.4) is used again. First, the pre-trained networks are fine-tuned with the data from the video database and then tested with the validation data (cf. Table 2).

Table 2. Prediction accuracy for fine-tuned networks.

Network	Accuracy
AlexNet - pretrained	94.91%
VGG16 - pretrained	96.78%

Videobased Networks. The two methods presented (tower and LSTM networks) are used to evaluate the video-based networks. The video sequences of twelve images are presented to the network. To check whether the networks have learned the temporal information for fire detection, the networks are additionally tested with a static sequence of images (e.g. tower networks: twelve times for the same frame). It should be noted that due to the small video dataset (compared to the large image dataset) the movement of the fire is not the significant criterion for the classification. But: The number of false-positive images can be further reduced by using video-based networks (cf. Table 3).

For the video-based networks, the classification accuracies are compared with the image-based networks in Table 4.

Table 3. Accuracy (Acc.) and False Positives (FP) image-based and video-based.

Network	Image-based		Tower networks		LSTM networks	
	Acc.	FP	Acc.	FP	Acc.	FP
AlexNet	94.41%	7.1%	95.61%	2.8%	95.14%	3.2%
VGG16	**96.78%**	7.1%	**96.82%**	**0.01%**	95.14%	0.01%

Table 4. Comparison of the accuracy of all networks.

Network	Single image	Tower networks	LSTM networks
AlexNet	94.91%	95.61%	95.14%
VGG16	**96.78%**	96.82%	95.14%
SqueezeNet	96.05%	-	-
Classic approach [11]	69.80%	-	-

6   Conclusion

In the process of developing a fire detector using images and video sequences, an image database with more than 160,000 images was created and expanded by 59 videos. It is large enough to be used for further research in this field. The methods developed for the analysis and detection of fires in video sequences can benefit from the additional temporal information (in contrast to single images). However, it is not possible to recognize fire only by its movement using the methods presented. Compared to the baseline networks, the tower networks have a better false negative rate at the same false positive rate.

7   Future Works

For future work on this topic it is recommended to expand the video database to further increase the influence of the temporal information. Furthermore, other architectures (such as ResNet [18] or DenseNet [19]) can be tested on the dataset. For example, a convolutional layer could not only spatially convolute the frames, but also extract the motion of the fire as a feature.

References

1. Bundesamt, S.: Todesursachen in Deutschland, Statistisches Bundesamt, 19 January 2017
2. AON plc: Weather, Climate & Catastrophe Insight - 2017 Annual report (2017)
3. Chen, T.H., Wu, P.H., Chiou, Y.C.: An early fire-detection method based on image processing. In: 2004 International Conference on Image Processing, ICIP 2004, vol. 3, pp. 1707–1710, October 2004

4. Cohen, G., Afshar, S., Tapson, J., van Schaik, A.: EMNIST: an extension of MNIST to handwritten letters. CoRR abs/1702.05373 (2017)
5. Deng, J., Dong, W., Socher, R., Li, L.J., Li, K., Fei-Fei, L.: ImageNet: a large-scale hierarchical image database. In: 2009 IEEE Conference on Computer Vision and Pattern Recognition, pp. 248–255, June 2009
6. Hochreiter, S., Schmidhuber, J.: Long short-term memory. Neural Comput. **9**(8), 1735–1780 (1997)
7. Iandola, F.N., Moskewicz, M.W., Ashraf, K., Han, S., Dally, W.J., Keutzer, K., SqueezeNet: Alexnet-level accuracy with 50x fewer parameters and <0.5MB model size. CoRR abs/1602.07360 (2016)
8. Karpathy, A., Toderici, G., Shetty, S., Leung, T., Sukthankar, R., Fei-Fei, L.: Large-scale video classification with convolutional neural networks. In: 2014 IEEE Conference on Computer Vision and Pattern Recognition, pp. 1725–1732, June 2014
9. Krizhevsky, A.: Learning multiple layers of features from tiny images. University of Toronto, May 2012
10. LeCun, Y., Cortes, C.: MNIST handwritten digit database (2010). http://yann.lecun.com/exdb/mnist/
11. Menevidis, Z., Ajami, M.: Vision based fire detection using colour variance, vol. 798–123, pp. 543–546 (2013)
12. Ng, J.Y., Hausknecht, M.J., Vijayanarasimhan, S., Vinyals, O., Monga, R., Toderici, G.: Beyond short snippets: deep networks for video classification. In: 2015 IEEE Conference on Computer Vision and Pattern Recognition (CVPR), pp. 4694–4702 (2015)
13. Rinsurongkawong, S., Ekpanyapong, M., Dailey, M.N.: Fire detection for early fire alarm based on optical flow video processing. In: 2012 9th International Conference on Electrical Engineering/Electronics, Computer, Telecommunications and Information Technology, pp. 1–4, May 2012
14. Tao, C., Zhang, J., Wang, P.: Smoke detection based on deep convolutional neural networks. In: 2016 International Conference on Industrial Informatics - Computing Technology, Intelligent Technology, Industrial Information Integration (ICIICII), pp. 150–153, December 2016
15. Zhang, Q., Xu, J., Xu, L., Guo, H.: Deep convolutional neural networks for forest fire detection. In: 2016 International Forum on Management, Education and Information Technology Application. Atlantis Press, January 2016
16. Muhammad, K., Ahmad, J., Mehmood, I., Rho, S., Baik, S.W.: Convolutional neural networks based fire detection in surveillance videos. IEEE Access **6**, 18174–18183 (2018)
17. Lee, W., Kim, S., Lee, Y., Lee, H., Choi, M.: Deep neural networks for wild fire detection with unmanned aerial vehicle. In: IEEE International Conference on Consumer Electronics (ICCE), Las Vegas, NV, pp. 252–253 (2017)
18. He, K., Zhang, X., Ren, S., Sun, J.: Deep residual learning for image recognition. In: IEEE Conference on Computer Vision and Pattern Recognition (CVPR), pp. 770–778 (2016)
19. Huang, G., Liu, Z., Maaten, L.V., Weinberger, K.Q.: Densely connected convolutional networks. In: IEEE Conference on Computer Vision and Pattern Recognition (CVPR), pp. 2261–2269 (2017)

Towards Automatic and Robust Particle Tracking in Microrheology Studies

Marina Castro[1,3(✉)] [iD], Ricardo J. Araújo[1,4] [iD], Laura Campo-Deaño[2,3] [iD], and Hélder P. Oliveira[1,4] [iD]

[1] INESC TEC, Porto, Portugal
up201404051@fe.up.pt
[2] Centro de Estudos de Fenómenos de Transporte, Porto, Portugal
[3] Faculdade de Engenharia da Universidade do Porto, Porto, Portugal
[4] Faculdade de Ciências da Universidade do Porto, Porto, Portugal

Abstract. Particle tracking applied to video passive microrheology is conventionally done through methods that are far from being automatic. Creating mechanisms that decode the image set properties and correctly detect the tracer beads, to find their trajectories, is fundamental to facilitate microrheology studies. In this work, the adequacy of two particle detection methods - a Radial Symmetry-based approach and Gaussian fitting - for microrheology setups is tested, both on a synthetic database and on real data. Results show that it is possible to automate the particle tracking process in this scope, while ensuring high detection accuracy and sub-pixel precision, crucial for an adequate characterization of microrheology studies.

Keywords: Particle tracking · Microrheology · Computer vision

1 Introduction

In rheology, the deformation and flow properties of matter are studied using a rheometer, a device that applies forces to the medium. However, fluids having low viscosity and/or elasticity values cannot be analyzed by this conventional technique, being assessed instead by microrheology. The latter consists in evaluating the behavior of tracer particles in the fluid via microscopic images. It can be passive, when the tracer particles move due to inherent thermal energy, or active, if external forces are applied to induce flow of the tracers. In this work, we consider only the case of passive microrheology, which involves analyzing the Brownian motion of the tracer particles, trough tracking techniques that are applied to microscopy image sequences.

Particle tracking in microrheology deals with specific motion conditions and particle types, but with a broad range of image properties. This means that, even though the tracer beads move randomly around a certain point, with low movement amplitude and, ideally, no drift, and have the same size and appearance, they can be immersed in fluids that create inhomogeneous backgrounds,

© Springer Nature Switzerland AG 2019
A. Morales et al. (Eds.): IbPRIA 2019, LNCS 11868, pp. 508–519, 2019.
https://doi.org/10.1007/978-3-030-31321-0_44

be temporarily occluded, form unwanted agglomerates with other particles, or even be affected by light artifacts that change their apparent shape and color.

Microrheology studies do not require that all the particles in a certain frame sequence are correctly identified and tracked, but their statistical relevance is enhanced if the mean squared displacements of many particles in tracks are taken into account. Moreover, particles that are agglomerated should not be considered in such studies, since the forces involved in their interaction have additional influence in their movement, which is not purely Brownian [1].

There are many available tools for particle tracking in microrheology problems, such as the various adaptations and improvements of Crocker and Grier's techniques [2], in programming languages and environments such as IDL (Exelis Visual Information Solutions, Boulder, Colorado, United States), Matlab (The MathWorks, Inc., Natick, Massachusetts, United States) and Python (Python Software Foundation, https://www.python.org/), or in tools such as the ImageJ plugin ParticleTracker [3]. These have a low level of automation, demanding the definition of multiple parameters about the dataset in use. Because researchers that work in microrheology studies are not necessarily proficient programmers, the task of particle tracking should be facilitated as much as possible, adding efficiency to the pipeline, but also avoiding error in the complex modulus computation, which is highly dependant on the accuracy of the tracking methods.

The main contributions of this work are the following: the design of more robust tracking methods for microscopic images; the creation of a pipeline that automatically deals with different microrheology settings; the extension of an existing synthetic database [4] by including more conditions that may appear in real images of microrheology, such as drift and large wavelength noise.

2   Particle Tracking

Particle tracking involves finding the trajectories of particles along a sequence of images, and is often divided into two steps - particle detection and linking - which we also address separately. In this work, we aim to automatize and standardize the process of particle tracking in microrheology image sequences coming from different settings, in order to facilitate the studies that experts of the field conduct.

To achieve this goal, we start by performing pre-processing to ensure that images have proper contrast and to attenuate background noise. Afterwards, calibration is made in one of the frames, in order to infer whether particles are darker or lighter than the background, and estimate their radius, a parameter that will be crucial in following steps. We also find a relatively small but robust set of search regions which are important when the method next described overlooks certain areas due to occlusion. To find particles at each image of the sequence, we start by applying Crocker and Grier's approach [2] to detect plausible searching regions. As this traditional algorithm tends to produce an overwhelming number of false positives with such limited knowledge of the data (as can be seen in Fig. 3), more robust approaches have to follow. In this work, we investigate how adequate and robust Gaussian fitting and Radial Symmetry are to produce the final detection of particles. In the end, we filter the particle aggregates, leaving out the particles that have an Euclidean distance smaller than 20 pixels between each-other, and the coordinates of the remaining particles are linked into trajectories. An illustration of the described pipeline can be seen in Fig. 1.

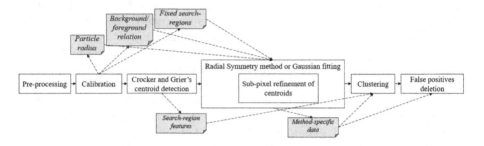

Fig. 1. Particle detection pipeline.

2.1 Pre-processing and Initial Calibration

The contrast of the images is enhanced via gamma adjustment [5]. This is followed by an estimation of the image noise variance [6], assuming that the noise is modeled by an additive zero mean Gaussian. A Gaussian filter [5] having the estimated variance is then applied to reduce noise. To revert the enlarging effect that this filter has on foreground objects, the erosion morphological operation is applied [7], having as kernel a circle with the calculated noise variance as radius.

Concerning calibration, an image of the sequence is used to estimate the radius of the tracer particles, required later in the Crocker and Grier's detection framework, and also to set some of the regions in which the particles are going to be searched for, in every frame. To do so, we start by computing the local image gradient as the difference between the local maximum and minimum in a 4-neighborhood. The gradient is computed with a rank filter [8], meaning that the local histogram is used for its calculation. The gradient image is denoised through a rolling-ball background subtraction [9]. A subsequent histogram of gradient values throughout the whole image is computed, since one wants to keep the edges of the particles, which are ideally represented by similar gradient values. The number of bins used in the histogram is given by the Stone rule, which resorts to the minimization of a loss function defined by the bin probabilities and number of bins [13]. Using Otsu's method [10] to establish a gradient threshold, one can separate the image gradient values into two classes, where the class encompassing the higher gradient values is the most likely to represent the particles and their steep intensity transitions. This method is invariant to the foreground being lighter or darker than the background, but is highly sensitive to high frequency noise, reason why previous and further denoising are required. Therefore, applying the morphological opening and closing to the thresholded (binary) gradient image, with a kernel that takes the 4-neighborhood of each pixel, one can remove salt and pepper noise. Then the particles' contours are filled in, creating solid particle regions. Through the analysis of histograms created for the region area values, the most common area value is defined as representative of true particles if the eccentricities of the regions associated with these areas are characteristic of circular regions. Regions are considered far from being circular, and therefore rejected, if their eccentricities surpass the value of 0.6 (in a scale from 0 to 1, being 0 the eccentricity of a perfect circle). An illustration of this pre-processing and initial segmentation pipeline can be seen in Fig. 2.

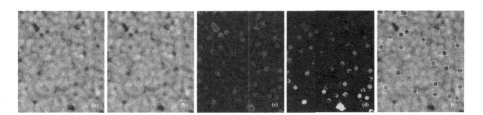

Fig. 2. Illustration of the pre-processing and initial segmentation pipeline. (a) Example image from blood sample video used in [15]. (b) Smoothed image according to the estimated noise variance. (c) Local image gradients. (d) Clustering through histogram analysis. (e) Identified particles.

The found particles allow to estimate an average radius and tell whether they are brighter or darker than the background, condition to which Gaussian fitting is not invariant. Also, as previously stated, the regions where the particles are found can be used as searching regions in the following frames, helping in the scenarios where Crocker and Grier's approach does not retrieve adequate results. Even though it could be used for particle detection, if followed by sub-pixel refinement, this method is only used for calibration, since it is not expected to detect all the particles in a certain frame, but only those which are highlighted by the gradient operation. Moreover, its computation time is prohibitive of its use for whole frame sequences, being of around 7 s per image.

2.2   Frame-by-Frame Detection Pipeline

In order to detect tracer particles in each image of a given sequence, we start by using the Crocker and Grier's algorithm. This method allows to tune several parameters in order to successfully detect particles in different microrheology settings by its own. However, the process of finding good parameters is very tedious, especially for the practitioners, as typically they are not deeply aware of how the algorithm works. Thus, in order to automatize as much as possible the procedure for different types of data, we only give as input the previously estimated particle radius and whether or not the background is darker than the particles. In such conditions, Crocker and Grier's method tends to produce many false positives (as can be seen in Fig. 3), which is not a problem, as we just use its outputs to obtain search areas that will be further processed using either Gaussian fitting or a Radial Symmetry-based approach and features of those search areas, such as integrated brightness, and size eccentricity of the particle region. These features are going to be used in combination with method-specific ones, either coming from the Gaussian profile that is fitted to the particle or from the radial symmetry of the region, to generate data clusters that will separate particles from non-particles.

Gaussian Fitting Through Least-Squares Estimation. The Gaussian fitting method for particle center estimation assumes that the intensity profile of particles can be modeled as a Gaussian distribution. The least-squares estimation minimizes the sum of squares of an error function defined for the fitting of Gaussian parameters to a given search region. This error function is the subtraction between a Gaussian with a set of parameters and the image region it is fitted to. The initial parameter estimates of the Gaussian that is fitted are given by the region's moments, namely the mean in each dimension (first moment) and the variance (second moment). The fitting process is aware of the particles being darker or lighter than the background, as inferred from the calibration step. To filter false positive search areas coming from the Crocker and Grier's algorithm, we start by neglecting all search areas where the fitting does not converge after 1000 iterations. Next, we cluster the fitted Gaussians through Fuzzy C-means [12], using as features their standard deviation and region features: the integrated brightness, size and eccentricity of the circular region in which the Gaussian is being fitted. A membership degree is assigned to each feature point, being higher for points that are closer to the center of the cluster. The "fuzzifier" coefficient, which defines how fuzzy the clusters will be (how small can the membership values become) is set to 2, a common value when there is little information about the features, and the stopping criteria are an error of 0.005 or a maximum number of iterations of 1000. Two data clusters are defined with this method. Then, cluster analysis is processed in a simple manner: the values of the integrated brightness that are placed in each cluster are compared and, if it was previously seen that the background is darker than the foreground/particle candidates, the cluster that is kept is the one with the higher integrated brightness values, and vice-versa. It must be highlighted that data clustering is only done if the number of points detected is superior to two times the number of points in the calibration phase or if the coefficient of variance of the features (given by the mean of the quotients between standard deviations and means of each feature) exceeds 0.5. An illustration of the Gaussian fitting method as a particle center estimation layer is presented in Fig. 4.

Radial Symmetry Method. The Radial Symmetry-based approach, as described in [4], tries to identify spherical particles and accurately compute their center by assuming that the intensity of a particle representation is radially symmetric in relation to the particle center. This algorithm estimates the particle

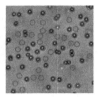

Fig. 3. Circled in red, the detections made by Crocker and Grier's method [2], when only the particle radius and whether the particles are brighter or darker than the background are specified. These detections are made in a frame of a microrheology setup for a blood analogue described in [15]. (Color figure online)

center as the point with the maximum value of radial symmetry, working for differently shaped particles, such as concentric rings and Gaussian intensity distributions. To do so, it assumes that the lines parallel to the intensity gradient of an image region that has perfect radial symmetry will be pointing, in any point, towards the origin, which is the center one wants to find. Therefore, the origin that minimizes the weighted sum of the squared distances between the lines parallel to the intensity gradient is the center of the particle representation. The mean squared distance between the lines parallel to the gradient and the center, weighted by intensity, are a measure of goodness of fit. If the coefficient of variation of the set of mean distance squared values of the detected particles is superior to 0.2, clustering is done through the method presented in the last subsection, but adding the mean squared distance to the feature space, instead of the Gaussian's standard deviation, which is not evaluated in this case. The cluster that is kept is the one that contains the lower mean distance squared values, denoting greater radial symmetry of the regions. An illustration of the Radial Symmetry method as a particle center estimation layer is presented in Fig. 4.

2.3   Linking

After detection, the coordinates of the particles are linked into trajectories, also following an approach by Crocker and Grier [2], which, being adapted to particles undergoing Brownian diffusion, looks for the particles near their most recent locations. The ideal case of particle motion is the one where the particle's velocity is uncorrelated from one frame to the next one, as is characteristic of Brownian diffusion. However, particles can drift together in a directed movement, getting carried away from their initial positions. Calculating and correcting the drift is possible through a simple post-processing routine, in which we average the particle velocities per frame and subtract the displacement caused by this average speed to the final tracks. A simulation of drifting particles through the synthetic database (detailed in the Experiments and Results section) and further drift correction is presented in Fig. 5.

Fig. 4. Illustration of the particle center detection through both approaches. (a) Synthetic 2-particle image generated with Poisson noise of SNR = 10 and fractal noise merged with a weight of 0.1. (b) Centers of all the defined search areas. (c) Detail of the XY-plane projection of the Gaussian that is fitted to the particle on the left. (e) Detail of a search region in which some of the lines parallel to the image gradient are drawn and their maximum convergence point identified. (d), (f) Refined positions of the centers through the Radial Symmetry method and Gaussian fitting, respectively, with the true centers in blue. (Color figure online)

Fig. 5. Illustration of drifting particles with Brownian motion, following a [2, 2] pixel drift vector (pointing down and right from the upper left corner) for 7 frames (a), and their trajectories after drift correction (b).

3   Experiments and Results

To validate the proposed methodology, we conducted experiments on an extension of the synthetic database of [4], and also on sequences of microrheology setups used in [15]. In this section, we start by describing how the database was extended by us, then we evaluate our detection steps in this synthetic data. Finally, we report the accuracy in terms of number of detected particles in the real datasets.

3.1   Synthetic Database Extension

Our simulated multi-particle images are generated from an extension of the generator described in [4] (Matlab code available in https://pages.uoregon.edu/raghu/particle_tracking.html), which models single fluorescent particles as point sources in random positions of the image, convolved with a point-spread-function (PSF):

$$PSF(r) = \left(\frac{2J_1(v)}{v} \right)^2 \tag{1}$$

that is characteristic of the representation of a particle in fluorescence microscopy, where J_1 is the Bessel function of the first kind, of order 1, $v = (2\pi NAr)/(\lambda n_w)$, λ is the wavelength of light, NA is the numerical aperture of the objective lens, n_w is the index of refraction of water and r is the radial coordinate [4]. The simulated particles were created by applying this function with $\lambda = 550$ nm and $NA = 0.55$. The multi-particle images are complemented with Poisson-distributed noise. For this type of noise, the signal-to-noise ratio is the square root of the number of photons detected at the brightest pixel. Therefore, the pixel intensities are scaled so that the peak intensity is equal to the square of the signal-to-noise ratio, and a constant background is replaced by pixels with a random intensity drawn from a Poisson distribution whose mean is the expected background intensity [4]. Because this image transformation only creates short wavelength noise, an additional source of large wavelength noise was added, to simulate the background effects that exist in inhomogeneous fluids, such as blood (see Fig. 8(a)). This noise layer has a cloud-like 2D texture generated by the superposition of increasingly upsampled portions of a white noise image, which creates a fractal noise pattern [14]. Asides from adding noise to the images, an optional gamma correction [5] is made. Besides mimicking Brownian motion, we also allow the existence of a drifting motion that is equally applied to all the particles, through the definition of a motion vector that is added to the previously defined random position of the particles. The pipeline for image generation is visually detailed in Fig. 6.

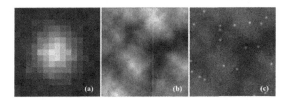

Fig. 6. Illustration of the image sequence generation pipeline. (a) Detail of particle generated with the previously detailed properties and applied Poisson noise (SNR = 8). (b) Cloud-like structure to be merged with a weight of 0.2, forming the final image, (c).

3.2 Performance Evaluation on Synthetic Data

We evaluated the detection capabilities of our methodology in this synthetic data. The Gaussian fitting method showed better accuracy and less occurrence of missed particles and false positives. However, the execution time of the method for each 50-particle image from a sequence, SNR = 10 and a cloud-like texture overlapped with a weight of 0.1 is of around 0.8 s, while the task is completed through the Radial Symmetry method in around 0.16 s.

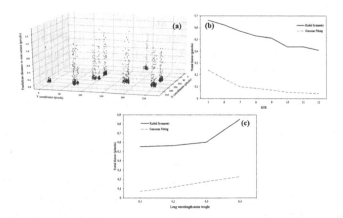

Fig. 7. Accuracy metrics of the detection methods on synthetic data. (a) Euclidean distance to true center in function of each center's true position, for Gaussian fitting (in blue), and Radial Symmetry method (in red). (b) Total error, calculated as the square root of the mean of the sum of the squares of the error in each dimension, in function of SNR. (d) Total error in function of long wavelength noise weight. (Color figure online)

In Fig. 7(a), one can see a three-dimensional representation of the euclidean distance to the true center in an image sequence of 100 frames with 10 particles, to which no drift is applied, but only Brownian motion. These images have a SNR of 10 and long wavelength noise with a merging weight of 0.1. It can be seen through this figure that the maximum Euclidean distance to the true center never exceeds 1.5 pixels, for any method. Also, in Fig. 7(b), the total error for both detection methods is plotted when varying the SNR from 5 to 12, and in (c) when varying the overlapping weight of fractal noise from 0.1 to 0.4, for 100 images with 10 particles. All plots denote the superior accuracy of Gaussian fitting over the Radial Symmetry method. One must, however, notice that the generated images have particle representations whose PSF forms an intensity profile that can be approximated to a Gaussian function, ensuring a good fit with the Gaussian fitting method.

3.3   Application to Real Data

To test the algorithms with real passive microrheology data, we used the videos of the study detailed in [15]. These include two videos of blood analogues, in which the tracer particles are highly contrasting with the background, but also a video of real blood, in which the background has intensity fluctuations, due to the heterogeneity of components, and the particles are often occluded and aggregated. The latter is challenging for this type of study (see Fig. 8(a)).

Since there is no manual annotation of the centroids, but only of the particles' their bounding boxes, the detection accuracy cannot be evaluated in terms of total error. Instead, it can be shown whether, for real case scenarios, the number

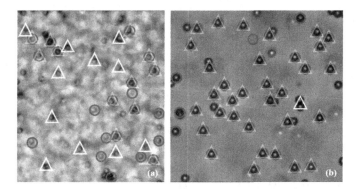

Fig. 8. Representation of the particle detections in a blood image, (a), with applied gamma correction (gamma = 2.5), and a blood analogue image, (b), with no pre-processing applied. Gaussian fitting detections are labeled with red circles and Radial Symmetry method detections are labeled with yellow triangles. (Color figure online)

of particles that were segmented by hand and the number of particles that could be detected by the automated methods match. Therefore, 1 in each 100 frames of three microrheology videos with 5000 frames were annotated and, for the blood analogue images (Fig. 9(c)), 92% of the particles could be identified with the Radial Symmetry method, and 93% with Gaussian fitting. 86% of the particles could be tracked continuously throughout the duration of the video, for both methods. As for true blood images, 53% of the particles could be identified with the Gaussian fitting method, and 56% of the particles were detected through the Radial Symmetry method. 60% of the detected particles could be tracked throughout the duration of the video, for both methods. A representation of the detected particles in both image cases can be seen in Fig. 8.

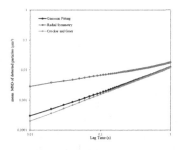

Fig. 9. Mean squared displacement in function of lag time for a blood analogue 5000 frame dataset, analyzed with the Radial Symmetry method and through Gaussian fitting. The MSD acquired for the same particles with the method that was used in the reference [15], Crocker and Grier's [2] approach for particle detection, is plotted in gray. The relative deviation between the MSD curves for Gaussian fitting and the Radial Symmetry method is 126%.

After particle tracking, drift correction, and removal of aggregates, we could calculate the mean square displacement of the particles in each video. Evaluating the mean squared displacement results for the particles in one of the blood analogue fluids detailed in [15], one could see that, even though the Radial Symmetry method has advantage over Gaussian fitting in terms of computational time, its slight inaccuracy at the sub-pixel level leads to the overestimation of the particles' movement, and, therefore, of their mean squared displacement, as can be seen in Fig. 9. Estimating the microrheology parameters with the Radial Symmetry method would lead to highly biased results.

4   Conclusion and Future Work

We have successfully created a way of automating the process for particle tracking in microrheology images. It was shown that the methods in use for this type of problem still have margin to improve, and that their combination and adaptation for different cases is possible. Nevertheless, it was shown that the Gaussian fitting method is adequate for the microrheology setup, since it detects the particle centroids with enough accuracy not to induce erroneous movement in the track calculations. The same cannot be said about the Radial Symmetry technique, which showed to be more inaccurate at the sub-pixel level. Increasing the number of tests on real microrheology scenarios is imperative, since the heterogeneity that is possible for images in these studies is not sufficiently represented in the used real data, and the synthetic database cannot model all the possible conditions.

Acknowledgments. This research was funded by FEDER (COMPETE 2020) and FCT/MCTES (PIDDAC), grant number POCI-01-0145-FEDER-030764. It was also financed by National Funds through the Portuguese funding agency, FCT - Fundação para a Ciência e a Tecnologia within project UID/EEA/50014/2019, and within PhD grant number SFRH/BD/126224/2016.

References

1. Wirtz, D.: Particle-tracking microrheology of living cells: principles and applications. Ann. Rev. Biophys. **38**, 301–326 (2009)
2. Crocker, J.C., Grier, D.G.: Methods of digital video microscopy for colloidal studies. J. Colloid Interface Sci. **179**(1), 298–310 (1996)
3. Sbalzarini, I.F., Koumoutsakos, P.: Feature point tracking and trajectory analysis for video imaging in cell biology. J. Struct. Biol. **151**(2), 182–195 (2005)
4. Parthasarathy, R.: Rapid, accurate particle tracking by calculation of radial symmetry centers. Nat. Methods **9**(7), 724 (2012)
5. Reinhard, E., et al.: High Dynamic Range Imaging: Acquisition, Display, and Image-Based Lighting. Morgan Kaufmann, Burlington (2010)
6. Immerkaer, J.: Fast noise variance estimation. Comput. Vis. Image Underst. **64**(2), 300–302 (1996)
7. Serra, J.: Image Analysis and Mathematical Morphology. Academic Press Inc., Orlando (1983)

8. Gonzalez, R.C., Woods, R.E.: Digital Image Processing. Publishing House of Electronics Industry 141.7 (2002)
9. Sternberg, S.R.: Biomedical image processing. Computer **1**, 22–34 (1983)
10. Otsu, N.: A threshold selection method from gray-level histograms. IEEE Trans. Syst. Man Cybern. **9**(1), 62–66 (1979)
11. Allan, D.B., et al.: Trackpy: Trackpy V0.4.1. v0.4.1, Zenodo, 21 April 2018. https://doi.org/10.5281/zenodo.1226458
12. Ross, T.J.: Fuzzy Logic with Engineering Applications, vol. 2. Wiley, New York (2004)
13. Stone, C.J.: An asymptotically optimal histogram selection rule. In: Proceedings of the Berkeley Conference in Honor of Jerzy Neyman and Jack Kiefer, Wadsworth, vol. 2 (1984)
14. Gardner, G.Y.: Visual simulation of clouds. ACM SIGGRAPH Comput. Graph. **19**(3), 297–304 (1985)
15. Campo-Deaño, L., et al.: Viscoelasticity of blood and viscoelastic blood analogues for use in polydymethylsiloxane in vitro models of the circulatory system. Biomicrofluidics **7**(3), 034102 (2013)

Study of the Impact of Pre-processing Applied to Images Acquired by the Cygno Experiment

G. S. P. Lopes[6](✉), E. Baracchini[2], F. Bellini[1,3], L. Benussi[4], S. Bianco[4], G. Cavoto[1,3], I. A. Costa[1,6], E. Di Marco[1], G. Maccarrone[4], M. Marafini[5], G. Mazzitelli[4], A. Messina[1,3], R. A. Nobrega[6], D. Piccolo[4], D. Pinci[1], F. Renga[1], F. Rosatelli[4], D. M. Souza[6], and S. Tomassini[4]

[1] Istituto Nazionale di Fisica Nucleare Sezione di Roma, 00185 Rome, Italy
davide.pinci@roma1.infn.it
[2] Gran Sasso Science Institute L'Aquila, 67100 L'Aquila, Italy
[3] Dipartimento di Fisica, Sapienza Università di Roma, 00185 Rome, Italy
[4] Laboratori Nazionali di Frascati,
Istituto Nazionale di Fisica Nucleare, 00040 Rome, Italy
[5] Museo Storico della Fisica e Centro Studi e Ricerche "Enrico Fermi",
Piazza del Viminale 1, 00184 Rome, Italy
[6] Electrical Engineering Department, Federal University of Juiz de Fora,
Juiz de Fora, Brazil
guilherme.lopes@engenharia.ufjf.br

Abstract. This work proposes to evaluate the effect of digital filters when applied to images acquired by the ORANGE prototype of the Cygno experiment. A preliminary analysis is presented in order to understand if filtering techniques can produce results that justify investing efforts in the pre-processing stage of those images. Such images come from a camera sensor based on CMOS technology installed in an appropriate gas detector. To perform the proposed work, a simulation environment was created and used to evaluate some of the classical filtering techniques known in the literature. The results showed that the signal-to-noise ratio of the images can be considerably improved, which may help in subsequent processing steps such as clustering and particles identification.

Keywords: Image processing · Digital filters · Particle physics experiment

1 Introduction

Digital image processing techniques emerged in the 1960s, being applied in fields such as medical science, observation of earth resources and astronomy. Since

This study was financed in part by the Coordenação de Aperfeiçoamento de Pessoal de Nível Superior - Brasil (CAPES) - Finance Code 001.

© Springer Nature Switzerland AG 2019
A. Morales et al. (Eds.): IbPRIA 2019, LNCS 11868, pp. 520–530, 2019.
https://doi.org/10.1007/978-3-030-31321-0_45

then, its field of application has grown considerably, and particle physics experiments have been making intensive use of them [1].

Currently, gas detectors have proven to be a choice for detecting particles with low emission of energy. In these detectors, the light produced by the de-excitation of the gas molecules during the multiplication process of electrons can be captured by a camera based on CMOS (Complementary Metal-Oxide-Semiconductor) technology [2]. The acquired images can be processed for better pictorial information and for a more effective human interpretation. Moreover, additional information can be extracted and processed by classifiers for applications in pattern recognition and machine learning.

For improving the detection and classification of patterns in images, it might be necessary a good pre-processing, mainly on images that have a low signal-noise ratio. To evaluate the impact of any pre-processing technique, a simulated data-set may be essential to assess the potential of any proposed algorithm and, in general, as an aid in the search for solutions to mitigate the effect of noise on images.

This work proposes a study aiming at understanding the importance of filtering for the CYGNO experiment [3] regarding the improvement of the signal-to-noise ratio of its captured images, evaluating, in a first approach, the efficiency of detection of straight tracks using spatial filters. To accomplish this task, a simulation tool is also proposed.

The present paper is organized as follows: Sect. 2 describes the used data-set and the analysis methodology. The image generation procedure is explained in Sect. 3 and the filters definitions are presented in Sect. 4. Section 5 shows the relevant results and discusses the applicability of the method. Section 6 concludes this work.

2   Data-Set Definition and Analysis Methodology

In this section an overview of the data-set used in this work and of the analysis methodology are given.

2.1   Data Set

The Cygno experiment aims to develop a triple-GEM (Gaz Electron Multiplier) detector with combined read-out system by using an optical readout structure employing high granularity and low noise CMOS sensors to obtain a good-enough tracking performance for measuring low energy particles to search for Dark Matter massive particles. The detector is developed to read the light produced by the de-excitation of gas molecules during the processes of electron multiplication.

In this work we are using the data acquired by the ORANGE (Optically Readout GEm) prototype, which is described in details in [4–6]. A $10 \times 10 \, cm^2$ Triple GEM structure, with a 1 cm high drift gap, using a binary gas mixture He/CF_4 60/40), was readout by an Orca Flash 4 CMOS-based camera[1] equipped with a large aperture ($f/0.95$) lens, as shown in Fig. 1.

---

[1] For more details visit the site www.hamamatsu.com.

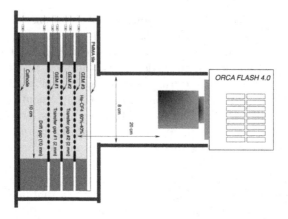

Fig. 1. Drawing (not to scale) of the triple-GEM stack with the lens and the CMOS camera. The drift and transfer gaps are shown.

The images composing the data-set were acquired in free-running mode, without any trigger and a $AmBe$ neutron source placed near to detector. Images produced by the GEM were recorded with an exposure of 2 s and all data were saved without any selection. Figure 2 shows two images generated by the detector for illustration.

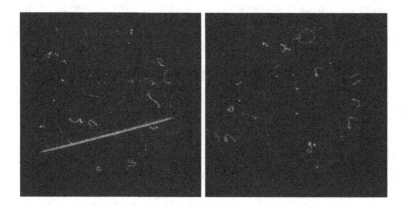

Fig. 2. Images acquired by the ORANGE prototype.

The tracks to be identified have different shapes and different intensities (luminosity). For this work, the main features of the straight tracks (signal) and of the noise have been extracted to study the images characteristics and to use them as input for the proposed simulation tool.

2.2 Analysis Methodology

The proposed analysis steps are summarized in Fig. 3. It can be divided into 4 steps, as follows:

- *Signal and background extraction*
  Using a real data-set acquired by ORANGE, the straight tracks and the image noise are studied and their main parameters characterized (see Sect. 3.1).
- *Image generation*
  In this step the tracks and noise are generated based on the parameters extracted from real data. For each track, its contrast is swept in order to produce a data-set which allows evaluating the filters performance for many levels of difficulty (see Sect. 3.2).
- *Image filtering*
  Filtering techniques are then applied to improve the signal-to-noise ratio of the images. For this work, some of the most used spatial filters (mean, median and Gaussian filters) are employed (see Sect. 4).
- *Performance evaluation*
  The last step is then, after performing a binarization of the image using a simple threshold level based on intensity, measure the number of signal pixels and the number of noise pixels with intensities above the threshold to produce two parameters: DE (detection efficiency) and FA (false alarm), respectively (false alarm might be seen as its counterpart known as background rejection, given by $BR = 1 - FA$).

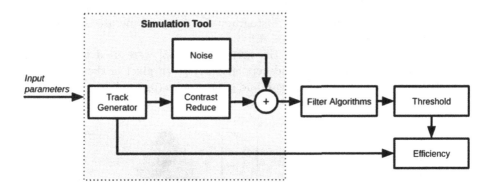

Fig. 3. Analysis steps.

3 Simulation Tool

In this section, the steps performed to obtain a first proposal of a simulation model based on the images acquired by the ORANGE prototype will be presented. The objective here would be only to create a tool that facilitates the evaluation of any method of image processing that could be considered by the Experiment.

An image signal can be modeled as in Eq. 1, where I_t is the binary image representing the track signal whose length and width are given using Monte Carlo method, i is the track intensity (or contrast), and η represents the additive noise. In particular, i will be swept over the entire intensity range of the image to generate data-sets with tracks of varying intensities in order to evaluate the filtering algorithms at different levels of difficulty.

$$I_g(x, y) = iI_t(x, y) + \eta(x, y) \tag{1}$$

For the moment, only straight tracks are considered. Subsections 3.1 and 3.2 describe the procedure used to extract some of the main characteristics of the noise η and tracks I_t, respectively, and explain how they were employed to generate events within the simulation tool.

3.1 Noise Features Extraction and Generation

As usual, for any detection based experiment, to create a simulation tool, signal and noise need to be characterized. To extract the noise features, a data-set acquired without making use of any radioactive source was used. A luminosity histogram built using such data-set is shown in Fig. 4a. Studies on the tracks characteristics have shown that their minimum and maximum intensities are 85 and 135, respectively. Figure 4b shows the histogram considering only that interval. To model such noise, it was used the method known as KDE (Kernel Density Estimator) [7]. The resulting model was then used to generate noise values for the simulation tool. A histogram of a set of generated values are shown in the same Fig. 4b.

The luminosity values beyond the interval [85–135] were used to model a noise component known as salt-and-pepper [8]. A given pixel in the image has the probability of being part of the salt and pepper noise ($p_{s\&p}$). If it is the case, it

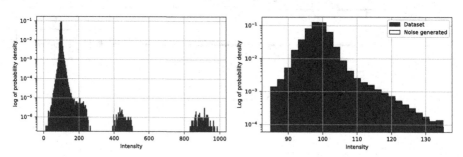

(a) Luminosity histogram from real data   (b) Luminosity histograms from real and simulated data

Fig. 4. Luminosity histograms from real and simulated images.

can be modeled by the Bernoulli distribution where the parameter p_s represents the probability of the pixel to be salt and $p_p = (1 - p_s)$ the probability to be pepper. When a pixel is salt (pepper), its intensity value assumes the maximum (minimum) value on the considered image scale. The probability of a pixel to be a salt-and-pepper noise was measured by summing the number of events out of the interval [85–135] divided by total number of events in the data-set. The probably of a pixel to be salt (pepper) was given by the number of events over (under) 135 (85) divided by the total number of events in the data-set, normalized so that the sum of the two components equals one. The estimated values were: $p_{s\&p} = 0.0083$, $p_s = 0.16$, $p_p = 0.84$, in other words, the probability of a given pixel being degraded by salt-and-pepper noise is 0.83%, and given that it has occurred, the probability of that pixel being salt is 16% and of being pepper is 84%.

3.2   Signal Features Extraction and Generation

As mentioned before, for a first approach only straight tracks have been considered. The considered features were length and width, to be used as input for the simulation tool. Contrast will be given by sweeping the tracks intensities within the range [85–135]. The length and width distributions are presented in Fig. 5. The marginal of the measurements were modeled using KDE, considering that the dependence between length and width is low enough not to justify, at this stage, a two-dimensional density estimation. A Gaussian kernel has been chosen and the KDE bandwidth was determined by Silverman's formula [9].

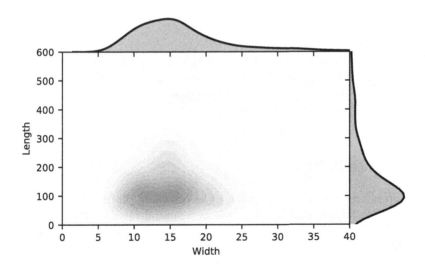

Fig. 5. Histograms of length and width of tracks

The next step is to generate the tracks using the length and width extracted from the estimated pdfs. The track generation is divided into 3 parts. The first is to define a straight segment with length obtained by Monte Carlo and divide it into n equally spaced points.

The second part is to rotate these points by a random angle and centralize it in a random position. This is necessary due to the random orientation and position of the tracks. It is assumed that they are uniformly distributed within the image area. Then, after having drawn an angle θ from 0 to 2π and a central position point (x_0, y_0), the rotation/translation matrix showed in Eq. 2 is applied.

$$
\begin{bmatrix} x' \\ y' \\ 1 \end{bmatrix} = \begin{bmatrix} cos(\theta) & -sen(\theta) & x_0 \\ sen(\theta) & cos(\theta) & y_0 \\ 0 & 0 & 1 \end{bmatrix} \begin{bmatrix} x \\ y \\ 1 \end{bmatrix}
\tag{2}
$$

The third part is to join these points to get a straight track. For that, a Gaussian radial base function [10] $K(\mathbf{x}, \mathbf{x}')$ given by Eq. 3 is used for each point defined previously.

$$
K(\mathbf{x}, \mathbf{x}') = e^{-\frac{||\mathbf{x}-\mathbf{x}'||^2}{2\sigma^2}}
\tag{3}
$$

where the vector $\mathbf{x}$ defines positions on a Cartesian plane (x, y) and the vector $\mathbf{x}'$ defines the position of each of the n points that form the straight segment. The output K are made of values from ≈ 0 (far from $\mathbf{x}'$) to 1 (near to $\mathbf{x}'$). Each point has a function K, so the final output, after processing all the n points, is given by the sum of all functions K as can be shown in Eq. 4. As the Gaussian function is infinite, a threshold is necessary to define the track region (defined as 4σ from $\mathbf{x}'$).

$$
\sum_{j=1}^{n} K_j(\mathbf{x}, \mathbf{x}'_j) = e^{-\frac{||\mathbf{x}-\mathbf{x}'_j||^2}{2\sigma^2}}
\tag{4}
$$

For simplifying the adjustment of track intensity, a variable normalized from 0 to 1 has been defined as shown in Eq. 5, where $i_{max} = 135$ (defined in Sect. 3.1) and $i_{med} = 99$ (which is the image noise baseline).

$$
i = \alpha(i_{max} - i_{med}) + i_{med}
\tag{5}
$$

4   Filters Definition

Spatial filters act directly on the pixels of an image [11] by moving a filter mask thought out its full region. The response of a spatial filter for each point (x, y) is obtained by the relationship between the central pixel and its neighbors. A spatial filter can be linear or non-linear. For this work, two linear filters (mean and Gaussian) and a non-linear filter (median) are considered.

Linear Filter. In general, the output image $g(x,y)$ is given by a 2D-convolution between the system, defined by its impulse response $w(x,y)$ (also known as filter mask), and the input image $f(x,y)$, as given by Eq. 6 [1]. The sizes of w and f are defined as $a \times b$ and $M \times N$, respectively.

$$g(x,y) = \sum_{s=-a+1}^{a-1} \sum_{t=-b+1}^{b-1} w(s,t)f(x+s,y+t) \tag{6}$$

For the mean filter, the mask is defined by Eq. 7, where W is the size of the window. This filter is used to soften the image, reducing the effects caused by the presence of high-frequency components, reducing the variance of the image.

$$w(x,y) = \frac{1}{W} \tag{7}$$

For the Gaussian filter, the mask is described by Eq. 8 and its size is usually defined as a function of σ (e.g. $= 5\sigma$).

$$w(x,y) = \frac{1}{2\pi\sigma^2} e^{-\frac{(x^2+y^2)}{2\sigma^2}} \tag{8}$$

Non-linear Filter. This type of filter uses non-linear operations between mask and image. Some examples are the *max* and *min* operators that get the maximum and minimum values of a pixel neighborhood, respectively. The median filter replaces a given pixel by the median of all pixels in its neighborhood w, as given by Eq. 9 [1].

$$g(x,y) = median\{f(x,y), (x,y) \in w\} \tag{9}$$

5   Results

The main objective of this work is to verify if digital filters have the potential to produce relevant improvement in the signal-to-noise ratio on the ORANGE images. To generate the results, the following items were considered:

- Simulation data will be generated as described in Sect. 3;
- Track intensity will be swept by varying the parameter α used in Eq. 5 through out the interval $[0, 1]$, allowing to measure the performance of the applied filters for various levels of difficulty.;
- Tracks were divided into three categories according to their width values: slim (width < 10 pixels), medium ($10 \leq$ width ≤ 20) and thick (width > 20 pixels);
- The filters known as mean, Gaussian and median will be tested for a given range of mask window size;
- The best mask window size will be found for each filter given a fixed false alarm level of 1% (or, equivalently, a background rejection of 99%).

– Finally, the results will be given in terms of detection efficiency for a background rejection fixed to 99%, as mentioned above.

The left plot of Fig. 6 shows the detection efficiency for many values of α and window size for slim tracks only. As it is possible to notice, when the windows size gets large, the linear filters have their performance degraded, not happening the same with the median filter.

The right plot of Fig. 6 shows the best detection efficiency curves achieved by the filters (which means that their best window sizes are used) and the one achieved when raw-data is directly used, without any image processing. The detection improvement offered by the filters are clear and the median filter has presented the best results since it is immune to outliers and salt-and-pepper noise, followed by the Gaussian filter.

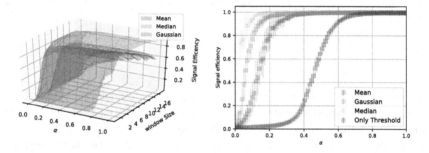

Fig. 6. Detection efficiency for slim tracks.

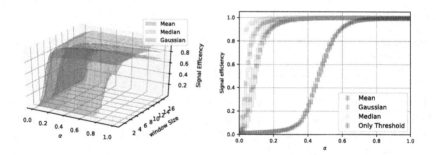

Fig. 7. Detection efficiency for medium-width tracks.

The same analysis has been done for the medium-width and thick-width tracks as shown by Figs. 7 and 8. Similar conclusions can be made with slight differences:

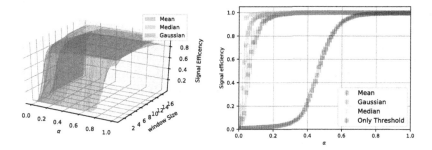

Fig. 8. Detection efficiency for thick-width tracks.

- The left plots show that the mean and Gaussian filters are less effected by the filter window size due to the thicker tracks width, but yet their performance are considerably degraded, differently from the median filter;
- The right plots show that the best curves offered by the filters get better (moving to the left and getting more pronounced) for larger track widths.
- The raw-data based method (only threshold) yields always the same performance, independent of track width.
- The detection efficiency curves achieved by the linear filters get closer.

6   Conclusions

This work presents an initial proposal for constructing a simulation tool for modeling the ORANGE images for evaluation of image processing algorithms. The simulation tool was used to verify the importance of applying digital filters to the CYGNO images for improving their signal-to-noise ratio.

Looking at the achieved results, it was possible to observe that the application of a pre-processing algorithm to the CYGNO image may be a way to improve the signal efficiency, even using a high background rejection requirement. These results justify further studies on the subject. Nonetheless, since this was a first attend, the simulation tool is been studied and upgraded to generated tracks even more closed to the reality of the experiment, in order to keep evaluating the already implemented filtering algorithm and add more complex algorithms.

Acknowledgment. The authors would like to thank the following Institutes/ Universities for their valuable work: INFN-Roma I, Gran Sasso Science Institute L'Aquila, Dipartimento di Fisica della Sapienza Università di Roma, INFN-LNF, Museo Storico della Fisica e Centro Studi e Ricerche "Enrico Fermi" and Post-Graduation Program in Electrical Engineering of UFJF.

References

1. Gonzalez, R.C., Woods, R.E., et al.: Digital Image Processing, vol. 141, no. 7. Publishing House of Electronics Industry (2002)
2. Marafini, M., Patera, V., Pinci, D., Sarti, A., Sciubba, A., Torchia, N.M.: Study of the performance of an optically readout triple-gem. IEEE Trans. Nucl. Sci. **65**(1), 604–608 (2018)
3. Pinci, D., et al.: Cygnus: development of a high resolution TPC for rare events. PoS, EPS-HEP2017, p. 077 (2017)
4. Marafini, M., Patera, V., Pinci, D., Sarti, A., Sciubba, A., Spiriti, E.: High granularity tracker based on a Triple-GEM optically read by a CMOS-based camera. JINST **10**(12), P12010 (2015)
5. Marafini, M., Patera, V., Pinci, D., Sarti, A., Sciubba, A., Spiriti, E.: ORANGE: a high sensitivity particle tracker based on optically read out GEM. Nucl. Instrum. Methods **A845**, 285–288 (2017)
6. Marafini, M., Patera, V., Pinci, D., Sarti, A., Sciubba, A., Spiriti, E.: Optical readout of a triple-GEM detector by means of a CMOS sensor. Nucl. Instrum. Methods **A824**, 562–564 (2016)
7. Scott, D.W., Sheather, S.J.: Kernel density estimation with binned data. Commun. Stat.-Theory Methods **14**(6), 1353–1359 (1985)
8. Chan, R.H., Ho, C.-W., Nikolova, M.: Salt-and-pepper noise removal by median-type noise detectors and detail-preserving regularization. IEEE Trans. Image Process. **14**(10), 1479–1485 (2005)
9. Silverman, B.W.: Density Estimation for Statistics and Data Analysis. Routledge, Boca Raton (2018)
10. Fornberg, B., Larsson, E., Flyer, N.: Stable computations with Gaussian radial basis functions. SIAM J. Sci. Comput. **33**(2), 869–892 (2011)
11. Lim, J.S.: Two-Dimensional Signal and Image Processing, 710 p. Prentice Hall, Englewood Cliffs (1990)

Author Index

Abass, Faycel II-177
Abdelbaset, Asmaa II-169
Abdel-Hakim, Alaa II-169
Abreu, Pedro H. II-322
Acevedo-Rodríguez, Francisco Javier I-77
Acien, Alejandro II-12
Adonias, Ana F. II-247
Agudo, Antonio I-423
Aguilar, Eduardo I-65
Ajami, Mohamad II-497
Alejo, R. I-216
Alén-Cordero, Cristina II-485
Alfaro-Contreras, María II-147
Almazán, Emilio J. I-137, I-329
Alonso, Raquel II-247
Amirneni, Satakarni I-589
Antunes, João I-194
Aouache, Mustapha II-86
Araújo, Ricardo J. II-473, II-508
Ardakani, Parichehr B. II-64
Arens, Michael I-101, I-302
Arevalillo, Jorge M. I-113
Arias-Rubio, Carlos I-161
Arista, Antonio II-239
Arroyo, Roberto I-329
Ávila, Mar I-498
Ayad, Mouloud II-177

Banerjee, Biplab I-472
Baptista-Ríos, Marcos I-77
Baracchini, E. II-520
Barata, Catarina I-239, II-3
Barrenechea, Edurne I-553
Barrere, Killian II-201
Baumela, Luis I-449
Becker, Stefan I-101
Bedoui, Mohamed Hédi II-260
Bellini, F. II-520
Ben Abdallah, Asma II-260
Bengherabi, Messaoud II-86
Benussi, L. II-520
Bernabeu, Marisa I-485
Bernardino, Alexandre I-194
Bessa, Sílvia I-355

Bianco, S. II-520
Bilen, Hakan I-508
Blaiech, Ahmed Ghazi II-260
Blanco, Saúl I-342
Bocanegra, Álvaro José II-359
Brandão, André I-225
Brando, Axel I-29
Brea, Víctor M. II-273
Buenaposada, José M. I-449
Bueno, Gloria I-317, I-342, II-189
Buhrmester, Vanessa I-302
Bulatov, Dimitri I-302
Bustince, Humberto I-553
Byra, Michal I-41

Caballero, Daniel I-498
Calvo-Zaragoza, Jorge II-135, II-147, II-159
Camacho, Camilo II-394
Campo-Deaño, Laura II-508
Cancino Suárez, Sandra Liliana II-465
Cancino, Sandra Liliana II-359
Canedo, Daniel I-620
Cardoso, Jaime S. II-38, II-247, II-473
Caro, Andrés I-498
Caro, Luis I-206
Carvalho, Pedro H. I-355
Casacuberta, Francisco I-16
Castillo-García, Fernando I-600
Castro, Francisco Manuel II-296
Castro, Marina II-508
Cavallaro, Andrea I-239
Cavoto, G. II-520
Cepeda, Karen II-465
Cherifi, Hocine I-170
Chetouani, Aladine I-170
Cleofas Sanchez, Laura II-239
Corroto, Juan Jose II-371
Costa, I. A. II-520
Costa, Joana II-383
Cristóbal, Gabriel I-317, I-342, II-189
Cunha, António II-335

de León, Pedro J. Ponce II-159
de la Calle, Alejandro I-137

Delgado, Francisco J.   I-329
Delgado-Escaño, Rubén   II-296
Demircan-Tureyen, Ezgi   I-89
Deniz, Oscar   I-342, II-371, II-441
Di Marco, E.   II-520
Dias, Catarina   II-335
Djeddi, Chawki   II-177
Dobruch-Sobczak, Katarzyna   I-41
Domingues, Inês   II-322
Domingues, José   II-217
Duarte, Hugo   II-322
Dutta, Deep   I-543

El Hassouni, Mohammed   I-170
El Haziti, Mohamed   I-170
El-Melegy, Moumen   I-270, II-169
ElMelegy, Tarek   I-270
El-Sayed, Gamal   II-169
Escalante-Ramirez, Boris   I-289
Esteves, F. F.   I-461

Fernandes, Kelwin   I-3
Fernandez-Carrobles, M. Milagro   II-441
Fernández-Sanjurjo, Mauro   II-273
Ferreira, Bárbara   II-3
Ferreira, Diogo Daniel   II-404
Ferreira-Gomes, Joana   II-247
Fierrez, Julian   II-12, II-108
Figueiredo, Rui   II-346
Fiori, Marcelo   I-148
Forcén, Juan Ignacio   I-553
Forero, Manuel Guillermo   I-161, I-378,
       I-577, I-600, II-416, II-465
Franco, Annalisa   II-25
Franco-Ceballos, Ricardo   I-398
Fusek, Radovan   II-76

Galeano-Zea, July   I-398
Gallego, Antonio-Javier   I-485, II-135
García, Vicente   I-249
García, Zaira   II-239
García-Vanegas, Andrés   I-600
Garzon, Johnson   I-398
Gaspar, José António   II-309
Gattal, Abdeljalil   II-177
Georgieva, Petia   I-225, II-217, II-404
Gerson, Christian   II-497
Ghosh, Tanmai K.   II-98
Giraldo, Beatriz F.   I-367

Gómez-Moreno, Hilario   II-485
Gonfaus, Josep M.   II-64
Gonzàlez, Jordi   II-64
González-Barcenas, V. M.   I-216
Gonzalez-Rodríguez, Antonio   I-600
Granda-Gutiérrez, E. E.   I-216
Grechikhin, Ivan   II-429
Gruber, Dieter P.   II-453
Guil, Nicolás   II-296
Gupta, Phalguni   II-50
Gutiérrez-Maestro, Eduardo   I-386

Hadjadj, Ismail   II-177
Hamidi, Mohamed   I-170
Heleno, Sandra   I-279
Hernandez, Andrea   II-465
Hernandez-Ortega, Javier   II-108
Herrero, Elias   I-437
Higuera, Carolina   II-394
Horta-Júnior, José de Anchieta C.   I-161
Hurtado, Antonio   I-329
Hussain, Mazahir   I-565

Iñesta, José M.   II-147
Islam, Sarder Tazul   I-543

Jacanamejoy Jamioy, Carlos   I-577
Jafari-Tabrizi, Atae   II-453

Kamasak, Mustafa E.   I-89
Kanagasingam, Yogesan   I-543, II-98
Kerkeni, Asma   II-260
Khatir, Nadjia   I-77
Kim, Eung-su   I-610
Kisku, Dakshina Ranjan   II-50
Korzinek, Danijel   I-41
Krüger, Jörg   II-497
Kurihara, Toru   II-229

Lafuente-Arroyo, Sergio   II-485
Lee, Junesuk   I-610
Lehr, Jan   II-497
Leira, Luís   II-404
Liberato-Tafur, Brhayan   I-600
Libreros, José A.   I-342
Lichtenegger, Hannah Luise   II-453
Lima, S. Q.   I-461
Lopes, Bernardo   II-217
Lopes, G. S. P.   II-520

Lopes, Gabriel   II-38
López, Dolores E.   I-161
López, Juan Manuel   I-367, II-359, II-465
Lopez-Lopez, Eric   II-25
López-Najera, Abraham   I-249
López-Santos, Oswaldo   I-378
López-Sastre, Roberto J.   I-77, I-386
Lumbreras, Felipe   I-423
Lumini, Alessandra   II-25

M. Pardo, Xosé   II-25
Maccarrone, G.   II-520
Maharjan, Samee   I-531
Maldonado-Bascón, Saturnino   I-386, II-485
Mamidibathula, Bharat   I-589
Mansour, Asma   II-260
Marafini, M.   II-520
Marasek, Krzysztof   I-41
Marcal, André R. S.   I-257, II-285
Marín-Jiménez, Manuel Jesús   II-296
Maroto, Fernando   II-441
Marques, Jorge S.   I-239, I-279
Martins, Aurora L. R.   I-257
Mateiu, Tudor N.   II-135
Mazzitelli, G.   II-520
Mejía, Juan S.   I-367
Mena, José   I-29
Meneses-Casas, Nohora   I-378, I-577
Messina, A.   II-520
Mihaylova, Petya   II-217, II-404
Mirza, Ali   I-565
Mohamed, Doaa   I-270
Monteiro, Nuno Barroso   II-309
Morales, Aythami   II-12, II-108
Moreno, Plinio   II-346
Moreno-Noguer, Francesc   I-423
Mucientes, Manuel   II-273
Münch, David   I-302
Murillo, Javier   I-398
Mustufa, Syed Ghulam   I-565

Nagae, Shigenori   II-108
Nait-Bahloul, Safia   I-77
Nakano, Mariko   II-239
Narotamo, Hemaxi   I-53
Neto, Fani   II-247
Neves, António J. R.   I-620
Nin, Jordi   I-182
Nobrega, R. A.   II-520

Nowicki, Andrzej   I-41
Nunes, Afonso   II-346
Nuñez-Alcover, Alicia   II-159

Ochoa-Ortiz, Alberto   I-249
Oliveira, Ana Catarina   II-322
Oliveira, Hélder P.   I-355, II-335, II-473, II-508
Olveres, Jimena   I-289
Ondo-Méndez, Alejandro   II-465
Orrite, Carlos   I-437
Oulefki, Adel   II-86

Pagola, Miguel   I-553
Pande, Shivam   I-472
Park, Soon-Yong   I-610
Patnam, Niharika   I-589
Pedraza, Anibal   I-342
Pedraza, César   II-394
Peixoto, Patrícia S.   I-355
Peralta, Billy   I-206
Perdomo, Sammy A.   II-416
Pereira, Francisco   I-279
Perez, Guillermo F.   II-416
Perez, Hector   II-239
Pérez-Palacios, Trinidad   I-498
Peris, Álvaro   I-16
Pertusa, Antonio   I-485
Phong, Nguyen Huu   I-521
Piccolo, D.   II-520
Pieringer, Christian   I-206
Pina, Pedro   I-279
Pinci, D.   II-520
Pinheiro, Gil   II-335
Pinto, João Ribeiro   II-38
Piotrzkowska-Wroblewska, Hanna   I-41
Pires, Pedro   I-225
Pissarra, José   I-257
Pižurica, Aleksandra   I-472
Portêlo, Ana   I-239
Pujol, Oriol   I-29, I-182
Pulido, Sergio David   II-359

Quimbaya, Mauricio A.   II-416
Quirós, Lorenzo   II-123

Radeva, Petia   I-65
Rahaman, G. M. Atiqur   I-543, II-98
Rakshit, Rinku Datta   II-50

Rendón, E.   I-216
Renga, F.   II-520
Reyes, Juan   I-206
Ribeiro, Bernardete   I-521, II-383
Rio-Torto, Isabel   I-3
Robledo, Sara M.   I-398
Rodriguez, Lizeth   II-465
Rodríguez, Pau   II-64
Rodríguez-Pardo, Carlos   I-508
Roldán, Nicolás   II-465
Rosatelli, F.   II-520
Rotger, Gemma   I-423
Ruiz, Angel D.   I-367
Ruiz-Santaquitaria, Jesús   I-342

S. Marques, Jorge   II-3
Saha, Sajib   I-543, II-98
Saini, Manisha   I-409
Sakpere, Wilson   I-125
Salazar-D'antonio, Diego   I-378
Salido, Jesús   I-342
Sanches, J. Miguel   I-53, I-461
Sánchez, Carlos   I-317, I-342, II-189
Sánchez, Josep Salvador   I-249
Santos, Elisabete M. D. S.   II-285
Santos, João   II-322
Sarbazvatan, S.   I-461
Savchenko, Andrey V.   II-429
Sayed, Md. Abu   II-98
Scherer-Negenborn, Norbert   I-101
Segundo, Marcela A.   I-355
Serrador, Diego G.   I-329
Sfeir, Ghesn   I-449
Siddiqi, Imran   I-565, II-177
Siewiorek, Daniel   I-194
Silva, Ana Rosa M.   I-355
Silva, Catarina   II-383
Silveira, Margarida   I-53
Singh, Harbinder   II-189
Sistla, Sai Shravani   I-589
Smailagic, Asim   I-194

Sojka, Eduard   II-76
Souza, D. M.   II-520
Suárez, Iago   I-449
Susan, Seba   I-409
Sznajder, Tomasz   I-41

Tavares, Fernando   II-285
Teixeira, Luís F.   I-3
Tistarelli, Massimo   II-50
Tomassini, S.   II-520
Torres, Fabian   I-289
Torres, Juan Pedro   I-498
Torres-Madronero, Maria C.   I-398
Toselli, Alejandro H.   II-123, II-201
Tovar, Javier   I-137, I-329
Trifan, Alina   I-620

Unceta, Irene   I-182

V. Regueiro, Carlos   II-25
Valdés, Matías   I-148
Valdovinos, R. M.   I-216
Valencia, Mauricio   I-437
Vállez, Noelia   I-317, II-371
Velasco-Mata, Alberto   II-371
Velazquez, Diego   II-64
Ventura, R.   I-461
Vera-Rodriguez, Ruben   II-12
Vidal, Enrique   II-123, II-201
Vitrià, Jordi   I-29
Vujasinović, Stéphane   I-101

Xavier Roca, F.   II-64

Yanai, Keiji   II-239
Yen, Ping-Lang   I-289
Yu, Jun   II-229

Zarzycki, Artur   I-398
Zhan, Shu   II-229

Printed in the United States
By Bookmasters